Françoise P. Coupal
1987

3495
oul
Rpt 1/67

D0809621

THE FIELD DIRECTORS' HANDBOOK

AN OXFAM MANUAL FOR DEVELOPMENT WORKERS

Edited by:

Brian Pratt, B.Soc.Sci.,Ph.D.(Cantab)
and
Jo Boyden, B.Sc.,Ph.D.(LSE Anthrop),F.R.A.I.

Published for Oxfam by
The Oxford University Press

Oxford University Press, Walton Street, Oxford OX2 6DP

London, Glasgow, New York, Toronto,
Delhi, Bombay, Calcutta, Madras, Karachi,
Kuala Lumpur, Singapore, Hong Kong, Tokyo,
Nairobi, Dar es Salaam, Cape Town,
Melbourne, Auckland

and associate companies in
Beirut, Berlin, Ibadan, Mexico City

© Oxfam 1985

The Field Directors' Handbook:
An Oxfam Manual for Development Workers — 4th ed.

    1. Economic development projects —
    — Developing countries
    — Management
    I. Pratt, Brian   II. Boyden, Jo
    338.9' 009172' 4   HC59.7

    ISBN 0-19-920153-6

Produced by Oxfam Publications, Oxford
Phototypeset by Getset (BTS) Ltd., Eynsham, Oxford
Printed by Alden Press, Oxford

ACKNOWLEDGEMENTS

Edited by   Jo Boyden and Brian Pratt

Part Editors:

Social Development:	Peter Oakley
Economic Development:	Anthony Hall
Agriculture:	Mary Cherry
Health:	Tim Lusty

Contributors included: Jeff Alderson, John Best, Jo Boyden, Margaret Bryer, Chris Daniell, Shafiq Dhanani, Pat Diskett, Marie Theresa Feuerstein, Brian Hartley, Jim Howard, Christopher Jones, Tim Lusty, David Marsden, Colin McKone, Michael Miller, Caroline Moser, Bob Plumtree, Brian Pratt, Lesley Roberts, Paul Shears, Marcus Thompson, Doug Thornton, Suzanne Williams, Peter Wood, Lincoln Young, plus many other members of Oxfam staff and from the Agricultural Extension and Rural Development Centre, Reading University.

Special thanks   to the following people who provided support in many different ways: David Bryer, Jill Bidie, Stephen Biggs, Eva Crane, Geoffrey Cuming, Bob Gibson, Jock Griffiths, David Hanson, Michael Harris, Caroline Heath, Rex Hudson, Tony Jackson, Christine Hugh Jones, Beryl Knotts, Adrian Moyes, Patrick Mulvaney, Ingrid Palmer, Philip Parker, Mark Pembleton, Sue Phillips, Sylvia Saunders, Gill Shepherd, Pat Simmons, Charles Skinner, Guy Stringer, Jeremy Swift, John Turner, David Turton, Tony Vaux, Tony Warner, Sandra Wilson and many others who volunteered their time. And those who contributed to earlier versions of the handbook.

And a very special mention for Mrs. Betty Hawkins who fought with various drafts over two years and filled dozens of discs on her word processor.

PREFACE

The Oxfam Field Director's Handbook was first issued as an internal document in 1974 and subsequently revised twice (1976 and 1980) before appearing in its present form. The Handbook was seen primarily as a set of guidelines for field staff working in the various Oxfam offices throughout the world. Few staff have access to sophisticated information systems, libraries, specialist consultants and the like. Moreover, most work with a large number of development programmes and could not be expected to be conversant with the many forms they take or the various technical problems they present. The Handbook provided guidance on the issues that had to be taken into account when appraising projects prior to funding, and was also a reference source on the problems commonly encountered by projects. We felt that by publishing our revised Handbook and making it widely available we would be able to share our experiences with others working in the field of development. Many field workers throughout the Third World are isolated and often obliged to make decisions based on the minimum of information.

Over the years Oxfam has supported many thousands of small projects in low-income areas. As with any agency, we have had our fair share of successes and failures, and it is hoped that some of these are reflected in the following pages. Resources for development are not so plentiful that we can afford to be wasteful and turn our backs on the learning process.

The manual reflects an underlying philosophy which is fundamental to Oxfam's work; that the most effective form of development is the development of people rather than technologies. Programme beneficiaries must be involved in all stages and aspects of development activity, both as individuals and communities, for there to be any real positive impact on their lives. Because of this basic philosophy we feel that although the manual is written from the perspective of non-governmental organisations the experiences are equally relevant to government and international agencies.

Too often development programmes have been designed on the basis of planning and management carried out by professionals (often foreign) in cities far removed from the people to be affected. In the present manual we have stressed small-scale, inexpensive options which can be implemented to a large extent by the communities themselves. Perhaps non-governmental organisations are fortunate in that they can work at this level relatively unimpeded by domestic and international politics and unfettered by large cumbersome bureaucracy. Where the political will exists, there is no reason why decision-making and implementation cannot be brought to the poor instead of remaining in the hands of politicians and development professionals.

Although Oxfam has chosen to publish this manual, there are many other agencies working along similar lines and it should not be thought that we have discovered a magical formula for development. Rather, we have sought to identify useful guidelines from our own work and that of our many project partners and those other agencies with whom we collaborate. The intention was not to produce the definitive work on development nor promote rigid models, but to give broad guidelines to some of the main practical issues. In many cases it will be necessary to refer for detailed information to other sources. Thus, we have tried to indicate in the Resources sections which books, journals, or institutions, could provide further, more specialised information. The aim has been, as far as possible, to include in the Resources sections accessible titles, rather than the more academic papers which can only be found in the larger university libraries.

Similarly, we have been at pains to avoid jargon and complex theoretical concepts. This is in part because Oxfam works in 80 countries throughout the world and we feel that certain terms can be interpreted in dramatically different ways from one country to another. We are not denying the importance of theoretical studies; but it is not the role of Oxfam to develop such a framework. On the other hand, most development workers would probably now accept that a wholistic approach to development is essential; for too long the technical approach has dominated development thinking and practice. Throughout the manual we stress that no programme should be commenced without prior investigation of its social context. Inappropriate and expensive technologies can be extremely wasteful and resources designated for poor communities may be of no benefit whatsoever if the will and capacity to manage them is lacking.

It is not intended that the manual should be read from cover to cover, but that readers should refer to those sections directly of relevance to their own interests. Some of the sub-sections may seem arbitrary, and some points are repeated. Our intention has been to cover all the main areas of development, and of course there is considerable overlap between them; it is somewhat artificial to separate agriculture from nutrition, for example, when the two are so intricately linked. Furthermore, some points are considered so important that they are included a number of times under different sub-sections. Elsewhere, we have relied on a system of cross referencing to help the reader identify complementary topics.

Our advice is that Parts One, Two and Three of the book are relevant to all development programmes and should be consulted by all readers. Whatever the scale of the programme, due cognisance must be taken of the historical, economic and social circumstances of the country in which it is located. The first two parts include a discussion of the major principles of development which are seen as crucial to all programmes. Part Two raises the issue of targeting in development and proposes that certain particularly disadvantaged groups should receive special priority. Part Three outlines some of the procedures for putting these principles into practice. The Parts concerned with Social and Economic Development, Agriculture, Health and Disasters give more detailed treatment of specific types of programme. Though, even here certain sections will have more general applicability (for example, those on training).

At the end of each Part is a basic checklist of questions which attempts to summarise the questions which must be addressed before commencing any programme. The specific checklists should be read in conjunction with the general checklist for project appraisal, and in some cases reference should be made to that of Part Two.

Rather than employ a system of academic footnotes, basic references can be found in the Resources sections. In a number of cases, where unusual or especially notable experiences have been drawn from direct project histories, we quote the Oxfam project number for reference. However, we have tried to avoid frustrating the reader with references to too many unpublished and confidential sources.

Brian Pratt/Jo Boyden
Oxford, 1985

CONTENTS

*For ease of reference a detailed Contents List of all Sections
has been included at the beginning of each PART.*

PART 1 INTRODUCTION

Oxfam

Part One INTRODUCTION

Section 1   Oxfam — An Interpretation

Oxfam believes in the essential dignity of people and their capacity to overcome the problems and pressures which can crush or exploit them. These may be rooted in climate and geography, or in the more complex areas of economics, politics and social conditions.

Oxfam is a partnership of people who share this belief, people who, regardless of race, sex, religion or politics, work together for the basic human rights of food, shelter and reasonable conditions of life. We believe that, if shared equitably, there are sufficient material resources in the world to enable all people to find a measure of fulfilment and to meet their basic human needs. We are committed, therefore, to a process of development by peaceful means which aims to help people, especially the poor and underprivileged overseas. This development will sometimes generate conflicts of choice both for us at home and for our partners overseas; but it must be a commitment to a process which encourages people to recognise and develop their potential and to decide their own values and priorities.

Oxfam's contribution is modest within the constraints of our limited resources. But we have learned that we can serve as a small-scale catalyst, helping small groups to become self-reliant and to combat the oppressive factors in their environment.

If we are to be effective, Oxfam staff, volunteers and supporters must function as an integrated movement. In our fundraising and trading activities, in the stewardship of our resources, in our patterns of consumption both personal and corporate, should be reflected the same aims for which we work in our overseas development programme.

Our public opinion forming and educational work should be rooted in the lessons learned from our overseas programme. We also recognise our responsibility as citizens to influence, where appropriate, the organisations and institutions in this country that are involved in the wider aspects of our relationships with the poor countries.

In all this we are aware that we live in a changing world. Our own organisation and the policies we pursue must keep pace, therefore, with new insights as they develop. We must be sensitive to the need to change ourselves.

All people, whether they be rich or poor, strong or weak, privileged or deprived, are interdependent and should share in the common task of seeking to achieve humanity's full potential. Oxfam provides people in the United Kingdom as well as overseas with the opportunity of playing a small part in a much larger struggle to eliminate poverty and to help human kind develop in a spirit of partnership.

The Objectives of Oxfam's Work Overseas

I   The non-governmental organisation

In its field operations the NGO cannot expect to create structures parallel to those of the State, nor act as a substitute in the provision of services which are recognised as a State responsibility. However, NGOs can be effective in working with those groups ignored or by-passed by large State development schemes. The main priority should be to try to reach the growing numbers of people not affected by liberal reforms or increases in wealth.

Perhaps the main problem for the NGO is that the scale of its operations is too small to effect direct structural change at a general level. It can, however, be effective in supporting small-scale self-help schemes and pilot projects, and can assist in the development of ideas. It must be recognised, though, that the self-help approach, while it is often the only way open to the poor in a world of shrinking resources, and can bring tangible benefits to those in need, is part of a trend in economic development that mainly benefits the wealthy. So long as the poor are forced to sustain their own development, without the assistance or even cooperation of society at large, they will remain marginalised both economically and politically. Therefore the NGO should encourage enlightened development education directed at the general populations of both Third World and industrialised nations and especially at governments. By exposing the problems caused by poverty to society at large, it is hoped to change attitudes towards the poor and affect policies so that those who are privileged will become more active in helping the poor and will direct funds and resources to that end.

Frequently in low-income areas many of the resources necessary for development exist, but are either under-utilised or appropriated by wealthy elites. Foremost among these is the intelligence, ingenuity and effort of labour itself. Self-help programmes funded by the NGO may well not bring about structural change on a large scale, but they can at least help ensure that reserves of capital, labour, appropriate technologies etc. are used to a more constructive end — an end which will benefit the poor and not the upper income groups.

Thus, the ability of the NGO to achieve its goals depends not only on its own resources and policies, but also on the constraints operating in the environment in which it works. Much of the NGO input may seem discouragingly fragmentary and yet the development agency must work within prevailing political conditions. With persistence and tact some solutions may be possible, even in the most intractable political situations. In many areas the NGO faces not so much an antipathetic government as a complete vacuum, where the development effort is circumscribed by problems other than the political. In such circumstances the government may well welcome the operations of the NGO.

However, even in areas where the need is very great, NGOs should never take for granted their impact, and their work must be monitored and assessed regularly. One of the shortcomings of much past NGO work has been the way programmes have been influenced by the need to be financially accountable in the short-term. Development in a true and profound sense is a slow process. One cannot expect people to change their life-styles and life-chances in the short time-span of the average NGO programme. In some ways funding and volunteer agencies can be blamed for a project-centred view of development whereby projects are expected to come to fruition according to a timetable. This pressure, which may cause considerable strain in a project, is compounded by the need felt by most NGOs to build structures to demonstrate a programme's success.

This is not to say that the NGO should not support small, short-term projects with limited, concrete goals, but that development should be seen as a gradual and sometimes radical process rather than a series of projects. Often it is not the outcome of a project — in terms of the material benefits accruing, the numbers of people affected etc. — which is most important. The very act of demanding a service can be a reflection of a people's awareness of their role in society, their needs and the obstacles they confront. And sometimes the development of this awareness can be the main goal of a programme. (See PART FOUR)

II   Oxfam and development

Oxfam's main objective is outlined in its Memorandum of Association:
"The relief of poverty, distress and suffering in every part of the world without regard to political and religious beliefs."

This involves, on the one hand, trying to relieve the suffering caused by poverty and, on the other, reducing the total number of poor. But Oxfam is trying to do something more: namely, to influence the process of development in such a way that the poorest are enabled to take charge of their own lives and to mount their own initiatives in the improvement of living standards. Thus, the main objective is to try to end bias against the poor in the development process.

It is hoped that the projects supported by Oxfam will be designed both for:

☐ the poorest to *have* more (particularly in terms of food and health care) and to gain control of a fair share of the world's resources, and

☐ for the poor to *be* more, in terms of self-confidence, ability to manage their own future, and improving their status in society at large.

We are not just concerned, therefore, with material improvement through rural or urban production schemes or health programmes, but also with the manner in which material change is organised; i.e., with the social institutions and organisations that accompany these programmes.

It follows that the main aim of Oxfam as a funding agency is not simply to improve access to resources or deliver essential services, but to assist the poor to gain increasing control over resources and to remove the obstacles to a more equitable distribution. To have any real impact, development must involve a more rational distribution of ownership; income transfer; institutional reform; the provision of services and the increased participation of the poor in political life.

Oxfam has identified three key objectives in its work:

a: To fund and support small-scale development programmes in certain priority areas (geographic and functional) which will enable the poor — as far as possible — to provide for their own needs, to obtain social justice and to secure their basic human rights.

b: To fund and support humanitarian welfare work among deprived people in certain priority areas.

c: To fund, plan and, where necessary, operate an effective relief response to those disasters identified as requiring Oxfam's assistance.

Each of these priorities is discussed in some detail in the following 'PARTS' of the Handbook. Often development, relief and welfare programmes will in reality be indivisible. And it is hoped that all programmes will have a strong development component, enabling people gradually to devise their own strategies for subsistence and survival.

It is the primary purpose of this Handbook to provide guidelines, grouped largely under the headings of Agriculture, Health, Social and Economic Development, to show how our aims may be pursued. But, however important such guidelines and the development framework they imply may be, we must always remember that Oxfam exists to serve people, not to promote an ideology or development model.

The first essential, then, is to respond to human need out of compassion and a sense of natural kinship with those who suffer. The guidelines are necessary so that this humanitarian impulse may be tempered with realism, so that it does not degenerate into mere sentimentality. Oxfam field staff should be flexible and imaginative in interpreting the guidelines and should respond to human need wherever there is a genuine opportunity to do so.

1 The Distinction between Development and 'Rescue' Aid

The main challenge for aid agencies is twofold:
☐ the urgent need to save the destitute from extreme deprivation and premature death, and
☐ the need to arrest and reverse the process of decline towards destitution.

To the extent that in any given situation the former dominates, Oxfam's response will be in terms of gifts in cash and kind, and will be a short-term measure until a more substantial change in the condition of the victims can be effected. Thus, even though Oxfam's main aim is to help people to help themselves, 'welfare' projects may in certain circumstances be necessary, especially where the people concerned are not in a position to benefit from a development programme.

In situations where the main priority is to reverse the process of destitution, it is important to understand not only present conditions, but also the past and likely future dynamics of change, and to seek a durable solution. This will be one which results in the rehabilitation of the poor in such a way that there is a good chance of continued improvement after Oxfam ceases to be involved.

The alternative strategies available to field staff, when set against the limited resources at their disposal, serve to underline the dilemma they constantly face between 'rescue' and development aid. Whereas rescue aid may ensure survival but solve none of the long-term problems, development aid, while holding the promise of a self-sustaining pattern of development, entails a high degree of risk and may easily fail.

2 Relief

Although the Handbook is primarily concerned with development strategy, it must be emphasised that the destitute who have no resources and little hope should receive assistance to secure their survival. Ideally, relief will give way to longer-term development programmes.

Disasters and emergencies present a challenge which Oxfam cannot ignore. The plight of people made homeless or afflicted by sudden disaster such as earthquakes, hurricanes, epidemics or forced migrations evokes a response from the public which has provided much of the impetus for the growth of Oxfam. To do justice to the compassion felt by our supporters and by the public as a whole, it is essential that we continue to assist in times of emergency and disaster (for detailed guidelines see PART EIGHT).

There are many communities in the world where people live in a chronic state of destitution, degradation and illness, for which development can offer no solutions. Compassion demands that Oxfam should respond in such circumstances.

3 Development

If it is accepted that Oxfam strives to arrest and reverse the processes that result in deprivation, the difficulties of designing appropriate and effective programmes should not be underestimated. It takes time, patience and imagination to gain a full understanding of current conditions and their causes, and to formulate and promote solutions. This is in many ways a new field and, unfortunately, little can be learned from orthodox development programmes designed from the top down; indeed, in some cases it is the unequal and unacceptable effects of such programmes that Oxfam will have to counter.

Disillusionment with development projects which stress narrow economic objectives has resulted in a search for other strategies which stress the social dynamics of change and the nature of the distribution of assets within society. The dominant, 'modernisation', approach to rural development may well have increased levels of agricultural production but has also increased inequalities and helped to perpetuate unjust and inegalitarian forms of organisation. Modernisation has benefited some, but the vast majority of people in the world continue to be marginalised and excluded from development projects.

Oxfam tries to avoid becoming involved in development work as an operational agency, but to retain its integrity as a funding agency. Whereas in the past funds were often channelled through large international agencies, missions and other expatriate groups, now the emphasis is on funding directly either grass-roots groups or local intermediate development agencies. Sometimes the operational approach to aid may seem less frustrating and more dramatic in its effect (especially if development is measured by material indices) than the less centralist approach. But a decentralised form of development is based on the view that the development initiative is more likely to be sustainable if there is strong support for, and a high degree of control of, the programme at the local level.

It is mainly for this reason that Oxfam's policy in development has moved away from the funding of large numbers of expatriate experts and technical advisers towards providing local people with the opportunity to put their own initiatives, ideas and enthusiasm to work for the benefit of their own communities. Thus, Oxfam's role in the development process is that of a partner. Although the organisation has benefited from the experience of working in many different countries and contrasting circumstances, each culture and socio-political context presents unique problems, and it is therefore neither possible nor desirable that the organisation should manage the process of change. Funding and intermediate agencies, grass-roots groups and all other interested parties should work together in a partnership, exchanging ideas, sharing their effort and taking risks towards achieving their goals.

Emphasis is placed on the individual and unique circumstances of each society. The era of 'blue-prints' for the development of the Third World, in which pre-packaged solutions are worked out in the capitals of the industrialised countries is over. Stress is placed on small-scale projects rather than on comprehensive planning strategies, and only secondary importance is given to quantitative results. Much more emphasis is placed on the *processes* by which results are achieved than on the resultant *forms*: for example, it is much more important to ensure that the content and continuity of educational provision are appropriate than merely to construct a school. Major emphasis is placed on increasing the opportunities of previously excluded groups to participate more fully in the development process.

If people are to 'be more', they must participate fully in their own social and economic development: they must make their own choices and not become the servants of an externally-devised grand design.

In the past ten years or so an approach has emerged which emphasises the *participation* of the poor in determining self-reliant development strategies. The paternalism of 'traditional' strategies is replaced by an emphasis on the provision of education and employment skills which will enable the poor to formulate their own programmes for change (see PART FOUR). Thus, an increasing number of development projects which Oxfam supports stress what we might term non-material objectives. Such objectives include the raising of people's consciousness, increasing solidarity and creating organisations through which the impoverished and marginalised groups within a society can build a more secure, less oppressed base from which to challenge established privilege. Oxfam's aim to relieve poverty, when translated into such terms, means supporting projects to increase awareness and self-determination and to remove the material barriers to self-reliant development.

This approach means directing support to particular groups within society: the poorest of the poor, those marginalised socially or politically and the unemployed. It is different from the traditional social welfare approach, which emphasises the care of the socially disabled. Firstly, it extends the meaning of this disability to include much larger sections of the population. Secondly, and more importantly, it sees disability as a product of the disabling processes of rapid social change; processes which encourage dependence and push more people into poverty and thus on to the periphery of society. Participatory democracy is identified as the most important mechanism for ensuring that a programme is appropriate to the problems presented in a given area. Participation is not merely an ideal; neither is it simply a question of consultation with project holders; nor is it defined by the acceptance and/or use of a particular service provided by the development agency. Participation is crucial to the identification of the goals of a programme, its implementation, organisation and evaluation, and is thus a vital factor affecting its potential for success. The participatory approach places strong emphasis on the need for social awareness; the identification by people of their common problems and goals; the importance of information, education and training to enable the effective use of local resources; and the necessity that all programmes be founded on collaboration between the funding agency, grass-roots organisations and, if applicable, the intermediate agency.

The stress on participatory democracy in development has certain practical implications.

a: The outcome will depend on the extent to which participation has occurred at the local level in all aspects of the project and on whether local resources have been truly mobilised.

b: The level of funding and the technological input must be appropriate to the local culture economy and the administrative and organisational capacity of the project holders. For example, in those rural areas where the need for assistance is very strong, but the population is only nominally incorporated in the market economy, it is all too easy to flood a zone with funds. In such circumstances, local initiative and participation in, or control of, a project may easily be destroyed.

c: The potential for full participation at the local level is very much determined by the degree of social differentiation and the interests of the local leadership. Charismatic leaders may be very persuasive and may have a strong influence on field staff and project holders alike, and yet they frequently inhibit the political development of the group they represent and marginalise the weaker members.

d: If the emphasis is on a democratic polity and collective effort, then it is important to examine the tradition of collective action in any given area so as to avoid imposing models. For example, in Latin America, despite extreme social differentiation, the pre-Columbian tradition of collective labour organises and reinforces many present day development programmes in both rural and urban communities. In India, on the other hand, strict caste observances would make this kind of community involvement impossible and collective effort would more likely be based on different principles of unity, such as kinship.

e: It may be that democratic participation is not simply a means of achieving a given end, but is seen as a goal in itself; a goal which helps focus social awareness and provides continuity in development.

2 Self-Reliance

Where there is full participation at all levels and stages of a development programme, it is hoped that one result will be self-reliance in financial, administrative, educational and social terms. However, in many instances — and particularly in remote rural areas — complete independence is not possible. Even in industrialised countries where extensive infrastructures have been established, the cost and organisational problems involved in creating a new service may be enormous; but in low-income countries the obstacles to improving health or education provision etc. will be even greater. A people occupied almost full-time with survival cannot be expected to have the time, energy or even the interest to sustain a programme without support of some kind from outside.

Perhaps in an ideal world all funding would be directed at grass-roots organisations — communities, cooperatives, associations etc. Even though this kind of work requires a high staff input from the funding agency, direct funding can bring considerable savings, since the institutional structures, staff and capital equipment of intermediate agencies can be extremely costly. However, it is often necessary to seek the support of local development institutions in the planning and implementation of projects or the provision of specialist services. Working through such organisations has the advantage of assisting in the formation of local development expertise. Further, the foreign funding agency is not capable of providing outreach to all groups that need help, nor is it desirable that it should do so. Therefore, where appropriate, local intermediate agencies should be identified and encouraged.

If a local development agency is involved in a programme, it is extremely important to examine its relationship with the grass-roots organisations. All too often it is assumed that the credibility of local professionals stems automatically from the very fact that they are local as opposed to expatriate. The marked social stratification characteristic of most low-income countries means that local middle-class professionals frequently have very little understanding of the problems faced by the poor. In most cases, the mandate to work in a given area, or with a given group of people, arises not because of a person's origins but because of their motives, their actions and their ability to foster a relationship of trust. The difference in status between the manual labourer and the local professional may be such that there is very little exchange of ideas between the two and very little chance of their working together on equal terms.

The achievement of self-reliance depends on a number of factors.

a: The constraints of the area: geographical, economic, social and political. Independent development initiatives are frequently undermined by forces external to the programme; income generation projects, for example, often increase dependence and exploitation.

b: The project timetable should not be planned simply to fit in with the requirements of the funding agency. The decision to end funding is an enormous responsibility. The progress of a project may be much slower than expected and therefore a lot of flexibility is needed in planning financial assistance. Constant monitoring and evaluation by all interested parties is the best way of determining when the funding agency should be able to withdraw without destroying a programme.

c: Where a project will require a permanent external input, for example, in the form of technologies which cannot be produced by project holders, then alternative sources of finance (other than the funding agency) must be identified from the outset. If there are sufficient reserves, the project itself may take over funding, but if not, arrangements must be made with the government or some other institution. It is too often the case that only when a project has already absorbed all the capital available from a funding agency does the search for an alternative source of finance begin.

③ Programme Scale

The scale of a programme can be measured by the level of funding, the area or population affected, the number of structures built, or the size of the implementing institution. Oxfam's commitment is mainly to small-scale development programmes, not simply because of the limited availability of resources, but more because larger programmes are less likely to guarantee a high level of participation or address problems at the local level.

In recent decades there has been world-wide criticism of many of the large-scale, 'top down', development programmes. Attention has been focused on certain shortcomings which seem often to result: alienation of project management from project holders; the lack of involvement of project holders in implementation and administration; bureaucratisation and the use of inappropriate technologies and policies. The impact of these programmes is rarely commensurate with the resources invested. Thus, throughout the world we see large, unwieldy development projects boasting massive funding, sophisticated equipment and expertise and large numbers of professional staff. And yet, despite exaggerated claims, they often prove unrealistic in their objectives, ill-advised in their methods and ineffective in terms of results; above all, failing to reach those most in need.

There are certain problems specific to funding large non-governmental operational agencies:

a: The institutionalisation of the agency. A commitment to working for/with the poor may rapidly be replaced by a commitment to maintaining the institution and providing jobs for the staff. This trend is accentuated in countries with high unemployment among middle-class professionals. Projects will often be devised by staff in order to secure their own future, and participation in project design and implementation will be severely restricted.

b: Large operational agencies will be seriously affected by changes in funding. The cost of maintaining institutional structures with inflated bureaucracies and staff numbers can be exorbitant — especially where they are based in urban areas — and it is unlikely that self-financing will be feasible.

Perhaps the most serious problem with large-scale 'top down' programmes is the tendency to create a level of dependence which leaves the project extremely vulnerable to changes in circumstance. If a project is beyond the control of project holders or is extremely expensive, then it is unlikely to be able to survive changes either in government or in funding policy.

However, small-scale development programmes also have a number of disadvantages. For example, small groups often have very limited skills and abilities; they are frequently dominated by one or two people; they have difficulty in planning a programme and in assessing the results of their efforts. The activities of small programmes, by definition, have limited impact and sometimes it is hard for the funding agency to detect any impact at all — much depends on faith in the organisations involved. Finally, their size frequently makes them very vulnerable to attack by the vested interests they offend.

Thus, there are no rigid rules regarding the size or scope of development programmes. The priority of field staff should be to support those programmes which with the minimum possible outlay reach the maximum possible numbers of people, the scale of funding or institutional development never undermining the feasibility of participation of, and control by, project holders.

4 Development as an Intervention

It must be recognised that however little field staff may wish to interfere in the development process, the role of Oxfam as a funding agency will always imply an intervention of some kind. But Oxfam at least has the advantage of not being forced to work to any specific directive and can support either single-or multi-purpose programmes, in response to local priorities. Continual evaluation and monitoring of both policy and practice is necessary to assess the impact of the funding agency in the countries in which it is active.

Perhaps the most controversial feature of intervention by the development agency is the way in which it may affect local cultural traditions. The wide range of cultural and institutional conditions in different parts of the world will affect radically what the funding agency is able to achieve in each region. Enormous tact and sensitivity are required when dealing with cultural attitudes which apparently obstruct educational or developmental activities. Yet one of the main challenges faced by the development agency is to try to reach those most deprived and marginalised: this implies that we cannot take local structures and attitudes at face value.

Rigid rules regarding marriage and kinship; taboos, prohibitions and other cultural practices; strict class and caste observances, etc., may all contribute to group security, but often at the expense of individual freedom. While these cultural traditions may be appropriate in many contexts, they frequently outlive their usefulness to the individual or community in changing circumstances. And yet they do not disappear easily, nor should a pattern of development which is destructive of a cultural tradition be encouraged. Sometimes the most violent social and psychological disruptions brought about by economic change can be those which result from the destruction of the integrity of a local culture and the removal of a people's links with their past (see PART TWO, Section 5).

5 Evaluation and Monitoring

It is recognised that all development programmes are experimental and entail some risk. Therefore, constant monitoring and evaluation is an essential component of ANY programme (see PART THREE). This kind of assessment not only helps to minimise detrimental effects but enables all interested parties (funding, intermediate agencies and grass-roots groups) to benefit more fully from the experience of development.

Evaluation should be integral to a programme and not added as an afterthought or as a requirement of the funding agency anxious to demonstrate success in order to justify its involvement. Evaluation and monitoring offer all concerned a chance of redefining goals or procedures where necessary and of applying the experience to future programmes.

Priority areas

In the field of development there are never sufficient resources to meet demand. Geographic priorities have to be established and resources consolidated if funding is to have any impact and projects are not to be dispersed and isolated one from the other. Whilst always remaining ready to respond to emergency needs, field staff are encouraged to concentrate their efforts as far as possible in those geographical areas identified as taking priority in Oxfam's overseas programme. However, to assess which areas, or indeed which countries, should receive priority in the programme of the funding agency is an extremely complex task. The problem is compounded by the competing demands of other priorities; the need to focus on particular disadvantaged groups or on certain kinds of programme.

The factors governing decisions as to which countries should receive assistance may differ from those determining which area within a country will take priority. When comparing countries and assessing relative need, factors such as GNP, poverty differentials, the percentage of total population in the low-income bracket, levels of infant mortality, incidence of fatal diseases, and the receptiveness of government, will all influence funding policy. While many of these factors may be equally as important in the assessment of priority areas within a country, field staff will also be expected to evaluate whether involvement in a given area — particularly if it is very isolated — can be justified in terms of travel costs and the time incurred by staff in visiting projects. It is also very important, of course, that programmes at the local level fit in with the priorities established for the country as a whole.

Ultimately, few programmes result simply from a decision about geographic priorities. The training, practical experience and perceptions of individual field staff have a strong influence on the way in which programmes develop and on programme priorities. Field staff may feel that a certain group of people should take precedence, regardless of whether or not they live in a priority area. Some of the more experimental programmes may merit funding even when they do not fall in with geographic priorities, simply because they can provide valuable experience for use in similar work elsewhere.

In the final analysis, many NGO development programmes are established not as a result of careful assessment of long-term need in a given area or the feasibility of sustaining a development initiative within that area, but are built up from an emergency intervention following a specific disaster. While it is important to work in hazard-prone areas, funding agencies must not harbour the illusion that an emergency response can, or should, *automatically* be converted into a long-term development programme.

Perhaps one of the main features distinguishing the geographic priority of one programme from another is whether it is set in the rural or urban context. Inequality in the production process and in the distribution of wealth characterises underdevelopment and is a condition of both rural and urban areas. For a number of reasons Oxfam has in the past tended to work mainly in rural areas. This has partly resulted from the effort to curb migration to the cities by strengthening the rural economy, but has also arisen because in most Third World countries the largest concentrations of poor have historically been in the rural sector and because the rural poor have seemed to be more isolated from essential services than city-dwellers.

Furthermore, it is often argued that the problems of the city in low-income countries are intractable: the urban poor are dependent for their livelihood and sustenance on people who take no interest in their welfare and on a State that does not have the capacity to meet their needs. It has been said that those urban poor not included in the permanent wage sector are unable to take charge of their own destiny. More important, it has frequently been argued that urban conditions inhibit collective action. Extreme heterogeneity in class and in sources of income challenge development initiatives, making it extremely difficult to find policies which would change prevailing conditions. The stress of urban life, it is claimed, encourages an attitude of individualism. Survival strategies in the highly competitive urban environment are felt to depend on entrepreneurship and are contrasted with the non-monetary, reciprocal exchanges which supposedly prevail in rural areas.

However, a more accurate picture of the state of affairs in most urban areas would be of households reinforced by ties of kinship, ethnicity, caste, tribe or friendship with other, neighbouring, households; ties which form the basis of association for the urban poor, despite a competitive and frequently hostile environment. The residents of a given neighbourhood will often mobilise for specific ends. Even if, once the goal has been achieved, the group appears to disband, this does not imply that the potential for future collective action is lost.

It is often believed that the NGO is able to have a more noticeable impact in the countryside and can elicit a more positive response among the rural population than in the city. Rural dwellers are seen to conserve a sense of community based on common values; fostering ancient traditions and favouring collective participation in development projects. It is argued that this sense of community is symbolised by structures such as the local church, the market, the washing area, or common lands.

But we should not assume that the concept of 'rural community' has any meaning to anyone other than the outside observer. In most rural settlements there will be at least a degree of differentiation, and many of the inhabitants will be marginalised either at the community level by their neighbours, or within the home by their kin. It is all too easy for a development project to benefit certain people while excluding, and maybe even harming, others. Rural leaders do not necessarily have credibility within their community — especially when they are appointed by outsiders — and may well be representing their own interests rather than those of the community as a whole.

In the following sub-sections some of the factors affecting the distribution of wealth in the Third World are discussed. These will have a strong influence on the selection of priority areas for funding and the assessment of the feasibility of mounting a development programme in a given area or country. However, it would not be appropriate in a manual such as this to name those countries to which Oxfam has chosen to give funding priority, since the situation of some individual countries can change radically in a very short period, as can the relative circumstances of different countries. What is more important is to highlight some of the principal problems and causes of poverty and inequitable distribution.

II Current trends

In a world which is at present suffering severe economic recession, a growing number of low-income countries face a deterioration in their economic prospects. On the one hand, as primary producers they are affected by deteriorating terms of trade in world markets. While the demand for most raw materials remains constant (and in the case of some minerals, is on the decline), the majority of producers rely heavily on exports to finance their external debt and are therefore forced to increase output in an already extremely competitive market. On the other hand, most Third World countries experience serious internal deficiencies in food and fuel which, despite the recent rise in the price of these goods internationally, force them to increase imports. They face a negative trade balance which in many instances can only be financed through massive loans.

Many Third World governments have responded to these problems by introducing deflationary policies — cutting back on government expenditure. It is the poor who are most affected by the removal of government subsidies on food and transport and the erosion of essential services such as education, health and agricultural extension. In recent decades we have seen a significant reduction in the real living standards of the poor and an increased polarisation in the distribution of wealth at both national and international level.

The Food and Agriculture Organisation has estimated that there are 400 million severely undernourished people in the world and possibly as many as 1,000 million who are significantly malnourished in some respect. These people's consumption of food and their access to other necessary goods and services is either at or below a subsistence minimum. They are deprived of the means to raise their standard of living; and they experience permanent hunger, squalid living conditions, endemic — and periodically epidemic — disease, a rate of infant mortality ten to twenty times greater than that experienced in industrialised countries, a reduced life expectancy and severe physical and mental weakness.

The process of impoverishment and increasing powerlessness is relentless. The victims of this process are frequently dispersed both geographically and socially and therefore find it extremely difficult to organise themselves in their own defence. Their isolation is compounded by the necessity to devote themselves full-time to the exigencies of mere survival. Nor is it sufficient to identify the poorest groups by the principal occupation of the head of the family. Family structures may be such that certain members suffer a great deal more than others — especially at times of maximum stress. In many societies women, female children or the aged may be deprived of adequate nourishment, even where there exists a margin over subsistence in production.

At the same time, scarce funds are being diverted into arms purchase and the maintenance of wealthy urban elites. The present world economic crisis and the struggle over the control of resources has in many areas contributed to a high degree of political instability and violence. Civilian populations in different areas throughout the world have been displaced, many becoming refugees and losing all ties with their homeland. Others have been detained as political prisoners or become the victims of arbitrary acts of politically motivated violence.

There is a very real risk that the populations of the industrialised countries will become complacent in the face of these conditions, accepting them as inevitable. Official aid programmes have been cut back dramatically in recent years as many of the world's wealthier nations also embark on a policy of retrenchment. But the reduction in aid above all reflects the loss of political will among industrialised nations to assist the governments of the Third World in the struggle against poverty.

1 **The Dynamics of Poverty**

Most Third World countries suffer disproportionately from the random impact of geography and climate. Many are faced with periodic — and in some cases frequent — natural disasters such as earthquakes and cyclones which cause deaths and are a constant drain on resources. In addition, the marked seasonality in the supplies of food and water has a dramatic effect — particularly in the tropics — on levels of nutrition and disease, afflicting those without reserves and leaving them even weaker and more vulnerable.

However, much of the poverty found in the world is attributable not to accidents of nature, but to the mechanics of human society and human interaction with the natural environment. The desire for short-term gain all too often outweighs any interest in securing resources for future generations. The indiscriminate removal of vegetation, over-production and the poor management of resources generally has effectively diminished economic opportunities relative to population growth. Ecological damage leads to the withdrawal of land from agriculture or drastically reduces its potential, thereby threatening subsistence.

The world's population is still increasing dramatically, even though in recent decades the rate of increase has been slowing down. In many areas this problem is compounded by the fact that while fertility rates remain high, there has been a reduction in mortality rates with improved health care and sanitation. Therefore, it is to be expected that, despite the reduction in rates of increase, in some areas the population will still double within the next 25 years. Clearly, a rate of population expansion of this magnitude will place severe stress on food stocks and public services. But another demographic trend — the significant increase in the proportion of the population under the age of 15 — presents special problems for certain services, such as education, as well as affecting employment strategies and placing a strain on social and family structures.

The demographic crisis can be felt particularly keenly in urban areas where the marked natural increase in population has corresponded with rapid growth resulting from migration. Population growth in rural areas has caused considerable pressure on land, but in many countries this has been lessened by the expulsion by property owners of large numbers of people. The whole question of population control has in recent years become an extremely controversial issue. Policies designed to reduce population growth have become the subject of much debate. It is felt by many that they evade the main issue of underdevelopment — that poverty is caused less by population size than by the inequitable distribution of resources (see PART SEVEN Section 6, Birth Control).

Indeed it is true that population growth only becomes a serious problem when the rate of growth exceeds the increase in food production, or where the distribution of food is very uneven. Very few countries are net exporters of agricultural produce and an increasing number rely heavily on food imports. The problem of declining per capita food production has been particularly serious in sub-Saharan Africa; it was falling in this area by about 1% a year in the 1970s, and by 1980 it was around 11% lower than in 1969/1970. Since agriculture is the sole source of income for the great majority of people in the region, a fall in production of this magnitude has had a very severe impact on incomes. Between the years 1971 and 1980 food imports to Africa multiplied in volume 2½ times. Projections of the continent's food requirements to 1990 show that without a sharp increase in domestic food output, the shortfall between production and consumption will become even more serious, resulting in an increase in imports to as much as two or three times the present level.

Various forces have come into play in sub-Saharan Africa which together undermine production by causing soil erosion and desertification. These include the 'parcelisation' of properties, inappropriate changes in agricultural practices, and overproduction in areas with fragile ecologies. But production has also fallen as a result of the abandonment of land because of economic and political pressure. Where there has been an increase in production this has often been in cash crops for export rather than foodstuffs for local consumption. Such a trend may earn foreign exchange, but will not solve the problem of food shortages.

Thus, population growth, although an important contributing factor, is not necessarily the main cause of poverty. Instead, we should look to the inequality in the control of productive assets and in the distribution of power and incomes and to people's relationship with the natural environment. Despite the tremendous shortage of resources, the allocation of them throughout the Third World discriminates heavily in favour of the owners of wealth: a small number of people are seen to prosper with economic development and a much larger proportion of society is deprived of any benefits. A growing number of people exist who have no prospect of even minimal support from national welfare programmes and to whom control of their own destiny is denied (see PART TWO and PART FOUR).

2 **Industrial Development and Investment**

The main sources of capital for economic development are foreign aid, commercial investment and domestic saving. The tendency in most low-income countries is for the greater burden of domestic saving to fall on the already impoverished agricultural sector. Despite the scarcity of capital in the Third World, domestic investment in the industrial sector is inhibited by low rates of return. The main deterrent to direct investment is the reduced size of national markets; but there are other problems such as the shortage of skilled labour and administration, as well as the inadequacy of internal transport facilities, communications and power supplies, all of which reduce the efficiency of capital. Only through substantial investment in supportive activities can improved rates of return on direct investment be achieved.

Since in most low-income countries national investors cannot supply sufficient capital to sustain economic expansion, there is considerable reliance on capital obtained through foreign exchange, or direct foreign investment. However, foreign investment represents a degree of foreign control in the economy. It also tends to distort economic development, favouring imported machinery and technical and administrative expertise and the concentration of indigenous capital into a few hands.

While most countries have surplus labour, development and investment policies derived from industrialised technologies stress labour-saving techniques and use scarce and expensive capital. Governments and business alike have encouraged the development of large industrial plants using imported technologies, which offer little benefit to the local population other than to a small elite of skilled labourers, who stand to gain employment. The accumulation of capital occurs almost exclusively in the urban sector. This is a particularly unfortunate trend for most African and Asian countries, since in these continents the urban sector contains only a small percentage of the total population.

Thus, the very process of development is seen to marginalise the poor further: industrial enclaves are established which are divorced from the national economy (or are even parasitic upon it) and the ill-conceived policies of governments result in technologically sophisticated development projects which increase the national debt without assisting the poor. Public capital formation tends to support the uneven development of the private sector, allowing the urban middle classes to appropriate an increasing proportion of resources. Those excluded from capital-intensive activities are forced to seek a livelihood in the urban informal sector, become unemployed or work in agriculture. The labour force entering these categories is growing faster than the rate of capital formation, hence the decline in real incomes.

③ New Patterns of Consumption

The progressive expansion in recent decades of economic and cultural ties linking the Third World with the industrialised nations has caused a noticeable shift in consumption patterns and expectations. New aspirations are now to be found in both the city and the countryside and are often completely inappropriate to the life-styles and life-chances of the national population.

Contemporary patterns of consumption reflect the concentration of wealth in a few enterprises and households. Those with political power also have the purchasing power, and so the development of the national economy is shaped by their demands. The demand for consumer goods by the rich contributes to an increase in imports which in turn compounds the balance of payments crisis. Further, the contrived cheapening of imports relative to domestic labour, quite apart from increasing local unemployment, reinforces economic dependence on the industrialised nations and undermines the integrity of the national polity. Given the world-wide depletion of natural resources, it is doubtful whether the exaggerated consumption patterns found in industrialised countries will ever be either appropriate or viable for the majority of the world's population.

III   Urban centres

Since the end of World War Two there has been an enormous increase in the rate of urbanisation in the Third World. However, the proportion of the population living in urban centres varies markedly from one continent to another. In Latin America today 60% to 70% of the population can be classified as urban, whereas in Africa the range is between 20% and 30%.

Often apparently comparable statistics are misleading. For example, in countries such as India and Egypt a population centre is only classified as urban if it contains more than 10,000 people — a number significantly higher than is used for such classifications elsewhere. Thus, it is possible to underestimate seriously the percentage of the urban population as a proportion of the whole and therefore to underestimate the numbers of urban poor. The problems faced by the urban poor are numerous. They include low incomes, or income insecurity, material deprivation, residential instability and isolation from essential services and the political process.

The considerable differences between countries and continents in the rate and extent of urbanisation can be explained by differences in the character of the local socio-economic structure and in the way in which each country has been integrated into the world economy. Perhaps the greatest determining force in urban development has been the expansion of the world market, which began with the colonial enterprise of the industrialised nations. The infant industrial sector in 19th-century Europe was fuelled by the export of primary products from the Third World. Urban centres were formed in the exporting countries as trans-shipment points along export routes, close to centres for the extraction of raw materials. The main problem with these early urban centres which were created primarily to further the colonial enterprise was that they developed no linkages with the hinterland and bore little relation to indigenous trading networks. Examples include many of the early cities of Hispanic Latin America, which were founded through the forced resettlement of the indigenous population to suit the economic goals of the conquerors.

However, much urban development is attributable not just to the exigencies of the world export economy, but to concentrations of population which were initiated in the pre-colonial period. For example, the Yoruba towns of Muslim Nigeria and the Mogul cities of the Gangetic plain of India all flourished in pre-colonial times. These earlier indigenous developments even today help to explain regionally differing rates of urbanisation and interior linkages.

It has become apparent that one of the main problems with the development of urban centres in the Third World is not so much the concentrations of people in towns, but more the rate of urbanisation. Urban expansion is attributable not, as many observers have claimed, simply to high rates of migration, but also to a rapid natural increase in population. In towns of ancient origin migration tends to be relatively unimportant. In Latin America — where many cities date from the 16th century, and some even earlier — migration accounts for less than 30% of the increase in urban population. On the other hand, in Africa — where the towns are fewer, much smaller and on the whole newer than in Latin America — migration is by far the most important factor in urbanisation. The high rate of natural population growth in many urban areas presents a challenge to those development agencies which have always concentrated on the rural sector on the assumption that this will halt migration and stabilise the rate of urbanisation. Third World cities continue to expand at an alarming rate, requiring of development agencies a reassessment of their priorities.

One of the main trends in urban development reflects the extreme centralisation of political control in most Third World countries: it is the concentration of populations, resources and industry into principal cities. This has the effect of both putting too much strain on the housing and labour markets of the principal city and undermining the development of provincial towns. A second trend is the increased suburbanisation of industrial location; the outlying districts in most cities now grow much faster than inner city areas.

Perhaps urban expansion would not present such a severe problem for development if it were in some way matched by the growth of the industrial sector and by income opportunities. The main characteristic of urban centres in low-income countries is that the majority of the population is either unemployed or underemployed. The formal labour market is very reduced in size and, so long as industrial development is capital-intensive, has little prospect of expanding.

Equally, there is enormous congestion in the housing market — with demand for permanent structures and fully serviced properties far outstripping supply. Rents are high and materials expensive, and so the majority of people are forced to depend on self-help schemes to solve their housing problems. Urban settlement in most Third World countries stretches far beyond the reach of potable water supplies, electricity, health and education provision and good communications.

The urban poor, then, are those people who are disadvantaged in the job market, in the political process and as regards their place of residence. They include those not engaged in full-time, permanent, paid employment, who are deprived of benefits and often live in squalid conditions without essential services.

1 The Urban Labour Market

Most urban employment is concentrated in the service sector, informal economic activities and artisan workshops. The majority of people are involved in intensely competitive, mobile and transient occupations such as casual street trades and construction. In most Third World towns and cities there is a multitude of people engaged in the sale of food and trinkets in the street and offering a variety of services, from car-washing to guarding property. However, not all the poor are concentrated in the informal sector, for they may include the workers in sweat-shops and other small-scale establishments and self-employed owners. The links between the formal wage market and the informal sector are numerous and diverse, and among the more exploited of the poor are those tied directly to the formal wage sector: the outworkers, sub-contractors and casual labourers. Furthermore, at the level of individual households it is common to find some members working in the formal wage market and some in the informal sector.

We have already seen that industrial development tends to be inappropriate to the needs of the local population in low-income countries. An increasing number of enterprises are capital-intensive, relying on sophisticated technologies which require little labour. Others produce modern consumer goods using labour-intensive processes; but these processes depend on the use of cheap labour and normally involve assembling the products of a technology based in the developed world. These types of enterprise rarely generate much local employment since their main ties are with production processes in the industrialised world. But also, industrial concentration of this nature will tend to destroy the basis of craft industries in small towns and rural areas.

The concentration of large numbers of people in the informal sector is vital for the interests of national business, multinational corporations and the State, because it ensures a large supply of cheap labour. It is not sufficient to concentrate development initiatives in urban areas on alleviating material deprivation without addressing the problem of exploitation. The informal sector exists beyond the limits of state regulations and social security systems. Workers in the informal sector are denied any legal protection or union membership and have to cover their own overheads. Where the State is controlled by an industrial bourgeoisie, even in the formal sector there will tend to be a repressive legal framework weakening labour organisation and lessening workers' benefits (see PART FOUR Section 4, Legal Aid).

2 The Problem of Residence

The impact of low incomes in urban areas is most clearly demonstrated in the problem of housing; the urban environment tends to be extremely expensive and housing provision hopelessly inadequate. Most of the urban poor live on the very outskirts of cities on land with little or no commercial value, above water or swamps, or on rubbish dumps. The materials used for housing are often impermanent and are rarely resistant to extreme weather conditions.

In some countries a very large proportion of urban settlement is the result of the illegal invasion of open spaces — usually land owned by the State. Thirty per cent of the housing in the urban centres of Asia and Latin America has resulted from land invasions. Invasion is one way in which urban residents can respond to inadequate housing provision and high rents. In Africa illegal sub-division of rented properties is a more common solution.

In the 1960s many Third World governments tried to overcome congestion in the housing market by destroying squatter settlements and building low-cost housing estates. This policy has, for a number of reasons, failed. The so-called low-cost houses proved too expensive for most poor people and, besides, the state schemes failed to keep up with population expansion. In effect, self-help schemes tend to benefit governments because they enable them to allocate fewer resources to housing.

Squatter settlements are often the outcome of an extremely well-organised invasion. A prospective site is located, an allowance for collective spaces made, streets planned and housing plots allocated — often all well in advance of the actual invasion. Commonly, committees are formed to defend the settlement and to organise residents to install sewage systems, pave roads and obtain supplies of electricity and water.

While self-construction may offer certain advantages, such as the opportunity to up-grade a property when possible, the residents of squatter settlements face many problems. Apart from the legal problems and the deficiency of services, most squatter settlements are very isolated from employment centres. They tend to generate only a very small amount of employment, and most residents work outside the settlement, frequently at a great distance from their home.

3 Programme Priorities

a: Employment. Although there may be many problems in the formal sector — poor working conditions, low wages, etc. — in general, because of the scale of the NGO input it is more likely that programmes in the informal sector will have greater impact. Further, those engaged in informal activities are likely to be worse off than workers in full wage employment.

One of the most common initiatives for tackling urban poverty is income-generation, providing funds for the purchase of capital equipment, training, etc.(see PART FIVE). However, such programmes should be approached with caution, and funding should only be considered when the market has been thoroughly tested and potential transport problems examined. Field staff should be particularly wary of programmes which propose to introduce new artefacts or skills or to produce goods in competition with industry. It is extremely difficult to assess whether there will be a real demand for goods which have not existed previously in that zone, and it is virtually impossible for small-scale craft enterprises to compete with organised industrial concerns. But, above all, income-generation projects carry a serious risk; they may well increase the levels of dependence and exploitation of the poor by bringing them further into a network of commercial relations over which they have no control.

Perhaps there is greater potential for assisting the formation of some kind of workers' association or union in specific sectors of the informal economy. Street traders, for example, have in some countries formed unions to negotiate with the government to secure certain rights and to protect themselves against harrassment. Thus, they have gained the right to use permanent licensed sites for marketing without the risk of being moved off by police. There are also many possibilities for improving essential services in the informal sector, with the provision of creches for working mothers, supplies of clean water for the sellers of cooked foods, etc.

The field of training and education provision to improve practical skills for the poor is enormous, and yet there are surprisingly few such projects in Third World urban areas. Clearly the most important feature of any training programme is that it should be appropriate to local technologies and job opportunities. (For further information see PART FOUR, Section 2.)

In general the mobile and transitory nature of the job market in Third World cities makes it very difficult to attack poverty at the place of work, and it is possibly easier to concentrate on problems presented at the place of residence.

b: Physical infrastructure. The poorer residential areas in Third World cities tend to be extremely heterogeneous in terms of levels of income, class and caste. Therefore it is frequently difficult to achieve solidarity in development programmes unless there are very specific narrow goals involving the improvement of living conditions (for example, the provision of street lighting, domestic electricity supplies, refuse collection, sewerage installations and water).

Equally, it may seem difficult to achieve continuity in urban development when collective action appears to be so sporadic in nature. However, even short-term mobilisation will have an impact on people's values and attitudes. It is important to take the history of a zone into account — is there a tradition of collective action or is there a tendency for factionalism to destroy local groups before goals have been achieved?

Among the urban poor in many areas there is a particularly high proportion of female household heads. Thus, an essential component of any urban development programme is to examine the kinds of strains on the family and to question, especially, the feasibility of increasing the contribution of women where they are already under severe stress (this problem is examined in PART TWO, Section 2).

The expense of improving living conditions and providing and maintaining services in low-income urban zones is considerable — particularly where the residential areas are distant from the city centre. Therefore it is not always realistic to expect project holders to become self-reliant in financial terms. Nor is it feasible for the NGO to maintain an indefinite presence. It follows that any pilot project of this nature — whether it be education or health provision, sanitation, etc. — must from the outset provide for a transfer to the State. (For information on urban development and housing see PART FIVE, Section 2.)

IV Rural areas

In most of the rural areas of the Third World the combined impact of colonial rule and the market economy has destroyed or distorted indigenous agrarian structures and methods of capital accumulation. However, underlying modern commercial relations are ancient systems of land tenure and stratifications of power and wealth which serve to reinforce further the inequalities characteristic of the market economy.

In recent decades the agricultural sector of the Third World as a whole has seen an appreciable increase in both productivity and income. Yet, paradoxically, at the same time there has been an increase in malnutrition, a worsening distribution of incomes and a decline in real incomes in the countryside. These trends indicate that a growing proportion of the rural population is living below the minimum of subsistence and a growing number are being alienated from the land. They face severe competition in the market place, appropriation of their assets, detrimental government policies and exploitation by powerful landowners and moneylenders.

This process prevails especially in those regions where an inequitable distribution of land ownership corresponds with a high rate of population growth. It is becoming all too evident that only those who control the productive assets (particularly land) stand to benefit from the expansion of the market economy and the introduction of new technologies in the countryside. Even in areas of low productivity the owners of large, inefficient feudal estates make handsome profits simply by virtue of the size of their properties.

The true extent of rural poverty is in many cases disguised by the remittances sent by migrant workers from towns and mining centres. One of the most devastating effects of the expansion of capitalist agriculture has been the expulsion of populations from the land. The most vulnerable sector consists of those rural inhabitants who possess no formal rights over their land, but farm it by virtue of centuries of occupation. Included in this category are many pastoralists, shifting cultivators, share-croppers and tenants. Few possess title to their land or have any security in law. They are subject to pressure from national settlement schemes and private business interests (lumber firms, property developers, mine and plantation owners, etc.).

There are various forces at play in the expanding market economy which serve to alienate people from the land. When land becomes profitable, owners either seek to reclaim their properties or increase rents. Either way, small-scale tenants are likely to lose control of their land. Another factor is the replacement of labour-intensive techniques with capital-intensive technologies: modernised agriculture requires far less resident labour, and the need for labour is no longer spread throughout the year.

Thus, many rural inhabitants are forced into seasonal, migrant labour on large plantations. Others move permanently from the countryside to towns. Even those families which retain control of some land are normally forced into the more marginal and least productive zones. They tend to inhabit areas prone to decline (through desertification, soil erosion and other processes) or to periodic natural disasters. One of the few ways of countering this process of destitution is by diversifying the source of income; hence the complex network of ties between rural producers and the members of their family working in the town.

The increase in cash-cropping has in many regions had an adverse effect on levels of nutrition. There has been a steady loss of food crops for local consumption — food crops which once not only sustained humans but also animals. The reduction in the numbers of animals has corresponded with a reduction in the proportion of protein in the human diet. The replacement of subsistence by cash crops has made farmers far more dependent on the market. Small-scale producers have no power to influence the market in the way that a large landowner might. Since they rarely possess storage facilities, they are forced to sell immediately after the harvest, at a time when prices are low. Neither can they produce in bulk, which would compensate for low prices. It cannot even be assumed that the increases in yield brought about by technological innovation will increase profit for the smallholder, because the greatest benefits will probably go to the moneylender or the intermediary. The increase in the cash-crop element in the domestic economy will often have a radical effect on family organisation, primarily undermining the role of women, who in many societies are traditionally excluded from the commercial sector. Where the produce is exported, the person who benefits will not necessarily be the producer but the one who can take advantage of the foreign exchange generated.

Technological innovation and the expansion of market relations will normally result in intensive production. But many of the ecologies of the Third World are extremely fragile. Whereas traditional agrarian practices preserved the environment for the future, intensive production without expertise in the maintenance of soil fertility can cause erosion or desertification. New 'Green Revolution' technologies require fertiliser, improved seed and regulated water supplies. All these factors will increase the dependence of the agricultural sector on purchased supplies and commercial sources of credit. Many of the farming supplies, since they are unavailable locally, have to be imported, and formal sources of credit are rarely available to the smallholder or tenant farmer.

Government policies tend to be directed towards increasing the output in the farming sector rather than the welfare of the rural population. The rural sector may contribute to the process of capital formation at a national level in various ways, such as direct taxation or investment by farmers in the non-farming sector. The expansion of rural production may be sufficient to reduce the prices for agricultural products, thereby increasing profits in the non-farming sector and favouring saving and investment.

However, the main problem with drawing on rural production for capital formation is that it is already an impoverished sector of the economy. Per capita incomes in rural areas are much lower than in urban areas. Besides, the long-term trend is for the prices of agricultural commodities to be in decline relative to industrial commodities. Therefore, the foreign exchange earnings from agriculture can only be maintained by increases in volume.

1 The Rural Community

It has generally been assumed that peasant systems of property are more favourable to the cultivator than plantation or other types of economy (since they give greater independence and are less exploitive) and that the peasant community is the ideal social unit for managing and implementing development work. Far more money has been allocated for the development of settled peasant communities than for assisting landless rural labourers or nomadic hunter-gatherers and pastoralists.

But the processes of population growth and market expansion have had an impact in peasant areas similar to that experienced at a national level. Many peasant communities now demonstrate a very high degree of internal differentiation. Differences in the domestic cycle and in diligence and luck have resulted in marked differences in the relation of production to consumption between families living in the same community. Those who consume more than they are able to produce become indebted and may be obliged to lease or sell part of their property. Those who make a profit may be able to add further to their income through rent and interest. In many peasant communities we find a complex network of leasing and sub-leasing, share-crop arrangements and the hiring and giving of labour. Sometimes a household will rent land from neighbours while also renting out a part of its own property. The property-owners may well be poorer than their tenants. Families deficient in labour, for example, may be forced to rent their property out to other, more prosperous households.

The sectoral development of agriculture places considerable pressure on the peasant household to supplement its farm income by wage labour elsewhere. This pattern suits the industrialist well because where workers control small properties, it minimises the need for housing and welfare provision and in some cases the labourers will even bring to work supplies of food from their land. Also, diversification and population mobility in rural areas in many cases expands the internal market for industrial goods.

Even where there are increases in prosperity in a rural community, population growth and prevailing inheritance practices often inhibit capital accumulation. Whatever the inheritance practices, so long as rights to land are attached to the family, growth in family size will reduce the amount of land available to each member. Population increases frequently lead to fragmentation of properties until eventually, regardless of the labour input, the holdings will no longer be viable. Unless there exists a mechanism for redistributing properties periodically, many farmers will be forced to withdraw from agriculture, often abandoning their land and migrating to the city.

2 Government Initiatives

The prime motive behind the intervention of Third World governments in the rural sector in recent decades has been to increase food output. Most of the Agrarian Reform programmes of the 1950s and 1960s were based on the premise that small-scale production is more intensive than large-scale and that the redistribution of ownership (giving land to the cultivator and achieving a more equitable distribution) would encourage output. It may be true that in a commercial system of agriculture, output will tend to be higher the more equal the distribution of ownership, but land reforms of this nature have been fraught with problems. Foremost among these has been the failure in the majority of cases to provide adequate credit facilities. Throughout the world new ownership groups in the countryside have gone bankrupt because there has been either a lack of political support or insufficient funds to secure a reliable source of credit. Many of those who benefited from land reform by gaining control of land have found themselves paying compensation to the original owners in addition to extortionate interest payments on loans. In many areas, land reform has served most the interests of those who were already wealthy, and isolated further the politically marginal.

The role played by national governments, the concentration on increasing food output, has led to a serious neglect of services in rural areas. Large areas of the countryside in the Third World are cut off from state services such as health and education, and in many countries little effort has been made to meet the challenge of rural poverty and deprivation.

3 Programme Priorities

a: Particularly in the rural context, field staff should be wary of funding community programmes which neither recognise nor address specifically the problem of inequality among their members. The truly cohesive, egalitarian rural community does not exist, even where it is represented as such to field staff. Each project should be examined closely to assess its impact on the more marginalised members of the community and the degree to which they participate and their interests are represented (see PART TWO).

b: In many rural areas there is a very real need to provide basic services. However, field staff should never underestimate the problems of providing services in the countryside. It can be far more costly, for example, to set up a primary health care programme in an isolated rural zone, where the problems of transport and supplies are acute, than in the city (see PART SEVEN, Section 1). It is frequently very difficult for professionals working in rural areas to sustain their commitment when salaries and career prospects are far more attractive in the city. Moreover, the provision of essential services implies a long-term financial commitment, and every effort should be made to secure the eventual transfer of funding to the State.

c: Income-generating programmes, such as those improving agricultural production or creating new sources of income, often appear to be the most suitable form of intervention in rural development. However, smallholder production presents a problem because each household or family will tend to produce a range of products and will therefore most likely be occupied the whole year round. It may be unrealistic to initiate a programme which increases the workload. Income generation may also increase the dependence of the producer on credit and therefore on the moneylender, increasing the burden of interest payments, rather than liberating him or her from such ties. Further, income-generation programmes are reliant on the existence of reliable transport facilities and a nearby market. All these factors must be considered when assessing the feasibility of income-generation in rural areas (see PART FIVE).

d: Development initiatives in rural areas must take account of the legal status of the community and of the land and other natural resources. Sometimes the main priority may be to convert *de-facto* rights to land acquired by centuries of occupation into *de-jure* rights. It is not possible to over-emphasise the importance of understanding the legal basis of rural populations, since many development programmes have floundered after their resources have been appropriated and their beneficiaries left unprotected in law (see PART FOUR, Section 4).

e: Despite the logistic difficulties presented by working with mobile groups, in many areas far greater priority should be given to working with nomadic or semi-nomadic peoples than sedentary farming populations. Too much emphasis has traditionally been placed on the rural peasant community, at the expense of other groups that are frequently much poorer and more vulnerable (for further details see PART TWO Section 5).

1

Section 4　Resources

The addresses of other Non-Government Organisations are to be found at the end of the book. References to other books, journals and organisations can also be found in the Resources sections at the end of each part.

For more information on OXFAM refer to the "Information Department", OXFAM, 274 Banbury Road, Oxford, OX2 7DZ, U.K. An independent history of OXFAM can be found in:

B. Whitaker, *A Bridge of People: a personal view of OXFAM's first forty years*. Heinemann, London, 1983.

I　Bibliography

[1] State of the Poor

UNICEF, *State of the World's Children*, Annual, Oxford University Press, Oxford.

World Bank, *World Development Report*, Annual, Oxford University Press, Oxford.

Development and Change, Institute of Social Studies, The Hague, Holland.

Development Dialogue, Dag Hammarskjold Centre, Ovreslottsgatan 2, 2-75220 Uppsala, Sweden.

Ecodevelopment News, International Research Centre on Environment and Development, 54 Boulevard Raspail, Room 311, 75270, Paris, Cedex 06, France.

Environment Liaison Centre, *Ecoforum*, PO Box 72461, Nairobi, Kenya.

Institute of Development Studies, *Bulletin*, University of Sussex, U.K.

New Internationalist, 42 Hythe Bridge Street, Oxford, U.K.

Reading Rural Development Bulletin, University of Reading, 16 London Road, Reading, RG1 5AQ.

Assignment Children. UNICEF, Palais des Nations, Geneva 10. 1983.

[2] Rural Development

General titles (for titles on agriculture see PART SIX, Section 14, for titles on rural peoples, PART TWO, Section 6, for social development PART FOUR, Section 7).

R. Chambers, *Rural Development, Putting the Last First*, Longman, Harlow, Essex, U.K., 1983.

U. Lele, *The Design of Rural Development: Lessons from Africa*, John Hopkins, Baltimore, U.S.A., 1975.

N. Long, *An Introduction to the Sociology of Rural Development*, London, 1977.

A. Pearse, *The Latin American Peasant*, Frank Cass, 1975.

E. Wolf, *Peasants*, Prentice Hall, New Jersey, U.S.A. 1966.

[3] Urban Centres

D. Drakakis-Smith, *Urbanisation, Housing and the Development Process*, Croom Helm, London. 1981.

D. Dwyer, *The City in the Third World,* Longman, Harlow, Essex. 1977.

A. Gilbert and J. Gugler, *Cities, Poverty and Development*, Oxford University Press, Oxford, U.K. 1983.

J. Hardy and D. Satterthwaite, *Shelter Need and Responses*, Wiley, Chichester, U.K. 1981.

P. Lloyd, *Slums of Hope*, Penguin, Harmondsworth, Middx., U.K. 1979.

J. Perlman, *The Myth of Marginality: Urban Poverty and Politics in Rio de Janeiro*, University of California Press, U.S.A. 1976.

B. Renaud, *National Urbanisation Policies in Developing Countries*, Oxford University Press, Oxford, U.K. 1981.

B. Roberts, *Cities of Peasants: The Political Economy of Urbanisation in the Third World*, Edward Arnold, 1978.

A. Turner, *Cities of the Poor*, Croom Helm, London, U.K. 1980.

The Urban Edge, published 10 times a year from the World Bank, c/o Publications Sales Unit, The World Bank, PO Box 37525, Washington DC, 20013 U.S.A.

1

PART 2 PRIORITY GROUPS

Jeremy Hartley

Part Two PRIORITY GROUPS

2

Section 1   Introduction

This part of the Handbook examines some of the special problems faced by the groups of people identified by Oxfam as taking priority in its development work. It is impossible to provide an exhaustive list of the poorest and most vulnerable groups throughout the world, since the most oppressed groups of people in certain contexts and in certain societies may not even exist elsewhere. However, a few principles of organisation can be identified as having universal significance in human society, and these are based on differences in gender, age, class and ethnic status. In any given situation, and in society as a whole, the system of classification will function to the advantage of some and the disadvantage of others. Women, children, disabled people and ethnic minorities have been singled out for special attention in this Handbook as sectors of society which almost everywhere in the world are poorly represented at a national level and in many situations are severely disadvantaged.

The priority groups are those with a 'silent voice', isolated by society from the decision-making process and taking no part in the decisions which most affect their lives. Even where they are the target of development programmes, these groups are rarely involved in the process of planning or implementation. They may play a crucial role in society and be fundamental to the economy, yet are usually treated as an afterthought in development planning. All too often, for example, a component part addressed specifically to women has been added to a major development programme simply to appease feminist pressure groups.

In some societies the most vulnerable groups have the legal status of minors and are not given the right to self-representation or independent expression. Often while they may be in a minority at the national level, in a given area they are in the majority — especially among the poor. They are disadvantaged in relation to the control of resources, socially, culturally and economically. The contribution they make to society is not matched by the benefits accruing to them: they do not benefit from the increases in wealth in society as a whole but are more often adversely affected by development, losing control of productive assets and becoming increasingly exploited by the more powerful.

The view that modernisation will inevitably prove advantageous to all, that its effects will eventually 'trickle down' even to the poorest, has now been discredited. There is ample evidence to show that benefits to the community or to the household are not only distributed unequally within it, but can actually worsen the situation of some members.

'The household', 'family' and 'community' are all fundamental concepts of the trickle down theory. When assessing projects, reference to these concepts should be treated with caution. Information must be gathered on the composition of these social units and the distribution of resources and relationships within them. The aim should be to ensure equality of opportunity and access and to recognise the rights of the weaker members of society. But the recognition of the importance of the weaker members of society is not just a humanitarian issue; it is an urgent practical issue. Women and children, for example, are the major food producers of the Third World and by denying them their rights the economies of whole communities are undermined. It is not sufficient to incorporate them in some artificial way into the development process — they are an integral part of that process.

Thus, there is a strong need to re-evaluate conventional thinking on development. Concepts such as community or family often disguise very real inequalities. New insights into the harmful effects of commercial farming on fragile ecologies should warn us of the need to re-examine the view that indigenous economies are 'simple' or 'primitive' — they are usually highly complex and extremely well adapted to the physical environment. Equally, western notions of childhood, or the idea that the prime role of women in society is as housewives, are highly inappropriate in most parts of the Third World. They often result in unsuitable programmes in which, for example, women are taught to embroider, knit or sew. While there is nothing wrong with these activities *per se*, they are frequently very far from women's real interests and work experience and fail to free them from exploitive relationships.

However, by focusing in this way on the vulnerable groups in society, we raise a serious methodological problem: Oxfam advocates an integrated approach to development, one which stresses the participation of all groups in the planning and implementation of projects. The danger with supporting programmes designed to cater specifically for a special sector of society is that they could promote a fragmented approach to development. Programmes for disabled people, for example, all too often use methods such as institutionalisation which increase their isolation from society. A physical or mental impairment need not stop a person from being an active member of his or her community. While handicapped people may not be able to undertake heavy manual labour in the fields or in the factories, they can undertake tasks such as primary health care work which are vital to the community. The aim should be to seek ways of integrating all vulnerable groups.

In some cases, where the level of prejudice or repression is extreme, a separatist programme of development may be necessary. However, it is not the main purpose of this part of the Handbook to promote such an approach. The aim is to alert field staff to the need not just to alleviate poverty, but more to identify those groups in society which are so often left out of development plans. Much of the traditional development work in the Third World has been extremely unsuccessful, even harmful. New ways must be sought to combat inequality and integrate the more vulnerable groups. The community or society as a whole must be encouraged to accept responsibility for all its members and to recognise each person's fundamental human rights.

Thus, the main aims of Oxfam's work with the especially disadvantaged sectors of society are outlined below.

a: To improve the life chances of those people who are vulnerable or at risk by supporting them in their struggle to overcome the barriers to the realisation of their full potential as social beings.

b: To ensure that programmes involving the poorest and most disadvantaged groups are always mounted with their full agreement and participation and are appropriate to their society and culture. As far as possible, the groups themselves should determine the pace and nature of their own development.

c: To assist disadvantaged groups, wherever possible, in ways that will reintegrate them into society, strengthening kinship, ethnic or community ties, and benefiting the community or society as a whole and not just its more vulnerable members.

d: To discourage any process which discriminates against the poorest and most vulnerable groups, placing them at greater risk and isolating them further from their society.

Field staff must make every effort to ensure that no development work supported by Oxfam adversely affects the more vulnerable members of society, directly or indirectly, and should, where possible, represent the interests of these groups when programmes run by the State or other agencies act against these.

2

Section 2　Women

I　Introduction

The report of the mid-decade conference of the UN Decade for Women held in Copenhagen in 1980 noted that while in some countries certain gains could be identified, in general the situation of women in poor countries had deteriorated markedly in recent years.

"In particular, it worsened with respect to the conditions of employment and education for women in the rural and so-called marginal urban sectors. In many countries the actual number of female illiterates is increasing. . . . . In many countries, transfer of inappropriate technology has worsened the employment and health conditions of women. In many countries, women have not been integrated into national development plans. Where special programmes have existed, they have failed for the most part in achieving significant results, owing to their narrow focus on stereotyped sex roles which have further increased segregation based on sex."

The causes of the worsening of the position of women are complex but not impossible to identify. A change in thinking about development is required if the situation is to improve. Some of the concepts used by development agencies must be closely examined and, in certain cases, abandoned.

The position of women and the problems they face vary according to cultural and religious factors (including caste stratification in countries such as India) and according to whether they live in rural or urban areas.

It is often extremely difficult for field staff to obtain the information necessary for an understanding of the real needs of poor women. In societies where women are not permitted by men to take part in discussions on a community level, it can be difficult even for female staff to gain access to the silent half of the community. It is virtually impossible for men. The same usually holds true for intermediate agencies. In addition, it is important to realise that women will often, by virtue of their role in society, accept without question prevailing norms and will not challenge the injustice they face.

Thus, the process of identifying women's real needs, their aspirations and the obstacles confronting them must involve sensitive consultation and careful assessment, based on a sound understanding of the culture and society in which they live. Where women are discriminated against and dominated by men, barriers are set up against their participation in political life, control of resources, control over child-bearing and fertility, access to education and new technologies and participation in religious life. These factors vary from society to society, but in all the work Oxfam supports they form the conditions of women's social and political life, and determine the effects of development not only on women, but also on children and communities as a whole.

Development programmes for whole communities should aim to benefit all members equally. Inequalities arising from gender differences within both society and the family make equal development difficult to achieve, and such inequalities must therefore be identified and specifically addressed. Since women are so often the victims of discrimination, there is a need for a special focus on women in development, and a need to support work which addresses their specific problems as well as those they share with other poor and exploited people.

1 Barriers to Decision-making

In many countries in which Oxfam works, women are excluded to a greater or lesser degree from decision-making in the community, and their participation in political life — by which is meant all social relations outside the family or household — is limited. In development terms, women are represented largely by agencies which are themselves dominated by men. Project design is in the hands of men, and cannot adequately address women's needs. In Brazil, for example, while it is not uncommon for women to be active in urban slum organisations, they often face considerable opposition from males in their own families. In other countries, such as India, women are brought up to play what has been called a 'life-long role of subservience and self-effacement', and in many cases accept a passive role in decision-making. Even fundamental decisions about child-bearing are usually taken by men, who often override women's desire to have fewer children. Decisions about the allocation of resources within the household frequently do not involve women, with resultant shortages of food and other commodities for them and for children.

In some societies the powerlessness of women is maintained by practices which isolate them not only from general society and political life, but also from contact with each other and from all but sporadic contact with kin. Fetching water, while a burdensome and time-consuming task, can often provide one of the few occasions when women can meet and talk together. Informal networks of this sort exist in many societies and help to counteract the isolation of women. Programmes attempting to ease the burden of water-carrying or affecting women's work in other ways should take this factor into account.

These informal networks function as support systems, and when women lose access to them through schemes which disperse communities, such as settlement programmes, they may suffer not only social isolation but also increased work-loads. Some kinds of economic activity, for example, rubber-tapping in Brazil and Bolivia, by their nature disperse households: while men may be drawn into marketing co-operatives, women are isolated in small jungle clearings and may pass months without seeing anyone other than their husband and children.

Seclusive practices such as purdah, found in areas of Muslim and Hindu influence, are extreme examples of women's isolation. Secluded women represent a captive and easily exploited labour force for the kinds of work which can be carried on at home. The lace makers of Narsapur in Andhra Pradesh, for example, are isolated in household compounds and thus unable to organise. Because they are constrained in terms of employment they accept work for very low wages.

Recommendations
☐ It is essential that contact is made with women directly;

☐ the attempts of women to organise should be supported and stimulated;

☐ educational work should be undertaken with women to help them form organisations, and with men to reduce their opposition where it exists;

☐ information should be sought on the nature of women's informal networks and labour exchange between kin and neighbours;

☐ where women work in isolation at home, they should be encouraged to form organisations and to value their labour at non-exploitive rates. This is particularly true for handicraft projects. (See PART FIVE, Section 2;)

☐ long-term educational work with men and women should deal with the problems of decision-making within the household as well as within the community.

2 Inequality Within the Household

Women suffer most from the failure of development projects to look at intra-household relations, for cash income which comes into the household is usually controlled by men, and spent on personal articles for themselves rather than on food for the women and children. This leads to the situation, often cited, where women and children may be malnourished and ragged, while the men of the household have new clothes or luxuries such as bicycles, wristwatches and radios. The women's status within the family can drop in relation to the men's with the introduction of new opportunities for the men, such as improved technologies or membership of co-operatives. Many Third World women have found that husband-wife relations have deteriorated with modernisation, and there has been an increase in conflict, wife-beating and other forms of violence within the home.

Recommendations

☐ When assessing a project application, references to 'the household' or 'the family' should not be taken at face value, but information should be sought on the composition of the households in question;

☐ the project should ensure that women participate directly in income-generating schemes, but not in schemes which increase their work-load;

☐ women should have equal access to membership of organisations such as cooperatives;

☐ patterns of distribution within the household should be identified, and measures undertaken to counteract unequal distribution.

3 Female-headed Households

One in three of all households in the world are headed by women, there being a strong correlation between poverty and households without men. In parts of India, 35% of destitute households are headed by women and in parts of Latin America 50% of households are female-headed. It is vital to stress this, for the concept of the household headed by a man still dominates development thinking.

The evidence showing the linkage between female family headship and poverty is compelling. It is women among the poor, whether in Central and South America, sub-Saharan and North Africa or Asia, who are increasingly becoming the main or sole economic provider for their families, and the majority of the world's poor today are women.

The number of female-headed households is growing with the migration of men to town seeking permanent or seasonal employment, with increasing warfare and with the abandonment of families because of poverty. Increased family impoverishment is often the result of reducing women's traditional ability to contribute to family income by concentrating development planning on men and disrupting traditional land-holding practices. A vicious circle is thus created: inappropriate development can increase the number of poor female-headed households through increased impoverishment of families. In periods of economic crisis, women are the first to lose out in the formal labour market, and are thus available to be hired informally at highly exploitive rates of pay.

In parts of rural Africa, where women are the main food producers, shortage of labour due to emigration of male kin has serious consequences for the health of female-headed households. Such households are characteristically short of food-processing equipment such as handmills, and the area of land they cultivate is smaller. In Latin America, female-headed households are common in urban slums, women often having migrated with their children from the rural areas.

Frequently under- or unemployed, they are forced to work as prostitutes. Some religions are particularly discriminatory against female-headed families. In the Hindu tradition, woman's true destiny is considered to be marriage, and her life should be devoted first to her father, then to her husband, and after his death, to her sons. Widows are shunned and excluded from participation in religious and social life.

Recommendations

☐ Special attention should be given to female-headed households with regard to access to income, training, technology, land, assistance with child-care, etc.;

☐ informal networks which help female heads of households should be strengthened, and new forms of organisation supported — e.g., co-operatives or trade unions;

☐ access to legal aid is vital for female household heads, especially in cases of expulsion from land or homes.

4 Child-care

In very few societies is child-care shared by men. In societies which still retain the fabric of traditional relationships, networks of female kin can help to take care of children. However, in many cases these networks have been disrupted, often by the development process itself. When women work outside as well as in the home, child-care can be a major problem. This is particularly true where the elder children, who might otherwise take care of the smaller ones, also have to work to supplement family income. A woman's participation in organisations, education, training and other activities, which may help to improve her own and her family's standard of living, is severely limited if she has no-one to care for the children.

As the health and well-being of the children depends largely on the health and resources of the mother, the welfare of future generations depends on the improvement of the conditions of life of women. Increasing numbers of abandoned children are the victims of the impoverishment of large sectors of the population in many countries, particularly of the destitution of women. Nutrition education for the improvement of children's health has its place, but it will not help women who have neither the income nor the resources to provide even the most basic foods for themselves and their families.

Recommendations

☐ Social education programmes which emphasise the importance of shared child-care should be encouraged;

☐ child-care facilities for working women, especially women heading households, should be supported;

☐ projects proposing new activities for women should take the problem of child-care into account;

☐ existing informal networks for child-care should be strengthened. (For further recommendations see Section 3.)

5 The Double Day

In most parts of the world women work longer hours than men, and often for no reward at all. Women are particularly important in food-processing, animal husbandry and agricultural (food) production. They are responsible for at least 50% of all food production in the world. The working day of the average peasant woman in the Third World begins at about four in the morning and ends at around nine in the evening. In addition to working in the fields women are expected to prepare food for their spouses and children, to gather water and fuel and to take food to their men at work. These tasks are made more arduous by the physical and emotional drain of frequent pregnancy, childbirth and breast-feeding.

Development often increases women's work-load by failing to take into account prior work commitments. Development initiatives also frequently perpetuate the myth that all farmers, or all workers in the city, are men. This means that benefit from training and technological innovation is geared exclusively to men. A tractor may well increase the area men can plough, but it will also increase the work undertaken by women, since it is they who usually do the weeding.

A shift in attitude is required in development planning, not just to provide equality of opportunity for women, but also because it makes sound economic sense. Greater recognition must be given to the economic value of the essential work carried out by women both inside and outside the home. Serious attempts must be made to ease women's work-load and stabilise food production through forms of organisation which involve men and women, and to introduce technology which facilitates the work women already do. This technology must be held firmly in the hands of women so that it is not taken over by men.

Recommendations

☐ Income-generating programmes for women should help them to derive income from work they already do, and not add to their workload;

☐ attention should be given to easing women's work-load by facilitating tasks such as water and fuel gathering and food processing;

☐ social education programmes which tackle the problem of women's excessive work-loads should be encouraged and supported;

☐ care must be taken that when technology is introduced to facilitate women's work, men do not assume control of it.

6 The Control of Fertility

At the centre of the problem of the control exercised by men over women lies the question of the control of women's fertility. In many societies men oppose women's use of contraceptives because it enhances their status to father a large number of children and because women in control of their own fertility could give sexual access to other men. Many women want to limit their families for economic reasons, but are prevented from doing so by their husbands, by local priests, or by social pressure in general. They fear being abandoned by their husbands if they cease to bear children each year, and worry about the side-effects of available contraceptives.

Sexual control over women is in most societies sustained by social charters which in one way or another link female sexuality with social disorder, wildness and danger to the community, and thus something over which social control must be exercised. Male control over female sexuality lies at the root of many proscriptive laws which are used to exclude women from religious and political life and maintain their social inferiority.

The question of fertility, and particularly the control of fertility, is fundamental to women's social and economic position in society and cannot be tackled outside the context of the sex-linked allocation of resources, food production and changing economic opportunities. Having fewer children will not necessarily mean that adequate income is allocated to women to feed themselves and their children. It is essential that social attitudes change, and that men assume equal responsibilities for the care of children.

All over the world, because of the social and religious restrictions on fertility control and because of the severe problems faced by large, closely-spaced families, infanticide is committed; alternatively, women have abortions. In many traditional societies abortion is accepted and practised effectively, but in most parts of the world it is unsafe, illegal and/or very expensive. Particularly in urban areas, poor women have to seek help from abortionists who may exploit them and damage them physically. When women have no access to post-operative care, infection may lead to infertility or death. To compound the problem further, aid agencies eager to support fertility control are often accused in the Third World of genocidal, imperialist plotting. Clearly, the whole issue is extremely sensitive and must be handled with both caution and tact.

Recommendations
(For recommendations on birth control see PART SEVEN, Section 6.)

7 Health

One of the most common health problems faced by women is anaemia — the combination, very often, of poor nutrition, frequent child-bearing and overwork. In many societies men eat first, and women and children afterwards: in times of food shortages, this discrimination is particularly damaging to women and children. In many areas traditional diets were varied and relatively nutritious, but the introduction of cash crops has reduced the range of available foods and altered diet. Where women are displaced from the land by cash-cropping and modernisation in agriculture, and lose control over the land and food production, their health suffers. In urban areas women frequently suffer from fewer employment opportunities than men, are often paid less for the same job or forced into sub-employment at minimal rates of pay, work longer hours than men, and live in slums where sanitation may be non-existent and water polluted.

In some countries it is women who get together in urban areas to press the authorities for sanitation, water, health posts, or who form health groups to disseminate information about simple home remedies like oral rehydration therapy, or collect money to buy essential medicines. In many societies, women are the traditional healers, and traditional midwives are still active. Women may be unwilling to go to male doctors or health workers, or their husbands may prevent them from doing so. Diseases and illnesses specific to women are often not understood by male doctors, nor taken seriously by them, leading to incorrect diagnosis and treatment, or simply to neglect. Women themselves are frequently ignorant about their own bodies — especially in societies where sexual ignorance is equated with virtue or where the religious constraints on physical self-knowledge are very powerful.

Recommendations
☐ Projects which aim to improve the health of women must take into account the full implications of women's socio-economic situation; training in health or better nutrition will be ineffective if women do not have sufficient access to land or cash income;
☐ women's organisations in urban areas which press for environmental health improvements should be supported;
☐ work which attempts to break down the barriers to women's knowledge of their own bodies should be encouraged;
(For further recommendations see PART SEVEN.)

8 Loss of Land and Earnings

In many countries, particularly in Africa, women had, and sometimes still have, customary rights over land and crops: land tenure is traditionally vested in the community or village, and allocated to both women and men, sometimes jointly and sometimes independently. Women are able to trade their surplus and gain income, and in parts of Africa today it is mainly the women who run the markets.

The introduction of a concept of land ownership vested in the individual not only paves the way for the loss of land of small rural communities in general, but has particularly affected women. On the assumption that land ownership should be vested in male heads of households, land registration schemes or land reform schemes often ignore women's rights to land. Land and the earnings from it may thus be transferred from women to men.

This is the beginning of impoverishment for women: land is usually security for credit, and land ownership often the criterion for access to agricultural extension services, irrigation, mechanisation and membership of cooperatives. In rural areas women's status suffers as a result in relation to men's, and women increasingly take over all the subsistence farming, which can result in an unmanageable workload, and loss of income opportunities.

Sometimes when women's economic activities (such as milk production in India) have been modernised and then taken over by men, women lose their only source of income. In other areas where co-operatives have been formed for the marketing of women's crops, but only men belong to the co-operatives, women lose their incentive to produce, and withdraw their labour altogether. All these developments will affect the nutrition of the family.

Recommendations

☐ Information on the exact nature of land tenure and use with regard to women and men should be obtained when considering agricultural/production projects;

☐ where technical innovation for women's work is proposed, there must be guarantees that women retain control of the work;

☐ projects should take into account what has happened to women's land and income in the past, in order to try to redress the balance;

☐ any land reform supported must deal equally with women and men, giving equality of opportunity as well as ownership;

☐ income-generating schemes to replace lost earnings should be supported on a collective basis in order to stimulate women's organisation and enable women to raise the capital for supplies and overheads which individuals could not manage.

9 Barriers to Knowledge, Appropriate Technology and Credit

In most societies, male children are given preferential education, while female children stay at home with their mothers and help with subsistence work. Women, already overburdened with work, tired, malnourished and lacking child-care facilities, are often unable to participate in education even when cultural constraints are not a major problem. Female illiteracy is increasing, and the gap between male and female illiterates is rising. While schooling and literacy do not necessarily provide a release from poverty and exploitation, they are important tools in the understanding of the social and political environment, without which the process of change is impeded. Poor women in Third World countries, when asked what they need most, typically reply — cash and education.

Training is usually offered to men and not to women, and in the field of agriculture in societies where women are the main producers, this makes no sense at all. Training is often given to women for completely inappropriate tasks, such as handicrafts which have little or no market value and are very labour-intensive, or homecraft/hygiene/nutrition where women lack even the most basic living conditions (such as accessible clean water or sufficient fuel to boil it). Training should preferably facilitate the work already undertaken by women, making it more productive and profitable.

Problems of access to appropriate technology are similar to those for training, and typically the introduction of technologies has favoured men, easing their work, while increasing the manual labour carried out by women. Technology in rural areas should be specifically designed to ease women's tasks in the fields and in the home.

Agricultural credit, or credit for income-generating schemes, is rarely available to women, particularly to illiterate women. In rural areas, this could be overcome by the recognition of women's land rights, and in towns, the registration of women's associations or work collectives.

Recommendations

- [ ] For women to be involved in education and training, work must be done with men to overcome their possible opposition;
- [ ] women's workloads and time available must be carefully considered;
- [ ] some complementary programme may have to be carried out to facilitate women's attendance at training and educational programmes — e.g., provision of child-care or organisation of food processing on a collective basis;
- [ ] new technologies must not fall under the sole control of men, for the likelihood is that women will suffer;
- [ ] programmes should be supported which rectify the bias in access to education and offer girls the same opportunity as boys.

2

I   Introduction

The Declaration of the Rights of the Child which was adopted in 1959 by the United Nations General Assembly embodies a number of important principles.

a: That the child should enjoy certain fundamental rights, regardless of the race, colour, sex, religion, political or other opinion, national or social origin, language, property, birth or other status of either the child or his/her family.

b: That the child should enjoy special protection and should be given the opportunities and facilities to enable him/her to develop physically, mentally, emotionally, morally, socially and spiritually in a healthy and normal manner.

c: The child should from birth be entitled to a name and a nationality.

d: The child should enjoy the benefits of social security and has the right to adequate nutrition, housing, recreation and medical services.

e: The child who is physically, mentally or socially handicapped should be given the special treatment, education and care required by his/her particular condition.

f: For the full development of his/her personality the child needs love and understanding. He/she should, wherever possible, grow up in the care and under the responsibility of his/her parents and in an atmosphere of affection and of moral and material security. Save in exceptional circumstances, the child should not be separated from his/her parents.

g: The child is entitled to receive education. The education should promote his/her general culture, enable the development of abilities, individual judgement and a sense of moral and social responsibility on a basis of equal opportunity, and should enable the child to become a useful member of society. The best interests of the child should always be foremost in the work of those responsible for his/her education and general guidance and the child should have full opportunity for play and recreation.

h: The child should in all circumstances be among the first to receive protection and relief.

i: The child should be protected against all forms of neglect, cruelty and exploitation. He/she should not be the subject of traffic in any form and should not be admitted to employment before an appropriate minimum age. He/she should not be caused or permitted to engage in any employment or occupation which is prejudicial to health or education, or interferes with physical, cognitive or moral development.

j: The child should be protected from practices which may foster racial, religious and any other form of discrimination and should be brought up in a spirit of understanding, tolerance, friendship, peace and universal kinship, fully conscious of the need to be devoted to the service of his/her fellow beings.

Unfortunately, in most of the areas where Oxfam works, few of these rights are upheld. Throughout the world the physical, emotional and intellectual development of children is impeded by poverty, malnutrition, neglect, ill health, discriminatory practices, ignorance, by the paucity of essential services and by natural and unnatural disasters.

Development efforts must focus on children — on protecting them, securing their health and normal development and upholding their rights — if the enormous waste of human life and potential is to be prevented. Most, if not all, projects funded by the NGO will affect children in one way or another, either directly or indirectly. Some of those concentrating on special problem areas (nutrition, education, vocational training, etc.) and all the more general programmes to do with technology, income-generation, organisation, etc. will have some impact on children's lives. However, children are almost never involved in the planning of projects and are rarely even consulted. In some countries they are not seen as having any special needs or, indeed, any rights. It is important not to be complacent about the likely effects of a development programme on their lives. It is, in fact, evident that the very process of development and the social and economic transformations that have accompanied it has harmed many children.

Children should take priority in all development initiatives because an investment in the wellbeing of the child is an investment in the future of the community as a whole.

However, it is hard to establish universal norms on child-rearing and child-care, all the more so because the very concept of childhood differs radically according to the social, cultural and economic context. In the industrialised countries few children are engaged in full-time paid employment, and most receive full-time education at least until the age of fifteen. In low-income countries on the other hand, children are often to be found working outside the home in paid employment from the age of eight onwards.

In most poor countries, little distinction is drawn between the various stages in life (childhood, adulthood, old age; the school years, the working life, retirement) which are given so much importance in the industrialised world. In many areas the most important distinctions are based on the division of labour according to gender and age differences, and retirement is a completely alien concept. In parts of Africa advancement through life is recognised by passage through a hierarchical structure of age grades, and these bear little similarity to the life stages recognised in Europe and North America. Therefore, any intervention designed to help children must be assessed in the context of local attitudes and practices.

II Principal problem areas

1 Obstacles to the Healthy Development of the Mind and Body

All children need love, security, enjoyment, stimulation and encouragement, if they are to develop a sense of responsibility for themselves and for others, to have confidence and to be able to express themselves in a manner appropriate to their culture and situation. They must be able to develop the skills necessary for adulthood, to copy roles, acquire a knowledge of their own culture and understand what forms of behaviour are acceptable in their society.

Certain needs will predominate at certain ages or stages of the child's development or because of different social or cultural factors. Parental and community aspirations for children are influenced by various factors: the values and attitudes of the wider society, individual experience and what is perceived as being desirable in other societies and cultures. There may be general agreement on the absolute minimum standard of physical well-being from below which any child should be rescued, but it is far harder to establish norms for emotional, spiritual or intellectual requirements.

Children face many hazards, especially in the earliest stages of their lives. Under-achievement and anti-social behaviour among adults can be linked directly with parental neglect, poor nutrition, poor health and inadequate early stimulation of children. In the Third World the rates of infant mortality and morbidity are extremely high: one in every four children dies before reaching school age. Diarrhoeal diseases alone cause between four and five million deaths worldwide each year among small children. Malnutrition and sickness impair children's capacity to learn or to function optimally, and those with a low birth weight are particularly vulnerable to early death or severe sickness.

Families under stress are rarely able to provide adequately for their children's welfare. They are often forced to surrender their children to employers (factory owners, wealthy urban families, etc.) in the hope that they will earn some money, or at least be provided for. These children are frequently neglected, abused physically and emotionally, and exploited. In only a few countries is the State able to curb this kind of abuse, and public services normally fail to reach the children who are most in need. In the majority of low-income countries there is no family counselling service, and state-sponsored child-care facilities and social work provision generally are extremely inadequate.

In any programme involving children there should be two main priorities;
☐ to protect the children's interests, securing their health and normal development;
☐ to help them to become integrated into society.

These two priorities are intricately linked. For example, nutritional programmes and health provision will improve children's physical well-being while at the same time enabling them to become more receptive to educational and general learning processes and to become fitter and stronger as adults. The focus of all programmes should be on the quality of child-care rather than the provision of a child-minding service (though this can also be important when both parents are working).

Even where family structures are intact, it should never be assumed that all children are receiving the love and care that they need. The value placed on individual children is involved with the values of society as a whole. Thus, in many societies female or handicapped children may be neglected because they are seen as a burden or because they are not able to contribute to the domestic economy. Inheritance practices may place great emphasis on certain children or on one particular child in the family, while isolating others. Where inequalities between children exist, discrimination is common and, especially in poor areas, infanticide may occur. All programmes should aim to protect the weaker, more vulnerable children and to change society's attitudes towards these children and towards birth control and child-spacing.

No project can be considered in isolation from customary practice and traditional values; these will determine priorities in child-care in any given society. Projects should be assessed in the context of attitudes and taboos regarding: family structure and size; the roles of father, mother and other relatives; the place of children in that structure (including differences between the sexes); preparation for adulthood; the role of education and child labour; what is seen as important in the rearing of children; what is regarded as normal and what as deviant behaviour; and how vulnerable children are cared for.

Recommendations

When the children's main requirements have been identified, it is necessary to examine whether any of these should be given priority in the programme: if the primary physical requirements are not being catered for, to at least a minimal level, then the rest must remain of secondary importance. A starving child will not immediately need exercise or education, but once he/she is being adequately fed, these requirements will become more urgent.

There are various types of service that can be provided for children.

Those that give support to the health and well-being of children who are living normally with their own families include:
☐ ante-natal and post-natal care
☐ health centres

- [ ] mother and child clinics
- [ ] well-baby clinics, pre-school and under 5s clinics
- [ ] day-care centres
- [ ] nursery schools, creches
- [ ] school health programmes
- [ ] child guidance clinics
- [ ] youth services, children's clubs and groups
- [ ] family assistance programmes
- [ ] probation.

Those services that provide for children who, for one reason or another, cannot be cared for by their own parents or relatives and for whom alternative residential care is necessary for a shorter or longer period include:

- [ ] foster home care
- [ ] adoption
- [ ] residential care for children: motherless babies' homes; transit or reception centres; remand homes; approved schools; homes for the mentally sick, maladjusted or subnormal children; homes for handicapped children (blind, deaf, crippled)
- [ ] special boarding education.

Many of these services exist only in the wealthier industrialised countries, and some are not suited to the work of the aid agency. However, certain types of project are very appropriate, and certain fundamental principles should always be followed.

Programmes concerned with food production should promote nutritional education and low-cost, high-quality foods appropriate to local conditions. Special emphasis should be given to maternal diet during pregnancy and lactation and to local food supplements and weaning foods for infants.

All programmes should encourage practices that provide appropriate mental stimulation for infants and young children. Primary health care programmes should aim to provide education on all aspects related to the health and development of the child: information should be given on child-spacing, child-rearing practices, ways of ensuring good health, harmful effects of practices such as smoking or drinking alcohol during pregnancy, ways of avoiding accidents in the home, appropriate diets and the advantages of breast-feeding. Emphasis should be given to preventive health measures: immunisation against disease, oral rehydration therapy, etc. (See PART SEVEN, Section 6.)

There are numerous ways in which assistance may be given to families to improve child-care. Support of this kind should be directed through infrastructures organised at the community level such as in a disaster aid programme, a health care intervention, communal shops or kitchens or a social work project. Advice and practical support can be given to families whose level of child-care is so inadequate that the children are at risk of being abandoned, neglected or abused. Support can also be given to people (relatives, foster-parents, etc.) who are looking after other people's children.

Children should only be helped in isolation from their family or community when these structures have really broken down. Unfortunately for many children the only possible alternative is institutional care. When assessing a project of this type it is vital to consider the purpose of institutional care and the means by which it is to be provided.

It is important to examine the position of the institution vis-a-vis the community: will it be part of the community or will the children be isolated from the community? Will the community help to fund the institution or provide volunteers to assist with the work? Are the children in care able to go to local schools or attend local health facilities?

Children may be socially or geographically isolated, and location can be critical. Does a support group exist in the community to help in the sponsorship/administration of the institution? Field staff must also assess:

- [ ] The quality of health care.
- [ ] The provision for play and stimulation.
- [ ] Staffing (the ratio of staff to children, the tasks required of staff, how staff are selected, whether they are trained, whether they are recruited locally, relationships between staff, and between staff and children, the level of continuity and the rate of turnover of staff).
- [ ] Links maintained by the children with their family/other relatives/the community.
- [ ] The buildings and equipment required/available; are they suited to local conditions?
- [ ] The status of the institution in the long term. Is it intended to become a permanent adjunct to community/family care, or to fulfil a temporary need to provide a refuge for just one generation of children after a disaster?
- [ ] The time-scale of care provided for individual children. Is it intended to take children into care for long or short periods?

The level of care provided by the institution should not differ radically from that provided by the community as a whole, or the children will find it hard to adjust when they leave. The size of the institution is also a critical factor; it must be large enough to justify the capital outlay, but never so large that the children cannot receive individual care and attention. Long-term institutional care is rarely the best solution for small children. Every effort should be made to find a way for them to be returned to their own family or, if this is not possible, to be brought up either in their own community or in another family or community. In most societies, when the immediate family is unable to look after a child, there exist alternatives — such as the care provided by members of the extended family. These customary methods of care should always be strengthened and not undermined by practices — such as institutional provision — which are essentially alien to the local culture.

Direct support may be given to children cut off from their families or community for whom institutional care would not be suitable. Groups of orphans, the destitute, street children and those employed in factories or on plantations may come into this category. Assistance can be given to children organised into some kind of group or association to improve the conditions in which they live and work. Where they are too young or unable to organise in this way, it may be possible to channel funds through a local welfare agency which can provide vocational training, night school facilities, or a refuge. Alternatively — and especially in the case of child workers — the most effective intervention might be to alert agencies concerned with the implementation of legislation on the rights of children.

Adoption (and sometimes fostering) — where permissible within the laws and customs of the country — is often one of the more satisfactory solutions for children who cannot remain with their own families. Adoption law and practice vary so much from country to country, however, that no generalisations can be made. It is wise to support only adoption schemes which provide homes for children within their own country, are organised by approved welfare agencies, have as their main priority the interests of the child and are run by those who are professionally qualified to do so. Oxfam should not become involved in any scheme by which children are removed from their country of origin for adoption, because of the political controversy surrounding such schemes.

2 The Abuse of Child Labour

Throughout the world children undertake domestic chores in the home, and are also to be found in rural areas working on the family farm or, in cities, assisting their parents by running errands, etc. These activities are frequently crucial to the domestic economy and can be an integral part of the process of learning in which the child gradually acquires the skills which will be necessary in adult life.

However, there is another widespread and growing phenomenon which is particularly common in the poorer countries of the Third World: the use of child labour in industrial and service establishments in urban areas (in both the formal and informal sectors) and also in commercial agriculture. Throughout the world working children are placed at the mercy of their employers, often working for long hours for low wages under conditions which are damaging to their health and to their physical and mental development, and deprived of appropriate opportunities for education and training.

The use of children to assist in domestic work or subsistence activities frees adults for other, directly productive tasks. Even though this kind of labour may present certain problems — the child may, for example, be deprived of a formal education — it is neither realistic, nor even necessarily desirable, to seek to eliminate child participation in a cohesive domestic economy.

Instead, our attention must be focused upon the exploitation of child labour in areas where the traditional system of work relations, based on the division of labour within the family, has been destroyed under the impact of commerce and urbanisation or the pressures of poverty. Child labour occurs in both rural and urban areas and abuses are common everywhere. India has perhaps the largest concentrations of child labour of any country in the world: the work-force of entire industries in some cities consists largely of low-paid juvenile labour.

Child labour is particularly common in countries with a narrow industrial base. One of the advantages of child labour is that it is cheap, and therefore helps the small enterprise to retain a competitive edge in the market: this is particularly true in countries where inexpensive imported goods are available. A high incidence of child labour is usually associated with high levels of adult unemployment and under-employment, together with extreme inequality of income.

In some industries children can be more efficient than adults. Their size and agility can be a great asset in certain activities. They are frequently used to clean machinery because they can crawl into small spaces which are inaccessible to adults. Child labour tends to prevail especially in those countries where the participation of women in non-domestic economic activities is prohibited by custom. Children may compensate for the absence of women by performing tasks such as street trading or marketing from which women are traditionally excluded.

Children are far more readily controlled than adults, and easier to exploit. They are usually neither aware of their rights, nor able to assert themselves sufficiently to protect those rights. Child labour legislation is usually ineffectual. In urban areas particularly, the sweat shop or small service establishment tends to be beyond the reach of the law. Government inspectors can often be persuaded with a bribe to turn a blind eye to the use of illegal child labour.

Because of poverty and unemployment, the household may rely heavily on the income-generating activities of children. Children may not only contribute to the household economy, but may also provide support for their parents in old age.

In the urban setting particularly, the breakdown of family and other supporting structures often forces children not merely to work outside the family in larger economic units, but also to live in isolation from the family and to be entirely dependent for survival on their own resources.

Child labour may take many forms:

a: Domestic work, including cooking, child-care, housework, etc.

b: Subsistence activities outside the home, including hunting, fishing, farm work, fuel and water collection (in rural areas) and the delivery of messages, guarding of goods, etc. (in urban areas).

c: Tied, or bonded, labour (usually illegal). This occurs in both rural and urban areas. Children are delivered by their parents to a job-placement agent in return for a cash payment. It is common for agents to take children from families in poor rural areas to the city, with false promises and claims that the child will be placed in a good job. Many children (both boys and girls) recruited in this way find themselves working without wages in brothels and massage parlours.

d: Wage employment. Children work as domestic servants, or as seasonal labourers on plantations; they are also found in mines, manufacturing industries and in service establishments.

e: Employment in the informal sector. This category includes a wide range of activities of varying intensity, but the work is normally either intermittent or short-term. Employment in the informal sector includes street-trading, shoe-shining, the collection and sorting of rubbish, the guarding of cars and other personal property, and carrying goods.

f: Marginal activities such as theft, drug trafficking and prostitution. Because of the illegality of these occupations, it is often difficult to identify the numbers of children involved, or the extent of their involvement. These children face special health and welfare risks.

g: Child soldiers. Whether recruited by government, guerrillas or outlawed political factions, there is a growing number of children in this category — especially in areas under constant pressure such as Iran and Palestine.

Child workers are frequently separated from their families and homes for long periods, or even permanently. Not only are they deprived of love and care, but their families may remain ignorant of their circumstances and whereabouts. Children are often forced to work under conditions which at the very least infringe their basic human rights but may also endanger their lives. The main areas of abuse are listed below.

a: Little or no remuneration. Often the wages are paid to parents, the child receiving only subsistence rations.

b: Excessive hours of work. Worldwide children work more than 40 hours a week. Many children working in service establishments are on night shift; girls in domestic service must be available at nearly all times of the day and night, and many children work without rest periods or holidays.

c: Safety hazards. In many factories and sweat shops even the minimum safety standards are not observed. Children work with machinery and tools designed for adult use, often without protective clothing. They are frequently assigned tasks which are totally unsuitable, given their lack of training or experience and their shorter span of concentration.

d: Health hazards. There are numerous health risks for working children and their vulnerability is increased by the high incidence of malnutrition and under-nourishment. Working children are more likely to suffer from malnutrition because of the additional energy requirements of work. Many work in appalling conditions with poor lighting and ventilation, fumes, dust, or high levels of humidity and heat which facilitate the transmission of communicable disease. They may be expected to work with toxic substances, or on manufacturing processes which cause the eventual loss of sight or hearing, or bone deformity.

e: Lack of ancillary services. There is frequently no provision for feeding, health care, education or training.

Recommendations

Many development programmes at the community level have the effect of preventing the worst abuses of child labour. For example, the provision of creches and nursery schools may enable women to work, thereby alleviating the pressures on older children to earn income. However, it should not automatically be assumed that all child labour is bad: the priority is to stop the exploitation of children and curb their involvement in activities that are harmful to their health and development.

Oxfam has been able to support few projects which alleviate the problems faced by juvenile industrial workers. It is difficult to work directly with children who are employed in large factories and mines, and intervention is a potential cause of political controversy. Trade unions tend to be opposed to child labour and therefore are reluctant to support moves to improve working conditions and rates of pay for children. A national trade union in Ecuador, however, is working to improve conditions for child labourers in industry.

Often the only resort will be to provide information on child workers to international bodies (such as the ILO) concerned with the ratification and application of international labour standards; to national organisations charged with the statutory obligation to protect children against abuse; and to government institutions and welfare agencies whose objectives embody the Declaration of the Rights of the Child. It is important that governments and other concerned bodies should be made aware that legislation prohibiting the use of child labour will merely force the problem underground and remove all protection for child workers. If implemented, it could exclude children from factories and other institutions, and force them into uncontrolled and dangerous occupations, in which exploitation and abuse are even more likely.

Children working on the streets may face a number of special problems. Not only are they trying to survive in an extremely competitive and volatile market, but many are also homeless. As far as income is concerned, various forms of assistance may be appropriate. Vocational education and training can be given to improve employment potential and literacy skills. Part-time night-school facilities can be made available to street children, and vocational training schools can been set up in industrial centres. In Ethiopia provision of small loans or low-cost capital equipment has been made to assist street children to earn a better living.

In some cases welfare agencies can provide food, shelter and health care for children living and working on the streets by establishing refuges in city centres or near markets and other places where casual employment may be obtained. An alternative (but one open to few children) is to seek to place the children in some form of permanent care. Before funding a programme for street children, it is important to establish whether they do in fact require shelter. Many children working in casual trades live with their families and would benefit most from education, training or provision for income generation.

Child labour cannot be eradicated where adult unemployment is high. Efforts to help the children must be based on the understanding that their income is usually vital for survival. Thus, education provision must fit in with their work routine (by being provided in shifts for example) and schools should be as near as possible to the place of work. Programmes assisting child workers are likely to be fairly experimental: a sound methodology for this type of project has yet to be found. Above all it has to be recognised that many working children experience minimal contact with, and control by, adults and are therefore unlikely to respond well to programmes involving institutionalisation, close supervision or a highly structured, inflexible routine. Street children tend to be particularly independent. Even those who live with their parents become used to making their own decisions and disposing of their own incomes at a very early age. Many work in trades controlled and regulated by gangs of young people. Programmes assisting these children should seek to foster and develop participatory organisation — giving the children considerable responsibility in administration, carrying our chore etc. — rather than impose authoritarian structures.

3 Children in Emergencies

More and more children are being caught up in natural and unnatural disasters. They may become the victims of war, monsoons, forced migration, earthquakes, etc., losing contact with their homeland, community, immediate family and other relatives. The assistance given to children in these circumstances will require a number of special features. Child victims of disasters may be under severe emotional and physical stress, isolated, perhaps, from their kin and living in unfamiliar and often hostile surroundings.

(For detailed recommendations of emergency provisions for children see PART EIGHT, Section 5.IV.)

Section 4   Disabled People

I   The problem

People afflicted by disablement are among the most disadvantaged sectors of any community, suffering not only physical discomfort through pain, incontinence or immobility, but mental anxiety as a result of dependence and rejection. There are some 340 million disabled people in the Third World. The major cause of their disabilities is poverty. Poor people are the most likely to become disabled and the least likely to receive treatment.

Mothers who cannot afford to eat enough food during pregnancy are far more likely to give birth to physically or mentally stunted children than those whose nutritional requirements are fulfilled. The ante-natal advice and care that is an essential prerequisite for a healthy baby is inaccessible to poor mothers. Dietary deficiencies such as a lack of Vitamin A can cause blindness in a growing child.

Social factors reinforce these medical problems. The type of work available to poor people increases the risk of disablement — carrying heavy loads on the head, wading in infected water, and a general failure to observe safety standards in places of work make poor workers more vulnerable. In the family, disabled people receive less food, less education and less medical care than others. Discriminatory practices are reinforced by political structures such as the caste system in India or the landlord system in Latin America, which serve to perpetuate the deprivation of the poor, and the hardest hit are disabled people.

The prejudice, superstition and ignorance encountered in everyday life are barriers just as hard to overcome as the wider socio-economic obstacles. Many disabled babies are killed or left to die; others die through lack of medical care. They may be mutilated or kept in the dark out of sight. Disabled people are often regarded as patients for life, forced into a dependent position where they have to accept and live with society's definition of their future. These negative attitudes can often be more of a handicap than the disability itself, since they prevent the disabled person from engaging in normal social activities.

340 million people is one in ten of the total population of the Third World, and this statistic remains surprisingly constant, regardless of the particular geographical area. This means that *any* project will necessarily include disabled people within its scope. Field staff should therefore assess the extent to which a particular project will benefit disabled people, as well as other disadvantaged groups. To this end it will be helpful to bear in mind two fundamental goals essential to any scheme designed either to alleviate the causes of disability or to help disabled people directly: prevention and integration.

1   Prevention

Prevention should, perhaps, be the highest priority. Since the major cause of disability in the Third World is poverty, all Oxfam work is, in a sense, aimed at the prevention of disability. Agriculture provides food to prevent malnutrition. Primary health care will reduce the risk of disabling diseases. There is, however, a vast body of medical knowledge specifically geared to the question of prevention. In many cases, no-cost or low-cost techniques are available which obviate the need for expensive institutional or hospital care (see PART SEVEN, Section 9).

Early identification and treatment is essential to stop the development of a disability, or at least to minimise its effect. Exercises and aids aimed at stimulating minds or muscles, increasing mobility, balance or reach are effective and very cheap. Training health workers in such techniques would reduce the number of disabled people as well as the seriousness of disabilities already developed.

Immunisation programmes are clearly a priority. It costs less to vaccinate 100 children against polio than to provide service for one disabled child. In the same way, it is cheaper to provide safe equipment to prevent home, agricultural, or industrial accidents than institutions to care for the victims.

Prevention is certainly the most cost-effective way of dealing with disablement. It must therefore be a priority. Nevertheless, it is a fact that one in ten do suffer from a disability of some kind; this does not mean that they are therefore unable to take an active role in community life.

2 Integration and Employment

Projects involving disabled people should not be regarded as 'humanitarian' or 'charitable' in nature — any more than those geared towards other disadvantaged groups such as women, children or the very poor. The conventional view that institutional care is the only way of treating people with disabilities is a dangerous notion and must be combated wherever possible. There are some advantages: institutions remove the burden from the very poor, for example, and provide a place for people who need care which could not be provided in their community. However, the disadvantages are numerous, for institutions:

☐ are usually based on a Euro-American model and are therefore very expensive;

☐ are highly selective, at present catering for the needs of a mere 2% of the disabled people in the Third World;

☐ tend to concentrate on the disability rather than the person's positive qualities;

☐ isolate a person from his/her community and therefore fail to make use of the considerable help that families can, and do, offer;

☐ reinforce negative or prejudiced attitudes, both in others who see 'homes' as the only acceptable place for people with disabilities, and in disabled people who may come to regard themselves as helpless patients.

Institutional care should therefore be considered only as a last resort, and then only for the shortest possible time.

Great emphasis has to be placed on allowing disabled people to live as normal a life as possible in their own community. The involvement of their own family is one of the most useful sources of help for disabled people. Health care that removes them from their own local environment for long periods should be avoided if at all possible; domiciliary services and day-care centres are to be preferred.

This emphasis on local care extends to artificial aids (prosthetics, wheelchairs, etc). Where possible, these should make use of local materials and techniques, to reduce costs. Disabled people themselves should be involved in their design and manufacture. (See PART SEVEN, Section 7.)

Employment for disabled people is another priority, and one that raises complex and controversial issues. It is not a question of charity, but simply one of equality of opportunity. Many jobs are performed better by people with a particular disability. Blind people are often good at massage, operating telephone switchboards, or as musicians. The Kerala Federation for the Blind in India has formed an orchestra to provide employment for blind people after they leave school or college (KER 30). The Federation's centre for blind adult males also has a variety of high-quality production lines including candle making, mat weaving and umbrella making, all products which compete on the local open market. Deaf people can work in noisy surroundings. Some mentally handicapped people can be happy doing simple repetitive jobs. According to the UN's International Labour Office, disabled workers in general tend to have higher productivity because they are more diligent, concentrate better and take more pride in their work. Perhaps they are also more apprehensive about losing their jobs.

The standard answer to the problem of employment is the provision of sheltered workshops. Ideally these should be avoided because:

- [ ] few people can be employed at any one time;
- [ ] workers compete with other workers;
- [ ] the goods produced are often of poor quality;
- [ ] they reinforce separatist notions about disabled people in general.

Frequently such workshops manufacture creative but useless objects. Embroidery in itself is of little long-term benefit to a disabled person unless there is a market for it. The making of redundant items, wanted and needed by no one, serves only to reinforce the view that people with a disability are peripheral and, like the objects they make, can serve no directly useful function.

Many such workshops are run on paternalist or even exploitive lines, thus degrading their workers. Cooperatives are preferable, such as the Cooperative Training Society formed by polio victims trained at the Government of Somalia's vocational training centre (SOM 35). Although there is a protected market for their garments, the standards compare very favourably with commercial businesses, and the society hopes eventually to become entirely independent. Where sheltered workshops are used, they should be subject to Oxfam's normal policies concerning employment and economic conditions (see PART FIVE).

Income-generating projects are clearly important, but social integration is just as important as economic integration. Those projects that lead to the rehabilitation of their workers in their home communities are preferable to those which set disabled workers apart as dependents who must be 'helped along'. The Olevagoya Hospital in Peru (PRU 217N) runs a farm for psychiatric patients. It is designed to prepare the patients for return to their families and to society in general. It is an innovative step, aimed at creating a new model of psychiatric assistance in Peru, based on the concept of a therapeutic community leading to reintegration of patients with society, rather than maintaining them in closed institutions. Part of the success of the project is due to the fact that the farm generates an income through the production and sale of chickens, rabbits and guinea-pigs (the pelts are used by the handicrafts workshop at the same hospital), potatoes, vegetables and maize.

Agricultural work by which disabled people can grow food for survival liberates them from reliance on charity. However, the shortage of cultivable land in any parts of the world, or its inaccessibility to poorer members of a community, limits the applicability of this form of employment.

Integration with their community often occurs spontaneously for disabled people in the Third World, for the extended family networks that operate in many areas place a strong emphasis on caring for all members of the family. In many cases, however, disabled people are rejected or pushed out of sight. Organisations and unions of disabled people are beginning to form in various parts of the world, for example, the Physically Handicapped Slum People Development Association in Bangalore, India. Groups like this provide invaluable mutual support and practical help for their members. They also bring non-disabled people's attention to their problems, which are often caused by ignorance and superstition. In Mauritania (MTA 6) the Union Nationale des Handicapes Physiques et Mentaux found that some taxi-drivers refused to pick up disabled people. Those affected simply took down the numbers of those taxis and reported them to the police. The drivers were not charged but had to report to the police station, thus wasting valuable taxi time. The problem no longer exists. The Union also runs training courses in a variety of skills and sells its products in the local market.

Recommendations

☐ The chief cause of disability in the Third World is poverty, and so the main form of prevention is the alleviation of poverty.

☐ Any project should take into account the needs of disabled people. If a development effort is designed to benefit the poorest, it must necessarily include disabled people.

☐ Prevention is far better than cure. Up to 50% of all disabilities can materially benefit from existing medical techniques, often costing little or nothing to implement. Health programmes designed to reduce the incidence of disabling communicable diseases and programmes designed to raise the nutritional status of the community should always be encouraged.

☐ Beware the isolating effect of institutional care; the proper place for people with disabilities is with other people in their own community. Care of the disabled should be neither centralised nor confined to professional workers.

☐ Disabled people are neither unemployable nor helpless, and paternalist or protectionist attitudes should be eradicated. Support should be given to income-generation projects — always ensuring that the project can compete well on the open market and that the products are in demand.

☐ Organisations of disabled people for disabled people should be supported and encouraged wherever possible. The active involvement of disabled people and their relatives should always be secured in any programme designed to help the disabled.

☐ Probably the greatest handicap for disabled people is the prejudice, superstition and ignorance they encounter in other people; educating society in general as to the true nature of disablement is vital, if disabled people are to achieve their full status. Support groups for disabled people should be encouraged, and social education promoting positive attitudes towards disability or handicap should receive priority.

☐ Simple devices should be promoted to help the handicapped to become independent and to earn their own living. All such devices should be appropriate to local socio-economic conditions, and wherever possible should be made by local artisans or by disabled people and their relatives.

(For further recommendations see PART SEVEN, Section 9.)

I   Introduction

Throughout the world economic oppression coincides with discrimination along ethnic lines. All societies contain groups who are consistently disadvantaged in relation to others in the population and distinguished by separate ethnic, religious, and/or linguistic characteristics. These distinctions not only categorise and rank people, but they also tend to foster a sense of solidarity in oppressed groups which is frequently directed towards preserving their unique culture and traditions.

However, ethnic groupings have no bearing in biological reality; ethnic categories are neither fixed nor permanent but are used subjectively according to the socio-economic context. At the very least, specific ethnic groups may be the subject of prejudicial attitudes. Many are also the victims of abrupt and insensitive policies and discriminatory legal codes imposed by repressive governments. Deprived of full citizenship rights, they are unable to gain access to adequate income and are denied power. Their resources are appropriated without compensation and they serve the national population by providing cheap labour or goods for urban markets. Their incorporation into the sovereign state is often achieved by using violence. In the most extreme cases they are the victims of systematic genocide. The growing concentrations of oppressed ethnic groups in urban ghettoes makes them even more vulnerable, particularly in times of economic depression and civil strife.

Such groups require Oxfam's special attention, though it is difficult to establish appropriate forms of assistance. Some live in isolated and inaccessible areas and a few are highly mobile. Moreover, their culture may seem elusive and their traditions obscure to people (donor agency staff, development workers and the like) who are more familiar with national society. Field staff will need to understand in some detail the internal characteristics of the society. Since they are unlikely to have the expertise, and will certainly lack the time, to undertake the necessary analysis themselves, they should make every effort to identify existing publications and research reports on the people in question. Information about such material is likely to be most easily obtained through contact with university departments and research institutions in the national capital. Anthropological research is likely to be particularly helpful in giving insight into the attitudes, preoccupations and, not least, knowledge of people at the local level. The aim must be to respond as far as possible to the needs of the people, as expressed by them, rather than introducing models of development devised by outsiders. All too often, without even consulting the groups concerned, development planners design interventions to improve economic security which are insensitive to culture and ignore crucial differences in values and attitudes. While competing economic interests and unequal commercial relations are usually at the root of ethnic discrimination, a programme addressing itself simply to the economy will most probably fail.

Among the peoples most commonly alienated from national society are those who live in rural areas and obtain their living by means of subsistence hunting, farming and/or herding. For want of a better term, these may be referred to as 'traditional' societies. Roughly 4% of the total population of the world (about 200 million people) fall into this category. For most traditional groups the consequences of contact with the modern nation states have been particularly severe: they have been the victims of centuries of land invasion; their populations have been decimated by Western diseases and warfare; their resources have been despoiled and their people disenfranchised and enslaved. Forced to live in ever smaller and more restricted, or marginal, territories, many groups have been compelled to change their whole life-style, their economic organisation and their traditional means of subsistence, making the transition from nomadic hunting and gathering or herding to settled agriculture, often combined with seasonal wage labour. Others drift into towns where they obtain casual work in the lowest and most menial of occupations.

Most traditional societies retain a number of pre-capitalist forms of social and economic organisation based on collective ownership and communal production. By comparison to the national society their social structure is relatively unhierarchical. Even though they normally have simple technologies, their knowledge of the environment is extensive and their methods of production based on highly specialised and complex adaptive mechanisms. Their relationship to the environment is rarely merely economic since certain tracts of land, or specific resources such as trees, lakes, or rocks, tend to have important religious or social significance.

It is sometimes argued that work with traditional groups should be discouraged because it is bound to undermine, in one way or another, and ultimately destroy, traditional values and ways of life. Although sincerely intended, the effect of this argument is to make a fetish of traditional culture — to treat it as a museum object. All societies, even the most remote or ancient, are continually developing and those who actually live in traditional cultures are usually only too well aware of ways in which their lives could be changed for the better. Nearly all such groups, furthermore, have long been in contact with national society and are therefore subject to forces of change that are beyond their control. Whilst striving to minimise the disruptive, unintended and undesired consequences of their intervention, field staff should aim to respond to the aspirations of such people for change. This can only be done on the basis of detailed information about, and respect for, local institutions, attitudes and knowledge.

What must be avoided if at all possible is a process of forced, accelerated change in line with the sole interest of the nation state. Indeed, many programmes with oppressed ethnic groups are initiated primarily as an attempt to reduce the impact of less compassionate forms of intervention. This has led some development workers to establish programmes revitalising a culture and preserving its most important symbols. However, unless such programmes also attempt to alleviate the economic problems of the group or secure its resources, they will not provide an effective weapon against domination and forced intervention. Instead, the risk is that they will merely confirm prejudice and reinforce discrimination.

The current concept of the nation state is particularly hostile to nomadic pastoralists, many of whom regularly cross national boundaries in their effort to feed animals on huge tracts of marginal land. Since animals are susceptible to theft, the nomads often carry weapons, thus increasing the chances of clashes with the authorities. Widespread settlement of land, destruction of dry-season habitats and many other factors, have combined to make pastoralists a particularly disadvantaged group in the modern world. Almost invariably they are ignored or even deliberately harassed by government schemes. In many cases efforts are made to restrict their nomadic way of life and ease them into a sedentary existence. Failure to understand the complexity of the pastoralists' life has led to disastrous results.

The root of pastoralists' problems is usually political. Governments dominated by sedentary peoples, long accustomed to fear the pastoralists, enact highly repressive measures provoking the nomads to violence. Refugee populations in Africa are replete with pastoralists who have been dislodged from their traditional way of life, partly by natural causes but very largely because of political intervention.

II   Principal problem areas

[1] The Denial of Civil Rights

Just as the position of different ethnic groups in national society varies enormously, so do the possible forms of assistance that can be given by the aid agency. For the majority of groups, though, development depends on first securing their fundamental human rights within the national context (see PART FOUR, Section 4). The legislation governing the exact status of different ethnic groups varies widely from country to country, some legal codes being systematically punitive, others far more liberal. Certain countries have consistently refused low-status ethnic groups the right of full citizenship — defining them as minors in law, under the guardianship of an appointed ministry or agency.

In many countries, even though full legal and civil rights are awarded, no specific provisions are made to protect the communal forms of land-holding, or special socio-political structures that are in many cases fundamental to minority society. Even where the law is fairly advanced — granting special dispensation to practitioners of minority religions, or the right to communal land-holding, or community-based political structures — often its application is piecemeal. Usually, where there exist strong opposing interests (as in most land disputes, for instance) the laws protecting the weaker ethnic group are not upheld, but are ignored and flouted by those more powerful.

Donor agencies can provide effective support for organisations that lobby governments and inform public opinion through the media. Equally important, assistance can be given to enable key documents such as legal codes to be translated into the language of the group concerned, thereby informing them of their rights. In some instances considerable pressure can be brought to bear on governments with the support of multilateral agencies, specialist organisations such as the Minority Rights Group and the international media. In countries with more favourable legal conditions, the priority should be to ensure compliance with the law — especially at the local level. Many ethnic groups can be encouraged to seek legal recognition, where appropriate, whether at the level of the tribe, sect or community. To this end, legal aid may be made available (see PART FOUR, Section 4). Legal recognition frequently facilitates the arrival of State services and is usually essential for securing title to land.

Some groups may opt for total integration with national society, on the premise that this is the most effective way of securing economic development and gaining access to essential services. However, as already indicated, integration schemes mounted by governments are more often than not thinly disguised attempts to draw different ethnic groups into the national economy as cheap labour.

Other groups have sought the guarantee of fundamental human rights and equality of access through self-determination, or co-existence, rather than integration. Autonomous political organisations have been formed in some countries which not only provide a mouthpiece for their people, lobbying for more humane and appropriate policies, but also control and administer development programmes. The growth of such organisations can be slow and difficult, and requires training in administration, legal procedures and accountancy (see PART FIVE, Section 4).

In the more extreme cases, where the level of oppression is severe, political secession is sometimes identified by the groups themselves as being the only viable alternative. This approach is, perhaps, the most problematic both for the aid agency and the group itself. The funding agency cannot become involved where a group is perpetuating acts of violence (see PART FOUR, Section 4). And furthermore, even though secession may give political independence, it is rarely the solution to poverty. Unless the group controls a sufficient area of fertile land, or other productive resources, independence is unlikely to ensure economic viability.

② The Threat to Resources

Many of the ethnic groups that have survived as distinct entities into the 20th century inhabit remote lands which formerly had no value for others. The spread of modern communications, population pressure, and industrial exploitation, have reduced the possibility of isolation. They find their land threatened by many forces, from spontaneous or planned settlement by farmers, to national development schemes, irrigation projects, highways, hydro-electric facilities, mines, agribusiness and timber extraction. The alienation of land is perhaps the most serious problem for the majority of oppressed ethnic groups and the greatest single cause of cultural disintegration and impoverishment.

The social relations within many groups are reflected in the spatial arrangement of villages and gardens and in rights to fishing, hunting and grazing lands. Sacred sites and burial grounds may mark out a group's history and provide social and cultural continuity. Thus, the loss of land not only threatens livelihood but in many cases undermines the social organisation and destroys the culture.

Recommendations

Under the more benign governments, a legal remedy for the problem of land alienation may be possible. Territorial demarcation and land titling, with the assistance of topographers and legal experts, should be a priority for any group under threat of encroachment. The members of the group must be involved in the procedure at every stage, advising on traditional landmarks, setting boundary stakes, and (where applicable) clearing undergrowth. The demarcation of land must respect traditional tenure patterns and production methods, and fragmentation of holdings should be avoided because it makes defence of property difficult. When a group is forced to compromise over boundaries, and accept a reduced territory, it is important to ensure that there will be sufficient land to sustain the population using traditional production methods.

Where a group has lost its lands altogether, two possibilities are land purchase, or, where all other channels have been tried, the payment of an indemnity (which is normally significantly lower than the free market price). These procedures are usually extremely costly, and rarely make available to the group sufficient land to sustain subsistence. They will, however, give a foothold to groups without other legal recourse. Another, less desirable, solution to land loss is relocation (see PART FIVE, Section 2). This strategy is used by some of the most repressive governments. Even so, it is still preferable to giving cash compensation for the expropriation of land, because this last alternative rarely provides long-term economic security.

Many of the more disadvantaged groups resort to invasion in the recovery of land. There are distinct risks, though, in provoking a hostile government and it is clearly difficult for the aid agency to become involved in such activities. However, it has proved possible under the more liberal regimes for Oxfam to assist groups to turn de facto into de jure rights on invaded land.

3 The Crisis in Production

For traditional societies the pressure on land can have particularly serious consequences: although the problem of over-exploitation of natural resources is common to most rural populations in the Third World, it has special implications for traditional hunter-gatherers and herders. For them, to survive crises in food production often means not simply adopting new production techniques, but also a whole new way of life.

Field staff should be wary of programmes designed to make radical changes in traditional production methods, and assess them carefully for their social and environmental impact. Most herding groups, for example, live in arid or semi-arid regions, at high altitudes and in areas of extreme temperature variation. The main feature common to pastoral areas is the seasonal availability of the pastures themselves, with extreme seasonal variation in the carrying capacity of individual tracts of land. Most herders have responded to this constraint by developing a nomadic pattern of exploitation. Nomadism permits a greater concentration of population and more complete utilisation of resources in marginal areas than would be possible under a sedentary system: it allows the pastures to rest and replenish. Attempts to increase production in pastoral areas have usually followed models drawn from North American and Australian experience. Beef production, commercial ranching, specialised breeding, growing and finishing enterprises, auction markets and processing facilities have been introduced into areas where local conditions are totally unsuitable. Such capital-intensive operations require, among other factors, stable markets, high meat prices, high rural wages; conditions that only rarely obtain in the Third World. Increased dependence on the market has for many herding groups only intensified prejudice and poverty.

Nutrition surveys carried out on traditional societies might in many cases show surprising results. For example, in most of the more isolated and marginal areas of the world, food-gathering is more reliable and maintains a higher nutritional status and wider dietary variation than cultivation. Furthermore, cultivated plants tend to be more vulnerable than wild ones to extreme weather conditions and disease, a very serious problem where producers do not have ready access to fertilisers, pesticides or insecticides.

Recommendations

In rural areas the problem of impoverished ethnic groups can be addressed in a number of ways, as indicated below.

a: The economy can be strengthened by developing traditional subsistence activities, improving production techniques, stock and yields.

b: A more radical approach is to introduce a range of new activities and products. This may entail cultivating new crops, breeding new animal species, or developing some form of settled or semi-migratory cultivation based on new techniques. Most such programmes are associated to a greater or lesser extent with production for sale.

c: Alternatively, traditional subsistence activities can be complemented by specialised income-generating programmes designed to capitalise on the particular skills and/or unique material culture of the group. Traditional handicrafts and other artefacts can be sold and profits ploughed into communal programmes (see PART FIVE, Section 2).

Care must be taken to ensure that economic change is gradual and is accompanied by training and by the use of demonstration plots and an appropriate level of capital. Among traditional societies particularly, variation in the rate of adoption of technological innovation by different people could lead to inequalities within a system once egalitarian, and individual modes of ownership might undermine collective activity. Cash cropping may be an effective way of raising funds to pay legal fees for land title, and of strengthening claims to land in the eyes of national society. Food production, though, should always take priority over cash crops, and particular caution should be exercised when highly profitable activities are introduced, since these will tend to attract people away from growing subsistence crops. It is important to assess whether cash cropping will have the effect of merely increasing the group's dependence on traders. Many ethnic groups are locked only tenuously into the market system, receiving very poor prices for their goods. Both marketing and supply can be very difficult for geographically isolated groups. Oxfam has supported a number of marketing programmes among rural groups which both distribute goods to communities (tools, clothing, etc.) at controlled prices and sell the agricultural produce and crafts made by them.

The development of subsistence crops with which people are acquainted may seem a far slower way of providing economic security than through cash crops, but it also tends to absorb far fewer external resources and is likely to be geared more effectively to people's understanding of the environment, its possibilities and constraints. Where new crops are introduced, they must be suited to local conditions; thus, for example, an appropriate innovation for a riverine people might well be fish farming.

Field staff must be particularly wary of the impact of economic programmes on the sexual division of labour customary to the group. Most ethnic groups in rural areas have a very marked pattern, in which men hunt, fish or herd the larger animals, and women cultivate subsistence crops, gather wild plants and tend the smaller domestic livestock. The introduction of cash crops can have considerable impact on the workload and general status of women in rural areas, and this problem must be examined before funding is agreed. (For more details, see Section 2, and PART SIX, Section 11.)

4 The Lack of Services

In poor countries, where essential services such as education and health provision are generally extremely inadequate, low-status ethnic groups are often denied access altogether. Those that do receive basic provision frequently find it is inappropriate to their culture and society. This tends to be true of official education programmes, for example, which are frequently used by governments as a tool of repression. Education is used to suppress minority languages and cultures and promote the national culture.

The assimilation of ethnic groups is indeed accelerated by education programmes based on national norms. But for the group concerned, even though education may be a major priority, what is normally sought is a system of teaching suited to their culture and history, and not an alien system using concepts imposed by representatives of the dominant society. Many groups do wish to learn the national language and understand the technologies and concepts used by national society, but few desire this knowledge at the expense of their own language and culture.

Although Oxfam does not normally give direct support for formal education, exceptions have been made in the case of ethnic groups trying to introduce bilingual teaching into their schools. Assistance has been given to popular education programmes, sometimes organised from special 'cultural' centres, where these have been seen to increase solidarity and strengthen communal organisation. It is always preferable that education programmes for ethnic groups be run by the groups themselves, using local promoters. Field staff should be cautious of programmes run by religious or political agencies, since they are generally more interested in proselytising than informing, promoting alien values rather than reinforcing local ones. Assistance has been given on occasion to groups seeking proper recognition of special educational needs within the official system.

Vocational training may be a higher priority for many groups than general education, especially where it equips them with specific practical skills (see PART FOUR, Section 2).

Health provision is another priority for most ethnic groups. Many ethnic groups will share the same health problems as the dominant national population, but others (such as those living in crowded refugee camps, or urban ghettoes) will face special health risks. A number of additional factors affect the health of traditional societies:

a: The disruption of the natural environment, population increase, loss of land, etc., means that many groups are no longer able to sustain adequate levels of food production. The social fabric and system of food-sharing has in many cases broken down; the normal risk-averting strategies no longer operate and the groups descend from relatively good nutritional levels to near-starvation in a very short time. These problems can only be solved by increasing food production and distribution.

b: The pressures to end dispersal and mobility of population and to concentrate population into nucleated settlements increase the risk of contagion.

c: The breakdown of the geographical isolation of many traditional societies has brought them into contact with new diseases (measles, tuberculosis, influenza, for example) which have had a devastating impact.

In most cases when traditional societies undergo a process of accelerated social change, traditional healing systems become distorted. Magic, witchcraft and sorcery tend to become more prevalent in times of social upheaval, and with the uncontrolled entry of Western drugs, much of the expertise and knowledge of diagnosis and healing may be lost. The role of the traditional healer is undermined by the attitudes of national society and the hard-sell of modern pharmaceutical companies. People who have been accustomed to a certain amount of autonomy, treating themselves with traditional remedies, either become dependent upon access to medical practitioners and pharmacists, or begin to treat themselves with powerful and often unsuitable drugs. (For guidelines on alternative healing systems, see PART SEVEN, Section 8.)

Although health programmes for ethnic groups should follow the general principles outlined in PART SEVEN, integrated health programmes always taking precedence over specialised ones, it may be necessary to assist those groups facing immediate decimation by establishing a specialist intervention (such as an emergency vaccination programme). One serious problem is that where prejudice and discrimination is strong, it can be extremely difficult to set up an effective system of referral. Many ethnic groups, if they are to benefit from a health service at all, will be forced to establish and maintain an entirely independent system.

The most successful programmes tend to be those that combine the indigenous healing system with allopathic medicine. In this way, traditional practitioners (such as birth attendants) can be incorporated into the programme and their work complemented by health promoters trained to deal with all the more common and straightforward ailments (see PART SEVEN, Section 8). In many traditional societies the problem of logistics is made far worse by their geographical isolation. It may be that the only practical solution is to create a small-scale, modest service which would never attempt to cope with the more serious health problems or the more complex interventions.

Section 6   Checklist of Questions

I   General questions

1. How is the group defined in relation to national society and what are the specific problems confronting it?

2. Do the objectives of the programme actually meet the real problems of the group concerned?

3. In what way could an outside agency provide the most appropriate assistance, is this in accord with the perceived needs of the group and the constraints acting upon them?

4. What are the relations of the group with the wider society, its culture and political and economic life?

5. What is the legal status of the group?

6. In general development programmes has sufficient attention been paid to priority social groups?

a) do these groups have access to any new opportunities resulting from the programme?

b) how are the benefits of the programme to be distributed?

c) is the programme likely to have adverse effects on certain priority groups?

7. What relationship is there between traditional cultural and religious practices and the project; is there any tension or contradiction between them, if so, how can this be resolved?

8. What support groups exist, how can these groups provide assistance, what approach do they use, are they truly participative or paternalistic in their work?

9. Does the programme include sufficient training to allow for the future autonomy of the group?

10. What provision is there for monitoring and evaluation?

II   Women

1. How does the project affect the division of labour and the use of women's time and labour?

2. How does it affect the distribution of resources within the domestic economy and community between men and women?

3. What provision is there for specialised health care for women especially those pregnant and mothers of young children?

4. Is there provision for educational programmes for women (literacy, marketable vocational training etc.)?

5. Are economic programmes merely adding to the work of women without really adding to their disposable income?

6. What child care is available for working women?

III   Children

1. Does the programme provide a welfare solution or longer term resolution of problems affecting children?

2. Who are the children to benefit, has the project distinguished sufficiently between different groups (handicapped, abandoned, working children etc.)?

3. Is the approach taken primarily for education, or vocational purposes, for income generation, residential care or the provision of a refuge?

4. Are the children themselves involved in running the programme? To what extent will relatives, the community, local welfare agencies or the State be involved?

5. What are the priority requirements that should be met immediately (food, shelter, immunisation, etc.)? To what extent can, or should the programme cater for secondary needs (education, vocational training, etc.) and to what extent can these be the responsibility of an alternative body?

6. What level of child care is intended? Is it appropriate to the local culture and socio-economic conditions? What is the minimum and maximum level of care possible in the circumstances, given the capacity of the project and local norms?

IV   Disabled

1. Are preventive low-cost techniques and aids (rather than institutional and hospital care) being implemented and health workers trained in the use of them?

2. Is the prevention of disablement a priority — are immunisation programmes, nutrition and health care adequate?

3. Is the disabled person living as normal a life as possible in their own community? Are employment opportunities being fully realised. Are productive activities commercially viable? Are the positive aspects of the disabled person's abilities being recognised, rather than just the handicap?

4. Are there adequate training courses for the handicapped and those who work and live with them? Are sufficient measures taken to integrate socially the handicapped and counteract any negative attitudes towards them held by the community?

V   Ethnic groups

1. What is the land history of the group? Has its territory been reduced? Is the present territory under threat, and does the group possess land titles? What forms of land tenure and usage are traditional to the group?

2. What economic activities are traditional to the group? How are products distributed? What are the dominant and secondary economic activities, how are risks spread between activities?

3. What is the division of labour in the group? Are the members of the group dependent on migrant labour or outside remittances? What

effects would the introduction of changes in the productive system have on the society and its culture?

4. Does the group have its own language? Is it officially recognised or taught in schools, is it a literate language?

5. What are the religious and other beliefs of the group especially in relation to land, animals, gender relations, kinship etc?

6. How is illness explained and what traditional methods of healing are available?

7. If the group is in danger of contracting diseases to which it has no resistance, is there a vaccination programme?

8. What are the preferred foods, how is food distributed?

Section 7   Resources

Bibliography

1 Women

E. Boserup, *Women's Role in Economic Development*, St. Martin's Press, New York, 1970.

A. Bronstein, *The Triple Struggle: Latin American Peasant Women*, War on Want Campaigns Ltd., London, 1982.

M. Buvinic and N. Youssef. *Women-Headed Households: The Ignored Factor in Development*, International Centre for Research on Women, WID USAID Bureau for Science and Technology AID, Washington DC, 20523 U.S.A. 1979.

Cain C. et al, *Sweet Ramparts, Women in Revolutionary Nicaragua*, War on Want and Nicaragua Solidarity Campaign, London, 1983.

R.B. Dixon, *Rural Women at Work: Strategies for Development in South Asia.* Johns Hopkins University Press, Baltimore and London, 1978.

P. Huston, *Third World Women Speak Out*, Praeger Publishers, New York, 1979.

M. Loutfi, *Rural Women, Unequal Partners in Development*, WEP Study, International Labour Organisation, Geneva, 1980.

M. Mukhopadhyay, *Silver Shackles: Women and Development in India*, Oxfam, Oxford, 1984.

B. Mass, *Population Target: The Political Economy of Population Control in Latin America*, Charters Publishing Co., Ontario, 1976.

N. May (compiled), *Of Conjuring and Caring Women in Development*, Change International Reports, London, 1983.

I. Palmer, *Rural Women and the Basic Needs Approach to Development*, International Labour Review 115.1, Geneva 1977.

B. Rogers, *The Domestication of Women, Discrimination in Developing Societies*, Tavistock Publications, London, 1980.

The Tribune, International Women's Tribune Centre. New York, U.S.A.

Women's International Bulletin: ISIS, Geneva, Switzerland.

2 Children

General

Child Survival/Fair Start, Ford Foundation Report, 1983. Obtainable from: Ford Foundation, Office of Reports, 320 East 43rd Street, New York, NY 10017.

International Children's Rights Monitor, Defence for Children International. Quarterly journal available from: Defence for Children International, P.O. Box 359, 1211 Geneva 4, Switzerland.

The Street Newsletter, Inter-NGO Programme on Street Children and Street Youth. Obtainable from: Inter-NGO Programme on Street Children and Street Youth, Rue de Lausanne 65, CH — 1202 Geneva, Switzerland. Also available: a series of papers presented at a sub-regional seminar for the Mediterranean on street children and street youth and profiles of projects for street children in various parts of the world.

Childright, Children's Legal Centre. Monthly review concentrating on

England and Wales available from Childright, 20 Compton Terrace, London N1 2UN, UK.

Alfred de Souza (ed.), *Children in India, Critical Issues in Human Development*.

Ideas Forum, UNICEF. Quarterly, with regular sections devoted to development education, NGO activities, UNICEF news, etc. Available from: UNICEF, Palais des Nations, 1211 Geneva 10, Switzerland.

International Child Welfare Review, International Union for Child Welfare. Journal on health education, family planning policy, development, handicap, etc., plus articles, information on meetings and a library supplement. Available from: International Union for Child Welfare, P.O. Box 41, 1211 Geneva 20, Switzerland.

Child Labour

Elias Mendelievich (ed.), *Children at Work*, 1979. Available from: International Labour Organisation, 1211 Geneva 22, Switzerland.

Exploitation of Child Labour, United Nations publications, sales no. E.82.SIV.2

Child Labour: a Threat to Health and Development, Ikesetsing Series no. 1. Available from Defence for Children International.

J. Challis and D. Elliman, *Child Workers Today*, Quatermaine House Ltd, London 1979.

Anti-Slavery Society Reports, Child Labour Series. Includes reports on child labour in India, Morocco, Spain, Thailand, Italy, Jamaica and South Africa. Available from Third World Publications, 151 Stratford Road, Birmingham B11 1RD, UK.

Benjamin White (ed.), *Development and Change*, Vol 13, No.4, Oct. 1982. SAGE Publications, 28 Banner Street, London EC1Y 8QE. Child workers' issue of the Journal containing case studies, e.g., Brazil, Senegal, and a bibliography of works on child labour in the Middle East and Europe.

3 **Disabled**

A. Moyes, *One in Ten: Disability and the Very Poor*, OXFAM, Oxford, 1981. £1.25 plus postage and packing.

WHO (ed.), *Disability Prevention and Rehabilitation*, World Health Organisation Technical Report Series 668.

D. Caston, *Low Cost Aids*, AHRTAG 1982, 85 Marylebone High Street, London W1M 3DE.

O. Shirley (ed.), *A Cry for Health: Poverty and Disability in the Third World*, The Third World Group for Disabled People, 16 Bath Street, Frome, Somerset, BA11 1DN, UK. 1983.

J. Wilson, Sir (ed.), *Disability Prevention: The Global Challenge*, Oxford University Press 1983.

4 **Blind**

J. Wilson, Sir, *World Blindness and its Prevention*, Oxford Medical Publications, 1980.

BBC, *In Touch: Aid and Services for the Blind and Partially-Sighted People*, BBC. (£1.00) Available from British Broadcasting Corporation, Portland Place, London W1, UK.

C. Blindenmission, *Without Holding Hands: Handbook of Approaches to Vocational Training and Rehabilitation Work with the Blind Arising from a CBM Seminar of Pioneers in 1978*. Available from Nibelungen Strasse 124, D-6140, Bensheim 4, West Germany. Recommended.

E.K. Chapman, *Visually Handicapped Children and Young People*, Routledge and Keegan Paul, 1978. (£5.50) Useful handbook for teachers working with the blind; includes good bibliography.

G. Bisley, *A Handbook of Ophthalmology for Developing Countries*, Oxford Medical Publication, Oxford University Press, 1976. (£1.50)

G. Salisbury, *Open Education Handbook for Teachers of the Blind*, 1974. Available from Royal Commonwealth Society for the Blind, Heath Road, Haywards Heath, Sussex. UK.

NB. The Society maintains a wide-ranging reference and information section.

5 Ethnic Minorities

M.G. Bicchieri, *Hunters and Gatherers Today*, Holt, Rinehart & Winston, N.Y., 1972.

J. Galaty, D. Aronson, P. Salzman, *The Future of Pastoral Peoples*, IDRC, Ottawa, Canada, 1981.

E. Leacock, R. Lee, *Politics and History in Band Society*, Cambridge University Press, 1982.

L. Mair, *Anthropology and Development*, Macmillan, London, 1984.

S. Sanford, *Management of Pastoral Development in the Third World*, ODI/Wileys, London and Chichester. 1983.

B. Whitaker, ed, *Minorities, A Question of Human Rights*, Pergamon Press, Oxford. 1984.

Organisations

Anti-Slavery Society, 180 Brixton Road, London.

Cultural Survival, 11 Divinity Avenue, Cambridge, Mass. 02138, USA.

International Work Group for Indigenous Affairs (IWGIA), Fiolstraede 10, DK- 1171, Copenhagen K,. Denmark.

Minority Rights Group, 29 Craven Street, London WC2.

Overseas Development Institute — Pastoral Development Network, 10-11 Percy Street, London W1.

Survival International, 29 Craven Street, London WC2.

PART 3 FIELD METHODOLOGIES

Karen Twining

Part Three FIELD METHODOLOGIES

3

Section 1   Introduction

Much of this part is concerned with management issues, but two points should always be kept in mind:

a: neither the objectives nor the strategy of a project should be taken for granted during monitoring or evaluation — their relevance and viability require continual reassessment;

b: social change, even in very small communities, is a complicated and infinitely varied process to which the NGO — through its project holders — can usually only make a modest contribution. While NGOs can participate in the process of social change, they certainly cannot manage it. The extent and nature of this contribution will depend on:

☐ a proper understanding of the social and economic forces affecting the area in question,

☐ an adequate and viable strategy for influencing the direction and impact of these forces and

☐ good organisation.

Above all, development is concerned with people: their aspirations, their lives and their problems. It is important to remember this when reading project proposals and applying to projects the criteria and analyses that are suggested in this section. *Projects are no more than a vehicle for development. They are not themselves development, and people can and do exist without them.* Good systems and efficient management are useful aids, but they are no substitute for a full understanding of the society in question.

The main priority in any system of appraisal or evaluation should be to indicate the likely impact of a project on people's welfare, their values and their capacity to take charge of their own lives. Unfortunately, the assessment of most development aid is restricted to measuring the likely material advantages to be gained from a given technical input.

For many projects, the following methodologies may be inappropriate. They may be too small to lend themselves to elaborate analysis, or, by their very nature, not be amenable to such methodologies. Nevertheless, field staff will have to make an assessment of all projects; this should be based on informal analysis, using, where appropriate, both these methodologies and their local knowledge and experience.

The idea of trying to measure the social impact of development work is relatively new, and still at an experimental stage as applied to the micro- or community level. It entails assessing the significance of the project and its relation to:

☐ the needs of the beneficiaries as expressed by them both at the outset and as the project progresses;

☐ the constraints on development operating in the area; and

☐ the project's objectives.

In some cases it may prove impossible to work exclusively with the poorest people in an area, and in others some of the poorest groups may be excluded because of limited resources. Field staff will face difficult decisions regarding the permissible extent of the involvement in a project of those people not living at starvation levels. For example, the involvement of the middle peasantry in a rural development programme may have an adverse effect on the poorer smallholders, quite apart from presenting a moral dilemma about the use of scarce resources to benefit groups that are already in some way advantaged, even if only marginally.

It can be seen from the foregoing that the proper training of new field staff in the techniques of project appraisal is critical. Not only must they be conversant with the principles involved, but they must also be readily able to apply them.

81

Section 2 Project Appraisal

I Appraisal of projects prior to making grants

1 Partnership

The idea has already been mentioned that the funding agency functions, not simply to underwrite material change, but more to enter into partnership to achieve social and economic development. *The achievement of mutual understanding and respect between field staff and project holders takes great sensitivity, understanding and perception, which cannot be substituted by any amount of paperwork.* A good relationship between field staff and project .holders is vital. Where the rapport does not exist and project holders are unwilling to discuss their programme, and where there is no understanding of the reciprocal obligations, it will not be possible to develop a working relationship.

2 Preliminary Discussions

Consultation between field staff and project holders may often start before the preparation of a project proposal. Such preliminary discussions frequently help avoid mistakes. If possible, when considering a new project the following should be incorporated:

a: Adequate baseline data. The information available should be sufficient to enable errors to be avoided in the decision-making process. Projects often collapse because the information on which certain assumptions have been based is either inadequate or false. It is difficult to assess a project if there is no baseline data against which to judge progress. Even in cases where it is hard to obtain information, the question should be asked: what is the minimum one needs to know before work can be initiated?

b: Tangible, and preferably quantifiable, objectives. Some projects may have objectives that do not lend themselves to assessment by statistical methods: for example, those concerned with social education. In these cases, it is vital to ensure that the objectives of the project are well defined and understood by all. Many programmes fail because project objectives are vague, ill-defined or over-ambitious. The objectives should also specify the group to benefit from the programme, its numbers and location.

c: Control groups. Where feasible, it is useful to identify a control group against which progress can be measured. Sometimes projects have resulted in apparently remarkable improvements, but further investigation has shown that the circumstances of neighbouring communities not exposed to the project have improved to the same degree. The use of a control group, therefore, can give a more modest and realistic impression of a project. It is unlikely that sufficient resources will be available for detailed monitoring of a control group; however, relatively informal comparisons with such a group may suffice.

d: Reporting and monitoring procedures should be discussed at the earliest possible moment: what information is to be included, by whom, how often and how to implement any changes indicated by the reports?

e: Provision for self-evaluation and for external evaluation studies, if appropriate, should also be discussed (see Section 4). A useful approach is to incorporate plans for a mid-term review in the life of a project, in which project staff and beneficiaries, field staff and outside consultants etc. meet to discuss the project and to measure achievements against goals and objectives. This should be done with a view to changing objectives and programmes in the second period of the project if necessary. Future funding should also be discussed at this stage.

If the above procedures are followed, the funding agency will be able to meet its obligations to donors and to recipients to obtain publicity material and assist in its own resource allocation and policy formulation. More importantly, these procedures provide a means whereby the project can stimulate and improve its own management. There is no doubt that good management is one of the critical elements affecting the outcome of projects.

Similarly, at this early stage it should be made clear to what extent grants will be subject to the receipt of satisfactory reports, and any other conditions should be specified.

3 Background

It is essential that in the appraisal of all projects due regard is paid to the national, regional and local socio-economic context. Even the most effectively designed programme will collapse if insufficient attention has been paid to the background against which it is working. National policies may easily run counter to the objectives of the programme. For example, a rural community may seek to improve agricultural production only to discover that the government has decided to cut the price paid to producers, thus rendering production unprofitable. Many political variables have a bearing on development initiatives and must be analysed before decisions are made. *Field staff may feel that the general political and economic circumstances of a country are beyond the immediate frame of reference; to ignore them, however, is to risk project failure.* It is rare for Oxfam to support surveys of national and regional social trends, because frequently there will be existing studies or specialists will be available who can advise on the context of a proposed project. (See Appendix III SOCIAL SURVEYS).

Many factors may affect the life of a development project, and it is impossible to predict all of these. There are, however, certain factors which should always be taken into account:

a: Government policies are of obvious relevance to any project; food prices, education policies and political outlook are some of the aspects of government which need to be assessed.

b: Development programmes. Before funding a project, field staff should find out about other development programmes in the area and programmes mounted in the past. It is not unknown for different development programmes to be antagonistic in their goals. For example, plans to settle jungle areas in South America are harmful to attempts to establish the rights of indigenous peoples inhabiting the zone. In other cases, development agencies have been found to be duplicating their efforts. This problem is particularly unfortunate given the limited resources available for development. A preliminary survey is especially important with service projects (e.g., health and education) where an NGO might duplicate work planned by the State or even give the State an excuse to withdraw from its programme, thereby depriving a population of a long-term service. Similarly, investigations should be made into the previous development experience of those groups intending to mount a project. Often the impact of previous aid programmes can be such that the potential for future success is severely restricted. For example, communities that have at some stage received large sums from official sources may well become dependent on external funding and in such circumstances small-scale, self-help schemes will probably fail. Investigation may reveal a poor use of funds in previous programmes. Or it may be found that a funding application has been made primarily to enable a group to pay off debts incurred through earlier programmes.

c: Local social and political hierarchies and structures have immediate relevance to a small development programme: care should be taken to ensure that local elites do not destroy the project in its infancy.

Large landowners fearful of competition, for instance, may well seek to undermine rural cooperatives or marketing groups. The first defence is to be aware of the possible dangers and to avoid a development plan that leaves gaps which could be exploited by outsiders. Many small marketing schemes, for example, fail because insufficient attention is paid to obtaining appropriate legal recognition or status for the group, thus leaving it open to harassment from local authorities. Similarly, there is little point in building a health post or digging a well if it is not clear who has legal title to the land on which the construction is to be sited.

d: Cultural norms and practices may impede a development project. Field staff must be sensitive to those practices and beliefs which could have an adverse effect on a project or isolate certain groups. Unfortunately, expatriate development workers have often been particularly remiss in this respect, not taking full account of local conditions. In some areas differences between classes, sexes, castes and tribes may be quite evident. In other areas social and cultural factors may not appear to be decisive; but it is all too easy to assume that all people engaged in the same kind of activity, dressed alike and speaking the same dialect have the same interests.

e: Long-term trends. Long-term climatic and social trends affecting a population are not always clear. Further, people's perceptions of these trends can sometimes be misleading. It is therefore vital to examine the history of the area in which the project is located. Extreme drought can be treated — mistakenly — as if it were an emergency which could be alleviated with short-term relief measures. However, if it is found that drought is becoming an annual event, then it might be more appropriate to think in terms of funding measures designed to have a long-term impact.

4 The Basic Questions

a: Field staff should assess projects mainly on the basis of:
- [ ] whether the people have a genuine need;
- [ ] whether the project is desired by the intended beneficiaries, both men and women;
- [ ] whether the objectives of the project fall within the scope of Oxfam's policies and terms of reference;
- [ ] whether the project is likely to achieve its objectives;
- [ ] whether the project is technically appropriate and economically viable;
- [ ] whether the project will survive the test of time (or, in other words, what likelihood there is that the benefits will continue once the project ends), and
- [ ] whether the constraints can be overcome.

b: Defining the objectives of a project is perhaps the single most important stage in project preparation. There will obviously be a hierarchy of objectives, and these should be listed in order of importance, always taking care not to confuse ends and means. Frequently neither project holders nor field staff distinguish between the purpose of a grant (e.g., to build a health post) and the objective of the project (to improve the general level of health or eradicate infectious disease, etc.). Once the principal objective has been defined, the secondary objectives should be established. Thus the main aim of an agricultural programme may be to increase production and incomes; and once this primary objective is clear, then the secondary objectives will indicate more immediate goals (such as the introduction of a new cropping system, reclaiming of land or building of an irrigation system). The next stage is to decide on the appropriate means to achieve these ends (e.g., to build an irrigation ditch, carry out educational and extension programmes, etc.). One of the main reasons why projects collapse is the failure to define objectives and, therefore, to relate project work clearly to these objectives.

c: There are certain guidelines which can assist in *planning objectives*:

☐ An objective should be realistic in terms of time, quantity, quality and cost. The beneficiaries should be clearly identified in number and location. If the objectives are not defined precisely, what remains is simply a vague statement of intent.

☐ The planning process must involve people, and objectives should be defined *with* people and not *for* them. Involvement of project holders in planning a programme generally results in their greater commitment to implementation.

☐ In some cases the main objective of a programme might result from an outsider's perception of a given community's needs, and might differ from those perceived by the community itself. However, the participants in the project must be able to see some objectives in the project which relate to their own experience; there will otherwise be no reason for them to continue working with the programme.

d: The following sequence can be used in *planning projects* in relation to their objectives:

☐ define objectives (both primary and secondary);

☐ consider the alternative ways of achieving these objectives;

☐ choose the best, or least costly, alternative, given the resources required and available;

☐ plan expenditure (see notes on project budgets);

☐ plan implementation of the project, and

☐ plan regular assessment of progress against objectives, allowing for the adjustment of objectives, methods and resources in the light of experience. (See Section 3, Monitoring.)

II Specific elements in project appraisal

The following notes draw attention to some of the problems likely to be encountered in project appraisal. Above all it must be remembered that its development is about people, and so is project appraisal. *However technically feasible a project, it is the participation of people which will determine whether it is a success or not.* The general atmosphere prevalent in a programme will indicate a great deal about the relations between the people involved in it. Are the people relaxed in their work? Are relationships within the group cordial? When entering a community in the company of project workers, do local people wave as you pass by? Do they even recognise the project workers? Do people stop and talk in the streets, or are they in awe of the project staff? Alternatively, are the staff respectful of local people? Do they keep them waiting for long periods? Do they treat locals as social inferiors, or as a hindrance to their work? On consecutive visits is the same small number of people always present, or is there evidence of a large number of people involved in the project? The appraiser should try to make sure that their visit is not treated as a special event since it is extremely important to establish, as far as possible, what a normal working day is like.

1 Leadership

People with qualities of leadership play an important role in all organisations. Many communities are fortunate enough to have a strong person to represent them, to lead the community in development and other initiatives. Inertia can be overcome by dynamic leadership. However, people may frequently agree with a leader, or be persuaded to take action during the excitement of a public meeting, but once away from the direct influence of the person's charisma may question the decision, and be unenthusiastic about implementing it. For example, people may promise to contribute labour to a community building project but fail to appear when required.

The charismatic leader may make appraisal of a project difficult, for his or her presence overshadows others, making it hard to assess with accuracy their attitudes and opinions. The leader may apparently know all the answers, communicate well with the appraisers, and always appear to speak on behalf of colleagues and neighbours. It is important to find out what the remainder of the group really thinks, or whether they even understand the project in question.

Many leaders work for long hours and put an enormous amount of effort into their work on behalf of other people. However, by taking so much onto themselves they do not allow others to participate fully in the project or to learn from their own experience. If leaders concentrate information, management and decision-making into their hands, others may fail to understand what is happening in a project.

It can be very difficult to explain to a hard working and committed person that their very immersion in their work may be detrimental to the people on whose behalf they are working. But for a group to be totally dependent upon one person is risky; leaders may obstruct new ideas or impede the participation of others, and their absence or removal can lead to the total collapse of the project.

There are people who will exploit power for their own ends. It is often all too easy for such persons to undermine morale by financial dishonesty or by encouraging a situation of permanent conflict in order to maintain their own dominant position.

2 Motives for Involvement

People's motives for becoming involved in development vary greatly, although there is little evidence that the nature of individual motivation has any direct relationship to a project. However, it is important to be aware of the possible problems that can be caused by particular sorts of motivation.

a: Religious conviction. Involvement in development work is sometimes seen as a path to individual salvation rather than a process of participation with a group in a common activity. The improvement of the individual ego becomes a major objective of daily work. Religion can also be used to exaggerate the righteousness of the adherents, and this can make them unreceptive to outside ideas. Finally, development and humanitarian work can be seen merely as a vehicle for proselytising.

b: Political. The same factors can also affect the work of those people with strong political motivations. Sometimes outsiders will attempt to use poor people for their own political ends. Thus, different factions may compete for the allegiance of a group of people by offering development programmes as an enticement. Any project risks coming into conflict with power elites, but certain political factions may have the expressed aim of inciting conflict, or using the poor as political 'cannon fodder'.

c: Professional. The field of development is increasingly becoming an independent profession and offers to many an interesting and challenging career. Greater professionalism in development can bring many advantages, but the danger is that projects will be written specifically with the aim of professional advancement in mind. Intermediate agencies for example may look more towards the survival of their own institution than to the interests of those for whom they supposedly work.

d: Expatriates. There are many committed, experienced and well qualified expatriates working in development programmes throughout the Third World. They often provide scarce skills and work in inhospitable areas neglected or abandoned by others. Furthermore, some expatriates make long-term commitments to their adopted country and cease to be regarded or to regard themselves as foreigners.

However, the relationship between expatriates and local people is complex and should be taken into account when assessing a project.

The advantages and disadvantages should be weighed carefully. Some of the more obvious advantages are that expatriates: may have professional experience and technical skills, and can train local people; have experience beyond national boundaries; are often highly motivated and committed and willing to work in difficult areas; often have greater objectivity and are able to identify more clearly local needs; can provide some protection against abuses of human rights by channelling information to the international press; can expose local prejudices and cross social boundaries, for example by working with women, low castes, tribal groups, minorities, refugees etc. The disadvantages are that: if expatriates are unfamiliar with local conditions and languages, a lot of time may be lost in training; they may be replacing local people in the job market; their participation continues the dependence on foreign knowledge; volunteers can be used as cheap labour, or if paid, expatriate labour can be very expensive; they may stay for a very short period and therefore destroy continuity; the project may collapse when they leave; they may push the programme too fast for local people in order to complete work before their contract ends; their lack of understanding of local society can lead to inappropriate responses; local agencies may seek to use expatriate assistance as a way of obtaining access to funds controlled by foreign agencies.

In an ideal world there would be little need for expatriate personnel. In certain areas, however, they are still required to assist in development and humanitarian work whether as paid staff or as volunteers. It is important to establish whether there is a real need to employ expatriate personnel or whether there are local people available with the necessary skills. Will the expatriate be teaching a counterpart or replacement? What will the role of the expatriate be? If it is as leader or project director, is someone being prepared to take over or will the project collapse on his or her departure? Are the expatriates encouraging a level of activity and funding which will be difficult to maintain in the future?

③ Funding

a: Support for individuals. In the absence of development agencies, individuals without any institutional backing sometimes become involved in development or humanitarian work. Some of the points raised above will be relevant in the assessment of such initiatives. For example, what will happen when the individual withdraws? Is management too centralised or is there allowance for community participation, etc.? What length of time is the person likely to continue their work?

It is highly inappropriate for agencies like Oxfam to provide funds in response to individual cases of hardship whether they be refugees, destitute people, the sick or anyone else. Not only is it an inefficient use of funds, but it entails many risks. It is not possible for the NGO to make a true assessment of individual cases. Individuals should be referred to an agency competent to undertake case-work. Other agencies specialise in providing grants for individual study. Where possible this is usually best carried out in the country of origin.

b: Overfunding. NGOs and governments alike constantly request an increase in resources available for development, but it must be remembered that the over-funding of projects can be very harmful. In part it is the uneven spread of assistance, which results in some groups receiving funds from many different sources whilst others are hardly touched by development activities. This imbalance is largely due to the deficiencies of funding agencies, which tend to concentrate their efforts in the same geographic areas, in those areas with good communications and with those people who are most articulate and able to present applications for funding. Unfortunately, many small groups have been destroyed by receiving excessive funds. Funding must be commensurate with a group's capacity to grow, and to cope with the management, administrative and personnel problems incurred in such growth. As a group grows, informal mechanisms of coping with decision-making and coordination between the participants begin to break down. The result may be inefficiency, an erosion of trust, and emergence of conflicts and tensions. Alternatively, some activities are possible on a small scale, but become more difficult if expanded; for example, handicraft production at the basic level requires little administration, but with many people involved may encounter problems of obtaining supplies and markets, and incur fiscal and legal obligations. A project should be allowed to grow at its own pace. New initiatives, good ideas and successful schemes deserve support but not at the risk of flooding them with funds.

c: Competition. Beware of the possibility of duplication of effort between agencies in the same area and, worse still, competition between them. Certain circumstances may exacerbate this problem. For example, political, religious or personality differences can cause communities or agencies to break up into smaller groups which continue their work in competition with each other.

d: Continuity and self-funding. All funding agencies are, or should be, preoccupied with the long-term management and financial arrangements of projects. Most recurrent costs are met by external funding agencies for only short periods, and therefore it is essential that before commencement some thought be given to possible sources of funding at the end of the initial grant.

Certain types of project should become self-sufficient as soon as possible. If, for example, an income-generating project is feasible, well-managed and well-planned, there should be no need to provide subsidies for running costs. Production should never be subsidised, although it might be acceptable to subsidise initial capital costs (or related social costs such as training). On the other hand, even though local people should make some kind of contribution to the costs of their services, it is unrealistic to expect all services to become self-funding,

One common problem is that many proposals for service projects are over-optimistic about the likelihood of a contribution from local sources, whether private or governmental. For example, the ideal health programme would allow for referral to a hospital and for laboratory services. However, the system of referral in many countries is very inefficient and in some areas is completely absent. Governments sometimes use the involvement of NGO funding agencies in health programmes as a pretext for neglecting the service. Allowance must be made for these possibilities.

If a service provided is more lavish than is the norm for the country or area, then the government is unlikely to be able or willing to maintain it. Many mission facilities transferred to state control have subsequently fallen into decay because they were more expensive to run than the average for the country. It may therefore be advisable to fund and plan for a programme which is less than the ideal, but which stands a better chance of obtaining finance for the future.

By not taking into account the longer-term future of a programme it is possible that services will be created and will disappear according to the availability of funds. It is important to plan for the longer term, to ensure follow-up, continuity and a lesser degree of dependence on external funds. It is unfortunately the case that many programmes only survive financially by being passed from one funding agency to another.

4 Assessing project proposals

a: Technical feasibility. Once the basic questions relating to a project proposal have been answered, it becomes necessary to analyse the technical feasibility of the project. Other parts of the Handbook analyse the technical details of agricultural, health, social and economic development, educational and disaster programmes. It is important to measure the objectives of the programme against its technical feasibility; people tend to be over-optimistic about the expected results of small projects and hence are disappointed when goals are not achieved. it is also important to ensure that the management and technical capacity of the project team is sufficient to meet the requirements of the programme.

b: Concentration on theory. The development process is highly complex, and in the attempt to understand often contradictory factors there is a tendency to formulate all-encompassing models. These may not only over-simplify reality but also restrict the implementation and achievement of goals. Any group which claims to have discovered the only true road to development must be treated with some circumspection, as it is unlikely to learn from its mistakes.

Project proposals which are over-theoretical in approach may obscure the empirical problems likely to be encountered in the project's implementation. Care should also be taken with plans for immediate replicability of a successful programme, as conditions may vary so much as to prohibit duplication of the model.

It is one thing to write a well thought-out proposal and provide all the necessary background information and analysis, but it is something else actually to implement the ideas. Excessive concentration on the theory of development can be as unproductive as extreme pragmatism.

c: Regional concentration. It is important to ascertain whether the proposed project is too concentrated geographically or, alternatively, too dispersed. Excessive concentration will lead to a reduction in the number of beneficiaries, and may increase costs per person as against benefits. Furthermore, neighbouring communities may become hostile if they are excluded from a programme: on the other hand, a programme taking in a larger area may become too thinly stretched and therefore lack impact. In dispersed programmes travel expenses may be high and a great deal of time can be wasted in travelling. There may be too little time available to spend in each locality for good relationships to develop.

Rather than concentrating on one locality, and rather than attempting to tackle many different problems in an integrated fashion, a programme may choose one single aspect of development. There have been some remarkable projects which have developed one type of expertise or pilot activity, sometimes as a complement to the development work of other agencies. However, such specialisation may lose the many advantages of integration.

d: Legal identity. This element of an organisation should not be taken for granted when assessing a project proposal. It is important to ensure that the legal identity of the project is appropriate to the nature of the activity proposed (e.g., if a production cooperative is envisaged, then in most countries the organisation should not be legally registered as a charitable body). Legal recognition may be essential for a programme to operate. For instance, an export licence may be a requisite for a handicraft group. Obviously the legal requirements will vary from country to country. Once legal status has been established, it is important to see how the control and ownership of assets are organised. Field staff must be cautious of projects which register ownership in the name of individuals. There is always a possibility that an institution may have only a short life-span; it should be clear how assets will be disposed of if the organisation dissolves (assets should preferably be given to a similar group to be used for the purposes for which they were originally acquired). The disposal of capital items will be of particular importance if expensive buildings are involved.

e: Organisation. The internal structure of an institution will have a marked impact on its operational success. A very small group with centralised management may in time run into trouble, since a highly centralised structure can become inadequate as the group grows. It is probably impossible to achieve a perfect balance between a hierarchy and democracy. While a hierarchical structure inhibits participation, total democracy impedes decision-making. But it is important to establish that an organisation does have a clear system of management, decision-making, procedures for reconciling conflicts etc. Is there, for example, a governing body to which staff and beneficiaries can appeal in case of conflict within the group?

However well a proposal is written, if the project holders are not sufficiently prepared technically, or committed personally to the project, then it cannot progress. Some people are excellent at preparing project proposals but lack the capacity for implementation. It is important to look at the project members as a team and not simply as a collection of individuals: a project director may be impressive in many ways but not work well with the team. The appraisal should include a study of group dynamics, leadership and, where funding is not aimed directly at grass roots groups, the relationship between the project team and beneficiaries.

f: Finance. There are two aspects of the project finance which need to be examined: firstly, the economic feasibility of the project — where this is appropriate — and secondly, the details of the operational budget. It must be ascertained whether the budget is realistic in terms of salaries, running costs and capital equipment; whether it takes due account of inflation and possible unforeseen costs, and whether it details additional contributions from the community and/or from other sources. It may also be important to examine plans for the attainment of self-financing or alternative funding at the end of the grant period.

III   Specialised appraisal techniques

1   Risk Assessment

A project which has admirable objectives may still be refused a grant if it appears to have a poor chance of reaching those objectives. However, in cases of urgent need, Oxfam may be willing to take much greater risks than would normally be acceptable, a willingness to take risks being one of the strengths of non-governmental organisations. Assessment of the degree of risk may depend mainly on intuition based on experience. However, certain questions may help in making this assessment:

☐ To what extent is the project likely to be interfered with by the government, local authorities, or by other power elites?

☐ How confident are you of the managerial ability of the project staff, and their ability to work together?

☐ In the long-term, what permanent improvements should the project bring about? Will the project have any adverse environmental effects? To what extent will the project be replicable?

A key part of the design of projects is the preparation of budgets covering anticipated expenditure and revenue. The failure to analyse budgets properly has often hidden from view potential risks or pitfalls. Field staff should pay close attention to budgets and income estimates and check especially economic forecasting and assumptions. Detailed advice on budgets is included in Section 6, Budgeting and Accountancy and in PART FIVE, Economic Development Guidelines.

2   Unit Costs and Cost Effectiveness

One simple measure for appraising a project is to estimate the *cost per person*: that is, divide the cost of the project by the number of people to benefit (it is important to define the exact number of beneficiaries, not the total population in the area). Where two alternative methods of reaching the same objective are presented, a decision between them may be made on the basis of cost effectiveness or cost per person.

This is a fairly crude measure, however, applicable only to clear-cut objectives where there are only 'direct' beneficiaries, all of whom stand to benefit equally; for example, with the provision of clean water to a village or construction of a health post. It is difficult to make any objective estimate in situations where the benefits resulting from a programme are indirect and unquantifiable.

The greatest *per capita* expenditures are probably justified in projects which train individuals to help others, for example extension workers or nutritionists, as these projects have a *multiplier effect*. In general, projects which invest in buildings or capital equipment have a low multiplier effect as compared with projects which train or educate people. In particular, investment in buildings is hard to justify because of the long period of use required to merit the high levels of capital investment entailed.

3 Cost Benefit Analysis and Investment Appraisal

The evaluation techniques of Cost Benefit Analysis (CBA) are now widely used in appraising development projects, and field staff should be aware of the principles involved in applying such techniques, the limitations and difficulties involved, and their relevance to the types of project likely to be supported by NGOs.

CBA is a project appraisal technique that helps decide whether or not a particular project is economically acceptable, or which out of a number of alternative projects is the most acceptable.

a: Concepts involved. It is important to see how the perspective of 'social CBA' differs from a purely financial analysis of a project. Both try to predict the profitability of a project by comparing the project's expected costs with its expected return. If the project's return exceeds its costs, then the project is acceptable.

In contrast to purely financial appraisals, social CBA takes a wide view of a project, and considers all the costs and benefits that may result. A businessman may not take into account the environmental consequences of a project, for example. The pollution of rivers or the preservation of wild life may be irrelevant to the economics of a project. Social CBA, however, views such effects as costs that may have a very important effect on a particular society, and they must therefore be considered if a project's wider implications are to be adequately assessed.

b: The three stages of social CBA.

☐ *Estimating future costs and benefits*. The procedure is first to identify all costs and benefits of a project, and then to put a value on them. All relevant costs and benefits should be included, whether direct or indirect. For example, an indirect cost of new industry might be the possible adverse effect on local artisans. Another important difference between financial appraisal and social CBA arises in estimating the monetary value of costs and benefits. A financial appraisal will assess the cost of materials, labour and equipment at just the market prices paid for them. Social CBA will assess these costs using the economic concept of the *opportunity cost*. For example, an industrial project decides to recruit a worker. He or she is at present earning a living in agriculture. The project offers a wage that is higher than the total value of his or her agricultural production. The cost to the economy is *not* the new wage that will be paid, but the cost of the agricultural production lost as a result of the worker's transferring from agriculture to the project. This is therefore the *opportunity cost* to the economy as a whole of the labour input of the new worker in the project; it is also known as the *shadow wage rate*, or the *accounting wage rate*.

☐ *'Discounting' future costs and benefits*. Since the benefits of costs do not appear until *after* the initial outlay, a technique known as discounting is used to take this into account.

Discounting occurs when the immediate prospect of receiving, say, £100 is more attractive than receiving the same sum, or even a greater amount, at some point in the future: £100 received *now* may be equivalent to £120 received next year (because of inflation, pressing needs etc.). Next year's £120 is discounted to this year's £100. The discount rate in this case is 20%, i.e. the difference expressed as a percentage of £100.

Typically, over the lifespan of a project the major cost will be incurred in the early stages and the benefits will accrue later. In an agricultural project, the cost of land improvement and irrigation demands an immediate outlay. The increase in agricultural output may not appear until a year or more later. So costs and benefits have to be identified and calculated year by year.

Once the future costs and benefits have been estimated over the life-span of the project, a second major step is therefore to reduce these costs and benefit flows to what are called *present day values*. This is done by the use of discount tables, after choosing a suitable discount rate over the time period.

In order to establish the suitable discount rate, the investor must decide upon a standard *required rate of return*. This is the lowest acceptable return on an investment, and is calculated in one of two ways: with an organisation like Oxfam, experience would suggest the lowest rate of return that makes a project a realistic proposition; alternatively, the required rate of return could be established by taking current interest rates into account.

Having discounted all future costs and benefits to obtain present day values, these can then be added to give *total discounted future costs* and *total discounted future benefits*. Appendix I includes project examples which illustrate the kind of calculations required to undertake a CBA.

☐ *Acceptance or rejection of a project*. Criteria need to be established with which to decide upon a project's economic acceptability. One or more of three decision rules may be used:

Net Present Value (NPV): this is the discounted total benefits less the discounted total costs. If this difference is positive, i.e. if the monetary value of benefits outweighs the monetary value of costs, the project is deemed acceptable.

Benefit-Cost Ratio (B/C): this is the ratio of the discounted benefits to the discounted costs. If this ratio is greater than one, the project is acceptable.

Internal Rate of Return (IRR): this is the most commonly used decision rule. It is rather like quoting a rate of return on investment: a project with an IRR of 30% is better than one with an IRR of 20%. The computation of the IRR is rather more difficult to understand and explain, and will not be attempted here. In practice, a project will be acceptable if its IRR is higher than the required rate of return (see 'Discounting' future costs, above). The required rate of return therefore becomes the cut-off rate for acceptance or rejection of the project.

c: Practical problems. The three main steps in conducting social CBA outlined above include a great number of theoretical and operational problems. Some of these are touched upon here:

☐ Predicting future benefits is clearly difficult or impossible in many projects. For example, in an agricultural project, the adoption of innovations by farmers depends upon a whole range of technical, economic, social and political factors operating in a given area. Most of these are difficult to predict; e.g., unexpected weather conditions, recession, crop failure.

☐ Estimating benefits of 'social development' projects (such as training schemes or social education) in money terms is highly controversial. In the case of an industrial or agricultural project, the main direct benefits will be the increased output attributable to the project, e.g., bags of fertiliser or tons of rice. However, in the case of a nutrition or rural health project, the benefits are not normally expressed in such terms. These can be estimated by considering the costs these projects will *avert* if they are implemented. For example, a malnourished population will impose costs on an economy because people do not work as well as they would if they were adequately fed. Needless to say, valuing benefits in such a way is a difficult and controversial matter, as it involves putting a quantitative value on the quality of life.

☐ The impact of inflation can only be ignored if all relevant resources will be subject to the same rate of inflation. For example, if labour rates are expected to be subject to a rate of inflation different from the general inflation rate, then the different rate for labour must be brought into the calculation.

☐ The treatment of risk and uncertainty can be done by carrying out a *sensitivity analysis*. This is to undertake a series of CBA calculations by modifying the original assumptions of costs and benefits (e.g., price of inputs, cost of labour, price and quantities of project output) by, say, plus or minus 10% or plus or minus 20%. This will produce a range of *internal rates of return*, each reflecting the assumptions or estimates made during the CBA calculation.

☐ The final internal rate of return of the project will have made use of a whole range of assumptions about opportunity costs, future prices and quantities of output, an appropriate discount rate, the life-span of the project, and so on. Many of the assumptions are likely to be based on subjective judgement alone. It is therefore advisable to test the more important assumptions by using sensitivity analysis as described above, particularly since the final rate of return may turn out to be disproportionately affected by a particular assumption or group of assumptions.

The many difficulties involved in valuing benefits has led to a growth of interest in developing alternative appraisal procedures, such as Cost-Effectiveness Analysis. It is possible that CBA, by emphasising monetary values, could distort the project's potential; however, this would not happen if the results were weighed against and compared with the results of other appraisal techniques. In other words, these techniques should not be used in isolation.

A simple list of all the relevant costs and benefits, and some indication of their magnitude, would aid the formulation of a suitably objective judgement of a particular project. Where CBA is considered appropriate, but where the project holders are unable to carry out a full CBA, this list could then be included in the project's proposal so that field staff can make these calculations. (See Appendix I for examples.)

4 Surveys

Field staff are unlikely to have to participate directly in surveys, although there are times when it will be necessary both to initiate a survey and to interpret its results. (A special note on nutrition surveys can be found in PART SEVEN, Section 1). All too often too much data is collected because the use of random sampling techniques has not been fully understood (see random sampling in Appendix III); excessive data collection can merely complicate the task of collating and interpreting the information.

There is always room for testing assumptions and hypotheses through the use of unstructured interviews and casual discussions; the more obvious information can often be overlooked by the use of complex questionnaires. However much thought goes into a questionnaire, the questions posed still determine to an extent the answers received. Important information may not be forthcoming because the questionnaire does not ask the appropriate question. Less structured questioning might well provide the information.

In planning any survey it is important to know:

- [ ] the types of situations in which a survey is/is not needed;
- [ ] where/how a group can get advice before it wastes time on the wrong kind of survey. What information is already available?
- [ ] the different approaches to surveys: participant observation, structured questionnaires, random sampling, participatory 'action'-styled investigation, etc.;
- [ ] the main types of sampling and statistical analysis available (ensuring that these are appropriate to the task in terms of sophistication, accuracy, and costs); and
- [ ] whether the survey will produce the sort of material necessary for future decision-making.

Further details are available in Appendix III.

Introduction

Monitoring is often seen as the poor relation of appraisal and evaluation, the assumption being that monitoring is not really necessary because everyone knows what is happening anyway — even without recourse to a structured information flow. In practice the pressures of the daily routine often obscure from view the general trends in a project's work. People may think that they know what is happening in their own project, but a concern with detail can impede progress towards attainment of long-term objectives. As participants become caught up in realising certain limited, short-term goals — such as the maintenance of plant, management of resources or the organisation of specific events — they may fail to monitor the impact of the project in the operational area (for example its effect on prices, mortality rates, hygiene standards etc.).

Furthermore, funding agencies such as Oxfam must be accountable to their donors and supporters. Regular reporting by those in receipt of grants is one of the most important ways in which the funding agency can account for the funds administered by the organisation on behalf of donors. It is vital that this point is understood by recipients of funds, as they may then be aware that the agency's responsibility is to both donor and recipient.

Equally important is the function of monitoring to improve project management. Often, project reports are not written as a genuine record of problems encountered, activities held or achievements. Instead, they are fabricated to fit the real or perceived priorities of the reader, usually the funding agency. Some funding agencies require the use of elaborate report-forms which tend to act as a strait-jacket on the monitoring process.

Reporting should be seen by groups as a positive experience, enabling them to maintain their programme in accordance with objectives set at the outset. Too often, reporting is seen as an activity external to the project, in that the reports are written for outsiders, whereas it should be seen as a part of the project's development and its management. By ensuring that reporting and monitoring become part of the structure of a development project, it is possible to strengthen the process of self- or participatory evaluation.

Participation in the monitoring process should also indicate to project members both the rate of progress of the programme and what mistakes have been made. This approach avoids a centralist attitude to management and encourages reflection and observation.

Because most reports are prepared for an external audience, to fulfil funding requirements etc., the monitoring process tends to concentrate upon the results of a development project. The quantifiable achievements and financial details of the work are reported. Monitoring, however, should also include reference to the procedures followed by the project. The monitoring and delivery of project inputs is as important as concern for the outputs, or achievements. By monitoring both the processes involved in development and the results, it should be possible to provide information which can be used as a tool by project managers. It is unrealistic to expect management to function effectively without recourse to basic information on the progress and direction of the project. The information gathered in the monitoring process will indicate what changes will have to be made to a project to bring initial plans in line with reality.

II   Initial considerations

1 Early Planning and Monitoring

To ensure good reporting and monitoring, the system of feedback, information collection, analysis and dissemination should be designed at the very beginning of a project's life. To impose, or introduce, a monitoring system at a later stage in the life of a project will lead to resistance by project workers, wary of being spied upon or afraid that the new system might be a means of control introduced because of lack of confidence in their performance. If, however, the reports are seen as an integral part of project work, they will in the last analysis absorb less work and time than if seen either as something imposed, or a burdensome duty to fulfil.

2 Base-Line Data

Ideally, no project should commence until it has discovered certain facts about the area in which it intends to work. The collection of base-line data is a prerequisite for both monitoring and evaluation, for without it the comparison of results against objectives can never be accurate. Many groups claim successes without being able to show what the pre-existing situation was in the project area; this is as true of projects measurable in technical terms (disease incidence, crop yield etc.) as of those with more general objectives. In many small projects it will not be appropriate to talk of obtaining base-line data, especially when its collection could entail more resources than the project is likely to have at its disposal. This will be the case, for instance, in small health projects. However, in small economic/production projects, the availability of basic data will be of crucial importance: without data on prices, levels of production, markets and transport costs, it is not possible to make an appraisal or judge the success of a project.

In sum, a degree of discretion will be required in establishing what level of information should be expected as a prerequisite for monitoring.

3 Objectives

Bad reports are partly the result of badly formulated objectives. In other sections of the Handbook, the importance is stressed of clearly defined objectives. Where a project has only a vague idea of its goals and its constraints, reports and other monitoring systems will not clearly reflect what is happening. Once the objectives are set, it is possible to decide on the information that should be included in the report.

The structure or system to be used in the monitoring programme can be chosen at this point to relate to a participatory form of evaluation, reports for external agencies, material for an evaluation or a mid-term review (see below).

III   Monitoring techniques

1 Structure of Monitoring

A formal progress report including a simple financial balance to illustrate how funds have been used should be the minimum reporting required of any project.

In many cases field staff neither can, nor should, expect much more from a group than the minimal report consistent with the need to account for the use of funds.

Financial reports should include all incomes and expenditures, and not be split artificially between specific grants. Unfortunately, some funding agencies now insist on separate accounts for each aspect of a project and in this case the financial reports give only a part of the picture, and can obscure rather than enlighten the reader. Details on financial reporting and accounts can be found in Section 6. From the financial reports it is often possible to make a rapid assessment of the fulfilment of practical achievements against original objectives and budgets. In this manner one can judge, for instance, the speed of implementation and how closely the programme is following the plan, and odd unforeseen expenditures will highlight those aspects of the programme which had not been envisaged in the original proposal. A financial report can also be used to check on other more descriptive reports; for example, by matching disbursements against other data.

In most large organisations it is now accepted that a review of staff is an integral aspect of monitoring. There is, however, a resistance to this type of monitoring by many groups involved in small-scale development and charitable programmes. It is felt that there is a contradiction between the process of monitoring staff or project-holders and the democratic ideals behind many development groups. However, monitoring staff performance can ensure that individuals are employed because they can fulfil a given task, and not merely because they have the right ideas.

In larger organisations, it is worth checking whether any attempt is made to monitor personnel, whether work cards are obtained from individuals, groups or teams, and whether these are discussed with the people involved. In general the degree of participation in discussions on the reports is a good guide to the type of management and how much feed-back and learning from experience there is on the project.

It is suggested that, ideally, all those employed in a project should meet regularly to discuss progress, match this against objectives, discuss possible changes in the light of experience etc. These discussions can also be used to inform and coordinate different groups within a project.

It is possible that the field staff will be asked to assist in constructing monitoring systems, especially where these require the collection of technical indices, such as price trends, water levels, health indices, production rates etc. When specialised technical information is required, it may be necessary to seek the advice of consultants to establish the best way of obtaining such information. Details on the technical aspects of monitoring and data collection can be found in Section 4 and Appendix III.

The size of the group involved in a project is obviously crucial in determining the potential for monitoring. Field staff should not expect a small handicraft group with handicapped workers to produce the same level of information as a large industrial cooperative. There may also be other unavoidable circumstances which work against adequate monitoring, for example civil strife, destruction of communications, death of key personnel etc. In this kind of situation discretion will always be called for. It is important to distinguish between those genuine cases where reporting is rendered difficult, if not impossible, and those trying to avoid what can often be a difficult, even if illuminating, process.

2 Reviews

One useful innovation, introduced recently by some groups, is that of a project review, usually initiated mid-way through the project life. This should be allowed for at the start of a project, and incorporated into the monitoring/evaluation system. The general idea is that the work of the project should be measured against its objectives, and that changes should be made in objectives or methodologies in the light of the review discussions. It is suggested that the review body should include most project holders where possible, and some outsiders, including perhaps the Oxfam field staff, and/or an independent consultant with a long-term brief to work with the project.

③ Some Types of Reports

a: Regular progress reports. Oxfam normally expects these every 6 months but in exceptional cases they might be produced either more or less frequently. This is really the minimum reporting that should be required of a group for it to be fully accountable.

b: Financial reports and audits. A basic financial report should be produced by all groups, whether required by external funding agencies or not. There is no better way to ensure that a group is open to an accusation of mismanagement of funds than for it to fail to produce financial statements. In many countries, one of the most common reasons for community groups to break up is a lack of confidence among members in their financial procedures. The tradition of auditing accounts varies between countries. There is much to be said for professionally audited accounts, as these help groups to defend themselves both against external accusations and internal conflict. An external auditor can also help a group to maintain a proper system of accounts. When supporting large projects, it is advisable to carry out an external audit in order to avoid difficulties over accounts at a later stage of a project's life. Field staff should be aware of the possibility of allowing auditors' costs in the budget for large projects. (PART THREE, Section 6.)

c: Tour reports by field staff. Often the most useful reports on a project will be those prepared by field staff; this is especially the case where the project is very small, and the participants may possess only low levels of education and literacy. The check list of questions at the end of this part will help field staff when visiting a group and writing reports.

d: Reports from visitors. Field staff should insist that all visitors to a project submit a short report giving their impressions; the more external comments available, the fuller the impression one can get of the project.

Where possible, comments from field staff and visitors should be fed back to the project holders.

e: Reports by external consultants. Where consultants are employed, they should maintain their contact with groups over a long period and participate in the reporting, review and evaluation processes, rather than be employed for short single visits. For this reason, local consultants have the advantage of being more readily accessible for consultations than expatriates brought in for the purpose.

Section 4   Evaluation

I   The meaning of evaluation for Oxfam

In industry evaluation has long been a highly developed and integrated element of the production process. More recently, among both the larger international development agencies, such as the World Bank and the OECD, and in many government agencies, such as USAID, it has become a well-documented science. However, among NGOs evaluation still has a tendency to be surrounded by mystique, to the exclusion of hard thinking and rational practice.

The concern of NGOs with people and their welfare means that evaluation will involve much more than, say, simply measuring the increase in agricultural production of a project. The concern is also with the surrounding infrastructure, such as marketing and credit possibilities, and above all, with the social consequences both to the immediate beneficiaries of a project and to the community as a whole. It is these considerations that render inappropriate the customary evaluation methods which are based principally on financial and economic returns with little regard to their social impact. We need to look for an evaluation style that recognises the dignity and validity of the local community and that does it justice. In particular, we should always be looking for better ways of evaluating the distributional effects of projects by asking check questions such as:

☐ has the project resulted in an increase in the incomes or prospects of the poorest?

☐ has the project raised the income of the poorest above a poverty line?

☐ has the range of income within the community been reduced?

☐ has the project helped the poor organise in a way which will allow them to reduce their own poverty?

It must be remembered that an evaluation study made in isolation will cure no ill unless based both on a firm personal relationship between the project and the funding agency and on adequate baseline data and/or quantifiable objectives that have been duly monitored. Evaluation should be an integral part of the development process. In its proper context, it is vital; it should never be made to stand alone. It is not the function of evaluation to make decisions, but it can present information and identify alternatives which may facilitate the making of better decisions, and help project holders to learn from their mistakes.

It is critical, therefore, that all concerned understand and accept from the outset:

☐ for whom the evaluation is intended, and

☐ for what purpose.

From an examination of large-scale evaluation studies undertaken for Oxfam in the late 1970s, it transpired that many provided conclusions that were irrelevant, incomprehensible or too lengthy. Many of the reports were received too late to be of use to anyone. Rarely have the results of a major study been brought together into concise, straightforward conclusions with realistic and pragmatic recommendations. In short, evaluations have frequently resulted in the postponement of important decisions, yet these decisions have usually been made with little regard for the results of the evaluation. Such a situation is clearly unsatisfactory.

In contrast to these large and unwieldy studies, a number of smaller, less ambitious evaluations have been undertaken which, for the most part, have been more analytical, objective and pertinent.

The objectives of evaluation can be summarised as:

a: demonstrating the success of the project in relation to its objectives, and the extent to which the intended beneficiaries have really benefited;

b: providing a check on the use of funds, especially in view of the responsibility of NGOs to their donors;

c: enhancing the work, both of similar development programmes in the same area and of the funding agency as a whole, through the dissemination of information about project experience and outcome. The optimum way of meeting the requirements of the project beneficiaries should always be a priority. How can an evaluation study serve their needs most effectively? Often an external evaluator (not necessarily expatriate) may prepare an excellent report which fully meets the funding agency's requirements; but by failing to have full discussion with the project holders during the investigations, and especially on their completion, a feeling of distrust may arise. The sensitive evaluator should aim to enhance the understanding between donor agencies, implementing agencies, and project holders, although there can be no rules on how to achieve this. The personality and experience of the evaluator will be critical. This puts a heavy responsibility on those setting up an evaluation and on the person who is selected for the task.

The following criteria upon which to base a decision whether or not to hold an evaluation have been established by the Oxfam Evaluation Panel:

☐ where there are doubts about the operation of the project;

☐ because of the large size of the financial commitment, taken in the context of the geographical conditions and type of project;

☐ because of a request for further funding;

☐ because of the experimental nature of the project, and

☐ at an appropriate stage in the project's development, e.g. at the end of a specific phase.

It is probable that more than one of the criteria will apply before the decision is made to hold an evaluation. Only in the case of pilot projects should evaluation be automatic.

The *cost* which will be incurred in obtaining information from an evaluation should always be set against the potential value which can be expected from it. To translate this into proportional terms can be misleading, but in general the cost of an evaluation should not be more than 5% of the total funding provided by the donor agency and in many cases will be much less.

There is much to be said for *mid-term evaluations*, i.e. those taking place part-way through a period of funding. At this stage it is possible to assess the progress of the project against the original objectives, and to change these as may be appropriate. Further, the pressure arising from mid-term evaluations is often less than that from the end-of-term evaluation when further funding may be at stake.

III   Methodologies

Development literature is filled with complex articles and books on evaluation design and methodologies. Perhaps fortunately for those associated with NGOs, much of this is inappropriate.

The methodologies listed below are among those that have been found to be most useful.

1	**Management Study Analysis**	The assessment is based on an analysis of: ☐ the infrastructure — personnel, building, equipment; ☐ the coverage — the number and range of beneficiaries; ☐ the quality — of the project services provided; ☐ the vital statistics — e.g. rates of infant mortality or cropping levels; ☐ the unit costs, and ☐ the significance — to the local community or certain sectors of the community.

1 **Management Study Analysis**

The assessment is based on an analysis of:

☐ the infrastructure — personnel, building, equipment;

☐ the coverage — the number and range of beneficiaries;

☐ the quality — of the project services provided;

☐ the vital statistics — e.g. rates of infant mortality or cropping levels;

☐ the unit costs, and

☐ the significance — to the local community or certain sectors of the community.

2 **Input/output Analysis**

☐ effort — a measure of the activity that has occurred, i.e. of the inputs;

☐ effect — the results of the effort;

☐ adequacy of effect — degree to which the effect is both adequate and appropriate to the needs of the project;

☐ efficiency — optimum use of resources, and the alternatives that are/were open;

☐ importance — largely self-analysis, leading to ranking the priority order, and

☐ impact — on the local community or certain sectors of the community.

3 **Rapid Rural Assessment**

A system has been developed for projects covering a number of activities. This arose from the problem of trying to evaluate a diversity of activities over a large area in the limited time normally available to evaluators.

It is based on forms prepared for each aspect of the project: on the left-hand side, the *objectives* of that project, and any other relevant factors, are itemised, while the right-hand side is left blank for *comments*. These comments are written in at prescribed intervals by the project holders as a form of monitoring, and then analysed by the evaluator.

This type of assessment is only effective where good baseline data are available together with a clear statement of objectives, and where every project holder within the scheme does a complete appraisal before funding begins.

It is therefore essential that the data thus gathered are carefully preserved and kept readily available. But at the same time beware of excessive data collection producing material which is never utilised.

A major problem in this approach can be to persuade project holders to identify their activities and for them to appreciate that their work is itself definable with specific objectives. Only when this identification has been done is it possible to quantify the various inputs and outputs.

Many people responsible for running development projects are reluctant to sit back and think in these terms, yet some of them must have done so initially when preparing the grant application.

Whereas project costs can and should normally be accurately recorded, *benefits and disadvantages* are more difficult to assess as:

☐ they should extend beyond the life of the project, and probably have a multiplier effect;

☐ there may be demonstration effects on neighbouring communities, and many of these are not readily quantifiable.

Nonetheless it is important to try to assess all these as they accrue to the people and to the environment for succeeding generations.

3

(See also PART FOUR Section 1)

1 Objectives

Evaluation seeks to understand the *effect* of a project, which is an important determinant of future project design and of the future allocation of resources. In this respect, Oxfam is concerned to understand the *effect* of the social development work it supports. The term 'social development', though, is open to broad interpretation. It can refer to activities involving traditional 'material' or 'physical' development, in which case evaluation is fairly straightforward and largely follows conventional procedures stressing *results* which are *quantifiable*.

However, the term 'social development' is also used to refer to projects which try to develop self-reliance among people, encourage their participation and generally seek to stimulate in them an awareness of their social reality and of their ability to influence and direct their own lives. This type of social development work is largely an educational process: it stresses *processes* which are *qualitative* and which cannot be evaluated by conventional evaluation techniques. Two examples, taken from Oxfam-supported projects, illustrate the kinds of objectives set by these social development projects and similarly indicate the difficulties of evaluation:

"to help the small farmers choose their own destiny and handle their own development;"

"to encourage a greater self-determination and greater awareness of the possibilities for development."

The objectives noted above are becoming increasingly common in development initiatives, and it is important to be able to judge whether such objectives have been achieved. The difficulty, however, is first, to establish satisfactory *criteria* by which such objectives may be better understood, and secondly, to determine what observations or *data* to collect which allow us to judge qualitatively the effect of these objectives. To illustrate the point it might be useful to note the kinds of social development objectives commonly found and by which these projects would be judged:

☐ increase participation
☐ increase organisation/solidarity (of client group)
☐ reduce dependence
☐ develop initiative/motivation/leadership
☐ create awareness.

These are the objectives that we need to understand more clearly in terms of how they are achieved. They are characteristically associated with projects directed towards the rural poor. The reasoning is that these objectives are indispensable prerequisites for such people to benefit from the more economic and material aspects of development.

2 Criteria and Indicators of Social Development

There are, of course, no lists or models which are universally applicable to all types of social development. Little research has been done to date on the evaluation of non-material objectives, and thus the literature offers little guidance. Much of the evidence to date is *subjective*, when project staff make their own judgement as to whether the project's social development objectives are being achieved or not. Similarly, self-evaluation or 'participatory evaluation' in which both project staff and beneficiaries participate are important aspects of social development evaluation.

a: Criteria. Social development is largely an *educational process*, and it is necessary to determine the *criteria* by which the success of this process can be judged. The social development process seeks to develop a number of critical *qualities* in the group such as 'solidarity', 'awareness', and 'participation'. Therefore, first the *qualities* which it is sought to achieve should be determined. They can then be used as the *criteria* by which some kind of judgement as to the outcome of the project can be made. One research project supported by Oxfam sought to identify criteria for judging social development projects. Project holders in a number of countries in India and Latin America were asked to characterise the qualities of the project group before and after the social development work.

The qualities of groups before a process of social development included:

☐ individualism (this was reflected in the predominance of private property, little interest in community projects, low participation in decision-making);

☐ lack of critical analysis of their situation, i.e. an inability to identify the causes of structural problems and work out solutions to them;

☐ economic, social and political dependence on, and exploitation by, intermediaries, large landowners, etc.;

☐ lack of confidence in their own ability to change the situation;

☐ absence of organisations which effectively represented their group interests and lack of cooperation between people in general;

☐ ignorance, suspicion and isolation with people afraid to talk/discuss and become involved.

The qualities of groups after a process of social development were suggested as:

☐ internal cohesion;

☐ sense of solidarity;

☐ critical consciousness/critical faculty;

☐ active and critical participation;

☐ reduced dependence, increased self-confidence;

☐ self-management capability;

☐ democratisation of power, collective responsibility;

☐ involvement in regular discussions with other, similar, groups and institutions;

☐ involvement in the creation of other similar groups;

☐ ability to deal with government officials.

These lists indicate the *qualitative* changes which can occur in project groups as a result of social development work. Qualities of this sort can serve as the criteria by which the social development work can be judged. They do not constitute any kind of universal model but merely serve to indicate *how* we might approach the evaluation of social development projects. The processes involved are complex and subject to both internal and external factors specific to the given context. The above is a framework which may prove useful in helping to understand social development initiatives.

b: Indicators. Having identified possible qualitative criteria, the next step is to identify the indicators which could be used to illustrate the qualitative changes taking place. Again these are not universal models in this respect, merely the results of some research on Oxfam-supported projects. This is a composite list of the kinds of indicators which could be used and which might indicate the kinds of qualitative change noted above:

☐ *activities/events* in which the group participates, e.g. collective economic activities, meetings, training sessions;

☐ *action* — specific action undertaken by the group to tackle a particular problem, e.g. action against river pollution, action to obtain social welfare, action to defend members' interests against exploitation by traders or action to demand the implementation of existing labour legislation;

☐ *changes in group behaviour* — the nature of group meetings, issues discussed, decisions taken, degree of participation, use of language, etc.;

□ *nature of intervention* — the relationship between the community development worker/*animateur*/project agent and the project group i.e. whether it is based on trust and mutual understanding or suspicion and intimidation etc.;

□ *relationship with other groups* — process of discussion, level of intergroup contact, joint action, etc.

These indicators could serve as a framework for understanding the qualitative changes taking place as a result of a social development project. Again the framework must be related to the specific context and adopted accordingly and is not intended to be universally applicable.

3 Operational Aspects

The most appropriate system for the evaluation of social development is one in which the major responsibility for data collection and analysis lies with the people engaged in the process. Ongoing evaluation should emphasise either self- or participatory evaluation. The great advantage is that the evaluation process is itself educative for the project team and project groups: through it they will develop their understanding of the complexities of social change and assess the outcome of their work.

In designing a monitoring and evaluation system, decisions have to be taken by the project team and target groups on the points below.

a: Type and amount of data to be collected. Here it is important to identify only the basic data required to:

□ assist the project in designing a programme and identifying its specific goals and objectives for work in each community on economic, educational and organisational levels;

□ enable the community or group to identify its problems, their causes and viable solutions, together with a time scale for implementation;

□ enable the project team and target group to monitor progress towards the solution of problems by the group, using criteria and indicators of social and institutional change.

b: Mode of collection to be adopted. Some data will be collected from secondary sources by the project team, but most will be a product of the ongoing process of education in which project team and groups are involved. The following methods can be considered:

□ community or group self-survey;

□ preparation of community profile by community assembly;

□ reports on visits to communities/groups by project team;

□ descriptive accounts or observations of important incidents, meetings, actions, activities and events involving the group. These could be written by group members or the project team, and observations should be made within the framework of indicators which have been previously planned;

□ interview of community members, group members, to assess attitude and behavioural changes;

□ collection of quantitative data on organisational activity: number of meetings, participants, frequency of changes in membership of key committees etc.

c: Procedures for processing and analysis of data. It will be necessary for the project team to develop a system for handling reports and recording data which ensures that prompt action is taken. Staff reporting forms are a means of providing regular monitoring of group activities and initiatives and ensuring that the right type of support is provided by the project team at the time it is required. They may also indicate that changes in methods and approach are called for.

Information coming into project headquarters from a continuous monitoring procedure can be analysed at regular evaluation meetings or seminars attended by all the members of the project team. In the case of larger projects it may be helpful to give one member the sole responsibility for collating the information for use at these meetings, and for the preparation of evaluation reports for external agencies. With the smaller projects, however, members can process and analyse the data for the communities and groups for which they are directly responsible. At these meetings the team can review the evaluation and monitoring system and if necessary propose changes to ensure that it continues to be both effective and useful.

4 **Conclusions**

Given the broad range of social development approaches, the differing and frequently changing political environments in which the projects have to operate, and the multiplicity of other economic, social, cultural and organisational variables, it is evident that the task of designing an appropriate evaluation methodology for social development work is a very difficult one. But Oxfam is concerned to understand the *effect* of this kind of work and is anxious to examine ways by which its effect may be understood. This section, therefore, is presented as a framework or guide to the processes involved which may help to structure experience in different contexts.

3

V   Practical issues in evaluation

1 **Evaluators**

From all that has been mentioned above, it will be clear that Oxfam gives every encouragement to the use of nationals to assist with evaluations. These may be experienced development workers or teams from local universities. However, while involvement of students in this kind of work may give them first-hand field experience, due regard must be given to their effect on the project personnel and beneficiaries; they must be carefully selected and supervised.

Expatriate evaluators, especially from the wealthy, industrialised countries, should be employed only when suitably experienced nationals are not available, except in the case of:

☐ projects of major importance which could be replicated elsewhere;
☐ large-scale projects which appear to have 'gone astray'; and
☐ situations where for political reasons a national would not be acceptable.

Evaluators should have *experience of the types of activity involved* in the project, *a good knowledge of the local culture − including language −* and *an appreciation of Oxfam's approach to development*. The normal run of evaluators tend to be experts in statistical analyses of profit-making activities and lack the qualifications or qualities necessary for a more profound study.

When providing the terms of reference for evaluators, it should be agreed both for whom their report is to be prepared, and who will decide on its distribution.

2 **Arrangements for an Evaluation**

When agreeing the details for an evaluation, field staff should ensure that the following points are covered:

☐ why and by whom the evaluation has been requested;
☐ the terms of reference, which should be sufficiently detailed to ensure that the information required is specified;
☐ arrangements for employing the evaluator(s);
☐ the budget for the evaluation;
☐ the methodology to be used, including the process by which the findings will be relayed back to the project team and beneficiaries, and how these will be taken up by them;
☐ the evaluation report: responsibility for the preparation and distribution of the evaluation report;

☐ payments to the evaluator. For example, this may be 50% when the evaluation starts, and the balance when a satisfactory report is received. Or if the evaluator(s) require payments to cover expenses during the course of the evaluation: 40% initially, 40% midway and 20% when the report is handed over.

③ Constraints on Ex-Post Evaluations

Oxfam has found that in many circumstances evaluations may not always be either feasible or appropriate. The opportunities for mounting evaluations are in practice often limited. The constraints include:

☐ resistance of projects to outside evaluations;

☐ the problem of finding evaluators who have the desired understanding and approach, who are acceptable to all the parties involved, are available when required and are willing to work for a reasonable fee;

☐ the difficulty of feeding back the results to the project team and beneficiaries, of their making proper use of its findings and of obtaining an adequate report;

☐ *it is, in any case, part of Oxfam's evaluation policy to try to reduce the overall number of ex-post evaluation studies.*

④ Participatory Evaluation
(see also Appendix II)

Participatory evaluation can be considered as a particular type of methodology. However, it has become so important to the work of NGOs that it warrants discussion in its own right.

The merit of participatory evaluation is the improved understanding and consequent increase in morale that it can bring to the project team and project beneficiaries alike.

Balanced against this is the fear that the results may be less satisfactory in terms of statistical analysis than evaluation by outsiders. However, this should not be a risk if adequate baseline data are recorded at the outset together with easily quantifiable objectives. Indeed, regular monitoring in this way can encourage improvement in basic administrative routines such as filing and data collection.

Where one of the main concerns in a development programme is to assist the poor to develop independently, *this type of assessment process should be built-in as a basic principle.* Projects are increasingly seeking advice on this procedure as they want to be able to assess their own experience.

The use of participatory evaluation has the added advantage for the donor agency of refuting possible accusations of trying to impose organisational and cultural values.

One of the best self-evaluations of this type in which Oxfam has been involved was of a community health project in Honduras (HON 3B). This project had a highly motivated group of health promoters who themselves decided that they wanted such an evaluation. 95% of the final report was written by the promoters themselves. Although an expatriate was involved, she acted simply as an adviser during the course of the evaluation, adding comments to the final report.

This evaluation proved successful in terms of both attitudes and motivation. The promoters were pleased that they could understand the evaluation and see more clearly the significance of their work. It was perhaps less successful in its statistical analysis — though in fairness some of the 'hard data' were not available.

This approach to evaluation is still largely in its infancy. However, such is the interest now shown in its use by projects, agencies and researchers, that an Appendix (Appendix II) is included on the subject. This is an outline of the main issues involved, and can be a useful reference for advising on the subject.

A variant of the self-evaluation process is the growing number of studies involving an advisory and training — or *'accompanamiento'/*

'appui technique' — element. In this, the consultant works alongside the project for a period of probably several months, and returns at intervals to investigate and advise. Oxfam has supported studies such as that of a community appropriate technology scheme in Brazil (BRZ 141), of village cereal banks (VOL 64) and of a rural development project (VOL 58) in Burkina Faso.

This arrangement has many advantages over the traditional ex-post evaluation study in that:

☐ the consultant works *with* the project team as opposed to looking *at* them;

☐ the findings can be fed back straightaway;

☐ it is a continuing process and the consultant gains a close understanding of the project and the team.

However, it is unrealistic to expect good results from participatory evaluation techniques in programmes where the general level of participation is very low and management is highly stratified and autocratic.

3

Section 5   Project Support

I   Non-financial assistance

1 Introduction

Rarely is the role of the donor agency restricted to financing development programmes. Most funding agencies devote considerable time and resources to giving advice, supplying information and providing general support for development initiatives. This is especially true of agencies such as Oxfam, which maintain permanent offices in developing countries.

Oxfam field staff are spending an increasing amount of time with individual projects. This extra effort is partly necessitated by the development in recent years of more thorough appraisal and monitoring. There has also been an increase in the priority given to providing non-financial assistance to groups, in helping to plan their projects and future development work, and in providing general and technical advice and information.

2 The Options

There are a number of ways in which field staff can give support to a project, other than financial provision. For example, they can assist with:

a: Inter-project Learning. People often learn best not from external 'experts' but from people like themselves facing similar problems. Field staff can help by putting such groups in contact with one another. People from projects funded by Oxfam and by other NGOs can be encouraged to visit one another, or attend seminars together. Groups for whom a development project is just an idea may be helped with an introduction to a group already successfully working in the field. Field staff are in the privileged position of having contact with a large number of groups, often in a number of regions or countries, and are able to assist in providing a link between them.

b: Discussions. Field staff can contribute a great deal simply by sitting down with people who are trying to promote development in their own locality and talking through their problems. Project holders are often highly competent in their own field, but may lack key skills, for example, in management or accountancy. Field staff may be able to point out pitfalls which might otherwise be overlooked, and relatively isolated people may be given the opportunity to try out their ideas on someone with a trained and critical mind.

c: Participation. The field staff's experience of a wide variety of projects, and their awareness of the need for community or group involvement and training, is especially important. Many projects with a technical, agricultural or medical content fail to achieve their objectives because these crucial organisational aspects are neglected. Field staff can assist technically qualified project planners to be more aware of the social, cultural and other needs of the group (see Section 4, EVALUATION, and PART FOUR, SOCIAL DEVELOPMENT GUIDELINES).

d: Local Experts. Where these exist, they may be mobilised to assist field staff and project holders alike. Apart from providing a useful service to projects they may help to bridge the gap between the intellectual elite of a country and local development agencies. Field staff should therefore enlist the support of local consultants and try to involve local universities and research institutes when seeking solutions to technical problems. Other local agencies which already possess tried experience in community development etc. may assist new groups struggling with the first stages of a development programme. It may also be possible to assist in strengthening local technical service centres, such as PATAC in

Brazil (BRZ 141) and the Kumasi Technology Consultancy Centre (GHA 22).

e: Information and Publications. Field staff often carry with them a variety of publications when they are visiting groups. New ideas and technical innovations can be spread with the aid of such material. Information spread by word of mouth can often be supplemented by booklets and leaflets either from Oxfam itself (e.g. memos on TB and leprosy), other foreign agencies (World Neighbors' 'In Action' series on soil conservation etc.), or other local groups. The Oxfam Book Token Scheme is another way in which books and periodicals can be distributed free of charge to those requiring information who lack library facilities or the foreign exchange to purchase such materials (details available from Oxfam Field Offices). Many Oxfam Field Directors now have separate grants to enable them to purchase essential texts for groups, whether Oxfam is funding their activities or not.

There are now many agencies offering technical advice, publications and consultancy services, often in very specialised fields. A list of some of these agencies may be found in Section 7.

II Inter-project learning: visits

It has been noted that people can learn effectively from their peers through visits and seminars. Various possibilities are open to field staff to assist in this process.

⒈ Short Visits

Short visits by project holders to other similar projects have often proved very effective. In Zaire, for example, one such visit demonstrated the feasibility of a comprehensive public health programme and showed how such programmes might be formed (ZAI 70). In Peru, groups of tribal Indians have visited projects run by neighbouring tribes and have learnt of the possibilities for independent development in their own communities (PRU 86). It is preferable that visits should be made to projects which have reached a certain degree of stability rather than to those which are still feeling their way to maturity, although there is a danger that the level of sophistication or degree of success of the project visited may prove intimidating. Short visits may be arranged simply to make initial contact or to show people around. Visits may also be arranged to further contacts made in seminars. Visitors may also contribute to the process of participatory evaluation by providing their hosts with new insights into their work.

Inter-project visits can be expensive when long distances are involved and frontiers have to be crossed. Lack of planning and of clear objectives can lead to problems. For example, when only short notice is given, project visits frequently amount to little more than a sight-seeing trip. It is unlikely that this kind of approach will be appreciated by those being visited. Similarly, it is difficult to organise events and translation facilities quickly. Often those visited without much prior warning have little idea of the visitors' interests or motives.

In view, therefore, of the need for careful arrangements it is suggested that the following measures be taken.

a: Visits should be planned several months in advance. The visitors should write directly to the project, not through Oxfam, indicating:
- [ ] their proposed programme,
- [ ] their special interests and qualifications/experience,
- [ ] the objectives of their visit,
- [ ] alternative dates.

b: A report should be prepared immediately after the visit on:
- [ ] activities observed at each of the locations visited,

☐ assessment of the project as a whole,
☐ lessons learnt from the project.
c: an approximate budget should be submitted before the visit (this prevents 'Holiday Inn'-style visiting). Public transport should be the norm wherever possible.
d: Projects visited should also prepare a report on the visitor(s).
e: If possible, a return visit should be arranged so as to enhance the learning process. In such circumstances, board and lodging may be provided at a reduced rate, or even free, by those being visited.
f: Try to plan and budget for inter-project visits at the initial stages of a project, thereby incorporating them into the general process of project monitoring and evaluation.

2 Training visits

These are longer visits, where people have been seconded to projects to learn about the techniques and methodology used in their work. In one such project, a group from Malawi (MAL 28) went to Zimbabwe to study savings clubs and rabbit rearing; someone from Burkina Faso was sent for three months to the Kumasi Technology Consultancy Centre in Ghana (GHA 22) to study appropriate technology, and Amuesha Indians in Peru studied agriculture at the *Instituto de Estudios Rurales* (PRU 141). For further information on training visits see PART FOUR, SOCIAL DEVELOPMENT GUIDELINES, Section 2, Training.).

3 Informal contacts

It is often sufficient for field staff to introduce project holders to others in similar institutions or programmes; it should never be assumed that people will already know one another. Fruitful collaboration can follow from the briefest introduction or reference.

III Inter-project learning: seminars

1 Introduction

Seminars may be arranged for development workers or grass roots organisations. If well organised and concerned with relevant topics, seminars can be extremely useful in providing a forum for the exchange of ideas and experience. It should be remembered that, despite being trained in development or technical matters, many field workers lack comparative experience and are not exposed to new ideas or methods, because of the isolation in which they normally work. In Brazil the members of several centres were brought together to discuss the use in social education work of a locally-made brick machine and an allied self-help housing system (BRZ 141). In India and Peru meetings have been held which have brought together people from a wide range of projects to discuss common problems. For example, small-scale artisans met in Peru to discuss the problems of marketing handicrafts, and water engineers from various countries in South America participated in a seminar on simple water pumps (PRU 158 and PRU 257).

2 Guidelines for organising seminars

a: Allow yourself plenty of time. Start planning at least eight months before the event.
b: Find someone (or an institution) who can organise it for you. Make sure that the organisers are people with whom you have close contact and in whom you have confidence; if you employ an institution, it should be a competent professional agency. Friends and/or projects who offer to run the seminar may be very helpful, but they may have many other commitments. It is also difficult to make demands of them if they are working free of charge; it is better to pay and get the job done properly.
c: Limit the participants to a reasonable number: sixty would in most cases be a maximum; forty would be better, and less than twenty hardly worthwhile.

d: If possible, live and eat together in one place. The location should preferably be some distance from the nearest town. Food and accommodation should be adequate but modest.

e: Decide whether the seminar is intended for professional development workers or for the beneficiaries of development projects. If the seminar is directed at peasant groups, for example, ensure that a good percentage of the participants are actually peasants, not professional advisers.

f: See that all participants are given the opportunity to put forward their views. For example, in a seminar involving both professional extension agents and farmers, it is vital that the presence of professionals should not inhibit the other participants from expressing their opinions. It may be preferable to invite professionals to attend in the capacity of advisers or observers rather than direct participants, so that they do not dominate the discussion.

g: Decide early on how much time is going to be spent in plenary sessions and how much in smaller workshops. The latter are more useful, but they should always report back to the general assembly for a plenary discussion at the conclusion of each workshop.

h: To keep costs down and to make sure that funds are used well, all travel — even journeys across frontiers — should be made by bus, train or car. Air fares are only paid to those who have to cross oceans or similar obstacles, or who have to travel great distances.

i: Leave as much of the day-to-day organisation as possible in the hands of the participants so that they can respond to people's needs as expressed in the general sessions.

j: The institution which you have employed to run the seminar for you should provide a competent person to chair the general sessions, but the participants might also like to appoint someone from among their own number. The institution should also provide a reporter, so that each day the minutes of the various meetings, including the points made by the workshops, may be written up and be available for distribution the following morning.

k: All participants should attend full-time. It is not advisable to agree to participants attending for only part of the allotted time, since it might be disruptive to the other participants, quite apart from the waste of resources.

l: Make sure participants bring examples of booklets, pamphlets, films, slides or any other educational material they may have produced. If they can provide sufficient quantities to enable them to sell to other participants, so much the better.

m: Those being asked to give a presentation should send at least a precis of what they are going to say to the organisers some weeks before the event, so that copies may be prepared for distribution.

n: Group discussions following presentations should be structured; a reporter, chairperson etc. should be chosen. These positions should rotate within the group. Groups should be representative of the participants. Rooms should be available for group meetings. Paper and pens should be available for notes. Time should be allocated for leisure so that people may have the opportunity to talk to whom they wish and, if necessary, to air views which they feel cannot be put in the general meetings, and to organise counter meetings.

o: The seminar should include at least one evening devoted to leisure activities.

p: To follow up contacts made during the seminar, groups should be encouraged to visit each other's projects, at Oxfam's expense if necessary.

q: A full report of the seminar should be prepared immediately after the event by the institution engaged to organise and run it. This should be sent to all participants as soon as possible, a copy being provided for each person who attended, not just for their organisation.

r: Write to the participants 6-12 months after the seminar, asking them what follow-up they undertook in their project or area as a result of what they learnt.

Section 6　　Budgeting and Accountancy

I　　Types and assessment of accounts

① Introduction

The aim of this section is to describe the types of accounts which should be used in small-scale development projects, and to indicate to field staff what they should look for when studying the accounts and proposed budgets of projects put forward for funding.

② Types of Accounts Suitable for Small-Scale Projects

a: For simple projects with a single purpose or activity, and without much in the way of fixed assets, a simple 'receipts and payments' account based on a cash book or on the procedures described below may suffice.

b: For projects involving several activities, full income and expenditure accounts and balance sheets are necessary; books should be kept using double-entry book-keeping.

c: Some kind of visual presentation of accounts is always worthwhile, especially for cooperatives, savings banks etc. For example, in Oxfam's annual reports, piles of coins of various sizes were at one time used to illustrate the way funds are apportioned for different purposes.

③ Sets of Accounts

The set of accounts presented by an organisation should show its financial position and structure. They can be an important aid when:

☐ deciding whether financial help is needed;

☐ deciding what kind of help would be most useful, e.g. a grant or cash, large or small, one-off or spread over a period, a piece of equipment, help in securing a loan on a commercial basis, etc;

☐ seeing what has happened to a grant or piece of equipment donated;

☐ evaluating the efficacy of a grant, etc.

A set of accounts may consist simply of a Receipts and Payments Account (see para (a) below), although an Income and Expenditure Account (see para (b) below) combined with a Balance Sheet (see para (c) below) is more useful. A Profit and Loss Account is the commercial equivalent of an Income and Expenditure Account. Accounts should be audited by a professional accountant, ensuring that they have been correctly prepared and on a basis consistent both with normal standards and with previous years, and that only cautious estimation has been applied.

a: Receipts and Payments Account. This is a summarised form of the organisation's cash book. Cash and bank transactions are amalgamated and balances at the beginning and end of the period are shown. The purchase and sale of fixed assets are shown along with expenses, and no adjustment is made for amounts owed or paid in advance. This is the most basic set of accounts and is sufficient for only the smallest organisations. Any organisation with a vehicle, premises or full-time employees should be asked to produce an Income and Expenditure Account (or Profit and Loss Account) and Balance Sheet.

In the assessment of the Receipts and Payments Account one must:

☐ check the period covered by the Account (usually one year). The older the Account, the less useful. A set of accounts for the previous period is useful for comparison;

☐ ask who prepared it and check whether it was audited (if not, it is worth inspecting the cash book, bank statements and invoices to confirm the figures);

☐ find out what was owed by and to the organisation at the end of the period; check what has happened since that time;

☐ check that any grant paid by the funding agency during the period is shown and has been applied in the way specified; ask about amounts received from other sources to see that they complement and do not duplicate the funding agency's contribution;

☐ find out what assets the organisation has, apart from the cash and bank balances shown on the Account.

b: Income and Expenditure (Profit and Loss) Account. This account shows the income and expenditure (adjusted for amounts owed and paid in advance) for the period. The purchase and sale of fixed assets are excluded. The excess of income over expenditure (profit) or excess of expenditure over income (loss) is also shown. Some items of income and expenditure will have been received or paid in cash; others, such as amounts owed and depreciation of fixed assets, are adjustments and will also be shown on the Balance Sheet. An Income and Expenditure Account for a period and a Balance Sheet dated at the end of the period together form a full set of accounts.

In the assessment of this account one must:

☐ check the period of the Account (usually one year), and whether it has been audited;

☐ go through the Account item by item, comparing it with the Account presented the previous year (period) and assessing the amounts given against knowledge of local costs;

☐ match expenses against grants from the funding agency and income from other sources (including other funding agencies in the case of co-funded projects).

c: Balance Sheet. On a Balance Sheet are listed the assets and liabilities of an organisation at a particular time. Like a photograph, a Balance Sheet shows a static view. The format varies, but it is usual to group items together under headings, e.g. fixed assets, current assets, long-term liabilities and current liabilities. Apart from the assets and amounts owed to outsiders (external liabilities), the remaining item to be listed may be described in various ways : capital fund, owner's equity, reserves, net worth, net/accumulated deficit. All these terms are ways of describing the estimated *net value of the organisation*.

In the assessment:

☐ check both the date of the balance sheet and also that it has been audited; compare it to the previous one;

☐ go through the items listed. Find out what *Fixed Assets* are shown; they should be both appropriate and reasonably valued. Most fixed assets are depreciated every year to take account of loss of value. *Current Assets* should all be realisable in cash within a short period. *Current Liabilities* should include all amounts due for payment and amounts which may be called in at any time (e.g. bank overdrafts);

☐ examine the items making up the net value of the organisation (see above). Where this is negative (that is, where external liabilities exceed assets), the organisation is said to be insolvent. If this is so, new inputs of funds are likely to be used to pay off old debts. Where assets are much greater than external liabilities, further funds may only be needed to overcome a shortage of immediately available cash (lack of liquidity) or to finance an expansion;

☐ look at the financial structure of the organisation by making comparisons of the larger figures. Fixed assets should be secure; this means being financed from the capital fund or long-term loans. The organisation should not be in danger of having to sell its vehicle to pay the electricity bill. There should be an adequate amount of liquid assets (cash, bank balances, easily saleable securities) to meet payments as they become due;

☐ if any fixed assets require replacement in the near future, check that either the necessary funds have been set aside or arrangements made to borrow them. (Note that the reducing of the book value of assets by depreciation means that yearly expenditure reflects the real cost of using the asset. It is a separate exercise to ensure that cash is available to buy a new asset.)

II   Financial management and budgeting

1 Accounting for funds on a day-to-day basis

A self-help farming project (KEN 81) has produced clear and useful instructions as to how funds should be moved between current accounts and petty cash etc. Funds were to be kept in the following four accounts:

☐ current account;
☐ savings account for depreciation of equipment (tractor etc.);
☐ petty cash account;
☐ account for tractor running costs.

All income, whether received in cash or by cheque, is paid into the current account, and receipts are issued for all sums. Money is banked as frequently as possible, so that large sums are not left in the office for any length of time. The money deposited in the bank on each occasion must tally with a specified set of receipts. All expenses are paid from the current account either by cheque, or via the petty cash account if by cash.

The function of the savings account is to accumulate depreciation so that expensive equipment can be replaced when this becomes necessary. At the end of each calendar month, money is transferred from the current account to the savings account. The amount will be a fixed sum each month for some items of equipment such as irrigation pumps; but for tractors, the depreciation payment is a fixed amount for each hour the tractor is used.

These savings must be deposited each month regardless of the financial position of the organisation, and even if it means that there is not enough money remaining to pay full member dividends.

The petty cash account is a revolving account of about £20 for cash payments of £2 or less. For each payment from this account, a receipt must be obtained and kept. When a purchase is made from a shop, a cash sale slip is usually provided as a receipt. In cases where no such slip is obtained, the treasurer should write the amount of the payment, the date and the expense on a piece of paper and sign it. These receipts are kept with the cash so that the total of cash and receipts always equals the original sum (£20 in this example). When it is necessary to replenish this account, add the amounts of all the receipts together and make out a cash cheque for the total. When the cash is then restored, the old receipts should be put together in an envelope and marked 'Petty Cash Receipts' with the date, the cheque number, and the total amount.

2 Depreciation

Eventually a pump, a plough or a building reaches the end of its working life, during which it reduces in value each year. This yearly reduction in value is known as depreciation, and is therefore regarded as a cost of owning that particular piece of equipment.

For example, if an ox-drawn planter costs £40 and its working life is 5 years, the depreciation or fall in value over 5 years is £8 per annum. An annual depreciation charge of £8 is therefore made in the accounts for each year of its working life; this will appear on the expenses side of the profit and loss account of a farm. Buildings, plant and machinery and all farm implements should be subject to a depreciation charge, whether they are new or second-hand, until the full purchase price has been recovered.

Note: With inflation now endemic in most parts of the world, this approach by itself is inadequate. Depreciation charges based on the original cost of a piece of equipment will usually be quite insufficient to replace that item when its useful life has expired. As a rough guide, in most countries depreciation should be charged at twice the normal rate quoted above.

When Oxfam gives equipment to a project which is expected to become self-sufficient, the usual policy is to insist that the project should set up a depreciation fund into which the annual or monthly depreciation charges are then paid. In theory the fund should allow the equipment to be replaced in due course. This is a good exercise for cooperative and other organisations which have to survive in real market conditions.

Oxfam policy in the past has been that depreciation funds set up by medical and welfare projects should receive annual payments equal to 10% of the cost of the equipment. For all other projects, a 15% payment should be made into the depreciation fund.

These are minimum rates, and even before current inflation began, it was recognised that a Land Rover (for example) would need to be depreciated at 25% of its initial cost each year. It was also recognised that depreciation payments at this rate are out of the question for many projects, so the setting up of these funds has not always been insisted on. If a fund could not be set up, the future budgets of the project would need to be considered in some other way when the time came, e.g. through another capital grant.

When assessing depreciation, the working life of equipment may be taken as the following:

a: land rovers/tractors, adverse conditions — 4 years
b: land rovers/tractors, normal conditions — 5 years
c: drilling rigs — 5 years
d: planters, maize shellers, etc. — 5 years
e: tractor-drawn farm implements in general — 8 years
f: ox-drawn planters — 5 years
g: other ox-drawn equipment — 10 years

3 Audits

In assessing projects, field staff should take care to note whether there has been a proper audit by a professional accountant. This ought to be an important feature, and should be insisted on in almost all cases. But where projects do present audited accounts, field staff should be willing to accept these, and not make further investigations except perhaps in matters of practical detail. Note that:

a: it is best if the accountant doing the audit can visit the project, though this is not absolutely necessary if he/she is given all the books and papers;

b: the audit will often reveal weaknesses in the financing of the project which may threaten its future. The auditor will not usually offer advice on this unless asked. Projects should therefore ask the auditor for his/her comments, indicating the strengths and weaknesses of the scheme.

4 Part-financing of Projects

Funding agencies often finance part of a project, leaving the people running it to seek the remaining funds from other sources. The danger of this is that attractive parts of a project get offers of funds from several agencies, while essential but unexciting aspects of the work can only be paid for if the project managers divert funds from other purposes.

Field staff should not give grants for part of a project without looking at the finances of the project as a whole. This would reveal whether, in fact, help with the running costs or in clearing a deficit or help with some relatively mundane work might be more urgently needed than the relatively spectacular capital works which may have been proposed as likely to attract a grant.

5 Budgets and Budgeting

Appraisal of a project for funding always involves studying estimates of future costs and sources of income to check that the project is viable.

This may include analysing capital budgets or cash flows, the latter being essential for checking that the project can always meet its payments and other outgoings. Farmers can use a simple cash flow approach to estimate what loans or credit they may need to carry them through from the time when payments have to be made for seed and fertiliser until harvest, when a return can be realised.

Budgets often go wrong because of *over-optimistic* assumptions: for instance, in tractor hire schemes, people have often budgeted on the basis that the tractors will each work 1,000 hours per year, whereas experience with this sort of project in developing countries indicates that, in most conditions, one can only hope for 400 hours.

Another common error in budgets for buildings is to include the costs of walls, roofs, windows and doors, but to forget that a building needs floor surfaces and furnishings. A budget is clearly no good if it is based on an inadequate bill of quantities.

3

Section 7   Checklist of Questions

I   General remarks

1. Does the project fall within the guidelines and priorities for the area?
2. Is the project known to you? If not, then have you allowed sufficient time to get to know the group, and for a relationship of trust and confidence to be developed between you and the group?
3. The size of the project will have some bearing on the visit, whether it will be possible to talk of past work, and on aspects such as evaluation, planning and management. Small grass-roots groups will be more concerned with the immediate project in hand than with more general discussions about development and its mechanics. Discretion will always be required when visiting projects and the people involved in them.

II   Specific points

It should be remembered that every project is to an extent unique, and that the relevance of particular questions will vary from project to project.

1. Objectives:
    Are the short-term aims and goals of the project compatible with its longer and medium-term objectives?
    ☐ Are they clear? Are they reasonable and practical?
    ☐ Are they based on reality as illustrated through experience and discussion, and on the results of study?
2. Design:
    ☐ Is the project feasible?
    ☐ Does the design take into account local traditions and customs?
    ☐ Is the project over-ambitious? Are the resources adequate to the requirements?
    ☐ Does the project design cater for sufficient feedback?
    ☐ Is the timing realistic?
    ☐ Would a short pilot project be advisable to gain experience before commencing the project?
3. Financial involvement:
    ☐ Do all the people aided by the project have a stake in it? Are they contributing in some way (in cash, in kind or in labour)? Is the project simply a source of income or career advancement for the professionals and other development workers involved?
    ☐ Could those involved be asked to contribute more?
4. Education:
    ☐ Are the people involved going through a process of 'critical appraisal'? Does the project enable them to assume greater control over their own lives?
    ☐ Have they reached a stage at which they are able to articulate their problems, aspirations and priorities?
    ☐ Is it felt necessary to mount a programme of 'self-awareness'? How is this to be achieved, and how does it relate to the rest of the programme?
5. Practical effects:
    ☐ Will the project improve people's lives by bringing them material benefits? Will they have more cash, food, animals etc.? Will they have better access to services such as health?
    ☐ Compare this with the answers on the educational process. Material and educational benefits should be complementary. Sometimes the

expansion of the economic process helps to build up people's critical awareness, but the reverse is not necessarily true.

6. Staff of project team and/or promoters:
- [ ] Are they reliable, realistic, capable and motivated?
- [ ] Have they sufficient experience? In what fields are they experienced? Is this experience appropriate to the task in hand?
- [ ] Are they well qualified or do they need more training?
- [ ] Are the leaders strong? Do they respect and/or reflect the opinions of others? What do other local groups and individuals say about the project leaders?
- [ ] Do the staff/promoters work well together? What relationship do they have with the poor? What do the local people say about them?
- [ ] How open to new ideas and suggestions are the staff/promoters?

7. Participation:
- [ ] How was the application drawn up? Who was involved in this process?
- [ ] To what extent are the people supposedly being helped able to influence the course of the project? What mechanisms are there for this? Have they already participated in the planning of the project?
- [ ] Is the feedback of information being properly heeded by the project managers? Is there a mechanism to allow participants to voice complaints or make suggestions to amend the project?

8. Organisation:
- [ ] What is the nature of the organisation? Does it have legal status? In practical terms what effect does the legal status have?
- [ ] Are there clearly set rules for running meetings, taking minutes, following up decisions, changing leadership, etc?
- [ ] How democratic is the organisation and how difficult is it for the membership to overturn decisions of the executive body?
- [ ] If things go wrong with the project, what allowance has been made for either resolving internal conflict, or disposing of donated assets?
- [ ] Is there an education programme for members of the organisation? This is especially important for cooperatives and similar institutions.
- [ ] What safeguards are there to ensure that the organisation is not taken over by an elite group?

9. Monitoring:
- [ ] Is there a system for regular self-evaluation? If so, how is this carried out and who participates?
- [ ] Are the conclusions of the monitoring process used to change objectives or plans? Does the project design allow for the incorporation of new ideas?
- [ ] Are any outside agencies or individuals invited to assist in the planning, monitoring or evaluation process?

10. Budgets and cost effectiveness:
- [ ] To what extent is the project, as planned, the cheapest way to achieve the objectives defined?
- [ ] Is the budget reasonable? How do staff/promoter salaries, for example, relate to those of the beneficiaries of the project? Are salaries compatible with local norms? Are equipment costs appropriate to local conditions?
- [ ] What previous aid has the project received? What were the results of this aid?
- [ ] Is there a danger that further aid will do more harm than good?

11. Measurability:
- [ ] Will it be possible to measure the progress of the project? If so, how?
- [ ] Is a form of measurement built in to the project design?
- [ ] How often will this monitoring be carried out? At regular intervals?

12. Replicability:
- [ ] If the project is successful, could it be repeated elsewhere? What is

the multiplier effect likely to be? Could the programme be expanded to neighbouring areas?

13. Continuity:
- [ ] When outside aid ceases, is it likely that the project would be able to continue on its own?
- [ ] What plans are there for continuation at the end of the project period?
- [ ] Is it likely that government and local authorities will encourage or oppose the project?
- [ ] Is the project so designed as to allow it to survive into future generations, i.e. are new techniques being passed on, and is there a teaching element?

14. Other institutions:
- [ ] Is the project linked to other similar institutions? Is it able to learn from other groups?
- [ ] Does an appropriate intermediate or representative group exist to which the project holders should belong, e.g. a federation of cooperatives?
- [ ] Is the agency responsive to requests for assistance from similar groups?

15. Risk assessment (see Section 2):
- [ ] Compared with other applications for funds, how much risk is involved in helping this particular project?
- [ ] Which applications are the most viable, with the greatest possibility of achieving results, of being able to expand and of having a multiplier effect?

III   Other considerations

---

1. When considering the funding of intermediate agencies it is essential to talk not only to the staff of the project but also to the people who are to be helped by the project, e.g. peasant groups, community committees, indigenous groups, women or handicapped people. It is also important to talk to other agencies and prominent local figures such as priests, development officers etc.

2. Even before a project has begun to receive funding, it can be useful to arrange for prospective project members to visit other similar programmes.

3. It can also be helpful to organise simple surveys and/or baseline data collecting. This should be done by an experienced agency. If at all possible, the people from the project area should help in the process.

4. Sometimes after visiting a group that is seeking funds, field staff will be forced to consider whether, in spite of its deficiencies, the group shows sufficient potential to merit receiving limited support, in the hope that — in time — the good points will outweigh the faults. This requires much thought, tact and a good relationship with the group.

5. In certain cases, where the needs of the community are great and where the prospect of a successful project seems minimal, it might be possible to give a small donation, on condition that an evaluation (internal or external) should be undertaken at once. If all goes well, a further application can be invited after this process has been completed.

6. Field staff should never try to impose forcibly their own, or the funding agency's, ideas or rules. Ultimately, both the applicants and the funding agency must feel that they are convinced about what they are undertaking. An imbalance in this respect is not only likely to result in disaster but is also an insult to the applicant's integrity. Hence the insistence that field staff should spend as much time as possible getting to know and understand the group concerned.

7. Funding agencies must be consistent in their work. They cannot say one thing and do another. Consistency is also important in the way agencies raise their money, deal with applications in their home office, alert their own nationals to the problems of the Third World, criticise their own government's policies towards development, and stand up for human rights issues, etc. Always try to explain to those seeking help how the funding agency is run, where the money comes from, what the organisation stands for, what it does at home in the way of educating the public on Third World issues, etc.

3

Section 8 Appendices

I Cost benefit analysis

1 **Examples of the Issues Arising from Changes in Agricultural Practices**

a: Replacing sorghum with groundnuts, i.e. budgeting to substitute one crop with another.

Table 3 – 1

	A	B
(i)	**Revenue lost from sorghum**	**Revenue gained on groundnuts**
	Value of grain	Value of groundnuts
	Value of by-products	Value of by-products
(ii)	**Costs incurred on groundnuts**	**Costs saved on sorghum**
	Seed	Seed
	Hired labour	Hired labour
	Annual charge for tools	Annual charge for tools
	Fertiliser	Fertiliser
	Sprays	Sprays

If B(i) + B(ii) is greater than A(i) + A(ii), the change may be worthwhile, but note:

☐ if increased expenditure is involved, there may be better ways of spending those funds;

☐ the calculation assumes that other things, like the availability of unpaid family labour, remain the same and that no opportunity cost is entailed in making the change;

☐ no allowance is made for carry-over effects; in this example, land fertility is likely to be higher after a legume like groundnuts than after sorghum (see PART SIX Section 2);

☐ if ex-farm prices are used even though the products are consumed at home, the differential contribution of the two crops to the dietary balance may be a factor influencing the choice.

b: Replacing bullocks by a tractor on a 10 ha farm, i.e. budgeting for a substitute form of tractive power where time is an important factor, see table 3-2.

Table 3-2		A	B
(i)		**Revenue lost on change from bullocks**	**Revenue gained on change to tractor**
		Custom work off-the-farm (net of concentrated feed and hired labour)	Custom work off-the-farm (net of running costs and hired labour)
		Fertiliser and fuel from dung	Net product from bullock-feeding land
			Net increase in production due to timeliness of land preparation
(ii)		**Costs incurred on tractors**	**Costs saved on bullocks**
		Annual capital cost *(purchase price less scrap value)* life of tractor	Annual capital cost *(purchase price less final sale price)* life of bullock
		Running costs	Concentrated feed purchased
		Hired labour	Hired labour
		Annual capital/running costs of equipment	Annual capital/running costs of equipment

If B(i) + B(ii) is greater than A(i) + A(ii), the change may be worthwhile, but note:

☐ if increased expenditure is involved there may be better ways of using those funds;

☐ for such a complex technological change the method of calculation is crude. It omits the effect on yields, which might arise from damage to soil of a heavy vehicle etc.; the financial costs of heavy initial capital investment in the tractor; and social losses and gains. Table 3-3 shows probable losses and gains:

Table 3-3	Social losses	Social gains
	Increase in unemployment	Increase in leisure and decrease in drudgery
	Polarisation of income distribution	Increase in time available to solve management problems other than labour
	Loss of community cooperation	
	Dependence on outside sources for spare parts	
	Dependence on small number of power units means higher risk of losing crops if machines break down	
	Deeper and more intensive cultivation may lead to detrimental effect on ecology and possible 'dustbowl' effect as well as disrupting established rotation.	

② Example of a Cost Benefit Calculation

The following example from an agricultural development programme in Guatemala indicates the kind of data required and the procedure involved. First of all the base data on the costs and benefits for the project are presented. Second, various assumptions are outlined which affect the analysis; and third, the subsequent calculations are given.

a: The Base Data

Table 3 – 4

Costs	YEAR	1	2	3	4	5
		Q	Q	Q	Q	Q
Salaries – Agricultural Director		975	975	975	975	975
Part-time Teacher		520	780	780	780	520
2 ploughs		300				
Small tools (revenue)		200	200	200	150	50
Operating expenses		10,000	8,000	8,000	8,000	5,000

Notes: **(i)** The 2 ploughs will have a joint residual value of £50 after 5 years.
(ii) Q = Quetzales (Guatemala currency).

Benefits after 5 years:
☐ Overall corn production doubles (see note (i));
☐ 30% of families with sloping fields will have made contour ditches which prevent rain eroding their fields (see note (ii));
☐ 25% of families with fruit trees will have increased their production because of better pruning and insect control methods.
Notes (i) The population of the programme area is 20,000. The average family size is 5 and the average salary per family per annum is 100 Q. The doubling of corn production will produce a 50% increase in salary by the end of 5 years. The benefits due to increased production per annum are shown in Table 3-5.

Table 3 – 5

Yr	1	2	3	4	5
Q per family	5	14	25	38	50

(ii) 80% of families have sloping fields and the percentage of those who will have made contour ditches, leading up to the 30%, and the money saved due to lack of erosion, is shown Table 3-6.

Table 3 – 6

Yr	1	2	3	4	5
%	3	7	14	22	30
Q	40	46	55	64	73

(iii) Half the families have fruit trees, and the increased production will give an extra 5 Q. Of those with fruit trees the following percentage will bear the extra fruit:

Yr	1	2	3	4	5
%	2	5	10	17	25

b: Further Information

(i) Inflation will be equal for all types of prices.

(ii) As costs are relatively certain and benefits relatively uncertain, assume a risk premium for benefits of 5% per year. This means that the benefit projections will be reduced by 5% each year.

(iii) Costs fall at the beginning of the year, and benefits accrue at the end of each year. This is a slight complication not often seen in economic appraisals. The implications of this are that different discount factors are used for costs and benefits within each year of the analysis. For example, from the following tables it can be seen that in year 1 the costs are discounted using a discount factor of 1 (i.e., effectively they are not discounted), while the benefits in year 1 are discounted using a discount factor of $1/1.15 = 0.87$.

(iv) Assume a required rate of return of 10%. This means that in calculating the Net Present Value (NPV), a discount factor of 10% must be used; alternatively, when using the Internal Rate of Return (IRR) rule, the project must have an IRR greater than 10% to be acceptable.

c: Problem

Calculate the net present value of net benefits and weigh up the solution against other evaluation techniques.

3

Table 3 – 7   The Calculations

(i) Costs

YEAR	1	2	3	4	5
Salaries	Q	Q	Q	Q	Q
a) Agricultural Director	975	975	975	975	975
b) Part-time Teacher	520	780	780	780	520
2 ploughs	300				
Small tools	200	200	200	150	50
Operating expenses	10,000	8,000	8,000	8,000	5,000
TOTAL	11,995	9,955	9,955	9,905	6,545
Discount factor	1 1	0.91 $1/1.1$	0.83 $1/(1.1)2$	0.75 $1/(1.1)3$	0.68 $1/(1.1)4$
Total NPV of costs	11,995 +	9,059 +	8,263 +	7,429 +	4,451

= 41,197Q

(ii) Benefits

Corn Production – increase in salary

Number of families in project area $\dfrac{20,000}{5} = 4,000$

Year 1	5Q × 4,000	=	20,000Q
2	14Q × 4,000	=	56,000Q
3	25Q × 4,000	=	100,000Q
4	38Q × 4,000	=	152,000Q
5	50Q × 4,000	=	200,000Q

continued overleaf

125

Cost of erosion saved

80% of 4,000 have sloping fields = 3,200

Year 1	3% × 3,200 × 40Q	=	3,840Q
2	7% × 3,200 × 46Q	=	10,304Q
3	14% × 3,200 × 55Q	=	24,640Q
4	22% × 3,200 × 64Q	=	45,056Q
5	30% × 3,200 × 73Q	=	70,080Q

Increased sales of fruit

Number of families with fruit trees $\frac{4,000}{2}$ = 2,000Q

Year 1	2% × 5Q × 2,000	=	200Q
2	5% × 5Q × 2,000	=	500Q
3	10% × 5Q × 2,000	=	1,000Q
4	17% × 5Q × 2,000	=	1,700Q
5	25% × 5Q × 2,000	=	2,500Q

(iii) Total Benefits

YEAR	1	2	3	4	5
	Q	Q	Q	Q	Q
Production – Salary Increase	20,000	56,000	100,000	152,000	200,000
Cost of erosion saved	3,840	10,304	24,640	45,056	70,080
Increased Fruit Sales	200	500	1,000	1,700	2,500
Residual value of plough					50
	24,040	66,804	125,640	198,756	272,630

Total benefits discounted to present value

Year 1	$24,040 \times \dfrac{1}{1.15}$	= 20,904 Q
		+
2	$66,804 \times \dfrac{1}{(1.15)(1.2)}$	= 48,409 Q
		+
3	$125,640 \times \dfrac{1}{(1.15)(1.2)(1.25)}$	= 72,624 Q
		+
4	$198,756 \times \dfrac{1}{(1.15)(1.2)(1.25)(1.3)}$	= 88,336 Q
		+
5	$272,630 \times \dfrac{1}{(1.15)(1.2)(1.25)(1.3)(1.35)}$	= 89,977 Q

Total net present value of benefits	320,250 Q
Less net present value of costs	41,197 Q
Net Present Value of net benefits	279,053 Q

This shows a positive NPV which means that the project is economically acceptable; this result should then be weighed up against other evaluation techniques.

Notes on participatory evaluation

1 Introduction

"Participation is a process in which a group or groups exercise initiative in taking action, stimulated by their own thinking and decision-making, and over which they have specific controls." M. T. Feuerstein.

Participation is sought at all stages of the project, from identification of goals to taking action based on the findings. This kind of 'full' participation is one of the most effective mechanisms for achieving the concept of a partnership in the development process — an approach which is fundamental to Oxfam's philosophy of development. Participatory Evaluation involves the funding agency, intermediate (or implementing) agency and grass roots organisations.

Providing that a pattern of constructive self-criticism has already been established in the project, the evaluation could indicate project strengths and weaknesses, show where changes are necessary, propose alternative strategies, increase mutual understanding between field staff and project holders, and perhaps boost morale. Where evaluation is not participatory, the news that a certain goal has been achieved may mean very little to project holders, since they have no understanding of the process involved.

2 Participation by the Whole Project

The project beneficiaries are not all expected to participate actively in every aspect of the evaluation. An existing team can be used to steer the evaluation or a team can be selected specially for the task. A prerequisite for success is the existence of a unified project leadership capable of carrying out group decisions. Whoever called for the evaluation, unless there is mutual agreement by all parties concerned it is useless to proceed further.

3 Determination of Evaluation Objectives

Participation should be sought in identifying the objectives of the evaluation. There may be difficulties in identifying project objectives where they have been stated only generally or where they have evolved gradually over time.

Participants should be helped to focus on particular problem areas, and to identify possible solutions. There should be enough flexibility to indicate more precisely the relationship between cause and effect.

4 The Problems of Determining Success or Failure

Sometimes success can be demonstrated in quantitative terms; but there exist other criteria for success which are less amenable to quantitative analysis. The social consequences of a project elude this type of approach: attitudes, relationships, fears, motivations, communication channels, goal priorities, leadership qualities and patterns can contribute to the success or otherwise of a project, and yet it is impossible to evaluate them precisely.

These factors may also determine success or failure and may be responsible for programme growth or decay. Evaluations are frequently a vehicle for conflict and tension, but the ability to prevent an evaluation from floundering or being torn apart by the more contentious issues depends on the project's capacity for self-analysis and its leadership skills.

5 Baseline Studies and Measuring of Objectives

Where few baseline studies exist, it is harder to quantify achievements. During the initial stages of the evaluation, baseline measurements can be taken, thus establishing a baseline for subsequent stages of the evaluation. This process may well indicate the need for more tangible and easily quantifiable objectives.

There may be problems in obtaining national data, such as census statistics and other reports, which may be considered confidential.

3

6 Evaluation Findings

The average evaluation report is replete with specialist terms and statistical interpretations too complex to be understood by the average participant. In a participatory approach, the evaluation findings emerge in terms understandable to the participants. The evaluation should be relatively simple, the priority being that it is understood and usable by those involved rather than that it should satisfy the standards of scientific research or external agencies. Methods employed in participatory evaluation are various and depend very much on the degree of literacy of the project holders, their organisation and leadership. Feedback to participants should be continuous as the evaluation progresses and complemented by a written report at the end of the evaluation. Where the report is written in a minority language, translation procedures must be designed to assure the participants that the translation is faithful to the original. Participants may wish to classify certain data and findings as confidential, especially if they disclose details of policy or finance. The evaluation process also aims to increase the participants' ability in decision-making, management skills and future evaluation expertise.

7 The Role of External Personnel

The role of external staff, advisers or researchers of various types is that of colleagues in a joint venture. Ideally, leadership is determined not so much by professional status as by the nature of the problem that presents itself. Future evaluations may well require less or no outside input. Even where evaluation procedures are in-built, there will be a need for periodic evaluation studies to prevent over-accumulation of data, to assess performance, to increase morale, and to indicate future directions. Where there are no in-built evaluation procedures, these can be established during the evaluation.

8 Times and Costs — the Implications

The length of the evaluation relates to such factors as evaluation objectives (large-scale or small-scale study), resources (personnel, finance etc.) and whether or not there is a need to obtain results in a hurry (for funding or policy-making purposes).

Projects generally continue functioning during an evaluation, but it may be necessary to slow the process down in order to accommodate the evaluation because of the deployment of staff in coordinating, collecting and assessing data for the evaluation etc.

The pace of the evaluation should never outstrip the participants' ability to comprehend what is taking place.

FINALLY, remember that participatory evaluation is not a panacea. Its big advantage, above all other forms of evaluation, is that it provides a far greater feedback to the project holders and helps foster confidence. It recognises the different perspectives of the different groups involved in the project — many evaluations done by external evaluators, paid for by funding agencies, with the reports going only to the funders, ignore completely the needs, problems or views of the project participants.

Participatory evaluation allows the subjects of the evaluation to define ways of assessing and measuring their own projects. It also helps to maintain understanding and motivation.

III   Social surveys

1 Introduction

Surveys are often useful either at the planning stage or after completing a project, because they help measure both the need for the scheme and its effects. This information should be supplemented, if possible, by careful record-keeping and observation throughout the duration of the project. However, it is vital to bear in mind that *survey results are only reliable to the extent that the data are competently collected and analysed*. Running a survey is a more complicated business than many amateurs realise, and is therefore best left to trained people. Beginners should only undertake a survey if a qualified adviser is willing to help. Many members of university sociology departments, college lecturers and secondary teachers have some experience in running surveys; as this is often part of training in the social sciences, lecturers may welcome a request for a small survey on some practical topic. Survey research centres can also be tried, though they tend to specialise in large-scale funded research and may find it difficult to fit a small localised survey into their programme. Governments are becoming increasingly wary of people carrying out surveys, and care must be taken that the necessary permission has been obtained at every level. Specify clearly what information is wanted; discuss the work fully with whoever will supervise it; and ensure that close supervision will be maintained at every stage of the work.

2 Sources of Information

Before planning a survey, it is best to find out what information is already available. Are there any local studies or reports on similar projects? What about ethnographic or community studies of the people and area concerned? Two major sources of information are local specialists and government reports.

a: Specialists tend to build up a wide range of contacts, whereas researchers working in one department (university or government) often have no knowledge of work going on in other departments. Small studies which have not reached publication stage are often useful because they are likely to be more up to date than published work. Specialists will also know about reports which can be found in libraries or in their own collections. Occasional attendance at conferences is useful for making contacts with people; correspondence can lead to information on topics of special interest which otherwise cannot be readily obtained.

b: Government. National censuses and regular or occasional statistics produced by government ministries are often useful sources of data on a specific area, e.g., composition of population, local leadership, development plans; or on a specific topic, e.g., education, agriculture, transportation, migration, housing. The adequacy of government reports varies considerably from one country to another and over time; beware of their limitations. Information is often inaccurate, so the data on certain subjects may be unreliable. Surveys are frequently restricted to large aggregates, so there may not be enough detail for your needs.

3 Surveys

Whether you commission surveys or rely on those conducted for other purposes, you need to be aware of how the methods used and/or the assumptions of the supervisors may bias the results. As a consumer of survey research, ask:

a: Purpose. What were the goals of this survey? What information did the author want? How were the major variables (income, aspirations, satisfaction) measured? Were the questions asked of the data meaningful in terms of the goals?

b: Sample. Was the sample adequate?

☐ *Choice*. A sample should be large enough to permit comparison of sub-groups, e.g., men and women, young and old, but a small well-chosen sample is usually much better than a large random one, such as interviews held with passers-by on the street. Interviewers must not be free to choose respondents, and they should be closely supervised to ensure that they interview the people who have been selected. The goal is a sample which is broadly representative of the population. Choice will depend on the availability of population lists and information on population composition (which may prove inadequate for the purpose of the study).

☐ *Generalisation from a sample*. How widely do the findings apply? Cultural, social, political and economic differences mean that findings in one place may not apply in another. A case study of one area or institution can provide useful information on some topics, such as community development or social relationships, but beware of claims that data are valid over a wide area. Data from several clusters, e.g., villages, schools and neighbourhoods, are usually more trustworthy than information from one place.

☐ *Response rates*. What proportion of the selected sample cooperated in the study? Were there some questions that almost no one answered? Low response rates, such as when less than 80-90% of those approached were willing to be interviewed, make the data suspect even if the original sample was well chosen. A high number of refusals or 'don't knows' on individual questions often implies that the interview techniques are poor or that the questions were badly structured.

c: Interviews. Did the interviewers establish rapport with the respondents or was there continuing suspicion of the sponsorship and/or motivation of the research? What checks were there on deliberate deception and the invention of answers to suit the occasion? Courtesy bias and the affirmation syndrome (being polite and saying 'yes') need to be watched for. Were the questions asked meaningful to the respondents? Were they clear? Were they within their competence and experience, so that the answers were not merely guesswork or statements made in order to satisfy a stranger? Were back-up questions used to find out what the respondent meant by a 'yes' or 'no' answer or by answers to other closed, 'fixed answer' questions? Open questions are also useful as a check on interviewers as it is harder to make up the range of answers which should result from such questions. Was a series of questions used for important topics so as not to be forced to rely on a single answer? How were problems of language handled, e.g., translation, the meaning of key words and concepts? How well were the interviewers trained and supervised? Who coded the data? Were the categories used for coding and analysis sharply defined and meaningful in terms of the respondents' answers?

d: Report. As you read a report, constantly ask whether it makes sense. Does the data force me to this conclusion? How would I have gone about it? What if the data are looked at from another point of view?

☐ *Method*. How well does the report cover the material collected? Is it related to other studies in this field? To what extent are assumptions substituted for evidence? Are the tables clear, so that they could stand without the text? Does the text summarise and analyse the tables rather than merely repeating or ignoring them? What checks are there on accuracy and reliability?

☐ *Timing*. Is this information up to date? It may take several years for a report to be written and published; readers many years later tend to take the report as an indication of present-day conditions, which are often quite different. New technology, new crops, migration and urbanisation, the spread of education, transportation and social services, changes in the national and local political and economic situation, may mean that research done in the 1970s is seriously out of date by the 1980s. We have to rely to some extent on what is available, but should be warned to check the date of the research as well as the date of publication, and try to work out the implications of changes over time.

This is a highly complex, specialist topic. Only an introduction to basic principles can be given here; these should help you to evaluate the results of the most commonly used measures. For more complicated statistics, it is usually better to refer to an expert rather than try to work from a book. Complicated analyses are sometimes applied to data of doubtful reliability, and impassioned argument is sometimes completely unsupported by facts. It is best to be able to draw your own conclusions from the tables provided.

Statistics seek to impose order on collections of diverse facts or data; to condense opinions, performances, and comparisons among them into summary numbers that can be understood immediately. Three major purposes are introduced here:

a: Statistics summarise information. There are two types of summary statistics

☐ measures of central tendency and measures of variability.

Measures of central tendency help you to portray in a single number the central message of a group of numerical scores. The mean, represented by the symbol $\bar{x}$, of a set of scores is the average score. The total of scores is divided by the number of cases. If some cases are way off at an extreme, the mean is pulled towards them and thus may be unrepresentative of the group; e.g., a few wealthy households raise the mean income disproportionately. Thus, for some data, the median (the middle case, with half the people on each side) may be a better measure. *Measures of variability* qualify measures of central tendency by showing how varied are the scores which contributed to the central tendency number. The variance is a measure of the spread around the mean. If most people are about the same, the variance is low. The standard deviation is the square root of the variance. A large standard deviation tells you that the mean which it accompanies is not to be considered a good summary number for the whole group of scores.

b: Statistics tell you how seriously to regard observed differences between results of two groups; one group being *the treatment group*, and the other group being *the control group* (where no change has taken place). For example, after dividing a group of 100 children into two groups of 50 children, and giving the children in the first group a special nutritional diet, we would like to assess the effect of this special diet on the weight increase of these children after six months, compared with the second group, called the control group. An average difference of 1 kg per child is observed between the two groups. How significant is this difference? Could it be due to chance? Significance tests calculate the likelihood of a real difference, as opposed to an observed difference, in the two groups. If the observed difference stands up to a significance test at 5% level, this means that we are 95% confident that a real difference exists; conversely, in stating that the observed difference of 1 kg is in fact due to the special diet, the probability of making an error is only 5%. A significance test at 10% level means that we are only 90% confident of a real difference between the two groups.

c: Statistics help you determine the degree of relationship among sets of data. In the example cited in (b), if, due to some quality in the special diet, 45 children out of the 50 in the treatment group all put on exactly the same weight, there is a strong correlation between quantity of special diet and weight increase. A correlation factor of 1 would have meant that all 50 children would have put on proportionately the same weight due to the same quantity of special diet.

Surveys are often best when they are combined with other methods, such as systematic records of progress, interviews with informants (structured or informal), and/or observation.

A survey may not be necessary if the information required can be

3

obtained by other means, but the reliability of this information should be assessed as rigorously as survey data.

a: Records. Field workers should keep a record, daily or weekly, of what has happened. What objections have been made to the project? Who has supported it, and what is their position in the community? Who benefits most and who least from the project? What are its potential implications? Are these realised and talked about by all, or by only a few? How has the work been organised? Has enthusiasm grown or decreased as the project has proceeded? Has it varied between groups? The act of keeping a record will encourage workers to be more aware of what is going on, and some difficulties can be handled before they become serious. When the project is finished, a summary of the record can be filed as a guide, for use in planning similar projects or other work in the same area. However, if the report is to be made available to outsiders, some consideration must be given to confidentiality, and if it is to be available to the public, names of people and places may need to be disguised.

b: Informants. Considerable local information can be obtained by talking to local leaders like the head teacher, imam or priest, chief or village head, or elders of the community. Their cooperation must be obtained for work in small communities; this is a good way for you to gain acceptance and credibility. In evaluating information supplied by informants, one should assess not only their knowledge and goals but also their credibility within the community.

☐ *Knowledge*. Partly because admitting ignorance may diminish self-respect, we often like to consider ourselves more knowledgeable than we are. Thus, we must continually question how well informed are those from whom we seek information. For instance, people in cities often know little about their neighbours, and elites frequently know little about ordinary people. Self-confidence is no substitute for experience. Watch for vague generalisations and information based on a single example.

☐ *Goals*. What are the motives of this informant? Does he or she hope to gain anything by providing this information? Does he or she expect increased power, prestige, access to resources, material reward? Is the interviewer seen as a representative of the government or as a neutral party? Communities are seldom homogeneous, and leaders are often affiliated to parties, which seek to assert their own views. Contacts with one group tend to invite suspicion from other groups; thus, it may be easy to get one side of a story but hard to get the other. Even if the community is united, its goal may be to keep strangers out or to mislead government agents; the informant may see his or her job as ensuring that the information wanted is not supplied.

c: Observation. Observation can provide a depth of understanding to complement the breadth of a survey; inadequately done, it can be just as misleading. It is inherently a much slower method than a survey. Observers may be participants or outsiders. Participants obtain additional information through experience and identification with the community; outsiders find it easier to remain neutral and thus examine the issues with less threat to themselves. Most observation is casual and unsystematic, but one can train oneself to observe systematically, watching how people relate to each other and do their tasks, and recording findings in an organised way. Interviewers should also observe: body positions, facial expressions and tone of voice can say as much as words. With practice, it becomes a habit. Getting started is the most difficult part of observation. Newcomers and strangers stand out, as they lack social contacts and a role which explains their presence in the community. The people need not know they are being observed if some acceptable role can be found, e.g., if the observer appears to be doing something else. Taking a preliminary census (for instance, of households, artisans, or anything which is visible from the street and found in reasonable numbers throughout the community) is a good way to get to know people, to watch daily activities and to talk about a planned project or survey.

(See Section 9, Resources, for Bibliography)

3

Section 9   Resources

I   Bibliography

1 Evaluation and monitoring

American Council of Voluntary Agencies for Foreign Service, *Evaluation Sourcebook*, 1983. Useful guide for voluntary agencies.

S.B. Anderson and S. Ball, *The Profession and Practice of Programme Evaluation*, 1978. (US $12.95) Gives particular emphasis to the ethics and principles of evaluation in the context of local politics and real-world concerns.

G.A. Bridger and J.T. Winpenny, *Planning Development Projects*, Overseas Development Administration (HMSO) London, 1983,

S.D. Briggs, "Rural Technology Checklist" in *Planning Rural Technologies*, IDS, University of Sussex, U.K.

I.D. Carruthers & E.S. Clayton, *Ex-post Evaluation of Agricultural Projects*, Wye College, 1977. Useful examination of approaches emphasising the practical rather than the mathematical.

D.J. Casley and D.A. Lucy, *A Handbook on Monitoring and Evaluation of Agriculture and Rural Development Projects*, World Bank, Washington D.C. November 1981.

R. Chambers, *Project Selection for Poverty-Focused Rural Development: Simple is Optimal*, IDS, University of Sussex, 1977. Much good sense stated without frills or jargon.

R. Chambers, *Rural Development Tourism*, IDS, University of Sussex, 1977. Helpful checklist to guide development workers visiting projects.

N. Clark and J. McCaffery, *Demystifying Evaluation*, World Education, 1979. A simple manual for training staff in the assessment of community-based programmes.

D. Conyers and P. Hills, *An Introduction to Development and Planning in the Third World*, 1984. John Wiley, U.K.

CMC, *Evaluation: Can It Become Collective Creativity?* Contact 48, Christian Medical Commission, December 1978. Includes a case study of the planning and evaluation of health care.

Evaluation of Social Development Programmes (with special reference to *youth programmes*), Commonwealth Secretariat, London, U.K. 1974.

W. Fernandes and R. Tandon (eds), *Participatory Research and Evaluation*, Indian Social Institute, New Delhi, 1981.

M.T. Feuerstein, "Self-evaluation: Educative Approach in Evaluation — an Appropriate Technology for a Rural Health Programme", in *International Journal of Health Education*, Vol. XXI, 1978/1.

Ministry of Overseas Development (UK), *Evaluation Activities*, 1976.

Oxfam and Evaluation, Drafts 1 and 2, March and May 1977. Internal documents.

Oxfam Evaluation Policy Paper, August 1977.

J. Pilgrim, *Guide to Development Project Design and Evaluation*, Crown Agents, London, U.K. 1977.

J.P. Gittinger, *Economic Analysis of Agricultural Projects*. (US $4.00) Relatively pragmatic and relevant. John Hopkins University of Pennsylvania, 1982.

Reading Rural Development Bulletin, *'Evaluation'*, Reading, U.K. No. 14 April 1982.

TAICH, *Criteria for Evaluation of Development Projects Involving Women*, New York, 1975.

TAICH, *Approaches to Appropriate Evaluation*, 1978. Workshop on Evaluation of American Voluntary Agencies.

A.J. Taylor and F.C. Cuny, *The Evaluation of Humanitarian Assistance*, Intertect, Texas, 1978. Useful and readable resumé relevant to most types of project.

USAID: *Programme Evaluation in Aid: Lessons Learnt*, 1976. *Project Evaluation Guidelines*, 3rd Edition, 1974. *Building Evaluation Elements into Project Design*, 1974. *Evaluation Handbook*, 2nd Edition, 1976.

2 Social Surveys

(NB. All but the last two are available in paperback.)

H.M. Blalock, *Social Statistics*, McGraw Hill, 1972. This text provides a basis for understanding a wide variety of statistics, as well as giving instructions for computing them.

3

J.A. Davis, *Elementary Survey Analysis*, Prentice-Hall, New Jersey, U.S.A. 1972. An easily readable explanation, with much good sense on the pitfalls of using data.

G. Hoinville et al, *Survey Research Practice*, Heinemann Educational Books, London 1978. A practical explanation of how to do research; based on British experience, but with much that would apply elsewhere.

M.J. Moroney, *Facts from Figures*, Penguin, Harmondsworth, Middlesex, U.K. 1952. An introduction to statistics, rather out of date, written before the calculator revolution.

C.A. Moser and G. Kalton, *Survey Methods in Social Investigation*, Heinemann Educational Books, London 1979 (2nd edn). This is a standard and widely available text, based on British practice; considerable coverage of sampling and many references to other sources of information on specific topics.

M. Peil et al, *Social Science Research Methods: An African Handbook*, Hodder and Stoughton, London 1971. Designed to provide all the practical details needed to conduct research in developing countries.

M.D. Shipman, *The Limitations of Social Research*, Longman, Harlow, Essex 1972. A guide to pitfalls.

J. Silvey, *Deciphering Data*, Longman, Harlow, Essex 1975. What to do with the data after collection.

C Seltiz et al, *Research Methods in Social Relations*, Holt, Rinehart and Winston, Eastbourne, Sussex 1976 (3rd edn). This American text gives a full treatment to most phases of research, including the use of available data and ethical considerations.

D.P. Warwick and S. Osherson (eds), *Comparative Research Methods*, Prentice-Hall, New Jersey, U.S.A. 1973. A set of papers on the problems of research in developing countries and how they were handled.

PART 4 SOCIAL DEVELOPMENT GUIDELINES

Mike Wells

Part Four SOCIAL DEVELOPMENT GUIDELINES

4

Section 1   Strategies of Social Development

Introduction

To have real impact on the development process the funding agency should not confine itself to supporting programmes with narrow economic objectives. The main aim must be the development of people, enabling them to lead fuller and more satisfying lives, rather than simply the delivery of material benefits. This approach is normally referred to as 'social development'. The term is common currency in the language of development studies, and most development projects could be presented as having certain 'social development' objectives. However, although the term is widely used, there is no one generally accepted statement of interpretation; just as there is much debate about what in practice is meant by phrases such as 'the development of people' and considerable disagreement as to what the process of enablement really entails.

Social development encompasses a very broad concern for the development of people, for an improvement in people's lives and for giving people some control over the forces, and some involvement in the decisions, which affect their lives. Field staff should encourage the people with whom they work to make their own decisions about programme objectives. The aim of both donor and operational agencies should be to help people to choose their own destiny and handle their own development.

Two very broad understandings of social development can be identified, both of which are important in the work of Oxfam and other NGOs. Although these approaches are not mutually exclusive, nor are they necessarily coherent models to be pursued.

a: The removal of material barriers to people's development, such as un/underemployment, ill-health and lack of education. In this understanding, social development is seen as the provision of services where they are lacking. Oxfam, therefore, supports both the government and non-government agencies' initiatives in providing an infrastructure of social services to those who previously had no access to such services.

b: Social development as an educational process designed to stimulate a greater self-determination and a greater awareness among people of the possibilities for development. This type of social development takes many forms and is not necessarily linked to a project which seeks some kind of material change. It stresses the need, first and foremost, for previously excluded groups to come to understand the causes of their exclusion and to consider forms of action which might overcome it.

Thus, the objectives of social development programmes may include:
- [ ] an increase in participation;
- [ ] the strengthening of communal organisation/solidarity;
- [ ] the reduction of dependence;
- [ ] the development of initiative/motivation/leadership;
- [ ] the heightening of critical awareness.

Social development programmes have historically been divided into two categories, *official* and *unofficial*. Official, or government-sponsored movements, are usually based on one of two strategies, Community Development and *Animation Rurale*. It is unlikely that the NGO would support an official CD or AR programme, as this would normally form an integral part of government administrative structure. It is more probable that NGOs would be asked to support projects which employ CD or AR techniques at the local level. Oxfam, though, has, on the whole, been more closely associated with the growth of unofficial programmes, referred to in this Handbook for convenience under the generic title 'Social Education' (but also known as 'Participative Education', 'Conscientisation', or 'Awareness Creation').

There are fundamental differences in ideology, basic assumptions and in the field methods employed by government and non-government schemes, although on occasion the two may share some features depending on particular circumstances. In order to make the major characteristics, advantages and drawbacks of the two main schools of social development a little clearer, they will be examined in more detail in the following sub-sections. (For further guidelines on the assessment of social development programmes see PART THREE, Section 4, IV, Evaluating Social Development; and for a more general discussion of the priorities in Oxfam's overseas work see PART ONE, Sections 2 and 3.)

II   Community Development and *Animation Rurale*

1 Background

Both Community Development (CD) and Animation Rurale (AR) programmes were elaborated by colonial powers, British and French respectively, to mobilise the labour of rural and urban communities in support of national government objectives to build social and physical infrastructures, and increase 'self-reliance'. The major difference between CD and AR is perhaps that the former was based on the British tradition of indirect rule, encouraging African local government as a preparation for post-independence, while the latter reflects more the French centralised policy of direct rule for implementing national plans. CD was the strategy of rural development of the English-speaking East African colonies in the 1940s, while AR was introduced into West Africa in the 1950s by the French colonial authorities. In the 1960s independent nations in West Africa adopted AR as the basic strategy for developing impoverished rural areas. CD has become widely implemented, the best documented being the Indian Community Development Programme. In recent years CD strategies have been adopted by governments with widely differing ideological bases, from socialist Tanzania, China and Cuba to capitalist Taiwan and South Korea. CD has even been explicitly incorporated into the development programmes of less redistributive and more overtly and aggressively capitalist nations such as Brazil, as a means of encouraging local self-reliance. For reasons which are explained in greater detail below, self-help strategies of this kind, implemented by non-socialist or non-redistributive governments, are unlikely to be successful in meeting their objectives. This is to some extent reflected in the waning of enthusiasm for CD and AR since the 1960s and 70s, and in the accompanying cutbacks in official aid for this purpose.

2 Methodology

CD and AR have been used mostly in rural communities, primarily as a method for educating adults to participate in economic development programmes. The aim is to 'animate' or activate the rural population into modernising their living conditions, means of agricultural production and economic and political institutions. Training is given in new farming techniques, marketing, health care, sanitation, etc.

However, with both these methods it is sometimes extremely difficult to understand precisely how they can be applied in practice and to what ends. Indeed, AR and CD programmes are largely unstructured, and usually take the form of a gradual and cumulative process rather than a series of specific, clearly identifiable activities. In many instances there may not be a lot of substance in the work: the programme may merely be a means of facilitating the intervention of a more technical service,

4

with the task of the *'animateur'*/community development worker being to mobilise people before this intervention.

Notwithstanding the fairly elusive nature of the methodology, certain concrete features of these strategies can be identified and should be carefully assessed when considering project applications.

a: The focal point of the programme will be the team organising and coordinating the AR or CD service at the local level. It will be necessary to examine the functioning and composition of the team and the way in which it undertakes the AR/CD work in the villages.

b: The change agent/*animateur*/community development worker is the key element in the whole process. Wherever possible the change agent/*animateur* should be local. He/she is normally a village representative selected as a result of an intervention by the AR/CD service. He/she is trained to analyse village needs and problems, design development initiatives, stimulate people's awareness of their collective capacity for self-improvement and facilitate the activities of technical agents. Various criteria are employed in the selection of change agents/*animateurs*, including age, availability for the work, position in the community, technical ability as a farmer, degree of confidence and respect he/she commands in the village. Clearly it is extremely important to examine the role and attitudes of the change agents/*animateurs* and to ensure that they do not abuse the power bestowed upon them.

c: The process of AR/CD functions through regular contact and dialogue between members of the animation/community development team and villagers. It is crucial to establish with whom the team are mostly in contact (i.e. the village as a whole or an elite group within the community such as the elders or the council), and whether or not any meaningful dialogue takes place. It takes a great deal of time and effort to establish contact and dialogue, and a relationship must be built up with the community as a whole rather than relying on the collaboration of local leaders.

d: Many AR/CD programmes are indeed to some extent structured, with the formation of local groups or cells at the village level to give the programme a more tangible form overall. With the help of these groups, technical goals, such as the construction of irrigation channels or health clinics, can be identified.

③ Major Criticisms of CD and AR Programmes

Several major criticisms may be levelled at both CD and AR as techniques for the alleviation of poverty and the promotion of autonomous development. Although their methods and structure differ somewhat, the philosophies of CD and AR have much in common and, indeed, can be found functioning side by side. The following observations may thus be directed at both types.

a: The language employed in these programmes, namely that of inspiring 'community initiative', 'cooperation' among villagers, 'self-reliance', 'solidarity' and 'participation', is largely rhetoric. Far from being a tool for encouraging local independence, CD and AR are frequently used as a means of extending government control into remote areas and even into the cities. Governments thus try to manipulate poor groups to suit their own political and national developmental goals, distributing benefits on a discretionary basis through systems of patronage. Under such programmes, 'participation', some critics argue, becomes a euphemism for conformity to officially defined objectives, leaving poor people little say in decision-making at any level or opportunity of questioning the development priorities established by the government.

b: CD and AR have not proved capable of alleviating poverty on any significant scale because they are based on the erroneous assumption that poverty is the result of fatalism and lack of self-confidence, and that the poor must simply be 'awakened' to their true potential for conducting their own development. However, most governments have not provided their people with the means to realise this potential. In other words, CD and AR exponents have ignored such causes of poverty as lack of access to land and services, the limited capacity of the modern sector in most Third World countries to provide adequate employment opportunities, and the whole structure of social, political and economic interests. This is hardly surprising, given the origins of CD and its dissemination during the 1950s Cold War period, when it was viewed by the West as a non-revolutionary means of achieving progress and change and of avoiding internal class conflicts which might encourage the spread of Communism. However, the cases of countries such as Taiwan and South Korea have shown that structural change within a democratic political system is feasible.

c: CD and AR also ignore the fundamental divisions within rural society, and the fact that conflicts of class, caste and lineage are far too great to be overcome by village level workers or *animateurs*. In most instances, therefore, village solidarity will always be undermined by serious conflicts inherent in rural life. Furthermore, the very selection of CD and AR personnel at local level is often made along elitist lines, precluding the empathy and identity of interests necessary for effective local action. The village level worker or *animateur* becomes a representative of the government rather than of the people's interests.

d: At the organisational level, the vast new bureaucracies created to establish CD and AR programmes have usually failed to overcome inter-ministerial rivalries, making it difficult to coordinate the provision of services and support for widely dispersed communities, particularly when many of the benefits have been monopolised by influential local elites rather than going to those in most need. These same problems have emerged in more recent strategies such as the Integrated Rural Development Programmes.

CD and AR not only failed to convey significant material benefits to the poor, but in terms of genuine self-reliance and independence of decision-making they have not been able to involve people more effectively in controlling the forces which affect their lives. Perhaps the major lesson from this experience is that popular participation cannot be created from the top by governments or populist politicians, but must spring from real needs as defined by the people themselves based on solidarity of interests among and within particular groups.

Although there are many problems with traditional CD and AR schemes, it should not be assumed that all officially-sponsored movements are entirely negative. There may be occasions in which NGOs have no option but to work through government-backed projects, either because there are no suitable alternatives, because the authorities may be reform-minded and open to innovative policies, or because of the nature of the particular problems faced. A disaster such as a drought or serious flooding may well be such an instance. Where the NGO does opt to work through government institutions, however, care should be taken to ensure that the people have a voice in the decision-making process and that their real needs, as defined by the people themselves, are being met by the measures adopted. But given the constraints that cooperation with official institutions must, by definition, impose there will always be limits to the extent which CD and AR programmes will be able to question established objectives or challenge authority. If, however, thought is given to the possibility of cooperating with such government schemes, field staff can ask some basic questions and adopt guidelines which will, it is hoped, help to ensure that the needs of the poor are being catered for.

4

4 Guidelines for Participation in Official Programmes

Questions which could be asked of an official programme are listed below.

a: Is the government agency in question really concerned with defining and satisfying the pressing needs of the communities in the programme?

b: Are the village level workers/*animateurs* selected by the local population from the area itself or have they been brought in from outside? They could be from an alien, urban environment with problems very different from those of the countryside, or could even include foreign volunteers who may have no knowledge of the language or culture of the people they are supposed to be helping. On the other hand, it must be remembered that locally-recruited field workers can sometimes be a disadvantage: for example, if they are closely tied to local elites and have a vested interest in resisting change or promoting prestige projects whose main beneficiaries will not be the poorer members of the community. Their attitudes and degree of commitment to the real needs of the poor should be examined.

c: Are local landlords and merchants so strong that they are able to appropriate for themselves most of the benefits accruing from the programme, thus accentuating the degree of social inequality?

d: Does the programme offer tangible benefits and does it provide the capital and technical inputs necessary for people to improve their production levels or health standards? For example, an agricultural programme may need to provide access to a whole package of inputs, from land to credit, and improved marketing facilities, which can be difficult given the political and bureaucratic obstacles mentioned above.

e: Does the programme merely attempt to bolster political support for the government? While this is to some degree unavoidable in officially-backed schemes, NGOs looking towards cooperation in such ventures should ask whether in addition there is an attempt to provide an educational input which encourages local initiative in assessing everyday problems and devising appropriate solutions; that is, a more honestly 'self-reliant' form of development.

III Social Education

1 Background

'Social Education' is a term that has been coined to denote a new and more progressive approach to development which holds that the main factors in poverty and under-development are the distribution of assets within society and the impact of rapid economic change (see PART ONE, Sections 2 and 3). By contrast, in the traditional approach to development it was the rural peasant who was perceived to be at fault; he/she was backward and conservative and needed to be changed if development efforts were to succeed. In more recent thinking under-development is attributed to the unequal distribution of resources, to exploitation and to the local, national and international systems which perpetuate conditions of unacceptable inequality and injustice, as well as to inappropriate development strategies which fail to take into account the true circumstances of the intended beneficiaries.

Social Education places emphasis on the provision of educational and employment skills which enable the poor to formulate their own programmes for change. Priority is given to self-determination and self-reliance and increasing the opportunities of the poor to participate more fully in the development process.

Although, like the government-sponsored AR and CD strategies, Social Education places considerable importance on the accumulation of concrete benefits to improve the living standards of the poor, it goes beyond advocating purely material progress. Social Education emphasises the promotion of certain features and values, in association with material advancement, which it is believed will be more effective in stimulating the longer-term development of particular groups, enabling those in society who are impoverished and marginalised to challenge established privilege and attain their fundamental human rights.

What might be called 'non-material' features include group independence from outside manipulation and control, the encouragement of local initiative in defining problems and in devising appropriate solutions, the value of association and group solidarity as a means of achieving project goals and the promotion of a general 'critical awareness' of local, national and international reality. This last feature owes much to the work of the Brazilian educationalist, Paulo Freire, and his concept of 'conscientisation'. It is an aspect, however, which should not be approached uncritically and is discussed in greater detail below. Unlike government-sponsored self-help schemes which seek to incorporate poorer groups into the State machinery as a means of controlling them and strengthening the ruling party's political support, Social Education programmes will try to avoid being controlled by government (although some contact is usually inevitable) and should maintain an independent stance in order to preserve their freedom of action. In contrast, officially-backed community development projects tend to be far less autonomous.

Social Education, then, is based on assumptions rather different from those of traditional CD and AR development strategies. The latter assume that the State is the only legitimate representative of the people's interests and should retain control over all participants in the development process. Social Education holds that many of the adverse social, economic and environmental effects of development can be ameliorated by local-level projects which remain essentially independent of the government and may indeed put pressure on it from time to time. This entails a process of promoting both the material and non-material aspects of change mentioned above, so that, although projects will have a sound economic base, certain attitudinal and organisational prerequisites are also necessary if they are to become to any extent self-perpetuating in the long term.

The very antithesis of Social Education would be groups depending upon government handouts distributed on the basis of political patronage and involving no local participation in decision-making. In such situations people become passive recipients of charity rather than active instigators of progressive change. The aims of Social Education differ radically from those of the conventional social welfare approach, which emphasises the care of those handicapped by poverty and powerlessness rather than their participation in the development process.

The increasing involvement with projects which have social as well as economic objectives necessitates a re-evaluation of the concepts used in conventional development strategies. No longer should it be seen as sufficient to deliver services to the community as a whole when higher priority should be given to particular groups within the community — the poorest of the poor, the marginalised, the unemployed (see PART TWO, Sections 1 to 5). Development efforts must focus on those groups worst affected by rapid social change and economic growth. As societies have undergone structural transformation, become less agrarian and more industrialised or urbanised, the more vulnerable sectors have suffered correspondingly. These include, for example, smallholder farmers forced off their land by a combination of market forces, property speculation and outright violence from larger interests; landless labourers; small fishermen and artisans whose sources of raw materials and markets have been monopolised by powerful concerns;

4

and of course the burgeoning army of un/underemployed in the sprawling cities of the Third World.

Oxfam works with a wide variety of groups and types of project under the general classification of Social Education, and it is only to be expected that the precise make-up and methodology adopted by projects will vary considerably. Indeed, emphasis is placed on the individual and unique circumstances of each society. Stress is placed on small-scale projects rather than on comprehensive planning strategies, and priority is given to the *processes* by which results are achieved rather than the resultant *forms* (it is much more important to ensure that the content and continuity of educational provision, for example, are appropriate than merely to construct a school). Nevertheless, it is possible to isolate certain basic features which are common to schemes approaching development problems from this standpoint.

a: An *economic base*, usually but by no means exclusively agricultural, which attempts to improve the productive capacity of the project group. The material content is fundamental, not simply in terms of providing a tangible sign of progress for materially-deprived groups, but also as a tool for strengthening group organisation and solidarity.

b: A system of *organisation* which may be formal (e.g. a cooperative) or informal, based on traditional arrangements (e.g. a community association). (For further guidelines on the assessment of organisational capacity of development programmes see PART THREE, and for economic institutions see PART FIVE, Section 4.)

c: A continuous *educational process* (in the non-formal sense) to propagate the values of group work, local independence in decision-making, critical discussion of the factors affecting the group's situation and other non-material aspects.

This is by no means a universally applicable model but merely an indication of the major features of Social Education. An examination of Oxfam's involvement with such projects in different regions of the world shows varying interpretations and applications of this form of project. Methods and goals will clearly have to be adapted to fit local conditions. However, although the emphasis placed on economic as opposed to organisational and educational objectives may vary, they will normally all be present to some extent.

In order to illustrate the kinds of objectives typical of Social Education schemes, the two following examples have been drawn from Latin America and Asia:

a: Project for poor fishermen and women in Brazil

☐ To stimulate the self-confidence of the fishermen and women and to make them aware of their value as human beings.

☐ To help the fishermen and women appreciate the fact that they can do much to solve their problems by acting together as a group rather than individually.

☐ To assist the fishermen and women to form their own association and thus avoid the need to market their catch through exploitive intermediaries.

b: Rural development project in India

☐ To help landless labourers and small farmers organise themselves into groups.

☐ To help them to an awareness of the underlying causes of their deprivation.

☐ To make them aware, through education, of their legal rights.

☐ To undertake projects with community involvement.

For more detailed examples see Appendix I.

2 Guidelines for Participation in Social Education Projects

With projects which have only material objectives it is a relatively easy task to draw up a checklist of relevant factors and critical issues for the field-worker. However, when we are concerned with more broadly-based schemes which emphasise non-material processes in addition to more concrete and quantifiable goals, assessment becomes more problematic. Yet it is certainly possible to identify key areas which should be examined either before funding or as part of a monitoring or evaluation exercise.

a: Overall objectives. Even if it is not possible to express all the goals of a Social Education project in precise, quantifiable terms, those concerned must have a fairly clear idea of the objectives towards which they are working, e.g. the formation of an organisation, the winning of concessions, improvements in wages and working conditions, etc. These goals should be committed to paper before the scheme gets off the ground in order to provide some yardstick against which to assess later achievement. Non-material objectives should not be vague or left 'in the air', awaiting an exact definition at some future stage.

b: Project activities. Most Social Education projects will need an economic base both to provide material gains for the group as well as to facilitate group activities. These activities should be examined for their technical and economic feasibility, for, if they fail, disillusionment can easily set in to undermine the broader educational process.

c: Educational work. The educational content of the work, whether undertaken by a group member, by an intermediary agency or an outside individual, should enhance the above-mentioned values of critical awareness, local initiative and group effort. It should be based on a regular pattern of meetings in which dialogue, constructive criticism and personal opinions are encouraged. Agents should be as non-directive as realistically possible and encourage as much independent group decision and action as those involved can cope with.

d: Organisation. The community group must eventually begin to organise itself more formally as the basis for longer-term operations when, it is hoped, any intermediary bodies or funding agency will no longer participate. Field staff should be wary of local development organisations which express the rhetoric of participating without in practice allowing groups independence of thought or action, and merely perpetuate relationships of dependence under the guise of Social Education.

e: Methodology. Examine the educational techniques used, such as group meetings, workshops, training sessions, drama, use of audio-visual aids, etc. Do they encourage free dialogue and stimulate the expression of people's ideas?

3 Common Problems with the Approach

It may have become apparent to the reader that there seems to be a contradiction between the general stress laid by Oxfam and other NGOs on the autonomous development of people on the one hand, and the activities of change agents in Social Education projects on the other. This is to a large extent inevitable, and intervention by an outside agency will often be the rule rather than the exception. This is because development efforts due entirely to local initiative are very few, dispose of too few resources and lack the specialised knowledge necessary for most forms of development. Although initial requests for assistance are often locally-inspired, contacts with funding agencies and locally-based development institutions will undoubtedly influence the ways in which poor groups perceive their problems as well as the likely solutions. We cannot pretend that it will be otherwise and should avoid using an over-idealised concept of local autonomy. What NGOs and their field representatives certainly can do is to build into projects as many safeguards as possible in order to minimise their dependence upon external support, especially in the longer term. The dilemma remains. The important thing, however, is that it be recognised as such and not simply ignored. Only then can active steps be taken to reduce the inherent and potentially harmful effects resulting from intervention by external change agents.

Taken at face value, Social Education appears to offer an almost perfect answer to development for the poor; a balanced prescription for material advancement, social justice and personal fulfilment. Yet this methodology is not without its disadvantages, some more obvious than others. The danger of overdependence has already been touched upon both in this and in other sections of the Handbook.

A perhaps more practical consideration is that of demonstrating signs of progress to those with an interest in seeing fast and clearly visible results. The donating public and governments engaged in co-funding agreements with NGOs, for example, need to be persuaded that their contributions are making an impact on the lives of the poor. Yet the dual emphasis placed by Social Education upon non-material changes as well as more concrete achievements makes such assessment more complex and problematic than with more conventional projects with more easily measurable targets (see PART THREE, Section 4).

It is not expected that Social Education projects will be able to list and quantify all their achievements, but careful analysis should be able to determine in what direction they are going and with what success. Modifications in attitudes and organisational forms often take far longer than straightforward quantitative increases. The approach is more demanding on fieldworkers whose methodology must be based upon a process of mutual exchange and feedback with target groups rather than didactic, top-down issuing of instructions as in traditional development models of the 'modernisation' school of thought. It may take time before changes in values and attitudes are reflected in the form of higher production figures or better health standards.

Yet once such material progress is made, it is far more likely to be self-sustaining if backed up by a process of Social Education which has taught people to do things and to think for themselves in their own interests. This methodology can, on the other hand, produce very fast results where the poor are threatened by external forces or events which endanger their livelihood. The threat of eviction, for example, has galvanised peasant groups all over the world into militant action in defence of their lands, groups which had previously known nothing about forming associations or joining syndicates.

This illustration also demonstrates another aspect of Social Education which may give rise to controversy. Since it entails the stimulation of group independence, it can often result in latent conflict becoming open. Oppressed smallholders may take steps to prevent their lands being stolen, labourers may press for better wages or working conditions, indigenous groups may try to stop prospecting mineral companies from encroaching on their territory. Confrontation with power and authority is far more likely to occur and is indeed frequently inevitable in situations where underprivileged groups take action to protect their families and sources of income. Funding agencies and field representatives must use their experience and discretion to decide what constitutes legitimate project activity, in the best interests of project holders, and within the law of the country concerned.

Another area of debate in Social Education methodology surrounds the concept of 'conscientisation', a less than satisfactory literal translation of the Portuguese *conscientizacao*, an expression coined by the Brazilian educationalist Paulo Freire in his pioneering work in Brazil during the 1950s and early 1960s. The closest definition would be 'consciousness-raising', and it has been much bandied about by development workers as if it offered a key to the enlightenment and liberation of the world's poor. Care must, however, be taken when employing such concepts. Who, for instance, is the authority who says one form of consciousness is superior to another? If we are not careful, we are in danger of substituting one form of oppressive consciousness with another. Critics have argued that it is a pretentious concept and that the poor are far more aware of their predicament and its immediate causes than they are given credit for. NGOs and middle-class fieldworkers, the argument goes, have no monopoly over true 'consciousness'. The mistake, of course, is to place too much emphasis on attempting simply to change attitudes without providing avenues through which these changes may take place. Thus, if it is to be implemented effectively, a process of Social Education must address the problem of inegalitarian forms of distribution, and involve the generation of material benefits and organisational structures, as well as accompanying changes of attitude.

4

Section 2   Training

I   Introduction

Training should not be confused with or be a substitute for education. Training has narrower, more immediate goals, related to enabling people to acquire specific skills which are usually transferable into a work or leisure activity. Education is a much broader process of inspiring human development. Many of its outcomes are not geared to any specific functions but are an inherent part of broadening the knowledge base upon which academic, social and cultural decisions may be more clearly understood or defined. All individuals require training and education, and they are both important, though there are many grey areas. Literacy training provided the basis for the emergence of the concept of conscientisation by Freire, which from the basis of literacy leads to a whole range of analytical skills, and aims to assist in personal development. (See PART FOUR, Section 1.)

Formal education has expanded in most countries of the world in this century, and increasing numbers of children attend school. With some exceptions, national governments have seen the provision of formal education as their responsibility and dedicate resources to its provision. Organisations such as Oxfam have not seen formal education as top priority for their work, and little support has been given to schools and colleges, which are considered the responsibility of government, although the quality of such education will ultimately affect other forms of training.

When assessing any training programme it is important to establish whether or not the problems to which the programme is addressing itself can in fact be resolved through the acquisition of new skills. Many training programmes attempt to address problems which in fact result from a lack of resources, not from inadequate skills. For example, nutrition education will not reduce malnutrition if the problem is not ignorance of which food to provide but simply lack of money to purchase nutritious food.

Everyone can benefit from further training and education, but, given the shortage of resources for such activities, decisions about priorities will have to be made. An assessment of the needs of individuals and groups of people should be made in order to establish priorities for the allocation of resources for training. There will be certain groups of people that because of their disadvantaged position will have special needs, including for example handicapped people (see PART TWO).

The educational methodology should be analysed in terms of appropriateness to the subject in hand, time available, level of previous education of trainees, their age, culture, gender, etc. It is not enough to know a subject to be able to teach it. The advantages and disadvantages of mechanical or 'rote' learning techniques should be weighed against those of more participatory approaches to training, especially important where practical skills are involved and 'doing is believing'.

The nature of the institution involved in training will also have an impact on the effectiveness of the programme. Loosely structured institutions with a high degree of outreach are likely, for instance, to provide a greater degree of community participation but a lower level of intensity in their training, whereas a tightly structured residential programme is going to provide a more intensive course of study for a smaller number of people.

Most training programmes fall into two categories, related either to production or to family welfare. There is a danger that such programmes will reinforce existing stereotypes by focusing the former on men and the latter on women. Examples are to be found elsewhere in the Handbook of the harm that can originate from failing to analyse adequately the training needs of a given population (PART TWO, Section 2; PART SIX Section 10).

Training programmes can be grouped into the following types:

a: Vocational, the aim of which is to help people to find employment by providing them with marketable and transferable skills.

b: In-service training, especially that designed for the advancement of economic activities. This is aimed at people already in employment to upgrade their skills. It can also include management and other training for members of cooperatives, for example (see PART FIVE, Section 4), and in a rural context could include extension programmes (PART SIX, Section 10).

c: Life skills: this tends to be training in activities related to subsistence agriculture (including production), but in a broader sense includes training in 'Do-It-Yourself' skills such as house building, sewing, knitting, basic nutrition and health care, which both the urban and the rural poor are obliged to carry out for themselves in the absence of sufficient income to pay for them.

d: Training in literacy: this is regarded as so important and crucial to the success of many other forms of training and general education that it should be considered as a special form of its own.

e: Training trainers: included in this category are the more formal approaches such as creating a group of '*animateurs*' or health promoters, or less formally structured 'leadership training' (see PART FOUR, Section 1).

4

II   Vocational training

This seeks to provide employment possibilities, thereby creating an advantage for the minority who receive the training. As a general rule, given the relative privilege provided by vocational training, it should be aimed at those with the greatest need. Hence vocational training programmes funded by the NGO should take in those from the poorest families and those least likely to find employment through other channels. A policy of positive discrimination should also work in favour of training for those with a physical or mental handicap.

Vocational training programmes should:

☐ be cost effective;

☐ avoid lengthy or expensive training;

☐ be in accord with local employment possibilities; too many schemes produce people with identical and not very marketable skills, e.g. carpenters and seamstresses;

☐ provide not only technical training but also an understanding of the management and accounting skills required, whether cooperative or self-employment is expected for the student;

☐ provide the skills technically appropriate to the local economy. Many schemes unfortunately use elaborate imported technology (often donated) which bears little relation to the technology to be found in local places of work;

☐ not be so long as to deter the poor. Scholarships might allow poor people to attend a course, as might an apprenticeship scheme in which students earn as they learn;

☐ be cautious of the tendency to over-estimate the full-time employment possibilities for artisans, especially in rural areas.

☐ indicate the sort of work rhythms and routines to be found in the work place. Too many trainers in training centres do not have practical work experience themselves;

☐ be wary of patronising and demoralising schemes aimed merely at keeping people off the street, rather than providing them with a genuinely marketable skill.

In a rural context it must be established whether there is a real need for advanced agricultural training, or whether there are local employment possibilities. The risk is that students will often use their training merely as a path to an urban job. In many schemes it is assumed that vocational agricultural training will help to keep young people in the countryside, whereas the opposite tends to be the case. There may be a great need for agricultural training, but at the farmer level. (PART SIX, Section 10).

III	In-service training

Many of the points made with reference to vocational training are relevant to in-service technical training. These programmes are aimed at those already working in defined areas, upgrading existing skills rather than introducing new ones.

Programmes will vary considerably, depending on whether people work for cooperative organisations, are self-employed or work in formal employment for large enterprises. Guidelines for training programmes within cooperatives are provided in PART FIVE, Section 4. Training for the self-employed may well include the improvement of management and accounting skills to enable the expansion of the small business. Training programmes for those formally employed should seek company sponsorship. Some countries have schemes paid for by firms either informally through levies or formally through taxation, although these are not normally the concern of NGOs. A few programmes will serve a mixed working population and may require external support.

It is important that in-service programmes should take into account existing work commitments, be in line with the needs of those working, and where possible provide some form of certificate and not be at variance with the national education programme. They should also provide an incentive to the recipient of the programme to take full advantage of the opportunity.

IV	Life skills

People living in a subsistence or semi-subsistence economy need to develop skills to provide certain basic services for themselves. Such 'life skills' tend to be confused with the training and extension work related to agricultural production. The confusion is understandable, as there is but a fine line distinguishing semi-subsistence farming from the everyday ancillary tasks undertaken by most poor people. The training in life skills will vary according to circumstances and may include everything from construction of buildings to maintenance of simple machinery (including hand-pumps, bicycles and the like). Such training will also include health-related subjects such as nutrition and child health.

By themselves, short-term life skills programmes may be frustrating for the participants, as they tend only to cover limited themes and fail to provide the capacity or potential to tackle the more profound problems. Life skills have their place in training, but most probably should be linked to general education, extension or other programmes rather than standing alone.

Over 50% of the adults in many developing countries are illiterate. Despite substantial reductions in the rate of illiteracy in some areas, the total number of illiterates, especially women, is still growing due to the increase in population.

Undoubtedly, the long-term solution to this problem is the expansion and extension of schooling to all sections of the population. However in the short term tailor-made adult literacy programmes are required to reach those who have had little or no schooling but who are currently productive. In this way more immediate results can be attained.

1 Approaches to Literacy Education

a: Functional literacy. This approach is based on the notion that literacy is best learned together with skill training for specific kinds of employment. Thus skills are learned through literacy education. This 'work-orientated' approach has tended to concentrate its efforts, therefore, on specific economic growth areas or development projects, rather than at national level.

b: Organisation. A reliable organisation is essential, supported by an adequate budget to prepare or acquire and distribute materials, train and pay instructors, and support learning groups with supplementary materials and activities. The organisation should be able to support a literacy programme for at least five years.

Literacy programmes tend to have greater impact when linked to national programmes of general social transformation, as in the mass campaigns. However, where such conditions do not exist, a well organised conscientisation or functional programme can be most effective at community level, providing the conditions set out above are met.

Continuity in literacy work is essential, but local volunteers, while useful in the short-term, have frequently proved unreliable in terms of attendance. However, in the mass campaigns where the climate is one of commitment to rapid, grass-roots social transformation, trained volunteers have been very effective. The general point is that literacy workers need to be well trained and well motivated whether by ideological fervour or by material incentives or both.

c: Propaganda and publicity for literacy campaigns have been generally noted as key ingredients of successful literacy work. Enthusiasm and interest in such work needs to be effectively communicated to and by the public to create an atmosphere of involvement in this crucial area of human development. (See Nicaragua (NIC 22A) and Grenada (GRE 14).)

d: Language. Functional literacy may be taught in the vernacular and may be achieved in about 300 hours, though further study should be encouraged. Where, however, the mother tongue is an unwritten vernacular or where teaching and reading materials are lacking in the vernacular, and instruction is in a second (for example, official or national) language, then the programme must be viewed as language teaching. Consequently, the time and effort involved must be multiplied perhaps tenfold.

e: Teaching methods and aids. Imaginative teaching methods and aids, while important in all aspects of education, are especially important with literacy. Students may be approaching formal learning for the first time and may be cautious and inclined to be easily bored and discouraged. Thus learning must be made attractive and worthwhile. Furthermore, literacy is not just an end in itself, but a key to other development and education activities. It therefore needs to be taught in such a way as to stimulate interest in moving on to such activities.

4

Although the three different approaches sketched above tend to favour different and often contradictory methods and aids (thus, for example, the mass campaign favours primers, to which conscientisation is opposed), there are some general guidelines which can be derived from the experience of all three approaches.

☐ Literacy learning, in order to make sense to learners, must be based on what people see as their needs. Such needs may range from very simple skills like letter writing and basic arithmetic to the skills necessary to form local welfare or political associations, with the literacy this will require.

☐ Aids and materials must be thoroughly pre-tested, and the instructors who use them must be adequately trained. These two processes can ideally be carried out at one and the same time with trained instructors being given teaching practice by doing the pre-testing, under guidance, among the local people.

☐ As much as possible, teaching methods should be student- rather than teacher-centred, with the teacher guiding the students through their learning rather than lecturing at them. Group work with students helping each other and self-learning activities, possibly through modules, should be encouraged.

☐ Good audio-visual aids can easily lose their value through dictatorial and verbose handling by teachers. They should be used to stimulate student activity and participation.

f: Costs and quality. A policy of quality rather than quantity, although expensive, is likely to be most cost-effective. Short, extensive, but poorly run and unprofessionally taught literacy programmes are rarely effective. There is no substitute for high quality teaching and good organisation and publicity, even though they are expensive.

The best guide to cost levels is to ascertain the cost of six years of primary schooling for a child in the country concerned. Per capita costs of adult literacy programmes should be slightly less than this.

2 **Oxfam's Involvement**

Field staff should consider supporting adult literacy programmes which are related to other development activities such as agriculture, rural industries, health and home economics (although such programmes should not be narrowly functional); also programmes which are concerned with social and political development. Many conscientisation programmes are relevant here. However, such programmes must be studied carefully to make sure they are not leading people into dangerous and futile conflict with powerful groups without due consideration of the consequences.

In addition to the above programmes, support should be given to mass campaigns, where the advance of literacy is perhaps the most rapid of all the programmes (See Nicaragua (NIC 22A)).

The following kinds of literacy programme activity should normally receive priority:

☐ the production of *teaching aids*, e.g. books and visual aids, charts, word and letter cards, to start a programme or to replace obsolete materials in an on-going programme;

☐ the training of teachers or study group leaders, including re-training;

☐ the production of *reading material*, relevant for adults, at different levels of readability. This again should be a continuous process.

☐ the reproduction of reading materials in collaboration with the production staff;

☐ *media programmes*, e.g. radio or film production to support and reinforce the literacy programmes as above;

☐ in Peru a *rural libraries' scheme* has given peasants access to useful and interesting books and pamphlets, thereby reinforcing literacy through continual use of the skill. (See Checklist of Questions.)

The multiplier effect of training trainers should be high, assuming that the trainers are used in programmes for the benefit of large numbers rather than small exclusive groups.

There will be as many types of trainers as training programmes, and it is difficult to generalise about them. There will be differences in programmes designed to assist people to work in strictly formal programmes including salaried extension agents, classroom teachers, and those working in the community as 'promoters', 'animateurs' and the like. Despite the different implications of 'bottom-up' and 'top-down' development, the sort of skills required by 'leaders' and 'staff' are similar. In other words, leadership training can learn from staff training and vice versa. Leaders need to be not only sensitive, trusted and trustworthy representatives of their community, but also skilled managers and administrators. A good trainer should have the ability to encourage and motivate people by conveying a genuine enthusiasm and, by communicating it, to draw their ideas out.

A great deal has been written about the qualities desired in a trainer; some seem to expect trainers to be superhuman beings. Unrealistic expectations of trainers should be avoided, in terms of both their personal attributes and the amount of work they can do after their training. If no natural candidates exist for training, the selection process must at least reflect the feeling of the local community. A candidate who is well qualified technically may not necessarily be acceptable to the people with whom he/she is expected to work. In particular it is necessary to be wary of local interest groups trying to decide the choice of candidate for their own purposes. Basic requirements for a candidate will include local standing and acceptability, sympathy, patience, sensitivity, ability to relate to people and to teach. A trainer with practical experience will carry more authority with students than one without. Many programmes accept only older people with families as trainers because they are less likely to leave the area or see the training as a means of career advancement.

The methodology used for training trainers should reflect the method which in turn the trainees are expected to use in their own work. Refresher courses and follow-up sessions should be included in all training programmes. This will allow problems to be resolved, new ideas to be incorporated and the trainees to receive support and confidence that they are not alone in their work.

4

Section 3   Communication

I

Introduction

Every social development project involves communication: communication between members of a community or group; communication between development workers and the people among whom they work; communication between funding agencies and field workers, and so on. Communication involves the sharing of ideas — of information, of emotions, of hopes and fears — and is part of every human activity.

This section is concerned with communication for the purpose of change, focusing upon the interaction betweeen those promoting change and the people with whom they work, and the use and role of communication in the context of the kinds of project Oxfam and other NGOs support. We only touch briefly upon interpersonal communication between villagers or between rural and urban dwellers, or communication within and between organisations.

Certain projects may have a separate communication element built in. This will usually involve funding for equipment or expertise to facilitate communication for a particular purpose: for example, a newspaper for people who have completed a literacy programme; audio-visual equipment for a training centre; the organisation of a workshop on community drama for development workers. In some cases, the NGO may be asked to fund a communication component of a project, other parts of which are being funded from other sources. But however the project is presented, communication activities should always be seen in the context within which they take place: they cannot be treated in isolation, as ends in themselves.

II

Communication channels

There is a wide variety of channels of communication currently used in social development projects. These range from 'modern' mass media, such as television (which can be broadcast from land-based stations, via satellites, or transmitted by cable) to traditional channels such as song, dance and drama. In between comes the whole range of printed material — newspapers, posters, leaflets, etc. — film and photographs, radio and audio-cassettes. A more recent addition, which is now being used imaginatively in development projects, is video.

Before including a communication component in a social development project, it is important to see what media, and media production facilities, are available in the area. There are three reasons for this.

a: It may be possible to make use of existing facilities without spending money on new ones. For example:

☐ *universities and colleges* may have audio-cassette copiers which they are prepared to make available for occasional use;

☐ *radio stations* (whether government-controlled or commercial stations) can be used to spread information over a wide area;

☐ *printing facilities* — ranging from simple duplicators to offset litho — are now widely available in the Third World.

b: When purchasing equipment, it is vital to make sure that it can be serviced and repaired within the country. Many projects have found themselves provided by donors with audio-visual equipment of a make for which there are no spares or repair facilities within several thousand miles.

c: It is important to know the channels of communication from which people are *used* to getting their information. Many modern media forms may be inappropriate for use with people who are not familiar with sophisticated methods of communication.

III   Communication strategies

Communication can be of importance in development for a number of reasons.

1 Information

The most obvious and commonest role of communication is to *provide information*. The purpose of providing information is usually to enable and encourage people to take action. The information must therefore be relevant, in the sense that it must be precisely the information that people need if they are to take the intended action. This may be information on the nature of a particular problem, the connection between wet ground and hookworm, for example, or on possibilities for action, such as the idea of constructing fire-breaks to protect homes and crops in areas prone to bush fires. The process of designing communication materials must therefore include an element of research: the current *knowledge*, *attitudes* and *practices* of the intended audience with respect to the topic under consideration must be assessed before decisions can be taken as to what information to provide.

It is essential to remember that the information provided by a project represents only a small fraction of the communications received by the people — communication that attempts to inform, reassure, manipulate or give pleasure. Communications come from many sources, from family and friends (and often from gods, spirits and ancestors), and from outside the community through radio, newspapers, television and films. All this is in addition to, and perhaps in conflict with, the contribution made by the development programme.

Whatever channel is used, the information must be presented in an attractive and comprehensible way. Radio programmes, posters, leaflets, video tapes, film-strips — these must all catch people's attention quickly, and then hold it: otherwise little information will be communicated. But they must also be comprehensible. Dissemination of information may be impeded by:

☐ the use of graphic conventions that are unfamiliar to the intended audience, e.g. arrows to indicate sequence, cross-section of drawings, disembodied heads and hands;

☐ the use of words and concepts that are not normally used by people in the area;

☐ the packing of too much information into a single programme.

It is therefore important to *pre-test* information materials wherever possible. This can be done by presenting the materials, in draft form, to a representative sample of the intended audience and then finding out from them whether they have understood it in the way the designers hoped they would. The information we put into such materials, and the information other people get from them, may differ considerably.

Information is more likely to be relevant, attractive and comprehensible when materials are designed and produced *locally*. Audio-cassettes, locally designed printed material, and the skills of local artists are useful in this context. Materials which rely on distant skills and technology are less likely to be suitable for local circumstances. But even locally produced materials should be pre-tested before the expense of reproducing and distributing them is incurred.

Information channels may be used to extend the impact of a single development worker, by bringing more people into contact with the

4

information than the worker can meet face to face. Or they may be used to complement or reinforce face to face communication. Flipcharts, instructional posters, leaflets that can be left behind after a visit, slide-tape programmes which illustrate a process or situation which is difficult to portray by the spoken word alone can all enhance face to face communication.

2 Discussion and Dialogue

Communication channels in development projects are increasingly being used to promote *discussion and dialogue*. This may be as part of a social education/conscientisation strategy, where the role of the communication medium is to contribute to the process of analysis of local problems, which should lead to an awareness of the structural conditions underlying them, then to discussion and, finally, action.

a: Drama is a powerful tool here: by dramatising a local situation — such as exploitation by money-lenders or oppression faced by low-caste groups — community workers and community groups can spark off a debate on issues that are normally simply accepted as a fact of life. At its most imaginative, drama has been used not only to present an issue but, by encouraging the 'audience' to direct the plot and even to join in, to explore the likely effects of alternative courses of action. Growing numbers of development workers are gaining experience in these uses of traditional and modern dramatic forms.

b: Video is also a useful, if expensive, tool in this context. Because it offers instant playback and re-filming (unlike cinematic film), a video tape exploring a particular local issue can be made by (or under the direction of) community groups and then used to generate discussion within the community. Video is a medium which requires no sophisticated skills and which therefore enables local people to express their point of view without relying on the professional skills of an outsider. Although such media can be used to present local views to the outside world, it is the *process* of producing the medium rather than the finished *product* that is usually considered important. The awareness, the analytical ability, the group cohesiveness, the self-confidence of those involved are all likely to be heightened by the process. It is important, however, to ensure that maintenance and repair facilities are locally available.

c: Drawings and photographs — within Freirian approaches to literacy training, these are frequently used as 'generative codes' which illustrate key words of local significance. The issues surrounding the word are discussed by the literacy group, after which the word itself is analysed into its component parts, which are then used to build up other new words.

As well as being an essential component of social education, discussion may also be a prerequisite for taking action in response to information from a project or agency. It is unrealistic to expect people to try a new activity just because someone on the radio said it was a good idea. The advantages and disadvantages have to be thrashed out; implications have to be explored and questions have to be answered. And so communication activities which include provision for discussion groups and for the feeding back of questions to the project or agency concerned are usually more effective than those that rely on a simple one-way flow of information.

3 The Sharing of Ideas and Experience between Communities

This is another valuable use of communication channels. Slide-tape or video programmes describing one community's response to a particular problem can be used to encourage others to tackle similar situations in their own locality. In a project in Ecuador (ECU 39), voluntary workers use simple tape recorders to produce programme items for regular agricultural radio broadcasts. With a small outlay and a few hours training, these broadcasts have been transformed from one-way, urban-based, channels of communication to a forum in which experiences can be shared by people who, in all senses, speak the same language.

4 Education and Training

Finally, most of the media referred to so far have been extensively used in education and training programmes, both formal and non-formal. Radio is used as the basis of many adult education projects in Latin America. Video, both in training centres and in mobile units, is a valuable medium for skill training: it can be used not only to show how a particular operation should be carried out, but also to record trainees' attempts with immediate playback and analysis. The development of lighter, more durable equipment, that will run on portable battery packs, is increasing the potential and flexibility of video. Most non-formal education projects rely heavily on printed materials of various kinds to reinforce and supplement the contributions of both development workers and radio broadcasts. *But as in all other applications, the media are merely tools: a communication project, or project component, is only as good as the social development strategy within which it takes place* (see Section 2 above).

IV   Function of communication

The purpose of communication in a development project input will almost inevitably be *to influence decisions that the people will make, decisions which often lead to changes in behaviour*.

People's use of, and need for, information differs according to the stage they have reached in the adoption process — the sequence of stages in mental and physical activity that leads to the adoption of new behaviour. This is briefly described below as straightforward and linear. In reality it is much more complex with many internal circulatory movements which it is hoped lead to a general progression forward.

☐ Awareness
☐ Interest
☐ Attitude change
☐ Behaviour change
☐ Consolidation into normal practices.

1 Awareness

Bringing new ideas to people is often seen as the major, and sometimes the sole, purpose of communication, but it is just the start of a sequence of communication activities.

Any communication that reaches an individual can create awareness, but only if it is attended to. Although everyone is constantly bombarded with information through all five senses, most is ignored or only superficially appraised. Only a tiny proportion of information available is selected for close attention. If our audience is not aware of the idea being promoted, it is unlikely that they will be interested in the topic, yet we have to get them to select our particular input from the mass of information, and give it their consideration. In advertisers' jargon it needs to have 'impact'. Three methods usually succeed in practice:

☐ to link the message to some topic in which the target group is interested and which it is known will be selected for attention;

☐ transmit it through some channel which will receive attention, e.g. through drama or radio serials (soap operas);

☐ to encourage the target groups to attend to the message through face to face discussion either on a one-to-one basis, or through group dialogue.

② Interest

Once an individual is aware of the idea, and seeds of doubt are sown concerning the appropriateness of his/her present activities (or lack of activity) he/she changes from being a relatively passive 'receiver of information' to an active 'seeker of information': information to clarify understanding, and seek reassurance (either reassurance that the information is wrong and can safely be ignored, or reassurance that the decisions his/her search seems to be leading toward will be the right ones and be locally accepted). The end result of this search depends very much on the knowledge and attitudes of those from whom he/she seeks information. If people know nothing about the new idea, or are not aware of its potential — or are frightened by the risk — the information they give will probably count against the new idea and simply reinforce and confirm the old behaviour.

This is most difficult to influence as the decision-maker will seek information from anyone on anything he/she feels might be useful. He/she might not even be seeking information without bias but may be seeking information to confirm that he/she can safely ignore this new information. In this context one of two communication strategies (or a combination of both) can be employed.

a: The whole community can be made aware of the idea and its potential through mass campaigns, radio serials (soap operas), village theatre or demonstrations. Before doing this it has to be confirmed that the community will appreciate the benefit of the proposal, that it does not conflict with present practices, knowledge, attitudes or values. The difficulty is that very seldom does an idea benefit the whole community, as some will inevitably see it as threatening. Will these people be included in those from whom the receiver seeks information and advice?

b: The number of people available to seek advice from can be temporarily reduced by introducing the topic to a discussion group. The topic may be a problem they are facing, or are likely to face, or only be a possible prescription for action. In either case dialogue within the group will help the articulation and clarification of ideas, definition of alternative solutions, and the consideration of possible scenarios of each if adopted.

When recommendations are to be discussed, one or more members or the extension worker may be briefed beforehand. The whole group explores the topic solely on the basis of past experience and logic, without any specific briefing, or, perhaps, as a result of a visit to a community practising the idea, or a demonstration.

One of the strengths of group work results from members seeking information from other members; people who know a great deal about the new idea and its potential, who have thought deeply and seriously about it, and clarified their ideas through discussion.

③ Attitude Change

All the foregoing activity is aimed at reaching decisions which together lead to the acceptance or rejection of the idea. If the idea is accepted, the receiver is assumed to have changed his/her attitudes. In practice, attitudes are probably changing throughout the process from 'It seems a good idea', to 'OK I'll give it a try', to 'Yes, it is good' (which will probably not occur until after the consolidation stage). This implies that it is easier for people to accept an idea at this stage if it can be tried out on a small scale with little risk.

After accepting that the new idea will be adopted or given a trial, two types of information are sought:

☐ confirmation and reassurance that the right decision has been made; and

☐ detailed information important for the actual implementation of the new idea, and training in the skills required for it.

Both these are best achieved through one-to-one extension methods. But visits to people already using the new idea are again useful, or talks (whether to a group, by tape recorder or over the radio) by people who have already adopted the idea or are trying it out. In this case they should describe in their talks what they did, what problems they encountered (and why) and how they overcame these problems.

4 Confirmation

After the activity has been put into practice, information is sought for reassurance and reinforcement. Personal assistance will also be needed to polish skills, to ensure that the new idea, and its implications, are completely understood, and that it fits the individual's unique circumstances.

4

Human rights

Oxfam is committed to the relief of suffering according to the needs of people overseas. We are concerned, however, with the causes, as well as the effects, of poverty. This inevitably leads us to support programmes whose objectives may include the attainment and improvement of basic human rights, whether economic and social, or political and civil, as defined in the UN Declaration of Human Rights, the denial or absence of which causes the suffering we aim to relieve.

In many areas of the world the inequitable distribution of resources is patently seen to be detrimental to the health and well-being of the poor. Therefore to meet people's basic needs for water, food, health and employment, Oxfam and other NGOs seek to increase resources through programmes facilitating greater levels of production of food, and the delivery of material benefits and services.

Through its developmental and humanitarian work, Oxfam seeks to help poor people to bring basic human rights within their reach. The methods used to achieve this are as varied as the problems encountered.

Work devoted to improving the conditions of the poor — and helping them to take effective steps to that end — will often raise issues of human rights. The leader of a village cooperative may, for example, face obstruction from corrupt or inefficient Government officials. His/her criticism of these officials might land him/her in prison. Legal aid to his/her family may secure his/her release and thereby strengthen the position of those in a similar situation.

It is not Oxfam's prime responsibility to seek full political or civil rights for people, but the organisation is increasingly unable to ignore the problems caused by the absence of respect for those rights, by conflicts of interest, civil war or armed strife. Arbitrary acts of violence, including torture and assassination, perpetrated in such circumstances oblige the NGO on humanitarian grounds, to offer help wherever possible. As a humanitarian movement of concerned people, with over four decades of experience in some of the most disturbed and troubled communities around the world, Oxfam must, and does, condemn the denial of basic human rights, not least when such rights are denied to people living in acute poverty and deprivation.

The defence and enhancement of human rights will be an important secondary objective of Oxfam's overseas work, particularly in emergency situations and in especially troubled parts of the world. The work Oxfam supports, although it may be described in terms of political or civil rights, takes the form of humanitarian aid to meet basic human needs; examples include assistance to families of political prisoners, food, shelter, medical treatment, water, supplies and, where necessary, legal aid and similar work.

In some cases, the decision may be taken to channel and monitor support through the only effective group working in a certain area, so as to alleviate the direct suffering of the dispossessed through medical or other humanitarian assistance, even though the particular group may have an explicit political objective to change certain policies in their country. That is not really our concern. In the same way, the fact that Oxfam collaborates on particular projects with a government does not imply agreement with all its policies.

The moral dilemma Oxfam may face in such situations of conflict, where human suffering is immense and widespread, is exacerbated when the organisation concerned uses violence to further its end. However, to quote from Oxfam's policy statement regarding the response to need in the context of human conflict and political repression:

"We reaffirm our belief in non-violence and deplore bloodshed, but we do not accept that a grant to, or at the request of, such an organisation (for humanitarian work) would of itself constitute support for all or indeed any of its objectives or methods ... Nor should we overlook the fact that failure to respond to human suffering may also be interpreted as a political decision."

The last point has been amply illustrated by the refusal of many international agencies to respond to need in areas of acutte political conflict, and conversely has borne out the justification of Oxfam's involvement in such countries. It is thus important to reiterate that Oxfam is committed in such circumstances to attempt to maintain a non-partisan policy of support to people in need. Over the years there are many examples of Oxfam working to meet human need irrespective of the political persuasion of a people, or of the government under which they live, and that will continue to be a policy objective. (See also PARTS ONE and TWO.)

II   Legal aid

1 Introduction

Legal aid is becoming increasingly important in the work of aid agencies. It has been identified as one of the most effective forms of assistance that can be given in areas where poverty and suffering result to a great extent from prevailing social and political injustices. In most countries there is a rule of law and a system of redress, but the problem is frequently one of *access* and *implementation*. The judicial and legal systems may appear to be satisfactory, but the poor are rarely in a position to use them, neither knowing of their existence nor having the resources or understanding to benefit from them.

Most poor people in the Third World are illiterate and completely cut off from the sources of learning and information concerning their entitlements. In such circumstances *knowledge becomes a basic need and a fundamental right, and lack of it can be a prime cause of poverty*. Many people will seek to take advantage of those who are unaware of their rights. For example, powerful landowners will make illegal demands of their tenants and moneylenders will encourage their clients into ever-increasing debt bondage. This is where the NGO can be of assistance, helping to redress violations of legal rights, providing information and equipping people to secure their entitlements — which in most cases are provided by existing laws. The success of a programme will not depend primarily on the number of favourable judgements in court but more on the confidence it fosters among the poor in using the national legal and bureaucratic systems to obtain justice.

However, it is vital that support for legal aid does not compromise the non-political and humanitarian aims of charitable organisations such as Oxfam. Even though it may be necessary to challenge the status quo in the struggle against poverty and deprivation, Oxfam, as other NGOs, stands apart from any overt political stance and always works within the rule of law in any country in which it is represented. Because of the delicate nature of legal aid programmes, they must be subject to greater scrutiny, monitoring and evaluation than other programmes.

2 Types of Programmes

The legal aid programmes funded by Oxfam generally fall into two broad categories:

a: Preventive/promotional work. This type of programme receives the highest priority and aims to assist organised groups or communities, educating them in their legal rights so that others will find it harder to exploit and violate them, and they will be able as far as possible to provide their own protection. The emphasis of this type of programme will be to avoid court action although legal precedents will be sought through test cases where necessary. The main emphasis, then, is on building awareness through education and counselling rather than pursuing cases in the courts. Educational work will include publishing simple guides to the law, training programmes, radio, and other popular means.

*b: **Curative work**. This type of programme is becoming less of a priority. It is more often than not directed at specific disputes and aims to redress individual cases of injustice by working with lawyers, both in and out of the courts.

These two main types can be further sub-divided according to their focus into:

i: General programmes. This type of work concentrates on the provision of advice and information on the civil and human rights of citizens. In many countries counselling of this sort will require the services of a lawyer or social workers to orientate people on the approach needed when dealing with public institutions, etc. The services commonly included cover general legal counselling in:

- [ ] civil law (obtaining legal documents and certificates etc.);
- [ ] social law (obtaining old age and/or sickness benefits, informing as to rights/obligations in debt bondage etc.);
- [ ] community law (legal assistance with authority procedures);
- [ ] labour law (labour injustices, union rights, etc.);
- [ ] family law (obtaining marriage and birth certificates, helping with divorces and establishing maintenance).

Although these more generalised programmes have considerable educational value, frequently too many resources can be taken up in addressing minor problems.

ii: Task-directed programmes. These often have a fairly dramatic impact. This type of programme offers a specialist service addressed to a specific problem/issue such as obtaining security of tenure or improving work conditions for a given group of people. Perhaps the one disadvantage is that task-directed programmes can become very narrow and inflexible, leaving aside many of the important educational elements of a general service.

iii: Group-directed programmes concentrate on a particular disadvantaged group requiring special attention, such as prisoners, disabled workers of indigenous peoples. This type of programme has the dual purpose of informing and educating public opinion with regard to the special problems and needs of this group and of providing legal aid to assist in redressing violations of rights.

3 Principles and Procedures of Legal Aid

Many poor people will be unable to afford to pay for legal advice; others often spend large sums they can ill afford on court cases. However, whenever possible legal aid programmes should *concentrate on supporting groups rather than individuals*. Efforts to supply legal aid on an individual basis — especially in disputes over land and other resources — are likely to be frustrated by the sheer numbers of people requiring help and the shortage of funds and lawyers. It is through collective efforts to establish legal resources within a community or group that the proper economies of scale and remedies may be developed. A more generalised and educational approach will produce less paternalistic and longer-lasting effects, giving people the confidence to pursue their legal rights.

The aim should always be to *avoid court cases*, since court procedures are extremely costly and may cause serious delays. On the whole the use of the court is only to be recommended when the issue at stake has wider implications, and the aim is to bring before the authorities a test case to establish a legal precedent. A favourable judgement may set to right the bias against the poor, discouraging the use of the courts as a weapon against them. Furthermore, a successful outcome could encourage other groups, similarly placed, to seek to secure their legal rights. To this extent, individual projects channelled through the court system may provide the first step in the development of a broader programme of legal assistance and education.

In many instances there may be a need for project holders to press for the implementation or repeal of legislation for the benefit of the poor and disadvantaged. *Lobbying, leaflets, the media and appeals to international bodies* have been used, and widespread campaigns have in the past prompted the authorities to reconsider a particular stance and rendered them generally more tolerant. It has been noted that not simply the legal possession of their land or other legal requirements are needed by the poor, but also credit and other inputs to work it, which shows the need for legal aid projects to be more integrated into a general development programme. Some legal aid schemes have formed part of an adult education programme, others have been included as a component of a more general programme of assistance involving technical, educational and administrative services, and this appears to have no detrimental effect on the quality of the legal aid received. A legal aid scheme should be a part of the general development programme and not be treated as an isolated service.

4 Community Outreach

A vital aspect of most legal aid programmes will be community involvement. It may be useful to train para-legal workers to assist lawyers, act as a referral service and a focus for educational work. *But it must be remembered that para-legal workers are not protected by professional status*, and therefore can be vulnerable to repression. Secondly, para-legal workers should not seek to replace traditional forums for resolving local disputes where these seem to be meeting local needs; rather they should be acting as a channel for those cases which cannot be handled locally. If the community acts together over a legal dispute, there is a far greater chance that it will be resolved in the communal interest than if they merely await the action of the lawyers working on their behalf. Communal support will also be important in obtaining implementation of the law, thus avoiding endlessly repetitious court appearances on the same issue.

As people become aware of their legal rights, and as the courts come to accept that poor people will call on these rights, a successful legal aid programme should slowly make itself redundant. Where there is already a State-aided legal aid programme with the potential to assist the poor, private legal aid programmes should seek to ensure that the government programme becomes more responsive to the needs of the poor rather than enter into competition with it. A legal aid programme which continues to run for many years has already demonstrated a degree of failure or government intransigence.

Section 5   Checklist of Questions

I   Literacy programmes

If the answers to the questions below are largely negative, it is likely that a literacy programme will not have useful and permanent results.

1. Does the potential audience suffer specific disadvantages through being innumerate and illiterate?

2. Are the people aware of these disadvantages? Is motivation high?

3. What applications will there be for literacy? Do farmers, for example, need to read instructions on pesticide cans?

4. Would numeracy and literacy give people more job opportunities?

5. Are there industries or agencies in the area which could support a literacy programme? Would it help other development activities?

6. Is there a sound organisation to run the programme?

7. Is there an adequate budget?

8. Does the responsible authority have clear and realistic aims?

II   Legal aid projects

1. *Organisation*. Personnel structure: how are lawyers hired and paid? Is there any evaluation of lawyers hired on a case by case basis? What is the comparative effectiveness of the lawyers?

2. *Programme area*. What is the relationship between the project and beneficiaries? How does the location of the office affect their work? Is the community involved in the decision making of the project? Is there sufficient outreach work to ensure that the community can fully benefit from the project? Is there a clear idea how the project can help to strengthen community organisations?

3. *Case identification*. In curative projects it is important to define which sorts of case will be dealt with. Action should be taken to exclude work involving inter- or intra-community disputes over land, personal problems or inter-family conflicts. Does the project have clear criteria for case selection? Are these criteria directed towards helping the poorest?

4. *Costs*. What level of costs is acceptable for individual cases? Obviously if there are spin-off effects from a single case, then a greater cost could be justified.

5. *Other legal aid schemes*. Are there other legal aid schemes in the area, particularly government schemes? Would it be better to strengthen these schemes rather than fund a completely new project?

6. *Oxfam policy*. Ensure that the project is in accord with Oxfam policy on Human Rights groups and political movements.

7. *General development*. How does the project fit in with other development projects? If land rights are being secured, are extension and credit also available? Does the agency take into account other needs related to the legal aspects of the project?

8. *Promotive work*. In curative legal aid projects has the project holder considered undertaking preventive and promotive work? The project holder should be encouraged to increase the project's outreach and

educational work where possible. If publications are produced, have the distribution and educational aspects been organised?

9. *Monitoring*. Constant monitoring and reporting are needed in all legal aid projects. Periodic evaluation of impact and effectiveness should be encouraged.

Section 6 Appendix

The examples 1. and 2. illustrate the different goals of Social Education.

1. Social organisation in India

Groups of young people in Bihar and other highly-polarised States have adopted a methodology based on organising village committees to represent the poorer people. The activities of these committees follow a progression roughly as follows:

☐ countering local practices regarded as harmful to solidarity and economic advance, such as alcoholism, dowry, oppression of women, etc;

☐ organising night classes for adult education and for children unable to attend conventional schools;

☐ taking up local issues against oppressive elements such as moneylenders;

☐ spreading information about Government schemes and advice on obtaining such benefits;

☐ organising demonstrations, rallies, etc. to put pressure on local vested interests and Government.

Generally these groups do not conduct economic programmes themselves since they feel that Government promises for such schemes should be honoured first.

However, in the less polarised areas (Gujarat, for example) a different methodology has been adopted. Informal local groups are given small grants for economic purposes in order to promote their cohesion and development. This may lead in turn to organised pressure on banks, Government, etc. Social organisation is not usually the immediate objective, but is developed later by exchanges between groups, training camps and other means.

In essence the first approach leads from educational work and local social issues to organisation, while the second approach reaches the same goal through economic development. Ultimately both approaches aim to take legitimate benefits from government (and other institutions) to protect civil rights and to raise wages for the landless and other wage-earners by organised pressure.

2. Social organisation in the Andes

The demand for services from local authorities has provided the issue around which many groups have sought to organise (such as the provision of water or roads); a single small project has also provided similar impetus (building a village school with communal labour, for example). The strengthening of community organisation by bringing people together for an issue of common concern is, however, not guaranteed to lead to longer-term social organisations emerging. Many groups wither, once the issue has been resolved or project completed. In some areas groups have found that success does contribute to the enhanced confidence of people in their own ability to resolve the problems confronting them, and a whole range of communal activities may emerge from a modest beginning. Where the community already demonstrates a degree of cohesion, some practical activity may lead to positive results in encouraging the community to act for its own benefit. Where the community is weak, any organisation set up for a short-term gain is likely not to withstand the test of time.

A very broad approach has been adopted by one peasant-run agency in the central Andes. It encompasses some 30 rural communities, and the agency is able to offer different services and support to communities, depending on their needs, although overall policy is decided by all the member committees in open forum. Thus fledgling groups of landless women have been assisted with literacy programmes and other basic training and discussion sessions. As the groups of women begin to work well together and problems are identified and solutions sought, they become more demanding of local authorities and the agency to provide practical and logistic support, in terms of land, credit or equipment purchase. The members of the groups also gain sufficient confidence to approach local authorities and to play an active role in community councils. When necessary, the agency is able to provide the technical support required (whether in fields of education, agriculture, water engineering or legal aid) as community groups identify their needs.

4

Section 7 Resources

I   **Bibliography** *(see also Part Two, Section 7, Resources)*

1 Community Development and *Animation Rurale*

D. Apter, "Community Development: Achievements and Deficiencies", *Internationales Asien Forum*, 97, 1968.

R. Armstrong, *Case Studies in Overseas Community Development*, University of Manchester, U.K. 1980.

T.R. Batten, *The Non-Directive Approach to Community Work*, Oxford University Press, Oxford, U.K. 1975.

R. Dore and Z. Mars (eds), *Community Development*, Croom Helm, London. 1981.

A. Meister, "Characteristics of Community Development and Rural Animation in Africa", *International Review of Community Development*, London, 27-8, 1972.

2 Social Education

P. Berger, *Pyramids of Sacrifice*, Penguin, Harmondsworth, Middlesex. 1977, Ch.4.

Commission on the Churches' Participation in Development, *People's Participation and People's Movements*, World Council of Churches, Geneva, 1981.

P.H. Coombs (ed), *Meeting the Basic Needs of the Rural Poor, the Integrated Community-based Approach*, Pergamon, New York, 1980.

P. Freire, *Pedagogy of the Oppressed*, Penguin, Harmondsworth, Middlesex, 1972.

B. Galjart, "Counterdevelopment; a Position Paper", in *Community Development Journal*, Vol. 16. No. 2, Oxford, 1981.

W. Haque *et al.*, "Towards a Theory of Rural Development", *Development Dialogue*, 1977, No. 1, Uppsala, Sweden.

J. Hoxeng, *Let Jorge Do It; an Approach to Rural Non-formal Education*, Centre for International Education, University of Massachusetts, U.S.A. 1979.

A. Norman, Articles on Freire's Methodology in Tanzania, in *Reading Rural Development Communications Bulletin*, Nos. 3 & 4, 1977-78.

P. Oakley & D. Winder, "The Concept and Practice of Rural Social Development", in *Manchester Paper on Development*, University of Manchester, Department of Administrative Studies, 1981.

G.V.S. de Silva *et al.*, "Bhoomi Sena: a Struggle for People's Power" in *Development Dialogue*, 1979, No. 2, Uppsala, Sweden.

W.A. Smith, *The Meaning of Conscientizacao; the Goal of Paulo Freire's Pedagogy*, Centre for International Education, University of Massachusetts, U.S.A. 1979.

3 Adult Literacy

P. Freire, *Cultural Action for Freedom*, Penguin, Harmondsworth, Middlesex. 1970.

P. Freire, *Education for Critical Consciousness*, Sheed and Ward, 1974.

INADES-Formation, *Cours d'Initiation au Developpement*. Series of booklets for literacy course in French. Available from: BP 8008, Abidjan, Ivory Coast.

A. Norman, *op cit*.

E.M. Rogers, *Modernisation among Peasants: the Impact of Communication*, Holt, Rinehardt and Winston, Eastbourne, Sussex. 1969.

UNESCO have many publications available, e.g. P. Furter, *Practical Guide to Functional Literacy* (7 Place de Fontenoy, 75700 Paris, France).

4 Technical/ Vocational/ Management Training

J. Boyd, *Equipment for Rural Workshops*, I.T. Publications, London. 1978. Not a training manual, but gives advice about setting up a workshop and suggests what equipment is required.

M. Culshaw, *Training for Village Renewal*, Lutheran World Service, 1977. Useful, short book on vocational and rural training. Interesting section about identifying rural skills.

M. Harper, *Consultancy for Small Businesses*, I.T. Publications, London. 1977. Useful guide for those running or advising on small-scale businesses.

M. Harper and T. Thiam Soon, *Small Enterprises in Developing Countries: Case Studies and Conclusions*, I.T. Publications, London. 1979. Illustrates many of the problems facing small businesses and analyses the attempts made to overcome them.

How to Start a Village Polytechnic, Youth Development Division, Dept. of Social Services, Ministry of Co-operatives and Social Services, Nairobi, Kenya (P.O. Box 30276).

Industrial Training Board (UK) publications. A catalogue of these publications is available from TETOC, 17-19 Dacre Street, London SW1H OD3, UK. The catalogue includes comprehensive list of training manuals. Many of these are also relevant to developing countries.

P. van Rensburg, *Report from Swaneng Hill*, Almquist and Wiksell (Stockholm), 1974. Story of Swaneng Hill School and the setting up of the brigade system in Botswana.

Village Polytechnic Handbook, National Christian Council of Kenya. Available from P.O. Box 45009, Nairobi, Kenya.

5 Training in Leadership and Life Skills

M. Carr, *Appropriate Technology for African Women*, UN Economic Commission for Africa, 1978.

FAO, *Training for Agriculture and Rural Development*, Rome. 1976. A cross-section of contemporary experiences.

Fr. McGrath, *Barefoot Management*, *Management Skills for All*. Available from Xavier Institute of Management and Labour Relations, Jamshedpur, Bihar, India.

6 Communication

a: General

A. Fugelsang, *About Understanding — Ideas and Observations on Cross-cultural Communication*, Uppsala, Sweden: Dag Hammarskjold Foundation, 1982 (paperback).

J. Diaz Bordenave, *Communication and Rural Development*, Paris, UNESCO, 1977 (paperback).

b: Information campaigns: theory and practice

R.E. Rice and W.J. Paisley (eds), *Public Communication Campaigns*, Beverly Hills and London, Sage, 25 Banner Street, London. 1981.

4

A. Leonard, *Developing Print Materials in Mexico for People Who do not Read*, Educational Broadcasting International, London. December 1980.

M. Byram and C. Garforth, *Researching and Testing Non-formal Education Materials: a Multi-media Extension Project in Botswana*, Educational Broadcasting International, London. December 1980, pp. 190-194.

R. Varma *et al*, *Action Research and the Production of Communication Media: Report of the All-India Field Workshop in Action Research in Agricultural Information Communication*, Reading, UK, Agricultural Extension & Rural Development Centre, 1973.

c: Communication for participation, discussion and action

F. Berrigan, *Community Communications: the Role of Community Media in Development*, UNESCO, Paris, 1981, p.50 (paperback).

P. Freire, *Extension or Communication?*, pp. 85-162 in *Education: the Practice of Freedom*, Writers & Readers Publishing Cooperative, London, 1976, p. 162 (paperback).

R. Kidd, *The Performing Arts, Non-formal Education and Social Change in the Third World: a Bibliography and a Review Essay*, The Hague, Centre for Study of Education in Developing Countries, 1982.

J. O'Sullivan-Ryan and M. Caplun (eds), *Communication Methods to Promote Grassroots Participation*, UNESCO report on Communication & Society, No. 6 Paris, 1981, p.155 (paperback).

H. Nigg and G. Wade, *Community Media: Community Communications in the U.K. — Video, Local TV, Film and Photography*, Regerbogen-Verlag, Zurich, Switzerland, 1980, p.269 (paperback).

d: On specific media

D. Crowley, A. Etherington and A. Kidd, *Mass Media Manual: How to Run a Radio Learning Group Campaign*, Friedrich-Ebert-Stiftung, Bonn, 1978, p. 197 (paperback).

R. Kidd and N. Colletta (eds), *Tradition for Development: Indigenous Structures and Folk Media in Non-formal Education*, German Foundation for International Development, 1982 (paperback). Available from Education & Science Division, Simrockstrasse, 15300 Bonn 1, W. Germany.

G. McBean (ed), *Illustrations for Development: a Manual for Cross-cultural Communication through Illustration and Workshops for Artists in Africa*, Nairobi, Afrolit Society, 1980, p.68 (paperback).

T. Peigh, D. Bogue, M. Maloney and R. Higgins, *The Use of Radio in Social Development*, Community and Family Study Centre, Chicago, 1979, pp.172 (paperback).

P.L. Spain, D.T. Jamison and E.G. McAnany, *Radio for Education and Development: Case Studies, Vols. I & II*, World Bank Staff Working Paper No. 266, Washington, World Bank, 1977, p.460 (paperback).

J. Zeitlyn, *Low Cost Printing for Development*, Delhi and London, CENDIT, (paperback). Available from the author at 51, Chetwynd Road, London NW5. 1983.

C. Fraser, *Video in the Field* — a novel approach to farmer training, Educational Broadcasting International, London, September 1980, pp.129-131.

A. Cooke and T. Gill, *Women's Rights*, Penguin, Harmondsworth, Middlesex, U.K. 1977.

M. Cranston, *What are Human Rights?*, Bodley Head, London, 1973.

M. Garling (ed), *The Human Rights Handbook*, Macmillan, London, 1979.

Minority Rights Group, *Reports*, 21 Craven Street, London WC2.

D. Owen, *Human Rights*, Jonathan Cape, London, 1978.

Sadruddin Aga Khan, *Human Rights and Massive Exoduses*, U.N. Economic and Social Council, Geneva. 1981.

B. Whitaker, *The Fourth World: Victims of Group Oppression*, Sidgwick and Jackson, London, 1977.

Development Rights and the Rule of Law. Report of a conference held in the Hague 27 April — 1 May, 1981, International Commission of Jurists, Pergamon Press, Oxford, 1981.

The Reporter, Anti-Slavery Society, London, (see below).

Human Rights Internet Reporter, 1338 G Street S.E., Washington D.C. 20003, U.S.A.

(*Human Rights Directories* for Latin America, Africa and Asia available from same address)

4

II Organisations

1 Communication

British Council Media Department, 10 Spring Gardens, London SW1A 2BN, publishes the journal *Media in Education and Development* quarterly (formerly *Education Broadcasting International*). The Department also offers consultancy services, placement of Third World students on media production courses, and a range of publications on media.

International Extension College, 18 Brookland Avenue, Cambridge CB2 2HN, provides information, training and advice on all aspects of "distance teaching" — the combined use of broadcasting, print and face-to-face tuition. The College publishes *About Distance Education* three times a year and a series of broadsheets on *Distance Learning*.

The Agricultural Extension and Rural Development Centre, The University, 16 London Road, Reading RG1 5AQ, publishes *The Rural Extension Bulletin*. The Bulletin contains articles on various aspects of field-level extension, including communication and the use of media.

The Third World Popular Theatre Newsletter is published by an informal network of popular theatre workers throughout the world. Contact Dickson Mwansa, DEMS/UNZA, P.O. Box 2035, Kitwe, Zambia.

UNESCO, Place de Fontenoy, 75700 Paris, France, publishes several series of monographs and reports on communication for development. Contact the Division of Library Archives and Documentation Services at UNESCO for full details.

UN Economic and Social Commission for Asia and the Pacific, Agriculture Division, United Nations Building, Rajdamnern Nok Avenue, Bangkok 2, Thailand, publishes an *Agricultural Information Development Bulletin*. This quarterly publication covers a range of topics from the role of media communication in promoting smallholder development, to information management in agricultural research and documentation.

World Health Organisation, ATH Unit, 20 Avenue Appia, 1211 Geneva 27, Switzerland. The WHO has set up a network of national and regional information centres, known as ATHIS (Appropriate Technology for Health Information System). A newsletter, a directory of participating institutions and further information are available from the above address.

Institute of Child Health, 30 Guilford Street, London WC1N 1EH, has established a unit called Teaching Aids at Low Cost (TALC) which makes available as cheaply as possible both books and teaching materials to developing country health workers.

Development Support Communication Branch, FAO, Via delle Terme di Caracalla, 00100 Rome, Italy, promotes the effective use of communication media in rural development. This is done by in-country technical advice and the distribution from FAO headquarters of media produced by the Branch itself. Catalogues of audio-visual aids available can be obtained from the Rome address.

The *Clearing House on Development Communication*, 1414 22nd Street N.W., Washington D.C. 20037, U.S.A., publishes a quarterly newsletter *Development Communication Report*. This usually contains short articles on rural development projects involving the production and use of communication media.

The David C. Cook Foundation, Elgin, Illinois 60120, U.S.A. is a "non-profit institution dedicated to Christian communication". It publishes a quarterly journal *INTERLIT* which is an excellent practical handbook covering the whole range of media skills but with an emphasis on print.

The *British Committee on Literacy* Documentation Centre, AERDC, University of Reading, 16 London Road, Reading RG1 5AQ, U.K.

2 Human Rights, Legal Aid

Amnesty International, 1 Easton Street, London WC1X 8DJ.

Anti-Slavery Society, 180 Brixton Road, London SW9.

Centre for the Independence of Judges and Lawyers (CIJL), PO Box 120, CH-1224, Chene-Bougeries, Geneva, Switzerland.

Human Rights Network, 26 Bedford Square, London WC1B 3HU.

International Commission of Jurists, (as for CIJL).

U.N. Economic and Social Council, Commission on Human Rights, Palais des Nations, Geneva, Switzerland.

PART 5 ECONOMIC DEVELOPMENT GUIDELINES

Brian Pratt

Part Five ECONOMIC DEVELOPMENT GUIDELINES

5

Section 1   Economic Development as a Goal

I   Introduction

Oxfam's work is inspired by a broad-ranging philosophy of development which embraces both economic and social objectives. Considerable emphasis is placed on providing the means for people to achieve greater social freedom, that is, enabling them to increase their capacity for self-determination. However, those involved in setting up or assisting aid projects should never lose sight of the fact that for the poor the main objective is usually to improve material living standards for themselves and their families. Therefore, in talking about the wider 'development of people', fieldworkers and aid administrators should remember that basic considerations of economic feasibility must form an integral part of development projects. Such schemes must be economically sound and convey long-term, concrete benefits if the activities being funded are to become financially self-sufficient and independent of donor agencies. Generally speaking, it will prove difficult, if not impossible, to achieve social development goals, such as increasing critical awareness among the poor, giving them more freedom to develop their own lives and to determine appropriate solutions to their problems, etc., unless the measures adopted are also economically viable. If this rule is not observed, projects may collapse as soon as foreign funding is withdrawn, and there is a serious risk that participants will become permanently disillusioned, thereby hindering future development efforts.

II   Objectives of economic development

Economic development should have certain fundamental objectives. These should be quite explicit, understood and agreed upon by everyone involved in a development programme.

a: To raise the income of project participants. Care must be taken, however, to ensure that such increased income is as equitably distributed as possible within the project, according to individual contributions and efforts. Profits accruing to farmers in an agricultural scheme, for example, should benefit all family members so that health and material living standards of men, women and children are improved. In order to avoid benefits being monopolised by one sector of the community, the project should incorporate mechanisms of mutual consultation and participation at all stages of design and execution (see PART THREE). This point cannot be stressed strongly enough.

b: To generate employment. Chronic unemployment and underemployment typify most situations in which Oxfam and other NGOs work. Wherever possible, therefore, projects should attempt to increase the number of paid jobs available to the local population. Job-creation schemes are, along with income-generation, sometimes accused of forming privileged elites among the poor, creating distortions which give rise to resentment among non-participants. To counter this, projects should try to avoid becoming closed economic enclaves, and should encourage as wide a participation in the scheme as possible. Dialogue between participants and non-members should be stimulated so that a 'spread effect' may be obtained and the economic base of the scheme gradually broadened. However, new recruits can only be taken on if the project holders feel able to manage efficiently with the resources at their disposal; for example, in the case of a cooperative, only if there is sufficient working capital, or in the case of a handicraft or agricultural scheme, if there are adequate market outlets.

c: To achieve self-reliance. One of the major stated goals of any NGO is to achieve self-sustained development within the projects it funds. Despite good intentions there is, nevertheless, a danger that situations of extreme dependency may be created, with projects collapsing soon after outside funding ends. Clearly, many organisations (especially those not directly engaged in producing goods and services) which carry heavy overheads and produce no cash income must remain, by their very nature, dependent upon external funding sources. In the medium to long term, projects involving the production of goods and services must be economically viable so that they may realistically expect to become autonomous. A funding agency which subsidises the economic operations of a given scheme too heavily may well sow the seeds of that project's destruction. For example, a cooperative should not be supplied with continual injections of working capital if this capital is being depleted through mismanagement. A careful examination should always be made of likely economic prospects for an intended project, which may be a simple or more complex exercise depending on the amounts to be granted in aid (see PART THREE, Section 2). In many instances it may prove more appropriate to provide technical expertise or other forms of non-monetary assistance rather than direct cash aid (see PART THREE, Section 5). Self-reliance is more likely to be achieved in situations where maximum use is made of local resources, both human and financial, in promoting development.

5

Section 2   Production of Goods and Services

I   Introduction

Most of the projects funded by Oxfam ultimately involve the production of goods and services in order to improve the livelihood of the poor. This may be possible through existing groups but may mean helping to set up new organisations created specifically for the purpose. Development workers should always take great care to avoid imposing inappropriate ideas and structures which recipients of aid may feel they must accept to be eligible for funding. Despite the best intentions of aid agencies, the relationship between giver and receiver is, in the final analysis, an unequal one in which the donor agency is the more powerful and is able to dictate terms if it so desires.

Granted the need for accountability at home and a certain degree of control over how funds are spent, locally-generated organisation and initiative should always be the major priority. Voluntary agency work may thus be able to counter-balance some of the effects of official 'top-down' development planning, which often fails to take into account real local needs. Oxfam attempts to minimise the likelihood of such mistakes by its policy of working, wherever possible, through indigenous agencies. Even in such cases, however, it is worth bearing in mind that the larger and more bureaucratised the indigenous organisation, the more difficult it becomes for target groups to take their own initiatives and exercise decision-making power. (See PART FOUR.)

A major drawback of much foreign aid is that it is frequently destined for capital-intensive industry and agriculture, which not only creates few jobs but may actually help to displace labour (as in the case of mechanised agriculture, for example). By concerning itself with local level, low-cost projects and techniques, the NGO is able to encourage labour-intensive activities in both rural and urban environments. The increasing use of appropriate technology adapted to local conditions is an instance of this.

Consideration of economic production is divided into two major categories, those of rural and of urban settlement and development, with an additional sub-section on handicrafts, which spans both sectors of the economy.

II   Rural development and settlement

⬚ Obstacles to Development

Most of Oxfam's work has, over the years, been located in the rural sector, where poverty has traditionally been concentrated. Although this may involve a variety of schemes in the fields of health, construction and education, most work funded in the countryside will be based on making agricultural improvements. Agricultural development may seem a relatively straightforward and simple technical task, but it is fraught with difficulties. Fieldworkers should never underestimate these problems, which range from technical questions such as soil and water deficiencies to basic institutional obstacles. The latter often appear insurmountable and frustrate many, if not most, attempts to effect substantial improvement in the agricultural performance and incomes of poor farmers. Poor cultivators are frequently denied access to sources of official credit (often subsidised), which tend to be monopolised by wealthier landowners, and are obliged to depend on informal sources, often at extremely high interest rates. The marketing chain is so structured as to minimise the profits of farmers who are not organised into associations or cooperatives, and thus have no bargaining power against merchants and intermediaries. In many countries, land-ownership is highly inequitable, and the majority of farmers have only small plots, barely adequate for subsistence, and are usually obliged to earn supplementary income elsewhere in temporary labouring jobs. In Latin America, for example, it is estimated that a mere 1½ % of land-owners occupy 65% of the land, and individual holdings can reach hundreds of thousands of hectares in size.

It is now generally accepted that inegalitarian land-tenure systems are one of the most intractable obstacles to promoting economic improvement among poor farmers. Furthermore, it appears that the problem is becoming worse. As farming becomes more modernised and highly capitalised, and increasingly dependent upon international agribusiness corporations, land is gradually becoming incorporated into larger and larger units, while small farms are either absorbed or fragmented further still, creating growing numbers of landless labourers and migrants who then swell the ranks of the urban poor.

It would be unrealistic to expect any one or a number of development agencies to stem this tide, but it is possible to mitigate the worst effects of these trends for some groups of people. For example, aid money can be used to produce economically sound agricultural projects (see PART SIX). Oxfam has made a positive contribution, through programmes of legal aid, towards preventing the illegal expulsion of small farmers from their lands by larger and politically more powerful interests (see PART FOUR, Section 4). Land-purchase schemes may offer an apparent solution in certain situations, but they have a number of disadvantages, and field staff should be cautious of this approach. Firstly, land prices may be artificially inflated by the presence of a foreign buyer in the market. Secondly, and perhaps more importantly, such schemes may create an 'enclave' mentality among the project-holders which could easily hinder attempts at stimulating more broadly-based development initiatives.

Another means of helping to overcome the land shortage problem is through directed rural settlement or colonisation projects.

2 Rural Settlement

A rural settlement scheme involves constructing all the necessary physical infrastructure to establish a self-sustaining community in a previously unpopulated area. It frequently involves moving people into a completely different ecological setting, often one with marginal rainfall and poor soils. (It is perhaps precisely for this reason that the region is available for settlement.) In addition, the community must develop an independent identity, adapting its traditional cultivation methods and modes of production to the new environment. New technologies that exploit local conditions should be available, so as to ensure self-sufficiency.

Settlement schemes should have the following major objectives:
- [ ] to facilitate successful adaptation to a new physical environment;
- [ ] to foster a communal approach in social organisation and means of agricultural production;
- [ ] to provide adequate training of settlers in their new life-style;
- [ ] to enable the communities to be self-reliant at an early stage;
- [ ] to initiate and encourage communal self-government by the settlers, so that all decision-making will eventually devolve upon the community as a whole.

Such projects are inevitably a costly solution, since they involve moving thousands of people and the rapid construction of buildings and amenities which would normally be established gradually over generations.

The rewards are unpredictable and slow to appear. Initially at least, the settlers are faced with considerable hardship, and the possibility of loss of interest leading to abandonment of the project must be taken into account. Settlement schemes frequently fail because the wishes of the settlers themselves are not sufficiently considered. They are often looked upon by regimes as ideal public relations exercises, chosen to generate positive publicity rather than to solve the problems of the settlers.

All settlement schemes must be placed in the socio-economic context of the relevant country and region; analysis of the relationship between political power and control over land will also be informative. A close examination of the distribution of land ownership and of the variety of forms of land tenure may reveal the need for a bold redistributive programme. This would release large areas of land already equipped with basic infrastructure for settlement at a cost significantly lower than that of new settlements. Reform of land-tenure arrangements to give ownership to those who till the land may not only encourage greater investment, but may also increase productivity per acre. Production would be for consumption rather than for export.

The motives of the government concerned should always be scrutinised: is it trying to reduce demands for land reform and quell agrarian protest movements among the landless? The landless status of many peasants may be far more a direct result of political and financial sponsorship of mono-culture, agri-industry (e.g. cotton and sugar cane) by the relevant government, than of a natural shortage of cultivable land. Landless peasants may be maintained as a readily available source of cheap, seasonal labour. Often, peasants have been dispossessed of their land precisely for this purpose. In fact, settlement schemes can even be a government tool to provide labour in newly-cleared tropical areas.

Alternatives such as land reform, or more intensive use of land by multi-cropping, irrigation, soil conservation and adequate credit facilities, may prove more beneficial.

The main argument in favour of directed projects is that, properly planned, they prevent adverse social and ecological effects, and provide for long-term viability. If there have been earlier spontaneous settlements in the region, then detailed analysis of these should be conducted to note: settlers' geographical origins; their socio-economic backgrounds; the degree of self-sufficiency attained; their relationship to the local power structure.

3 Settlement, Siting and Structure

A good deal of anthropological investigation into the peoples already living in the proposed area, as well as geographical research, will provide invaluable guides to choosing the most suitable settlement site. Some or all of the techniques listed below could be used.

☐ Satellite remote-sensing imagery, with resolution down to contiguous areas of 60 x 80m. This enables close examination of the natural resource base, physical features, the degree of human habitation, the mapping of current river courses, and the presence of forest tracks.

☐ Small aerial photographs should be taken at different times of the year to reveal flood-prone areas and changing river courses.

☐ Soil surveys should also be carried out, as settlement is inadvisable in areas where soil is lower than grades 1 or 2.

☐ Weather records should be consulted with special note taken of the frequency of droughts and flooding.

☐ The degree of isolation from markets and large urban areas is also crucial.

☐ Legal title to the land must be guaranteed for all settlers, with full rights of ownership.

The settlement should be designed in such a way as to facilitate community cooperation in as many activities as possible. Two viable alternatives are the *linear* and the *nuclear* arrangement.

a: A *linear* development along feeder roads allows for larger holdings and basic rectangular plots, and has frequently been the pattern of spontaneous, unorganised settlements. This structure typifies an individualistic, pioneering approach and enables large tracts of land to be cleared by each settler.

b: The benefits of a *nuclear* settlement design are easier and cheaper provision of services such as roads, water, electricity, health care and education. A nuclear settlement with a ring of houses and plots of land radiating out from the homes gives greater cohesion and acceptance of responsibility for communal activities. It is important that homes should be as nearly as possible equi-distant from the cleared land, and that land is allocated with due consideration of family size and varying agricultural potential of the different parcels of land.

In order that settlement schemes can be self-reliant in the long term, an agricultural service centre should be planned. If possible, the settlement should be situated near a large urban centre which would provide industrial opportunities outside the farming sector. A commonly suggested basic settlement unit is forty families, and the larger the overall settlement, the greater the variety of services that can be provided economically by the settlers. This also means that there are more multiplier effects in the region, and more employment is created.

If a settlement is to be integrated into a new area, care must be taken to avoid creating an elite with better services, easier access to credit, etc. As much as possible must be done by the settlers themselves, if necessary with a monetary incentive.

4 Selection of Settlers

The most effective form of selection is self-selection. Settlers who seek such opportunities in the face of great physical hardship and economic insecurity are usually those most in need, and also those who are most determined and idealistic. It is commonly true that these settlers are also more individualistically minded and not great believers in communal production. The most successful settler groups are those which have an agricultural background in a similar ecological setting.

However, if selection is necessary, the *need* is the primary criterion, and settlers should be drawn from the landless and from the most overcrowded regions, or from urban areas where unemployment is highest. A communal approach and a commitment to work for the common good is more likely to exist if settlers come from the same region, social or ethnic group, even more so if they have been particularly oppressed or discriminated against as a group.

In many instances, the initial settlement is undertaken by young males who are committed to an extended family elsewhere. However, the smallest structural unit for work purposes is not generally the individual. Most economic tasks are traditionally undertaken by families as a group, or are at least divided among different members of the family. Settlement schemes should therefore encourage the whole family to settle, so as to achieve the most effective use of land and labour. The problem of many schemes is that each settler has been seen as an individual, and not as a member of a social group or class, a part of a larger structural framework. Field staff should also determine how the new settlers fit into the regional context, and relate to older settlers, other ethnic groups, traders and large landowners. A settler's attitude to his/her new opportunities may depend to a large extent on his/her position within the familial power structure.

The influence of the regional cash economy in the form of agri-industrial concerns may lure settlers away from the settlement at harvest

5

time, as paid employment is available elsewhere. This may also result in settlers using paid labour on their own plots of land.

Government selection of settlers on political or strategic grounds must be identified as such. Settlement schemes in Ethiopia, for example, have been used to stabilise disputed frontiers, to quell secessionist groups, and to settle pastoralists who were traditionally rebellious and very loosely tied to the core of the country. In Bolivia, political capital was made from the settlement of exiles from Chile after 1973, even though they had no experience or commitment to subsistence agriculture.

Information about the previous cultural and agricultural practices of prospective settlers can give an indication of the likelihood of communal participation.

5 Orientation and Self-Management

Prospective settlers should be made familiar with the scheme before migration. A programme of this kind should:
☐ prepare potential settlers;
☐ identify the most suitable candidates;
☐ provide an accurate and objective picture of the new environment, and of the problems to be confronted.

Although it is advantageous to have a balanced intake of skilled settlers in all the necessary trades, training courses can be set up, preferably staffed by former settlers from older settlements in a similar setting. Settlers should be trained from the outset to become future course-organisers and instructors.

Orientation courses should be comprehensive enough to provide a working basis for self-reliance. The methods used should be democratic, as an authoritarian relationship between teacher and learner tends to work against the acquisition of essential skills for self-management. Withdrawal of advisers should be gradual. Courses should cover: agricultural techniques, education, house building, health care, forestry, animal husbandry, accounting, book-keeping, self-management, and the setting up of cooperatives. Examples of topics that should be included in these courses are given below.

a: Agricultural techniques. Settlers should be instructed in land conservation methods: this is especially important when tropical, semi-tropical or dry lands are being opened up. Tropical soil fertility is finely balanced, and for the first few seasons of cropping, careful attention must be paid to the retention of fertility (see PART SIX, Section 2). Since in many settlement schemes most settlers originate from more temperate zones, where the breakdown of minerals and soil nutrients is much slower, practical methods to control leaching, hard pan development, laterite production and waterlogging must be undertaken. The use of soil conservation, gully preparation, and silt resulting from flood waters must be a high priority.

The new environment may necessitate the use of new tools and a new range of animals. Farmers should therefore be instructed in the use of any new technology. Continual reference should be made to local research stations and older settlements. Information must also be provided on wild fruits, plants and vegetables that can be harvested, as well as animals that can be hunted. New cooking and eating practices should be initiated with due regard to traditional custom, taking into account the religious or ritual significance of certain foods.

The most important priority is a diversified base of production to cope with nutritional needs, climatic vicissitudes, seasonal labour requirements and market fluctuations in a regional, national, and international context. Also, a blend of subsistence and cash crops should be grown. The crop varieties needed are those that:
☐ are resistant to the pests and diseases encountered in a newly settled area;

☐ provide economic returns under less sophisticated management systems;

☐ are least dependent on purchased inputs in areas that have poor social and physical infrastructures;

☐ enable planting and harvesting to be spread over as many months as possible, with multi-cropping a possible option.

The previous division of labour between men and women should be integrated into the new pattern of production, if appropriate.

b: Nutritional requirements. A balanced diet should be the aim, based on a combination of the settlers' previous cultural tradition and the new ecological setting. Crops which have low nutritional value could be slowly phased out, e.g. cassava/manioc. In newly settled areas, the provision of meat is impossible, so soya beans or groundnuts could be replacements. Training courses explaining and demonstrating nutritional planning can be useful.

c: House building. Previous cultural practices and life rhythms must be catered for in new houses, so a blend of old styles with new building materials in a new climatic setting must be undertaken. Instruction should be given in possible construction techniques, but community decision should be left with the settlers, once the possibilities, advantages, and benefits have been outlined. A course of instruction in the use of newly available timber and the construction of furniture should be given where appropriate.

d: Forestry. Information should be given to aid identification and selection of timber for different uses, together with techniques for working with new wood. (For further details see PART SIX, Section 5.)

The original clearing of land should be carried out by the community as a whole, and until the first harvest, food will need to be provided from outside. The distribution of this, however, must as far as possible be by the settlers themselves, in order to promote a sense of self-government. A community spirit could be further enhanced by cooking and eating in the common space in the centre of the ring of houses. A communal vegetable garden should be set up along with a cooperatively-run consumers' shop. The well, clinic, grain mill, and latrines should all be communally used, so that frequent interaction takes place, and channels of communication are plentiful. Group projects should be quickly established and group recreational activities fostered.

Medical examinations should be conducted on all new settlers; natural immunity will not be present, as a whole new range of diseases and parasites may be found in the new ecological zone. An intensive immunisation programme is needed, and communal clinics should be set up in each settlement, staffed by local health workers. All settlers should be instructed in basic hygiene, and primary health care should gradually become the province of trained settlers.

6 Marketing

Detailed market research, including the consideration of expected local and regional demand for cash crops, ease of transportation, the likely surplus in adjacent areas, and an estimation of product supply, should be carried out at an early stage. Crops with a high retail value may seem a good prospect, but the market may be highly limited. The processing of cash crops can provide more employment after the harvest and increase the value of and demand for the crop, as well as reducing its perishability. Crops that are particularly susceptible to inclement weather must be carefully appraised for variability of supply .

Good road communications with nearby urban centres may allow for ease of marketing, but may also facilitate exploitation by local traders. Detailed research should be conducted into settlers' previous experience of marketing. Even if a cooperative marketing organisation is set up with cooperative transport, settlers may not use it although they provide a much more equitable forum within which to conduct business with local traders. The change from individual production to large-scale communal production may require too much adaptation from traditional practices. The availability of transport is crucial, as this allows for the selling of large amounts of produce in far-off urban centres, and the buying in bulk of necessary supplies. Collection points, storage facilities and processing industries should all be set up.

7 Evaluation

Evaluation of settlement projects can only be meaningful in the long term, as the capital outlay on infrastructure is so great. The project can be evaluated by determining whether the sustainable income of beneficiaries is equal to the average of farmers in a similar ecological setting. A cost-benefit analysis would be inappropriate in view of the scale of the project. More important than an objective economic appraisal is participatory evaluation of the project by the settlers themselves at an early stage. Long-term generation of employment in a wide variety of processes, self-reliance, and the creation of a large marketable surplus, are the best criteria for evaluation.

III   Urban development and housing

1 Introduction

Increasingly large proportions of the populations of Third World countries are becoming urban-based rather than rural. The resulting pressure on urban resources often means that people are effectively worse off, poorly nourished, living in insanitary conditions and unemployed. In response to these problems, therefore, NGOs have tended to devote an increasing proportion of their limited funds to measures aimed at alleviating the worst symptoms of urban squalor. This has taken a variety of forms dealt with in other parts of the Handbook, ranging, for example, from humanitarian aid to encouraging small, employment-generating businesses, handicrafts, and health care at community level.

Another area is that of physical infrastructure and *housing*. Building new houses and improving urban infrastructure may be viewed in purely physical terms of better dwellings, roads and water supplies, etc. Alternatively, such construction activities may be just one part of a wider development effort. This might mean, for example, building houses in concrete to replace wattle and daub as a means of helping to guarantee land tenure for slum dwellers (BRZ 86), or providing educational and legal guidance to prevent unfair evictions of poor inhabitants by property speculators (BRZ 172). These issues are dealt with at greater length in PART FOUR, Social Development Guidelines. In this section, however, we are concerned with providing guidelines for building and improving homes and neighbourhoods. For shelter in post-disaster situations, reference should be made to PART EIGHT, Section 4.

2 Assessing the Situation

In order to assess the need and potential for housing action, one must know about:
☐ the people and the organisations involved;
☐ their material, financial and human resources;
☐ their interests, expectations and motives;
☐ the operations they normally carry out, and are willing to carry out in the particular situation;
☐ how the various jobs or operations are carried out, and how they relate to each other; and
☐ the actual results, physical, financial and social.
It is impossible to plan realistically and to manage any situation effectively without reasonable understanding of all these factors. What matters in the longer term is the impact of the whole process and its

products, but initially this can only be considered in the form of the participants' own expectations. To summarise: one must have a clear idea as to who is likely to do what, why and with what results in the particular circumstances.

The background to any neighbourhood activity normally includes the following:

☐ the geographic or natural environment, especially the soil and subsoil, climate, vegetation, and the materials they provide;

☐ the man-made environment, especially the relative location of settled areas, communications, utility networks, surrounding land uses and buildings;

☐ property-ownership and tenure in the area;

☐ laws, customs and the administration of controls over land tenure, uses and improvements;

☐ the local building industry, including the supply of materials and components, labour and craftsmanship, tools and equipment, forms of contracting, and types of building;

☐ the local systems of exchange and credit, including those in kind, such as mutual aid;

☐ the composition and interrelationships of the local community, and its relations with the larger society of which it is a part, e.g. the ethnic or religious groupings, socio-economic classes, household and family types, age and gender groups.

The existence of local grass-roots organisations through which people can approach and negotiate with external organisations, especially government agencies, is extremely important. When these local mediating structures are lacking or too weak to be effective, their substitution by outsiders who mediate on their behalf can increase local dependence and subjection, undermining the goals of community development. The nurturing and strengthening of locally based and directly representative organisations must always be a central concern.

As the success of any local development activity depends on the active participation of those who stand to benefit, there must be a locally acceptable match between the goods and services supplied and the needs and priorities of those concerned. It is commonly and erroneously supposed by professionals and administrators that needs for housing and local infrastructures are similar for everyone everywhere. Highly variable factors determine what individuals in different circumstances will consider top-priority needs.

It is also commonly and quite wrongly assumed by those in charge of housing programmes that land, utilities and dwellings are equally essential parts of a package which must always be delivered as a whole. Sometimes people above all else need land, and are better off if they set up temporary camps while building their own homes, and obtain the utilities at later stages.

To be habitable, a place must have a number of qualities and material components. It must be accessible; that is, the residents must be able to get to and from their sources of social and material support. It must provide an acceptable degree of shelter and privacy, and be accompanied by sufficient rights of occupancy or tenure to make habitation feasible.

The great majority of people with low incomes see security as their highest priority, especially if their future expectations are low; they also give high priority to new or better employment, better education and an improved social environment. The only practical way of determining priorities in a particular situation is to consult the people concerned, if they have not already articulated their own programme.

The key criteria for deciding whether or not to take action should be:

☐ where people are insecure and threatened, and are therefore reluctant to change their present way of life;

☐ where people are reasonably secure but are poorly housed and serviced;

☐ where people have no home of their own or do not want to stay where they are.

Thus the criteria are based on the needs of the people, and are not necessarily the same as those of officially promoted programmes.

3 Settlement, Siting and Structure

The elements below are independent variables and so must be separately appraised.

a: Location. The quality of the land, its soil and subsoil, together with its orientation and tenure systems, are often crucial for decisions on what is to be built and how.

In the many countries where cities are largely new and the majority of the inhabitants are in the lower income bracket, there is often a lack of appropriate designs. This is a task which professionals and local builders and their customers can and must tackle together. Most traditional housing forms and neighbourhood lay-outs are much better than most modern ones, both socially and financially.

The essential criteria in the appraisal of neighbourhood layout designs are:

☐ the ratio of public to private land, i.e. land open for public use and that which is reserved for private and semi-private use;

☐ the ratio of circulation to area, that is, the length of public streets and paths to the area served;

☐ the gross density, i.e. the number of persons per hectare for the whole area, including all streets and open spaces. Private land should not be less than 55%, and public streets should not be more than 25%, leaving at least 20% for recreational, cultural and other semi-public uses. The circulation length for single-household developments should not exceed 250 metres per hectare, and the number of persons per hectare should not be less than 300 in most urban circumstances.

Small low-cost dwellings must always be designed to permit future improvement. One test of a good design is the ease with which it is modified and adapted to the changing uses and circumstances of the householders. Spaces must be laid out to provide the widest possible variety of uses; this is in direct opposition to the modern tendency of architects to design every space for a specific purpose. Structures must be designed so that they can be added to without significant waste of work already done. And of course, householders must be able to build as slowly as they wish.

In most low-income contexts little more than a roof is needed at first, along with sufficient screening for privacy. In some contexts, low-income householders prefer to start with a perimeter wall around their plot, and set up within this a shack from cheap provisional or waste materials. Others may prefer to start with one more or less finished room, and yet others with a temporary dwelling from materials they will use later in the permanent construction. In any case, designs must allow for the widest possible variety of individual house-building schedules. Proper attention to design can result in very considerable savings in material costs, as well as improvements to the health and social relations of the householders.

b: Public services and utilities

☐ *Drainage*. Whether surface and/or piped, drainage may be crucial in areas liable to flooding or erosion, or where a water source providing the domestic water supply may be contaminated.

☐ *Water supply*. This will be essential where it is not already available from uncontaminated wells or nearby streams.

☐ *Permanent roads and paths* are especially important in areas subject to heavy rainfall and/or where dust is generated by traffic. Roads and paths are both for access to places and for circulation within them. The impact of increased accessibility, especially on more or less isolated or peripheral areas, can be considerable: villages which become connected with markets by road may increase their trade and wealth, or may be more effectively exploited by intermediaries; peri-urban land with improved communications may become either more accessible for settlement and housing or, especially to those with low incomes, less accessible, owing to increased market values and prices.

☐ *Electric light and power* can be a vital service for the generation of local production and for educational and recreational activities. It can also greatly reduce costs of construction and enhance domestic economies and comfort.

☐ *Public transport* can be essential for neighbourhoods in isolated or peripheral locations. Improved services can do as much as, or more than, any other component to increase the use and market value of an existing or a new settlement.

☐ *Telecommunications and postal services*. The existence of a public telephone can be an immense benefit to low-income neighbourhoods, especially in helping to overcome physical or social isolation and the attendant risks to the sick and injured. Adequate postal services are obviously as important for low-income people as for anyone else, even if individual use is much less frequent.

c: Community facilities. A meeting-place is essential in order to organise and carry out cooperative activities of all kinds; in many cultures this is provided by a church, temple or mosque, where officiating members may also provide educational and medical services as well as the performance of rites and celebrations. In any case schools, especially for small children too young to commute, often have very high priority, together with clinics and/or pharmacies where medical advice can be obtained. Market stalls or shops for subsistence goods are often essential when alternatives involve expensive travel. Garbage collection may be especially important in some circumstances, and fire protection is vital in densely built-up areas or where flammable materials have been used.

d: Buildings. As in the case of services, dwellings are generally built by stages in low-income contexts. All more or less permanent dwellings have the following components, which are often independently variable:

☐ *Foundations*. Whether piles driven into mud, footings in trenches, reinforced slabs or other types, foundations must be designed to carry the loads that may be placed upon them.

☐ *Load-bearing walls and frames*. The structure carrying the roof and upper storeys may be a frame of wood or reinforced concrete, and monolithic walls from a variety of different materials, or a combination of both. Walls are to protect the users from extremes of temperatures and to provide visual and acoustic privacy, as well as to hold up the roof and upper floors. They must be capable of doing the latter under stresses of high winds, heavy snows or earthquakes without undue danger of collapsing on to the occupiers or allowing the roofs to do so.

☐ *Roofs*. These are the most expensive and technically the most difficult components in the majority of cases. Many traditional roofing materials, such as thatch from rushes or palm leaves, are inappropriate in densely built-up areas, or when cooking styles and fuels have so changed that smoke no longer eliminates insects. Alternatives, such as corrugated steel sheets, are often costly and have poor insulating properties. Since roofs may account for more than half the total cost of simple dwellings, economies in their construction are especially important. Roofs not only protect the residents from rain, wind, heat and cold, but may also be important for the protection of walls and foundations in weathering.

☐ *Carpentry*. Windows and doors especially may have very important social as well as physical functions; a large proportion of total expenditure is often lavished upon those that face on to public ways. These and other components, such as built-in fittings, are generally prefabricated in workshops and brought to the site. They are, therefore, an important source of local employment.

☐ *Installations*. Demands for internal plumbing and wiring are common, even in low-income homes. However, the quality of such installations is usually poor, and, in the case of electrical work, highly dangerous as well as wasteful. Because of a general lack of skilled workers in these trades, the latent demand is relatively high. This opens up substantial opportunities for on-the-job training in local development projects, probably much greater than those in trades which are already well-known and quite highly developed, such as masonry and carpentry.

4 Self-sufficiency, Dependency and Self-management

There are an infinite number of ways in which homes and neighbourhoods may be built, improved and maintained, but it is essential to make some assumptions with regard to the most significant and influential factors. It is commonly assumed, for instance, that the difference between contracted building work and self-help is crucial, as is the difference between subsidised and unsubsidised housing schemes, or the difference between sponsorship by private, public and non-governmental agencies. All these differences are indeed important factors. A few general categories may be based on principles by which any particular factor or combination can be usefully distinguished.

The three suggested below are justified by the assumption that the most important factor of all is who decides what for whom. The significant difference between contracted and self-built housing, for example, is who decides what shall be done by self-help and what by contractors. When householders are the general contractors, they are likely to help themselves to save money to a much greater extent than when they build their own homes in a sponsored project administered by an agency over which they have no control. People in most circumstances can save as much, or more, by being their own contractors as by being their own labourers.

The simplest useful distinctions in the field of decision and control are:

a: Self-sufficiency. The householders are the builders. They own resources without either support or contributions from outside agents or sectors. The poorest urban dwellers must usually settle wherever they can, even if very temporarily, and shelter themselves with whatever they can find or afford from their own very small incomes. While they may contribute a great deal to industry and commerce, they cannot afford the goods or services produced and they get no assistance from government: indeed they are fortunate if they are not harassed.

b: Dependence on others for the supply of housing over which the householders have no control and in the production of which they have no say is the opposite of self-sufficiency. Conventional housing projects and programmes providing centrally planned goods and services to categories of households, often decided by the same agency, are organised pyramidally with decisions flowing from the top down. Although government policies are now changing, assisted and sometimes pressured by international agencies, centrally administered 'categorical programmes' are almost universally assumed to be the alternative to the impoverished self-sufficiency of shanty towns.

However, shanty town dwellers are often unable to afford the new public housing. Alternatively, many poor people are ineligible for such amenities because, as unregistered independent workers, they cannot provide documentary proof of a minimum monthly wage to qualify for a mortgage. Consequently, much new public housing is occupied by minor government employees rather than by the poorest groups.

A more cost-effective and politically acceptable alternative to brand-new housing schemes is to upgrade unauthorised settlements, where such potential exists. This avoids large-scale displacement of urban populations from centrally-placed locations in big cities, valuable not only to property developers but also to low-income groups which must minimise costs of transport to their places of work. In Brazil, for example, the government is upgrading large areas of shanty town in the city centres of Rio de Janeiro, Salvador and Recife, rather than resorting to the age-old solution of moving populations *en masse* 20 or 30 miles away to the urban periphery. A long-term danger is that, if land tenure rights are not guaranteed, central locations with vastly improved infrastructures (houses, roads, drainage, electricity, etc.) may become subject to illegal pressures from property speculators anxious to take over these prime sites. Community associations can help to avoid this outcome by forming pressure groups and taking legal action to gain official protection. Community participation in determining the future of the area is more easily incorporated into such improvement schemes.

It is still generally assumed that centrally administered projects and programmes are the only ways of ensuring adequate housing for all. It is therefore supposed that the task is to reduce the cost of products to levels which the great majority can afford. Consequently, priorities are often given to technological innovations that promise to increase productivity or reduce costs, to self-help by which people can provide unpaid labour, or to sites and services projects in which recipients are left to build their own houses. All these innovations — which are sometimes no more than gimmicks — may well increase the provision of housing, albeit in ways that also increase dependence.

During the 1970s, the concept of 'progressive development' in housing arose in response to the failures of more traditional housing schemes. The concept of sites and services projects implied flexible programmes which would enable participants to make improvements at their own pace, and included self-help components such as mutual aid groups and the use of households' own labour in house-building. However, it was found in large projects in El Salvador, for instance, that middle-income groups were the major beneficiaries, while poorer groups continued to rely on the informal or illegal housing markets. A major reason for this was said to lie in the selection criteria, which excluded the very poorest households. The principle will only work in practice if official housing authorities actively favour the poorest groups rather than allow economic and political criteria to govern the selection procedures.

c: Self-management is quite different from self-sufficiency; people can manage their own housing and other affairs as long as they have access to the necessary resources. Where industry and government do provide people with sufficient resources for managing their own affairs, there is the opportunity for this to be done by people and organisations negotiating with each other on more or less equal terms and according to commonly respected rules.

The importance of these principles of decision-making and organisation lies in their relative capacities to obtain and use resources economically, and to enable people to fulfil themselves through responsible and creative activity. Although self-management is obviously the principle that most directly embodies these qualities, it is equally clear that it depends on access to fair shares of locally scarce

5

resources, and these can only be guaranteed by central planning and control. On the other hand, self-sufficiency can make major contributions to personal and social wellbeing, as it can relieve burdens on scarce resources, reduce pollution and increase personal security and independence. In almost all situations, the building and maintenance of homes and neighbourhoods demands a mixture and appropriate balance of all three principles of organisation.

IV   Handicrafts

1 The Role of Handicrafts

Handicrafts are just that — objects made by hand. However, the word usually also implies some or all of the following:
- [ ] the skill is a traditional one, handed down from one generation to another;
- [ ] the raw materials and labour are available locally;
- [ ] the finished product is of use in the local context, where art and craft are normally undifferentiated. The local use of handicrafts can be utilitarian (for cooking, marketing, dressing, working in the fields, etc.) or ritual, where the form, materials and decoration sometimes have symbolic meaning;
- [ ] the techniques of production are labour-intensive;
- [ ] the emphasis in the production process is on quality rather than quantity.

There is great potential for improving people's economic circumstances through the sale of handicrafts in those areas where there is a tradition of making hand-crafted goods and a market for them. The sale of agricultural surplus in low-income rural areas is extremely risky — prices tend to be very low and may fluctuate violently, and weather conditions are often very unreliable. Equally, the concentration of land-ownership in many rural areas leads to serious shortages of food. Thus, handicraft sales may become crucial to sustaining a rural population.

Handicraft projects abound in urban areas because of the high levels of unemployment and underemployment. However, in the city there is likely to be stiff competition in the market. Even where there are sufficient outlets, the necessary traditions and skills may have been lost. For this reason many urban handicraft programmes have stressed training — often not taking enough account of employment opportunities or market possibilities.

Training artisans without first assessing the real commercial potential of a handicraft project can do more harm than good. For example, in cities throughout the world women are being trained in sewing and embroidery. They are led to believe that with these skills they will be able to earn good incomes. The reality may be very different. If self-employed, a woman can usually only compete with factories by charging unrealistically low prices for her goods. If in employment, she will find herself relegated to a low-status, poorly-paid job.

Handicraft projects tend to be set up primarily for the purpose of income-generation. The making of handicrafts can, however, fulfil a number of other important functions. Where there is a strong tradition of craftsmanship a handicraft project can help to maintain the cultural heritage. This may be especially important in areas subject to migration or rapid social change.

Handicraft projects can also serve as a basis for communal organisation. The organisation of artisans into some kind of cooperative or other group can be a convenient tool for educational or social reform. Handicraft projects may be linked with educational programmes, benefiting members and their families; teaching basic skills such as literacy or, at a higher level, accountancy. Some handicraft programmes have been tied in with health posts, and others help to finance communal stores (for example PRU 101 and PRU 200).

2 Production

In recent decades the term handicraft has been broadened to include artefacts that have been modified to suit an outside market. In such instances the dimensions, pattern, colour and function of the article may change. Adaptations of the traditional craft to suit new market conditions and the introduction (when expedient) of labour-saving techniques and modern materials are controversial isssues. Frequently the response of the artisan to changes in demand is to lower the quality in order to be competitive — thereby often failing to supply the quality demanded by the market. And yet considerable economic benefits have accrued to some communities where local materials and skills have been applied to new or modified products. Mass production by capital-intensive methods can result in cheaper products and a more uniform quality. Artisans can now copy some of the traditional handicrafts, using semi-industrial techniques, and the market often does not differentiate between the two. This is true, for example, of handicrafts found in trade catalogues.

However, there are a number of dangers underlying this approach. By modifying the production process and the product, the artisan becomes more dependent on the market — not only is the artefact no longer suitable for domestic use and must therefore be sold, but the materials and other requirements must also be purchased. These developments will inevitably bring artisans into closer competition with large industrial concerns, and will also imply a considerable commitment of capital.

The mass production of handicrafts is also risky because the market is notoriously fickle — what is in demand one year may be out of fashion the next. Traditional handicrafts are less subject to these trends, and the production of relatively few, high-quality pieces may enable the artisan to set his/her own price. The use of traditional handicraft techniques will also help to preserve cultural values and secure ancient skills for future generations. However, this kind of product will probably not sustain large numbers of artisans, and the sales outlets may be very specialised (museums, for example) and relatively inaccessible to the producer.

Where artefacts are being produced exclusively for sale, there is a very real danger of a reduction in quality. The problem of *quality control* may be especially difficult for artisans to understand because they often have no access to large stores, catalogues, magazines or trade fairs where they might learn of the finer points that distinguish one product from another. They therefore often fail to understand why it is that one product may be acceptable while another is not.

a: Standards and safety regulations are being established and implemented in many countries. These particularly affect toys and ceramics, where locally made paint and glazes contain unacceptable levels of lead, chromium etc.

b: Packaging and presentation. Packaging is very important, to ensure that the goods arrive safely after sometimes hazardous journeys. Goods have to be packed dry; if not, fabric and leather will develop mildew, and wood develops mould and cracks. This is particularly difficult to avoid in a humid climate and in the rainy season. Packing materials are difficult to come by, and often artisans are forced to use straw, which presents its own problems. The provision of cardboard or wooden boxes or baskets, let alone plastic bags, shrink-wrapping or blister packing, can be a project in itself. Packing is vital in aiding presentation and must include careful labelling. The handmade product benefits from a small label describing where it comes from, how it was made and its cultural significance.

c: Pricing is perhaps the most difficult problem. Often articles have

5

traditionally been produced for the home or bartered within the community. Frequently the raw material is produced locally, not brought in; the tools are minimal; and the time allocated would not be used to carry out another productive task. This is especially true in the 'dead' agricultural periods when most handicraft items are produced. At first sight, then, it may appear that the products will be cheap. However, if the product is successfully marketed externally, then the costs will rise dramatically:

☐ raw materials have to be brought in or specially cultivated;

☐ time has to be diverted from other tasks;

☐ new tools and machinery must be purchased, replacing old ones which are no longer appropriate;

☐ someone has to keep accounts, sell, and, especially when several artisans are involved, administer the project. All this can be extremely costly.

Conversely, increased sales can make handcrafted goods cheaper, since raw materials bought in bulk are less expensive, and profits can be spent on the purchase of small machines to speed up certain processes such as polishing or pumping.

Many projects falter on the question of prices; in their efforts to secure a 'fair' rate for the work, their prices are set too high to maintain a reasonable level of sales. Projects with vast inventories of unsold and probably unsaleable stocks are numerous. Occasionally project holders have direct access (usually through churches) to the expatriate community or to an export market where high prices can be maintained, but this kind of project tends to be very vulnerable, and a very limited growth of sales must be expected. Besides, these kinds of market opportunity are usually not available to small groups.

It is worth noting that a 10% damage or spoilage rate effectively increases the price by 11%, and a 25% damage or spoilage rate increases prices by a third.

It is especially important that artisans should be organised into some kind of *legal grouping*, since in so many countries it is necessary to obtain a licence to sell products. Obviously it is even more important to have legal status if direct export is being considered. Once the group has some kind of legal standing, it is easier to obtain loans, exchange information, rent premises, etc. However, membership of a group should not necessitate working in a workshop or similar centre if it is more convenient for people to work on a flexible timetable in their homes. A well-organised group should not only be concerned with securing sales outlets but also with improving the conditions of work for members and generally having an impact on the quality of their lives through the provision of education, health and other services.

3 Women and Refugees

In rural areas most development projects tend to concentrate on agriculture. Handicraft projects tend to be initiated as a sideline, and are usually tagged on as 'something for the women'. This approach fails to take into account a number of factors, for example, the participation of women in many agricultural tasks, the responsibility for children of women in most societies and the often incredible work-load they bear. However, the need for women to receive an income is often great. Most handicraft projects involving women operate as a 'cottage industry', which allows for caring of children and other household tasks. It is perhaps the most exploited section of the labour market, since women have so little choice of occupation; the wage is traditionally very low, as these tasks are assumed to have little or no value. Access to income can be a contentious issue in societies where men control all finances. It is notable that in many women's cooperatives the officers are men, not simply because men are usually more literate but because tradition so dictates.

It would be very easy to conclude that handicraft projects involving women should be approached with extreme caution. However, handicrafts can be their only source of income and can also be valuable as a way of bringing them together to share problems and experiences.

Similarly, handicrafts are often the only option available to refugees for income-generation. Skills are often all that refugees have to offer even though they may not have ready access to raw materials and tools or marketing facilities either locally or further afield. In the short run it may be possible to find a market overseas, but there are a number of problems with sustaining a handicraft project for refugees.

a: Handicraft production for income may be a new skill, and the producers may not appreciate quality control and market preferences.

b: The raw materials may differ from those in the refugees' area of origin, and the products therefore may not turn out as expected. The raw materials may be of poor quality and their supply irregular.

c: The artisans may not be able to control their market or, worse, may have little knowledge of it.

d: It is frequently their only source of income in an often costly environment. Prices therefore tend to be high.

e: Interest from overseas will tend to wane as another trouble spot takes world attention.

V Marketing Schemes

1 Oxfam Practice

In general terms, marketing is a process which includes all discussions and transactions between buyers and sellers, as well as the transport, storage, processing, packaging and advertising which occur in trading. In one way or another, a large number of projects funded by Oxfam incorporate a marketing element. These may range in size and complexity and include agricultural and industrial cooperatives and informal artisans' associations. Marketing is crucial to all production of goods and services offered subsequently for sale, and is designed to generate income and employment. The improvement of marketing facilities on Oxfam-funded projects is based on the premise that poor marketing/transport facilities act as a major constraint on increasing production and on raising incomes. Poor marketing and bad producer organisation is viewed as a major cause of the exploitation of small producers, both farmers and artisans. (For details on export procedures, and especially the report of handicrafts, see Appendix I.)

5

2 The Marketing Chain

The economic and political weakness of the small producer in the 'marketing chain' (the label given to the sequence of transactions which links the producer of a particular commodity to the consumer) makes him/her vulnerable to exploitation by those with whom he/she transacts. Most marketing chains include various intermediaries between producer and consumer, who are normally divided into wholesalers and retailers. Intermediary activities can be divided as follows.

a: Bulking or collecting. Wholesalers with capital are more likely to deal with the stages of marketing in which relatively large loads are transported over long distances and economies of scale are possible in transport and storage.

b: Distributing. Both the distribution and production stages are marked by high inputs of labour and low inputs of capital, major economies of scale not usually being possible.

Minor processing operations such as washing or packing may take place at almost any stage in the marketing chain, but major operations are almost always conducted with bulked quantities.

3 Employment and Marketing

Poorer countries are renowned for their long and complex marketing chains, the profusion of intermediaries and the large number of petty retailers, particularly in the streets and markets of large cities. These are a reflection of the underlying social and economic structure of the countries in question. This situation is usually characterised by an abundance of labour, high levels of unemployment and underemployment, and an extreme poverty that forces many into petty trading activities. This is an occupation which can be entered or left relatively easily, requires little or no capital and which is therefore especially suited to rural-urban migrants.

Depending on their points of view, observers tend to see intermediaries either as useful, functional and necessary elements in commerce, or as villains who are out to squeeze every last penny from helpless producers who are forced on to the breadline by their activities. The truth normally lies somewhere in between these views. In most cases, particularly in isolated rural areas, and where there are corrupt or inefficient local officials, powerful intermediaries may collaborate to form monopolies, paying producers extremely low prices and charging retailers/consumers prices which are too high. This situation is exacerbated when intermediaries also act as suppliers of informal credit to producers who are denied access to bank loans. Under such circumstances, wholesalers can virtually dictate their own prices to individual producers. Landowners who engage in marketing are also able to exercise monopolistic power over their tenants. The intermediary is often able, therefore, to exploit the ignorance or powerlessness of clients to his/her own advantage in any situation where free competition among intermediaries does not occur. However, it is possible to exaggerate the inefficiencies and excess profits of existing marketing systems; in reality there may be less margin for improvement than at first sight appears to be the case. The systems may not be so much exploitive as simply the product of prevailing economic conditions. Marketing ventures must therefore be studied carefully to make sure that new projects can improve on those already in existence.

4 Objectives of Marketing Schemes

Marketing schemes should aim to improve existing marketing channels or to create new outlets. They generally have one or more of the following aims:

- [ ] to increase the prices received by producers;
- [ ] to reduce the prices paid by consumers;
- [ ] to increase the efficiency of the marketing process, by reducing the number of intermediaries, or by improving transport and handling;
- [ ] to reduce wastage by the application of appropriate preservation techniques to storage, packaging and transport;
- [ ] to stabilise prices;
- [ ] to standardise and control weights and measures and commercial practices in order to reduce corruption, discrimination, etc;
- [ ] to reduce trading in impure, unhealthy or contraband goods;
- [ ] to increase the efficiency of taxation on commercial activities and/or to augment the revenue arising from it;
- [ ] to reorganise the locations or types of trading institutions, for example through the provision of regulated markets as in India, or through the adoption of 'rural growth centre' strategies;
- [ ] to disseminate information on prevailing prices through the radio, handbills, etc. in order to reduce the likelihood of underpayment or overcharging by traders;
- [ ] to provide or encourage the development of credit and savings facilities;
- [ ] to assist producers to control marketing activities through a variety of economic institutions, ranging from formal cooperatives to

unofficial associations, which are likely to increase their bargaining power *vis-a-vis* other competing interests in the market.

Clearly, several of these aims are incompatible with each other, and the project holders must decide which are most appropriate to their particular needs and circumstances. Reforms on a large scale in any single area will have repercussions elsewhere and may prejudice the interests of other groups. For example, measures designed to reduce the prices paid by urban consumers for food may adversely affect small farmers and weaker intermediaries, possibly leading to increased underemployment. Many marketing projects are intended to modernise marketing processes through increasing efficiency and by substituting capital-intensive methods for traditional labour-saving techniques. Apart from the danger of creating unemployment, such new marketing chains are likely to prove most beneficial to the large-scale producers or consumers, because these people are generally better informed, have more capital, and can act more swiftly to take advantage of new opportunities.

Oxfam aims to assist small-scale producers and to protect the poorest consumers. Projects of this type are generally based on the formation of cooperatives, or may involve the development of savings groups. A simple solution for small producers in certain circumstances is for them to agree among themselves on the prices at which they will sell, so that traders cannot force prices down by playing one producer off against another (BOL 5 and BOL 13). However, a note of caution must be sounded on marketing schemes in general. Poor producers are often up against powerful forces which are not easily swept aside. Thus, for a farming or fishing cooperative to set up its own marketing facilities (storage, transport, etc.) may involve heavy capital investment and possibly a disproportionate outside input to make the scheme viable. The danger exists that without adequate management, the substantial resources put into such schemes may in the end benefit relatively few people, or that project members may lack the expertise necessary to maintain the project independently of further foreign aid. Projects should therefore try to look at the multi-dimensional nature of the problems faced, avoid simplistic solutions and consider the total package of reforms needed to make the scheme feasible as an economic exercise which will bring long-lasting benefits to participants.

5

5 The Role of Rural Roads

Road building is often assumed to be a relatively simple, though expensive, method of assisting small farmers by providing easier access to markets for their produce. Official development programmes frequently include the building of both major and feeder roads in rural areas. Studies have shown, however, that new roads on their own do little to increase agricultural production by the small farmer. In order to realise the full potential of new roads, it has been found necessary to provide complementary services such as agricultural extension, farm cooperatives, credit, transport facilities and even land reform (see PART SIX).

Even so, increases in production tend to consist mostly of more perishable crops, which cannot be marketed before the building of new roads, and a switch to cash crops such as cocoa, oil palm, rubber and coffee. As such, new rural roads generally lead to an increase in land values, particularly for land adjacent to the roads themselves, and to competition for land. Where small farmers have no access to complementary services nor security of tenure, they are easily displaced by larger farmers, and concentration of land-ownership takes place. In such areas of increasing penetration by new rural roads, therefore, projects to assist farmers by providing additional marketing and related services through informal groups or cooperatives (storage, transport, information dissemination, land titling, etc.) can help to counter these adverse effects.

I   Savings and credit

1 General Principles

For the most part, Oxfam's involvement with savings and credit is in the rural sector (primarily in agriculture), although there have been a few instances of support given to savings and credit programmes connected with housing, handicrafts and small-scale commercial and industrial undertakings.

In virtually all societies, households and individuals both *save* and *borrow* money and real assets. *Saving* takes place during periods when income exceeds expenditure; money and/or assets are then accumulated (and may be available for lending to others). *Borrowing* normally takes place for one of the following reasons:

☐ necessity, e.g. crop failure resulting in inadequate food supply, or the need to spend money on, say, a wedding or funeral;

☐ the expectation that the funds can be used to generate an income, e.g. improving seed by buying and using fertiliser.

In many parts of the world, recurring seasonal scarcity in the period before the main harvest forces people to borrow in order to buy food and other necessities. This gives rise to a *cycle of indebtedness* in which debts incurred before each year's harvest are paid off after the harvest, but households' remaining stocks of food and cash are not sufficient to carry them through the year; further borrowing is therefore unavoidable. A severe crop failure which makes it impossible to pay off existing debts in any year can turn a cycle of indebtedness into a state of *perpetual indebtedness*. Many of the people with whom Oxfam works are in one of these two situations.

Rural people in the Third World most commonly save and borrow through indigenous or 'informal' institutions in the village or local town. The most common of these are listed below.

☐ *For saving*: hiding money; buying valuables or livestock; depositing money/valuables with a friend or broker; reciprocal arrangements with neighbours (i.e. lending in the expectation of a reciprocal loan at a later date); rotating savings groups.

☐ *For credit*: friends and neighbours (as part of a reciprocal arrangement); rotating savings groups; moneylenders and traders; landlords.

In addition, people may have access, in varying degrees, to the following formal institutions.

☐ *For saving*: post office; commercial bank; cooperative/credit union.

☐ *For credit*: commercial bank; cooperative/credit union; specialist government credit agency.

It is very widely believed that informal lenders (such as specialist moneylenders, traders and landlords) exploit poor rural people. Oxfam has plenty of experience of cases where this is so, and of supporting projects which attempt to liberate debtors from informal lenders. However, the points below must be recognised.

a: The informal lenders *may* serve borrowers reasonably well in that they provide them with loans which would not be available from formal sources because: they are for very small amounts; they are made without security; and they are for purposes which may be urgent to the borrower (e.g. medical fees) but which would be regarded as 'unproductive' by a formal lender.

b: The informal lenders' rates of interest may not be excessive in terms of the real cost of capital in poor communities, and informal lenders also disburse speedily and with a minimum of formalities.

c: To attempt to replace informal sources of credit by, say, a credit cooperative using external funds may not be an effective way of meeting real needs.

d: People can also run into difficulties as a result of taking loans from a formal credit institution.

2 Conventional Credit and an Alternative Approach

It is today a very widely accepted theory that small farmers can only become more productive if given the opportunity to borrow money to buy improved seeds, to apply the necessary fertiliser and pesticides, or to develop water resources.

As a result of this theory, many rural development programmes incorporate credit schemes which disburse quite large sums to small farmers, often tied to a 'package' of inputs which enables them to adopt a particular agricultural innovation such as a new crop variety. Common results of this type of programme include:

- [ ] the 'misuse' of loans for buying consumer goods, paying rent and taxes, or meeting existing debts;
- [ ] failure to recover loans, because they have not been applied to the purpose for which they were made;
- [ ] the swamping of traditional savings and credit arrangements (e.g. rotating savings groups) because external loans (without saving) appear an attractive alternative to the smaller amounts available through traditional arrangements;
- [ ] inaccessibility of loans to the smallest and poorest farmers (because they are unable to offer security, are illiterate, are regarded as technically 'backward', etc.);
- [ ] positive damage to small farmers where the technology tied to the loan is a risky one and where the benefits of adoption may not, in the case of a crop failure, be enough to meet loan servicing and repayment requirements.

The experience of Oxfam leads it to reject the conventional approach to credit outlined above, since it tends to be both inefficient and inequitable. Oxfam's experience calls in question the assumption on which conventional credit is based, namely that the rural sector can generate so little in the way of savings that there is no real alternative to the large-scale injection of funds from outside. Village savings clubs have shown the contrary, that many small farmers can mobilise savings through appropriate institutions.

The approach now favoured by Oxfam can be regarded as having three elements:

- [ ] identifying the community's credit needs (including the different needs of different individuals and groups);
- [ ] identifying how these are met by existing savings and credit institutions;
- [ ] devising programmes based on a close assessment of these needs and how they are met. Such programmes may encourage the growth of new institutions which more adequately meet the community's needs, or they may attempt to devise ways of making existing institutions work better.

Three main types of project or programme can be identified:

- [ ] those which encourage more effective saving;
- [ ] those which help poor people to gain access to existing sources of credit which would otherwise be denied them;
- [ ] those which lend agency funds direct to individuals and groups.

The purpose of any savings or credit programme is to enable people to gain access on reasonable terms to assets which they can use to improve their livelihood. Which of the above approaches will be appropriate in any situation depends on a close assessment of that situation.

Households make many demands on ready cash, including any that becomes available through credit. The most crucial demands are those made by:
- [ ] the need for basic commodities (especially food);
- [ ] the need to pay rents and taxes;
- [ ] the need to pay off existing debts.

Others are ceremonies (weddings and funerals), school fees, medicines and medical care, durable consumer goods (e.g. cooking utensils, bicycles, radios) and gifts and loans to other households (towards whom customary obligations exist and who may themselves have particularly acute needs).

Productive investment (e.g. in fertiliser or high-yielding seed) often comes *low* on the list of priorities for a poor household, and indeed the distinction between 'production' and 'consumption' credit is not always clear to such a household. Where money is needed very urgently for a particular purpose (to pay school fees or taxes or to pay off existing debts, for example), then any funds coming into the household will be used for that purpose first, regardless of whether they were intended for an agricultural project or not. In other words, a lender cannot possibly ensure that any loan is applied to a particular purpose unless this purpose accords with the borrower's own priorities. Poor households are best helped by credit which enables them to meet their most pressing needs and, at the same time, results in an increase in income large enough to enable them to repay their loan without hardship.

In both these ways, 'liberation loans' to pay off existing debts have been a valuable form of assistance to landless artisans in Cuddapphah, Andhra Pradesh (AP 22) and in the Oxfam West Orissa Project (OXWORP — ORS 20). Traditionally the artisans in West Orissa depended on advances from moneylenders at a high interest rate (over 125%) to obtain both working materials and consumption necessities, and they were in perpetual debt. A revolving loan fund charging a high rate of interest, but with less stringent repayment conditions than the moneylenders, enabled the artisans to free themselves from dependence on the moneylenders, establish a revolving fund for loans to a similar group in a neighbouring village, and eventually reduce interest rates.

4 **Existing Savings and Credit Institutions**

The distinction between informal and formal institutions has already been made, and some examples of each type have been given. Any community will be served by some, or possibly all, of these institutions. Before attempting to put into effect a programme aimed at improving people's access to credit, it is important to ask:
- [ ] which formal and informal institutions are used?
- [ ] by which groups of people are they used, and whom do they benefit?
- [ ] do they function efficiently?

An investigation aimed at answering these questions might yield the results below.

a: Traders and moneylenders *may* offer a reasonably efficient and non-exploitive means for small farmers to borrow, because they understand their clients' needs, are able to assess accurately individual clients' credit-worthiness, and lend without requiring security; on the other hand, they may be highly exploitive.

b: Traditional rotating savings groups *may* function effectively, but the savings made by members are always used to buy consumer goods because the quantity and timing of payouts makes it difficult for them to be used for investment.

c: Saving may still be achieved largely by hoarding, and the encouragement of small informal groups may be an important means of breaking away from this.

d: Small savers *may* use formal institutions such as the post office and commercial banks; on the other hand, they may be prevented from this by a minimum deposit requirement, by physical distance or by mistrust of these institutions.

e: Small borrowers are unable to make use of credit funds on offer by formal institutions (e.g. rural banks, cooperatives, credit unions) because they are unable to provide security or guarantees, or because the amounts they need to borrow are below the institution's minimum.

⑤ Savings and Credit Programmes

*a: **Encouraging more effective saving***. In some communities, access to credit may best be obtained by encouraging saving. One important way of achieving this is through savings clubs. In Indonesia a traditional form of rotating savings club, the *arisan*, is widespread. The club meets weekly and each member makes a small payment into a central fund, the whole of which is handed over to one of the members (usually chosen by drawing lots). Adaptations of the *arisan* have been:

☐ to allocate the accumulated funds to individual members of the group in rotation rather than by lot, so that the recipient knows when he/she is to receive it and is able to plan how to invest it (IDS 16E);

☐ banking the accumulated funds and so building up a deposit from which any member of the group can borrow (IDS 62).

An inventive version of the rotating savings group, the 'Saveway Club' (ZIM 22), has been pioneered in Central Africa with Oxfam support (see Section 4). Initially devised by workers at the Institute of Adult Education at the University of Zimbabwe, the Saveway Club was, unlike the *arisan* in Indonesia, a new institution to the rural people to whom it was introduced. It was adopted with some success, however, because it was related to their needs and capacities.

The *credit union* is a type of formal cooperative which can facilitate saving by members, make available the savings of some of its members to others as credit, and channel external funds to its members as credit. The credit union is described more fully below (see Section 4).

*b: **Making existing institutions work better***. Formal commercial financial institutions (i.e. commercial banks as distinct from government agricultural development banks) find major barriers to making loans to the rural sector in that:

☐ they cannot give loans without security;

☐ it is very costly, in relation to the size of the loans, to administer loans to small farmers and other isolated or scattered borrowers in villages.

One method of giving small rural borrowers access to the funds of commercial banks (and in the process often liberating them from moneylenders) involves the placing of deposits with banks to guarantee loans which the banks then make to approved groups. Projects receiving bank loans in this way include the Ponnur and Dindi schemes in Andhra Pradesh (AP 24N) to assist marginal farmers and landless labourers to recover from the cyclones of 1976 and 1977. All participants in these schemes have holdings of 2.5 acres or less and would otherwise, because of their small landholdings, have been totally excluded from the formal credit market. These schemes have to date achieved a 90% rate of repayment. In Gujarat a deposit made with a commercial bank by the Rajpipla Social Service Society (GUJ 53C) has secured a loan equal to three times the amount of the deposit for a group of Adivasi (scheduled caste) dairy farmers to enable them to buy milk buffaloes and to establish a marketing cooperative. The nationalised Syndicate Bank of Manipal, Karnataka, has established a subsidiary, the Syndicate Agricultural Foundation, as a means of increasing its involvement in rural development ('barefoot banking'), and Oxfam has funded (KN 46) audio-visual equipment for the Foundation to use in extension work.

5

c: NGOs and Credit. It may be necessary for NGOs to provide funds for credit where the two types of programme described above — those which encourage saving and those which open existing sources of credit — are inappropriate. This is likely to be so where the capacity for saving is negligible, or credit from informal or formal sources is not available on equitable terms.

These conditions applied where liberation loans were given to artisans in West Orissa. This type of project will almost always involve advancing money (as a loan or as a grant) from a revolving fund and using a formal (cooperative) or informal group as a means of disbursing loans to individuals.

The following points about the use of agency funds need to be stressed:

☐ the purpose for which the loan is made must be income-generating;
☐ repayment within a maximum period of 2-3 years should be possible;
☐ the groups themselves should contribute a sizeable proportion of the total cost of the project (possibly in the form of labour);
☐ a reasonable rate of repayment is set, within the capacity of the recipients, possibly by discussion within the groups themselves;
☐ loans should *not* be made free of interest or at a subsidised rate of interest (e.g. one below the national rate of inflation).

These conditions should ensure effective use of the funds and a good repayment record. Where the investment made with a credit loan shows a high rate of return, there is no *need* to subsidise the rate of interest; where the investment results in a significant increase in the income of the borrowers, they will have the capacity to repay the loan without great difficulty. Where the borrowers have themselves been involved in decisions about the terms of the loan and where they have committed resources of their own to the investment, they are not likely to regard the loan as a hand-out or to try to exploit the agency providing the funds by failing to repay. Where these conditions are not met, it is probably unwise to commit agency funds directly to a credit programme.

6 Key Issues on Savings and Credit

Faced with the recognition that conventional credit programmes often have the effect of intensifying inequality in that they either exclude poor farmers or lead them into commitments which they are unable to meet, voluntary agencies are concerned to implement programmes which help the poor.

To encourage savings through groups for the purpose of making a specific investment is an attractive option but will hardly meet the needs of those who are heavily indebted; to serve them may require 'liberation loans', which may not be for a purpose normally thought of as 'productive', but which may well be repaid without difficulty by recipients released from heavy interest payments.

To make loans or grants to set up revolving loan funds for informal groups is an exercise fraught with difficulty, in spite of some notable successes. For example, borrowers have sometimes treated voluntary agency loans as a 'soft option', possibly in the knowledge that the agency will not go to extreme lengths to recover debts, partly for practical, partly for ethical reasons. Experiences of this type raise the question of whether it is possible to act as both development agency and 'mini-banker'.

One possible way out of this dilemma is to help the poor to gain access to the commercial banking system by guaranteeing loans. However, the fact that commercial banks are necessarily profit-oriented must raise the question of how widely they can be used as a means of providing credit for the poor.

Revolving Loan Funds

1 Oxfam's Experience

Revolving loan funds (RLFs) are designed to provide credit for those engaged in the production of goods and services, usually farmers or craftsmen. A grant is made to the project holder who, in turn, lends the money to participants. On repayment, the cash stays with the project for further lending to finance additional production, rather than being returned to the donor agency (as in the case of formal loans).

Grants are normally made for RLFs where projects experience difficulty in obtaining access to institutionalised credit. Such conditions are quite common in developing countries where formal organisations such as cooperatives embrace only a small percentage of producers or where, either for political or logistic reasons, poor farmers and artisans are denied access to bank credit. For example, in some countries a farmer is ineligible for bank loans unless he/she has a legal land title. In fact, most Oxfam grants for RLFs have been made in agriculture (more often than not through informal associations), with less frequent examples in the fields of craftwork and women's training.

Oxfam's experience with projects involving RLFs has not been wholly satisfactory, and a large proportion of such schemes has not lived up to expectations. A common situation is where a project has allowed its RLF to diminish gradually over time, repayments from members having failed to maintain the value of the working capital. Eventually, either the scheme folds or the project is obliged to request another injection of funds.

2 Why RLFs Go Wrong

Administering an RLF is indeed a far more complex exercise than many people imagine, and a variety of factors may be responsible for failures. These should be carefully considered when investigating new requests.

a: A major reason for failure is that the income derived from sales financed by the RLF rarely meet the producer's expectation in terms of higher earnings or profits as well as giving a satisfactory margin to cover loan repayments. In the case of a small farmer, for example, prevailing market forces may well squeeze his/her profit margin down to the absolute minimum, the larger surplus being extracted by more powerful landowners and merchant intermediaries. The poor producer is then forced either to default on his/her repayments to the central fund or to delay them for such a long time that inflation significantly diminishes their real value.

b: A number of factors may mean that the project holder (cooperative, association, etc.) is, despite the best intentions and a realistic assessment of the financial situation, unable to charge rates of interest which would make the RLF economically viable as a long-term proposition. For example, the project may be competing with other producers who are in receipt of heavily subsidised official credit. An NGO-funded scheme cannot subsidise its members without regular fresh inputs of outside cash — a situation which Oxfam tries to avoid since it defeats the very purpose of the grant, which is to make the project financially independent. The sad truth is that, in many cases, projects are caught in a vicious circle of dependence which is the product of the economic and political weaknesses in society, and which an outside agency, no matter how large, simply cannot break. In such situations, therefore, the wisdom of making grants for RLFs should be called into question. An element of permanent subsidy may be the only option in certain circumstances. This would then have to be weighed up against the possibility of having to forfeit the wider educational goals that so often result from financial self-sufficiency.

c: Of course, many disappointments in this field can be attributed to a

5

lack of experience in handling loans. Just as many recipients of loans fail to see possible difficulties they may have with repayments, so RLF administrators themselves frequently underestimate the hazards of handling such schemes. They may, for example, fail when calculating repayments, to make allowance for the effects of inflation, which can depreciate the real value of working capital in a very short space of time. Indeed, unpredictable and unstable markets and prices may make such a task virtually impossible for small organisations.

d: Often administrative ability in the handling agency is lacking, and the simple failure to keep adequate records makes proper reporting/accounting impossible and leads to confusion.

e: Easy access to loan funds often encourages the farmer to progress faster than his/her managerial abilities allow, resulting in poor usage of the items provided by the loan.

f: The economic feasibility of the scheme may not have been adequately appraised. This need not necessarily imply a formal cost-benefit analysis, but should at least include some sort of systematic attempt to examine the long-term viability of the project. For example, unless there is adequate investigation before a loan is disbursed, a farmer may find him-/herself with only some of the inputs required to benefit from the loan. A good example of this is a loan to purchase an irrigation pump without provision being made to buy the necessary improved seed, fertiliser and insecticide. A form of package loan seems to be the answer in these cases.

g: Loans are often made in times of stress such as drought or flood, to enable farmers to make good losses of seed, livestock, etc. At such times any form of assistance tends to be accepted without understanding, and there is rarely enough time to make the recipient aware of his/her other liabilities.

h: Loans for the purchase of draught animals must take into consideration the farmer's need to earn a much larger income to cover the repayments, usually entailing the cultivation of a larger area of land.

i: Settlement schemes often include a large loan element to cover the cost of housing, bush clearance, seeds, etc. Repayments are normally made through a single-channel marketing system, but are often so phased that an excessive burden falls on settlers in the early years, leading them to abandon schemes. This suggests the need for simpler, less capital-intensive housing, roads and water supplies, and differently phased repayment schedules.

j: Finally, it should not be forgotten that many of the agencies handling loan funds are seen as soft options by the recipients, who do not therefore take loan repayments seriously. Any educational pretensions that the development agency may have will thus be seriously endangered.

3 Conditions for Success

Where RLFs have been successful, they have included certain common ingredients.

a: The loans have been carefully administered with adequate supervision, record-keeping and a realistic loan repayment timetable backed up by some degree of law enforcement.

b: There has been a face-to-face relationship between the lender and the borrower which keeps the situation on a more personal level. This severely restricts the operation both in size and geographical area but results in much greater mutual understanding. Of course, a situation could arise in which such familiarity, without adequate financial controls, could lead to laxity in repayments and to general inefficiency.

c: The programme has increased the borrower's income, enabling him/her to repay the loan while at the same time increasing his/her own personal benefit. It is unrealistic to expect the borrower to live at a poorer level than before, in order to pay off a loan.

4 Administration of Small-scale Credit

A formal loan agreement in the local language must be made with the borrowers to remind them of their obligations to the central RLF.

Periodic statements of money owed should be circulated to borrowers.

Frequent visits by responsible staff must be made to build up a personal relationship.

Where possible, single-channel marketing of produce should be encouraged to enable a stop-order system to be implemented. Participation of borrowers in the marketing system is very important to prevent accusations of price exploitation.

Village-level loan committees, made up of trusted locals chosen by the communities, should act as a primary vetting organisation which can also put pressure on defaulters, and have considerable educational value.

If a project feels that it has inadequate managerial and/or administrative capacity, it should approach the funding body to discuss the possibility of further training in this field.

5 Points for Field Staff

Firstly, consider carefully the long-term economic feasibility of the scheme to be funded, and ask whether or not it is likely to generate enough income to enable the RLF to be maintained at a sufficiently high level, independent of further financial aid from outside.

In a high-risk project, a smaller initial amount with a restricted group would lessen the danger of large-scale collapse and act as a pilot scheme for a larger initiative later on.

Make sure an efficient reporting schedule is set up to give a complete list of all the borrowers, showing the amount borrowed, the purpose, the terms of the loan and the actual repayments made.

Examine whether the project has adequate managerial and/or administrative capacity to cope with the grant being applied for and the subsequent monitoring of members' loans. Will a realistic rate of interest be charged, or will the project's working capital simply evaporate with inflation after a year or so?

Could the activities for which the loan is needed be better financed by the formation of a savings group? This would take longer but has some advantages.

5

I   Cooperatives

1 Introduction

A cooperative is defined here as a legally registered organisation with a formal structure including membership, board of directors and official accounts, etc. This is to distinguish cooperatives from informal groups or associations, which have their own characteristics and are considered separately (see IV. below).

The cooperative idea originated in 19th-century Europe. It has been widely promoted in the Third World by colonial governments and, more recently, by independent national governments and international organisations. However, cooperatives in the Third World based on the Western model have more often been failures than successes. Failure of a cooperative is commonly associated with some or all of the features listed in sub-section 2 below.

a: The *objectives* of a cooperative may be a single purpose, or some combination of the following:

☐ to enable small-scale producers, consumers and traders to buy and sell with the same economies of scale and bargaining power as their large-scale competitors, or to process their produce to add value;

☐ to enable several individuals to share a major resource, e.g. an irrigation pump, a weaving loom or a piece of land;

☐ to help a group of individuals to save money which may then be used for individual purchases, or to provide credit for members of the group, or to further other aims of the cooperative;

☐ to create employment by organising the production or marketing of handicrafts, furniture, etc., or providing a service;

☐ to overcome exploitation of people in specific situations.

Some cooperatives are established for a particular limited purpose (single purpose) while others combine a series of functions and have several objectives. Experience shows that single purpose cooperatives are often more stable and effective because their management structure and operations are simpler. Multipurpose cooperatives are complex and need to ensure that their various activities are compatible. Difficulties may occur in determining priorities for allocation of resources, both in finance for development, and in staff time for management.

b: *Cooperative principles* include the following:

☐ membership should be voluntary and available without restriction to all persons who can make use of its services and are willing to accept the responsibilities of membership;

☐ the organisation should be democratic — administered by persons elected or appointed by their members and accountable to them. One member should have one vote, and should be able to participate in discussions affecting their cooperative;

☐ the share capital should only receive a strictly limited rate of interest, if any;

☐ any profit arising out of the operations belongs to the members. If distributed to the members it must be in proportion to their transactions;

☐ cooperative societies should make provision for educating their members, employees and others in cooperative principles;

☐ in the interest of their members all cooperatives should actively cooperate with others at local and national level.

Most countries have legislation governing the establishment and running of cooperatives. Under legislation of this type, all cooperative societies must register with a central Department of Cooperatives, and are then compelled to comply with a set of highly specific regulations governing such matters as the issuing of shares, the keeping of accounts, admission to membership, and the election of officers. These regulations can be, and often are, circumvented in practice, resulting in corrupt management and lack of democratic control of cooperatives. On the other hand, their existence can severely restrict the development of groups which may attempt to achieve some or all of the cooperative functions outlined above, using a simpler form of organisation, which may be a 'non-formal' cooperative or pre-cooperative.

2 Problems of Cooperatives

Experience shows that the failure of cooperatives is often associated with the factors below.

a: Failure to involve the members adequately in decision-making. This may be a direct effect of the complexity of cooperative regulations, which members may find difficult to comprehend, with the result that they are dependent on professional staff. The situation is exacerbated if the staff concerned leave and are not replaced by others of similar competence. Typically as a result, the cooperative gets into a vicious circle of poor member participation — poor benefits to members — poorer member participation.

b: Use of the cooperative for a *purpose determined not by the members* but by politicians or officials: for example, as a vehicle for a government credit or input distribution scheme, or as a monopoly buying organisation for cash export crops. Cooperatives are sometimes set up as instruments of government policy. Although it may be possible to use existing cooperatives to implement specific aspects of government policy, this depends on whether the cooperative members accept given policy measures as a legitimate objective of their society.

c: Lack of administrative skills. A 'mismatch' between the organisational requirements of the cooperative, as laid down by cooperative law, and the skills/capacity of members of the community. For example, effective participation in a formal cooperative requires literacy and numeracy skills which, in many Third World communities, a large proportion of adults may not possess. Without these skills, members cannot check, and participate in, management through the committee.

d: Structural conflicts between the cooperatives as an institution and the society into which it is introduced. For example, democratic elections to committee membership are likely to be meaningless in a highly stratified society. The responsibility of a cooperative committee member or official, for example, to control/speak out against corruption and inefficiency, may conflict with his/her general community role and status which require him/her to be deferential towards those older or more powerful.

e: Poor or corrupt management is often caused by the failure or inability of members to control managers. Lack of management skills and ineffective board members who do not fulfil their responsibilities contribute to it.

f: Breakdown of democratic control and the taking over of the cooperative from within by narrow interest groups such as large farmers or traders in order to gain for themselves the benefits of cooperative membership (e.g. access to input supplies, control of pumps, processing equipment), or prevent poorer members of the cooperative from increasing their economic power.

g: Direct *attempts to destroy* the cooperative:
☐ from within, by those opposed to the cooperative who gain membership and then sabotage it, e.g. by defaulting on large loans;
☐ from outside, by those who withhold supplies, deny employment to cooperative members, or use physical violence.

h: Failure to become self-reliant because of lack of participation by members, and insufficient contribution or usage, can lead to the cooperative being kept alive artificially by aid agencies.

i: Misappropriation of funds can occur when staff lack training in financial control and budgeting. This problem may be compounded by poor or non-existent auditing procedures. Where possible, copies of accounts should be available in local languages.

j: Appointment of *managers from outside* the local area, or even other cooperatives, or of government officials whose loyalty and motivation is not given to, or motivation derived from, the members but their paymaster.

k: Dependence is often created by well-meaning NGOs or a number of *donors acting independently* to fund the cooperative.

3 Setting Up and Advising Cooperatives

In many circumstances a formally-structured cooperative may be unnecessary or even disadvantageous. Donor agencies should therefore ask whether a cooperative is the most appropriate organisational form in relation both to the objectives desired by the particular group or community, and to its capacity to cope with a relatively complex organisational structure. The advantages of alternatives are examined in greater detail later in this section.

Formal cooperatives often fail to benefit the poorest members of the community. The very nature of a cooperative requires it to pursue specific objectives which are in the interests of its members. Thus, a cooperative whose members are small and medium farmers is unlikely to pursue objectives which would serve the interests of poorer groups, such as sharecroppers or landless labourers. For such a cooperative to try to reach the poor by, for example, offering membership on favourable terms, would tend to be counter-productive for the cooperative members, with no real benefits for the poor.

The Multipurpose Cooperative Association in Hyderabad, Andhra Pradesh (AP 27) assists the development of village multi-cooperatives that embrace a number of activities of interest to the poor in the local area. These include shops which provide goods and services at lowest cost. They can also encourage employment-generating activities based on production of goods or provision of services. These are likely to be off-farm activities and appropriate for those without land.

It is also useful to identify specific needs of particular disadvantaged or poor groups, and to establish cooperatives appropriate in structure and purpose to meet these. The benefits which have been achieved for the poor by cooperatives or cooperative-type groups are outlined in the examples given below.

a: New economic and employment opportunities have been created by opening markets, training in necessary skills, obtaining materials and equipment, e.g. Fotrama, a Quechua Indian women's knitting and weaving cooperative (BOL 22).

b: The poor have gained in economic power *vis-a-vis* suppliers, purchasers and landlords. For example, in Brazil, a chain of cooperative stores — *minipostos* — exchanges food and basic consumer essentials for agricultural produce and handicrafts, so keeping both selling and purchasing away from intermediaries who are otherwise in a position to squeeze peasants from two sides (BRZ 148).

c: The poor have gained access to staple products at fair prices through consumer cooperatives. Although a much larger amount of Oxfam aid goes to marketing and production cooperatives, consumer cooperatives are in some situations an effective way of assisting the poor and preventing their exploitation (BOL 32, COL 33 and ETH 88).

d: The poor have been helped to gain access to productive resources which are denied to them except as members of a group. Among cooperatives and groups with this aim are:

□ landless farmers' cultivation groups, giving access to the key resource of land (ORS 22, BD 96 and ZIM 84);

□ irrigation groups giving access to, and — equally important — control of, irrigation water (BIH 10, ORS 7 and ORS 21);

□ industrial and handicraft production cooperatives, giving access to capital equipment (BRZ 111, BRZ 141 and BOL 22).

In order that a cooperative may function effectively, education of members is almost always a necessary part of its activities. In some cases the educational role of a cooperative goes well beyond what is needed simply for the cooperative's functioning; it takes on an importance at least equal to that of its commercial activities and makes a wider-ranging impact.

A cooperative's educational programme may include the following:

□ training in specific skills required by cooperative activities, e.g. crop and/or animal husbandry, weaving;

□ literacy and numeracy training which may be either 'functional', i.e. linked specifically to the requirements of running the cooperative, or part of a separate programme. The programme should be aimed at helping members towards an increased awareness of their identity and their social and political role, as well as a general understanding of their physical environment, and of ways of extending control over it through knowledge of health care, hygiene and nutrition.

An example of a wide-ranging programme of education connected with women's cooperative groups and developed in response to the expressed needs of members is that of the Women's Self-Reliance Movement in Faridpur District, Bangladesh (BD 76). The Fotrama Knitting and Weaving Cooperative in Bolivia (BOL 22) has an educational programme aimed at the specific need of its Indian women members to develop an awareness of their ethnic and personal identity.

4 Key Features of a Successful Cooperative

One of the most successful formal cooperatives helped by Oxfam, an agricultural marketing federation in Central America (GUA 24), has put forward some basic principles for the organisation and management of cooperatives.

a: Members. Member participation is crucial to the democratic principles of cooperation. It depends on the extent to which members have received concrete, identifiable benefits which in turn are directly proportional to the time, effort and money they invest.

b: Board of Directors. The Board should be composed of two independent committees, an administrative committee and a vigilance committee. Prevention of mismanagement and/or corruption depends on these committees carrying out their duties. Officers of the Board should be elected on the basis of their honesty, capacity and participation in the organisation.

c: Manager. Although the manager should not usurp the functions of the Board of Directors, the selection of a competent manager is the most important step a cooperative can take in its organisational development.

d: Commercial enterprises. The strength of a cooperative is measured in terms of the success of its commercial enterprise(s); this success requires that a useful service is provided to members, and that at the same time a profit is generated. This is most likely to be achieved if:

□ the cooperative begins on a small-scale with one or a small number of enterprises, and aims to build up the scale and scope of its activities gradually;

□ the skills of the management are equal to the scope and complexity of the commercial enterprise;

□ credit is not extended to members until an effective system for ensuring repayment is established.

e: Financing. The shares which members buy in the cooperative are both the foundation for the sound financial growth of the organisation, and the basis of the spirit of self-help by which members maintain their dignity. The amount a cooperative can borrow and effectively manage is directly proportional to the amount its members have invested. The size of grant a cooperative can manage efficiently is directly proportional to the amount the members are willing to put into the project for which the grant is given, so that at no time does the cooperative become dependent on a grant. In general the cooperative must be aware of the intrinsic liabilities of grants and loans.

f: Accounting. The sophistication of the accounting procedures and controls needs to keep pace with the growth of the cooperative's operations. The financial statements of the cooperatives need to be written in a manner which members can understand.

g: Educational services. The success of a cooperative's educational programme depends on its helping members to meet their needs. This is likely to be achieved where the programme has practical, measurable goals, of value to both members and the organisation. A relevant educational programme encourages a high level of member participation, which in turn may have a proliferating effect as members teach others what they have learned.

⑤ Ways in which Field Staff can Assist Cooperatives

Advise project leaders of the potential problems associated with cooperatives. Try to ensure that they do not plan to foist a preconceived economic enterprise on a group.

Insist that the people themselves, *including the women*, are involved in the planning, implementation, monitoring and evaluation of the activity.

Look for ways in which the voice of the people can be heard locally and regionally so as to influence the environment in which they operate.

Ensure that the economic activity has a good chance of viability during the project's life. It is helpful to strive for early success in order to reinforce commitment.

Emphasise the importance of training in skills such as how to participate, conduct meetings, keep simple accounts and do budgeting and costing, etc. Grants for training organisers, members and managers are an effective way of helping the cooperative.

The strength of many individual cooperatives or groups lies in their links with other similar organisations. One small voice can be joined with others to make a large voice in order to influence development policy, and to try to ensure that the rural poor have access to material resources and opportunity to acquire power. Grants for creating linkages or unions of cooperatives in order to take advantage of scale could be useful (see PART THREE, Section 5).

Support for central cooperative organisations like the Multi-cooperative Association in Andhra Pradesh (AP 27) could be valuable if they are trying to improve the effectiveness of member cooperatives with identifiable policies and training programmes.

Try to match the aid input to the level of support and commitment shown by the people. Avoid creating dependence. Take note of other donors and try to assess the effect of a financial input in terms of encouraging self-reliance.

Consider support for consumer shops at the other end of the food chain. It can be an effective way of helping all the population including the poorest.

Examine carefully requests for the balance between expected economic and social benefits. Social benefits flow from economic benefits. Cooperatives intending to provide social benefits only can never become self-reliant. Some other form of organisation may be more appropriate.

II Credit Unions

Credit unions are a type of formal cooperative based on a nineteenth-century German model. The International Credit Union Movement has its headquarters in the USA, and this form of organisation has been promoted in a large number of developing countries through national credit union 'parent bodies'. Oxfam has assisted the Credit Unions and Savings Association (CUSA) of Zambia (ZAM 21), mainly with promotion of its educational activities.

Credit Unions tend to face similar problems to formal cooperatives. They are, compared with Saveway clubs (see below), a relatively sophisticated form of organisation, requiring literacy on the part of members and professional help with the running of groups. They differ from Saveway clubs also in that they enable credit from *outside* the group to be channelled into it, once the group has established itself by a period of saving. This carries risks of bad debts and the consequent failure of the group — a risk which does not exist in Saveway clubs. One of the reasons suggested for the failure of the Saveway movement in Africa is that Saveway clubs find it hard to compete with credit unions, partly because the latter offer credit as a 'carrot', and partly because national cooperative movements and credit union associations are usually keen to assimilate Saveway groups.

III Saveway Clubs

Saveway clubs are a particular model for Savings groups. The basis of the Saveway system, first developed during the early 1970s in Zimbabwe (ZIM 22), is the use of stamps to record deposits made by club members at weekly meetings. The stamps are stuck onto a card held by each member and the cash is collected and placed in a post office or bank account in the name of the club. When a member wishes to withdraw his/her savings he/she uses a card of stamps (or a certificate issued in exchange for a full card) for cash equal to the face value of the stamps or the certificate.

a: The advantages of the Saveway club are various. For example, it performs a function similar to that of a savings bank in situations where the savings which people are able to make are too small, or where they are otherwise prevented (by illiteracy or by distance from the bank) from opening accounts. It needs only one literate member per group (to hold the group's post office or bank account). Through group membership and group meetings, members encourage each other to save; or alternatively, the group can be used to provide an object for the activity of saving, as when members decide to use their accumulated savings to buy fertiliser or building materials. Materials can be bought by all members at the same time, at the end of a saving 'cycle' — this may lead in turn to the group performing functions other than that of saving, such as collective buying of materials (perhaps negotiating a discount with the supplier) and hiring transport. The Saveway club is a means of encouraging people to meet regularly, an event which can be used for extension/instruction in agriculture, health and child care, nutrition, etc. Alternatively, savings clubs can be 'grafted on' to groups already meeting for such purposes, including women's groups.

An evaluation of the 'Saveways' scheme in Kenya, Malawi, Tanzania and Zambia, sees the scheme as a first step in the following 'developmental pattern' of use of credit and savings institutions:

☐ a simple savings club requiring minimum literacy by members, leading naturally to

☐ a savings and credit-union type organisation where a greater degree of literacy is required, leading eventually to

☐ possession of a deposit and current account with a commercial or national bank.

b: Saveway clubs are not appropriate, however, in situations where people are in perpetual debt, so that any cash available, once the most basic subsistence needs have been met, is already committed to payment or partial repayment of loans. Furthermore, a high rate of inflation results in the real value of savings falling while they are being accumulated. Nor would Saveway clubs be suitable where there is no realistic investment opportunity for the savings which individuals or groups can make within a reasonably short period of time (10-12 months); in other words, where savings capacity is low in relation to investment opportunity.

Nonetheless, where people are not being seriously exploited and where the time-value of money is not unstable, savings clubs have shown that even poor people have a significant capacity for saving. Even in the poorest communities, savings of the equivalent of £67 per year have been made, even though the value of a stamp has been as low as 3-4 pence.

IV   Associations and informal groups

Many of the projects funded by NGOs comprise informal groups or associations rather than formally structured institutions such as cooperatives or labour unions. The development of informal group work is dealt with in detail in PART FOUR, Social Development Guidelines, as are the principles and educational philosophies which underlie it. It is enough to say here that informal groups have a number of advantages over formal organisations. They have no official links with government which might place them in a dependent or subservient position, nor obligations such as audited accounts or payment of taxes, etc. Consequently, they are relatively easy to set up, and it may even be possible to use or adapt existing groups established for traditional purposes such as saving, house-building or farmwork. An example is the network of cereal banks (VOL 53) encouraged by a local voluntary agency in Burkina Faso. Setting up a cooperative, on the other hand, frequently entails grappling with complex legal, organisational and managerial problems, which may well be beyond the capacity of inexperienced producers.

Such groups have a wide range of purposes, but all of them contain an element of group saving and/or use of an external loan to further their objectives. A common start is a small grant to set up a revolving loan fund because such informal groups, by their very nature, may have no access to institutionalised credit such as cooperative banks. These arrangements have had mixed degrees of success, and extreme care must be taken when setting up RLFs to ensure that community groups are able to manage them efficiently (see Section 3.II on RLFs). Oxfam has funded hundreds of such groups pursuing a variety of activities from farming and house-building to handicrafts and health care. Occasionally grants have been made directly to village or community groups but, more often than not, it is done through intermediary, preferably indigenous agencies, which are better placed to monitor progress and provide educational assistance in the broadest sense.

Informal groups are usually confined to isolated communities or, at best, small numbers of villages in the same geographical area. One of the disadvantages of informal groups is that it may be difficult to create a unified movement from a collection of scattered groups. With effective central direction, however, this obstacle can be overcome. A case in point is the Sarvodaya Shramadana Movement (SRI 37), with 100,000 participant families undertaking a vast range of activities including road- and well-building, sanitation, pre-school education, training programmes, community shops and village industries. Apparently unstructured village group work can also lay the foundations for specific institution-building at a later stage. The Community Movement stimulated by the Brazilian church and other local development organisations, for example, has helped to build a stronger rural syndicate movement better able to defend the interests of poor farmers.

V Small businesses

In order to distinguish small businesses from other types of income-generating activities (associations, cooperatives, etc.) funded by NGOs, they may be defined as those concerns which are not linked to broader educational goals, but are engaged in the production of goods and services simply for a cash profit. This is a low priority for most NGOs, since their projects are normally intended to have a wider social, as well as an economic, impact (see PART FOUR). However, there may well be occasions when the funding of small enterprises is justified. This could be, for example:

☐ in situations of extreme poverty, where a large proportion of the population depends on independent activities, e.g. small traders and artisans;

☐ where social or educational goals could be effectively pursued after a minimum level of economic achievement had been reached.

In practice, it is not always easy to find groups predisposed to work on a collective basis, or to pursue the types of educational objectives set by Oxfam as prerequisites for funding. In highly fragmented and divided societies such as India, for example, caste divisions among the poor frequently render such goals impossible. In these cases, the only feasible alternative may be simple economic enterprises. It may subsequently be possible to build upon the economic accomplishments, but experience has shown that this is unlikely unless there exists at least the germ of a collective movement in the first instance. On the contrary, it is quite possible that concentration on increasing personal profit could, rather than provide the foundation for cooperative initiatives, merely perpetuate isolationist attitudes.

Nonetheless, it would be unwise to dismiss small businesses out of hand on ideological or philosophical grounds. In India, many grants are made to small businesses in order to free them from the grip of merchants and moneylenders, as in the case of rickshaw pullers in Nagpur (MAH 53) and fishermen in Orissa (OR 41). The Brazilian organisation UNO has also had marked success in the North-eastern city of Recife (BRZ 130). With a population of two million, Recife has 40,000 small enterprises employing 120,000 workers. UNO has helped over 700 of these businesses, creating 1400 jobs at a tenth of the cost of government-created posts. UNO achieved this by acting as an intermediary between commercial banks and small borrowers who would normally be denied access to credit, as well as by providing courses in accounting, business management and labour laws etc.

VI Labour unions

The success of labour unions in defending and promoting the interests of the poor and underprivileged is, historically speaking,

5

indisputable. Oxfam's involvement with unions has grown steadily over the years as social development has been given a higher priority, in response to changing needs. This may be described as increasing people's awareness of the factors which condition their lives and providing them with a greater capacity for self-determination. These aspects have been dealt with at length elsewhere in the Handbook (see PART FOUR). Here it is worthwhile to re-emphasise the economic role of unions and the fact that, in the final analysis, they are concerned with improving the material benefits accruing to their members. Therefore, social development objectives are, to a large extent, although not exclusively, those of obtaining decent living and working conditions.

In most projects which involve union-related activities, the benefits are normally only apparent in the long term, and are the result of educational programmes and campaigns against unfair practices. However, persistence by project-holders has in a number of cases led to tangible economic results benefiting disadvantaged groups. In India, for example, the Industrial Services Institute (MAH 53) has assisted the rickshaw-pullers' union of Nagpur to get new laws passed allowing them to buy and own their own vehicles, rather than having to hire them at prohibitive rates. In North-east Brazil the public protests of rural syndicates have led to wider and fairer distribution of official drought relief aid (BRZ 337). In the same region, rural unions have been instrumental in challenging the potentially disastrous side-effects of planned government dam-building schemes, and have obtained fairer compensation and alternative land and housing for dispossessed farmers (BRZ 224).

VII   Educational support and training groups

1 Background

Oxfam is primarily a financing rather than an operational agency. In certain circumstances, for example, where complex procedures require sophisticated technical advice on a long-term basis, it may be advisable to fund an intermediary organisation which has the relevant expertise, rather than to allocate funds directly. This may be the case particularly with cooperative and other economic enterprises. It is important to recognise the basic characteristics of the intermediary organisation, to ensure that it is capable of achieving its stated objectives in a way that allows for and encourages the autonomy and long-term viability of the cooperative it serves. The decision to fund such groups depends on their professional expertise, the kind of services they offer, their relevance to the recipients and most importantly, their overall methodology. It is vital to recognise whether they are directive or non-directive: that is, do they adopt a democratic or an authoritarian approach to training and education? For many religious groups or political parties, for example, the missionary impulse to recruit adherents may well be stronger than the desire to grant autonomy and promote justice and equality. Directive approaches tend to create tension and mistrust, whereas non-directive organisations promote confidence and motivation, thereby creating the conditions for the eventual self-reliance of a project.

Oxfam has been involved in a number of projects where funds have been sought by groups which aim to advise upon and encourage the setting up of cooperatives and self-managed businesses. The methods and problems of these support organisations have a wider application to all groups that seek to supply the continuous, consistent, detailed and knowledgeable support that is a necessary complement to financial resources in any development effort.

In a training programme involved with cooperatives, a balance should be sought between *ideological* and *technical* training, in order to take into account the participants' dual roles of worker and member.

Since cooperatives challenge and question the values and assumptions which explicitly or implicitly underlie the workings of standard firms or companies, it is important to orientate members to give them an overview of the aims and objectives of cooperatives in general. However, these wider ideological implications must be integrated with values and ideas that bear directly on their work as members of a particular organisation. Excessive intellectual diversity and complexity can be detrimental and destructive, since it can often lead to apathy and disintegration.

This theme of relevance applies equally to the technical training given. Participants will be better motivated if the practical and immediate usefulness of information and techniques imparted to them can easily be demonstrated.

Both the ideological and technical training are related to the wider problem of propriety. Information should be correctly pitched in order to achieve its maximum effect. A Peruvian group, INPET (PRU 144), has established a tripartite structure of training designed to give basic instruction to workers, intermediate training to potential leaders, and advanced tuition to administrators. With less sophisticated cooperatives, particularly small groups in rural areas, all members can be given a chance to experience the various roles of the different sections. GUP in Bangladesh, for example (BD 51), has a policy of changing the leaders regularly (see PART FOUR Section 2).

2 Teaching Methods

The *teaching methods* adopted ought to reinforce the overall aim, which is to teach workers self-management. The techniques must therefore be appropriate, involving participation at all levels (see PART FOUR, Section 2). Authoritarian training methodologies would not only clash with the ideology, but would fail to prepare workers for the independent role expected of them. To this end, the teaching should foster useful intellectual skills and habits, such as:

☐ mutual respect and democratic behaviour;
☐ learning to identify and analyse problems as a group;
☐ how to collect and assess information;
☐ how to make decisions, implement them, and control and evaluate the results.

Differences between the classroom and the workplace should be minimised, as should any hierarchical demarcation between teacher and learner. Of primary importance is the teaching of future trainers by educating both technical and managerial leaders in group-leading techniques within a participatory environment. Ideally, the approach should promote a constant process of self-education through observation and reflection. This is best achieved by the use of experimental approaches involving participation, on-the-job training, and action programmes. An entirely theoretical approach was attempted by the Cooperative Activity Management Support Project (CAMS) operating in northern and upper Ghana (GHA 34). However, it soon became apparent that far more on-the-job training was essential, as people would then learn through experience.

3 Services

The services offered by support groups vary according to the scale and resources of the support group and the particular needs of the organisation. It is vital that the groups should not try to do too much. In Peru, INPET makes a point of not expanding their services to a particular group unless they are absolutely certain that they can provide a full service, rather than merely superficial coverage. Services such as those listed overleaf may be offered.

a: General training and education to provide an overview of the cooperative movement and its aims, together with specific training on special subjects. The Andhra Pradesh project, for example, runs special courses on the legal recovery of dues and the operation of a fertiliser business.

b: A consultancy service offering advice on specific problems, as well as support and encouragement through regular attendance at meetings, aimed at achieving participation of all members — with particular regard to vulnerable groups and sectors liable to exploitation, e.g. women, poorer members, workers lacking formal education and business expertise.

c: Technical services, such as small-scale feasibility studies, particularly where this is a legal requirement.

d: Accounting and book-keeping instruction. Organisations should be encouraged to adopt and maintain high standards of accounting, for both outside and inside information. Tension and mutual suspicion is far less likely to build up if all the members have a clear, precise knowledge of how the business side is being operated.

e: Legal advice can be vital, particularly where formal status must be acquired in order to give access to formal credit institutions and bank loans, as in Latin America.

f: Financial aid is always needed at the beginning of a project, and can be provided in a number of ways (see subsections II. and III. above), but care must be taken to avoid creating dependence.

g: Diffusion and dissemination of information and ideas. CIDIAG (PRU 207) publishes pamphlets, manufactures audiovisual aids, provides videos and oral histories on important events in the country's labour history.

4 Problems

In any group of this nature problems will arise.

a: Where the group is at variance with prevailing political trends, it may be subjected to outside interference and consequently need protection.

b: Some bureaucratic support organisations tend to develop their own needs independently of, or even in spite of, the organisations they are designed to benefit. Some groups extend loans and guarantees below market prices. Thus, the funds gradually disappear and, in order to prevent their total elimination, the funding groups are led into speculative ventures such as applying fines and commissions to raise the real cost of the credit given. This leads to a conflict of interest between the support organisations and their clients and it illustrates the profound need for adequate communication.

c: Volunteer groups are often staffed either by ex-government professionals, whose schemes tend often to be grandiose and bureaucratic, or by radical professionals, often aggressively ideological, impractical, and narrow-minded.

d: If a support organisation exists to create cooperative projects, it may be difficult for its members, as outsiders, to catalyse the formation of a group, particularly in rural communities with strong hierarchies and political currents.

The role of the funding agency here is multifaceted. As an external agency it can play an important corrective role, especially in reinforcing the non-directive tendencies within the organisation. Its objectivity provides it with an excellent standpoint from which to watch for political dominance by a particular faction that may obstruct the group's initial objectives. As an outside organisation, it should also encourage cooperation between the institution and other groups seeking to serve the poor.

I   Settlement schemes

1. Have all the alternative projects that set out to achieve the same results been considered and evaluated?

2. Have all the motives of the government been considered and compared with the expressed desires of the settlers and the development needs specifically of the poorest sectors of the region and country?

3. Do the orientation courses cover all the basic needs of settlers? Are they of sufficient duration? Do they foster self-reliance rather than dependence?

4. How are settlers selected, and will a balanced community be created in terms of skill requirements, social cohesion, and communal labour relations?

5. Will the crop and animal base be sufficiently diversified in terms of nutrition, marketability and climatic/ecological requirements? Will they be balanced between cash and subsistence crops?

6. Have plans been made for the development of non-farm employment and training, together with processing industries to enhance retail value of products?

7. What credit facilities will be included? How will they be operated and financed? Is the rescheduling of loans flexible enough?

II   Marketing

1. Who are the people who will benefit from the project? Are they all aware of the likely benefits they will obtain?

2. Who are the people likely to suffer as a direct result of the project, e.g. traditional intermediaries, competing producers etc? What action are they likely to take, and what can be done to neutralise, by-pass or counter this? Do they have legitimate interests which ought to be safeguarded?

3. Do the project holders have sufficient capital resources and, perhaps most importantly, the expertise necessary to deal with the complex problems involved — storage, transport, processing, packaging, advertising, etc?

4. Does the project have sufficient contacts with officials who can deal with possible complexities of permits, legal disputes and other proceedings?

5. Are there any representatives to deal with buying and selling in distant places? A permanent representative may be needed in a large city to sell the products and raw materials on behalf of a rural agricultural cooperative. Occasional visits by inexperienced farmers are unlikely to be successful when major transactions are being negotiated.

6. How efficient is the communications network, and how flexible is the production and distribution pattern to deal with unexpected changes in supply, demand and prices?

7. Is there an adequate system of internal control to avoid corruption and

inefficiency? Project leadership should always be accountable to members, who should be kept well informed of events and participate in decision-making.

8. Finally, has a careful study been made, both of the existing marketing situation and of the proposed project, to ensure that there is adequate margin for improvement?

III   Savings and credit

1. What institutions for savings and credit (formal and informal) exist in the community?

2. Are any particular groups (including the project's target group) badly served by these institutions (e.g. charged high interest by a moneylender, denied membership of a credit union)?

3. What are the specific needs of these groups for credit?

4. Can these needs be met by a savings group which mobilises community savings — or is the need *larger* or more *urgent* than this?

5. Is there a way in which these needs can be met by helping groups or individuals to gain access to existing institutions, especially formal institutions (e.g. by guaranteeing a commercial bank loan)? Or is it necessary to set up a new institution (e.g. an informal cooperative group) with a revolving loan fund?

IV   Cooperatives

1. What are the objectives of the cooperative? Can these be related to identifiable needs of the community or of specific groups within the community?

2. What is the actual or potential membership? Does this form a coherent group with a close identity of interest? Have the potential members themselves participated in determining their needs and assisted in drawing up the plan for the cooperative?

3. What national legislation governs the establishment and organisation of cooperatives? Are groups established under this legislation likely to be within the capacity of the membership? (i.e. is a cooperative really necessary?)

4. Is a 'non-formal' cooperative likely to be preferable to a full cooperative? What is the legal status of such a group? Would the group have access to resources if not formally registered?

5. Is the objective of the cooperative both commercially viable *and* feasible within the social and economic structure in which the cooperative must operate?

6. Are the actual or potential leaders and professional managers of a calibre to make the cooperative a success?

7. What is the educational role of the cooperative? Are the activities of the cooperative likely to have any favourable impact on non-members?

8. Is the request to Oxfam a contribution towards self-reliance or is it likely to increase dependence?

V	**Small businesses**

1. Is the activity in question a viable economic proposition?

2. Are the employees being given satisfactory wages and working conditions?

3. Will the new or expanded business create more new jobs for the currently unemployed or underemployed?

4. Will it put others, in competing small businesses, out of work?

5. Is there any potential for pursuing social or educational development, either simultaneously or at a later stage?

VI	**Labour unions**

1. Will the project increase the incomes of participants?

2. Alternatively, are there other economic benefits which will improve the lot of the group(s) in question: greater security of land tenure, improved access to social security benefits, fairer labour laws, compensation for losses incurred as the result of badly planned official development schemes, etc.?

3. Is there an educational programme designed to disseminate information about the issues involved to those affected?

4. Do the means exist for systematically consulting the membership and obtaining its participation in decision-making?

5. Does the union leadership accurately represent the views of the mass membership?

6. Will the union be able to become financially independent later on, or must it remain dependent on outside assistance? Every attempt should be made to avoid such dependence at the earliest stages of project negotiation.

5

VII	**Support group**

1. What is the feasibility of the institution's proposed programme? Is the group capable of taking on the full range of services that it desires?

2. Is there sufficient expertise and professionalism in the group? Does idealistic enthusiasm take the place of practical skill?

3. Are their methods suitable and appropriate? Are the services offered useful and relevant?

4. Does the support group have the skill and understanding required to cope with possible tensions in the cooperatives between the political and administrative leaders and about goals? Or between cooperative members and less privileged members of the community?

Section 6   Appendix

1. Oxfam Trading

Oxfam Trading is the commercial arm of Oxfam, which imports craft work from overseas producers under its "BRIDGE" scheme

a: Advances. Many smaller BRIDGE producers, if they accepted a large Oxfam Trading order, could be faced with a chronic shortage of working capital. Raw materials have to be purchased, and wages paid for several months before payment for the goods is received. Therefore, as part of its producer assistance function, Oxfam Trading can pay an advance of up to 40% of the value of an order, and in exceptional circumstances the Managing Director can approve an advance of 50%. Such advances are not automatically paid, and with a new producer Oxfam Trading will seek the advice of the field staff or local OT staff member, as to what size of advance, if any, would be appropriate.

If an advance is to be paid this will be shown on the purchase order sent to the producer: it will be paid immediately following confirmation from the producer that the order has been accepted. Other alternative marketing organisations also provide advances and the supplier should ask whether these are available when accepting an order.

b: Goods. Payment for goods can be made by any of the methods used for paying grants. In addition payment can be made by a letter of credit, or a bill of exchange.

c: Letter of credit. This is a document made out by the customer when the order is placed. It authorises the producer's bankers to pay the producer on the production of the stipulated invoices and shipping documents etc. Letters of credit must be exactly adhered to, and therefore become complicated and expensive if there are alterations to prices, shipping dates etc. However, they are the standard form of international commercial payment because they guarantee the suppliers their money. The addition of a 'Red Clause' can authorise the bank to pay the producer a stipulated advance payment, and even without this the producer can often negotiate a bank overdraft on the security of a letter of credit.

d: Bill of exchange. This is a document made out by the producer when preparing the invoice and shipping documents. It is basically a cheque, and by signing it the customer authorises the bank to pay the supplier. The producer should consign all the shipping documents to the bank, and therefore the customer cannot obtain possession of the goods until he or she has signed the bill of exchange.

A producer who has been sent an advance will often have the remainder paid by bill of exchange, and must therefore remember to deduct the advance in calculating the value of the bill of exchange.

Oxfam Trading is prepared to give advice to a producer on the most appropriate form of payment. However, in most cases, the producer can obtain this advice locally, and Oxfam Trading will then use the method preferred by the producer.

It is very rare for Oxfam Trading to refuse payment because a consignment is either short or faulty. The standard practice is to pay the amount indicated in the invoice and then to ask the producer to send additional products free of charge in the next consignment.

2. Export requirements

If exporting goods from one community to another is complicated, exporting to other countries is even more so. If at all possible, projects should concentrate on local marketing, meeting local needs. The second option would be to market to urban areas. The third would be to sell to wealthy nationals and tourists. However, success in catering to the tourist trade is not a guideline for success in exporting; tourists are notoriously poor judges of products.

It is extremely difficult for small groups outside the urban centres to handle the correspondence, documents and financial transactions necessary for a successful exporting operation. Often there are language difficulties in addition to problems of export licences, import quotas, exchange control and so on. Many countries now have state agencies who will act for small groups, but in the main their operations are not beneficial to the producers. The umbrella marketing organisations set up to serve small groups, often with those groups as members, have been more successful. Oxfam has helped in the formation of such organisations in Haiti, Kenya, Indonesia, Guatemala, Peru, India, Bangladesh and other countries.

For export, all goods must be acceptable under the *customs regulations* of the country of destination; the supplier must produce documents for the buyer to present to customs so as to obtain the goods. Certain *documents* are universally required, whatever country is exporting; others pertain only to that country or are required only for certain goods, whether textiles or handicrafts.

a: All goods must have an *invoice* stating how many goods are being sent and at what price. The invoice must show under what terms the goods have been purchased:

☐ Ex-works means that the customer pays for all expenses from the factory onwards.

☐ FOB means that the supplier is liable for all charges incurred until the goods are on a plane or boat.

☐ C + F means that the supplier is responsible for all expenses, apart from insurance and clearing the goods in the country of destination.

☐ CIF means that the supplier is responsible for all expenses, except clearing the goods through the customer's customs office.

☐ Franco Domicile means that the supplier is responsible for all charges to the customer's door.

b: Each consignment must have a *packing list* of the contents of each box/crate. This is used by customs if they wish to examine the goods; it also helps the buyer to check that he/she has all the goods paid for and can help to identify which items are missing if some have been stolen or mislaid.

c: Britain and other EEC countries have *fixed quotas* on each imported item. A quota denotes a certain quantity legally allowed into the country. It is set up to regulate trade and, in certain instances, to protect home production. There are several quotas, the most common being the generalised system of preference. This entitles certain goods from certain Third World countries to preferential rates of duty; i.e. none or minimal. As long as the quota has not been met, goods accompanied by a *GSP certificate* on a green form (in the case of the Philippines or India) or an *EUR 1 certification* on a yellow form (from countries in the African Continent, the Caribbean and Pacific territories) can be accepted at a lower rate of duty.

d: If the goods are coming by air, then an *airway bill* is drawn up by the airline in question. This lists the goods, weight, destination, etc. It can often be made out to the customer's bank so that the customer must pay the bank for the goods before taking possession of them.

e: If the goods are coming by sea, then a *bill of lading* is drawn up by the shipping line involved. It is a document of title, and the customer can only obtain the goods on production of the original bill.

f: There are various rules and regulations regarding textiles. All textiles

5

must be accompanied by a *certificate of origin* (yellow); without it, the customer cannot remove the goods from customs. Certain goods, if handloom-made, must be accompanied by a *handloom certificate* (blue), whilst millmade goods must be accompanied by an *export certificate* (yellow), so that the customer can obtain a licence from the Department of Trade to enable him/her to receive the goods. Certain items, notably cotton textiles such as shirts and skirts from India, are strictly monitored by the British Government, and a licence must be granted before the goods can be accepted through customs.

g: Some countries (India and the Philippines, but not Thailand or Burma) issue a *handicrafts certificate* through their Handicrafts Board covering goods considered to be handmade. This is a white certificate written in 5 languages, hence its name, '5-language handicraft certificate'.

Once all the paper work is assembled it can be forwarded to the customer well in advance of the goods to prevent any delay (especially important when licences must be obtained in the country of destination), or sent with the goods if they are being air freighted.

Delay in presenting the correct documents to customs can cost the supplier a lot of money in rent. Both docks and airport allow two days' free storage and then begin to charge per day by weight. The rent incurred at an airport is always more punitive, as storage space is more limited; people must therefore be dissuaded from leaving their goods there for a long time. In the case of Britain, with both air and sea consignments, if the goods have not been cleared at the end of three months, the customs department is legally bound to confiscate them.

Finally, it is most important to find a reliable agent to handle the goods; this prevents delay. If the supplier is packing his/her own goods or hires someone else to do so, it is vital that everything is well packed, well protected in the case of breakables and well boxed or crated in every instance. Every bale/crate must be well and clearly marked with the bale/box number and the destination. The suppliers' names and any airway bill details or shipping marks must also be shown. In the case of goods going to French Canada, details must be in both English and French. Too often, time and money is wasted with goods being broken into or parcels breaking open. It is also most important to *check insurance*, use a trustworthy insurance company if doing it yourself (in the case of CIF goods), or have the purchaser arrange it (in the case of ex-Works, FOB and C + F goods).

Exporting can be a lucrative business but may lead to a maze of documents and agents. If any queries arise, it is better to approach the local Boards of Commerce or Export to clarify any problems before undertaking exportation, as incorrect documents or, at worst, lack of necessary certificates waste both time and money for supplier and customer alike and do not make for good trade relations.

Section 7   Resources

I   Bibliography

1 Rural Development

R. Burbach and Patricia Flynn, *Agribusiness in the Americas*, Monthly Review Press, 1980.

R. Chambers, *Managing Rural Development: Ideas and Experience from East Africa*, Scandinavian Institute of African Affairs, Uppsala, Sweden, 1974.

B. Dinham and C. Hines, *Agribusiness in Africa*, Earth Resources Research Ltd, 1983.

S. George, *How the Other Half Dies*, Penguin, Harmondsworth, Middlesex, U.K. 1977.

J. Harris (ed), *Rural Development: Theories of Peasant Economy and Agrarian Change*, Hutchinson, London, 1982.

U. Lele, *The Design of Rural Development: Lessons from Africa*, John Hopkins University Press, Baltimore, Maryland, 1975.

N. Long, *An Introduction to the Sociology of Rural Development*, Tavistock, U.K. 1980.

A. Pearse, *The Latin American Peasant*, Frank Cass, London, 1975.

2 Rural Settlement

R. Chambers, *Settlement Schemes in Tropical Africa*, 1969. A study of organisation and development. International Library of Sociology and Social Reconstruction, Routledge, Kegan & Paul, London.

B.H. Farmer, *Agricultural Colonisation in India since Independence*, Oxford University Press, Oxford, U.K. 1974.

A. Gaitskell, *Gezira: a Story of Development in the Sudan*, Faber, London, 1959.

ILO, *Systeme de colonisation le long de la route transamazonienne du Bresil*, ILO D.20, Geneva, 1974.

A. Klempin, *Contribution of State-Directed Land Settlement to Agricultural Development*, Research Centre for International Agrarian Development, Publication No. 9, 1979. A comprehensive case study of the land settlement and Thai-German Agricultural Development Project, Sara Buri, Thailand.

C. MacAndrews, *Land Settlement Policies in Malaysia and Indonesia: A Preliminary Analysis*, Institute of S.E. Asian Studies, Singapore, 1978.

M. Nelson, *The Development of Tropical Lands: Policy Issues in Latin America*, John Hopkins University Press, Baltimore, Maryland, 1973.

M. Patterson, *Settlement of Pastoral Nomads: A Case Study of the New Halfa Irrigation Scheme in East Sudan*, University of East Anglia, Norwich, Development Studies Occasional Paper, 1980.

3 Physical Infrastructure

C. Abrams, *Housing in the Modern World*, Faber and Faber, London, 1966. Still a classic, and one of the best single references to practical policy issues.

V. Borremans, *The Librarians' Guide to Convivial Tools*, Bowker, 1980. Complete list of sources. (Details from the author at: Tecnopolitica, Apartado 479, Cuernavaca, Morelos, Mexico).

5

G. Boyle and P. Harper, *Radical Technology*, Wildwood House, London, 1976 (£3.95). Many short articles, well illustrated.

H. and C.H. Caminos, *El Precio de la Dispersion Urbana*, Universidad de los Andes, 1977. Useful plans and tables. (Facultad de Arquitectura, Universidad de los Andes, Merida, Venezuela).

H. Caminos and R. Goethert, *Urbanization Primer*, MIT Press, Massachusetts, 1979 (US $27.50). This manual analyses designs and layouts in great detail. Valuable for those who can afford it.

M. Clinard, *Slums and Community Development: Experiments in Self Help*. New York & Co., 1966.

H.K. Daucy, *A Manual on Building Construction*, I.T. Publications, London, 1975.

D.J. Dwyer, *People and Housing in Third World Cities*, Longman, Harlow, Essex, 1975. (£6.50). Focuses on cases from Asia and Latin America. Good reference source.

H. Fathy *Architecture for the Poor*, University of Chicago, 1973, (paperback £4.10) Excellent case study based on traditional rural housing in Egypt.

I. Illich, *Tools for Conviviality*, Calder and Boyars, London, 1973. (Paperback £1.50) Of his writings, the most directly relevant to planning and building.

National Academy of Science, *Ferrocement: Applications in Developing Countries*, Washington, D.C., 1973.

RAIN, *The Rainbook: Resources for Appropriate Technology*, Schoken Books, New York, 1977 (US $7.75). Useful source book.

Shankland Cox Partnership, *Third World Urban Housing*, Building Research Establishment, London, 1977 (Paperback £5.40). Represents the approach of sensitive architects and planners.

J.F.C. Turner, *Housing by People*, Marion Boyars, London, 1976 (Paperback £2.25). Based on experience in Latin America. A classic which has been translated into Spanish, French and many other languages.

UN. *Non-Conventional Financing of Housing for Low Income Households*, Department of International Economic and Social Affairs, United Nations, New York, 1978. Useful, and good bibliography.

Building Blocks with the Cinva Ram Block Press, Volunteers in Technical Assistance, Arlington, Virginia, U.S.A., 1966.

S.B. Watt, *Ferrocement Water Tanks and their Construction*, I.T. Publications, King Street, London, 1978.

P. Wilsher and R. Richter, *The Exploding Cities*, Andre Deutsch, London, 1975 (£5.25). Provides general overview.

AHAS Housing Advisory Service has set up an information retrieval system of case studies, references and organisations relating to urbanisation and urban neighbourhoods. AHAS hopes to provide an information service at a modest fee. Enquiries should be directed to:

AHAS, 5 Dryden Street, London WC2E 9NW, UK.

4 Marketing

W. Jones, *Marketing Staple Food Crops in Tropical Africa*, Cornell University Press, Ithaca, N.Y., 1982.

U. Lele, *Food Grain Marketing in India*, Cornell University Press, Ithaca, N.Y., 1972.

A. Mosher, *Creating a Progressive Rural Structure to Serve a Modern Agriculture*, Agricultural Development Council of New York, 1969.

E. Whetham, *Agricultural Marketing in Africa*, Oxford University Press, Oxford, U.K. 1972.

5 **Cooperatives**

R. Apthorpe, *The Cooperatives' Poor Harvest*, in New Internationalist, February 1977. (New Internationalist, 43 Hythe Bridge Street, Oxford, Oxfordshire, UK). A revealing and clear analysis of the reasons for the failure of so many cooperatives.

M. Bakuramwtsa, *Why do our Cooperatives in Africa fail?* Ideas and Action, Paris, FFHC/AD 146, 1982/83.

T. Bottomley, *An Introduction to Cooperatives: A Programme Learning Text*, I.T. Publications, King Street, London, 1979 (£2.95).

J. Launder, *Office Management for Cooperatives: A Prepared Learning Text*, I.T. Publications, King Street, London, 1980 (£2.95).

OXFAM Information Sheets are available on the following projects:- *The Chimaltenango and San Martin*, Guatemala (GUA 1 & 12) *Orientation Programme for Colonists, San Julian*, Bolivia (BOL 32 & 13) *Fotrama, Knitting and Weaving Cooperative, and Education Programme*, Cochabamba, Bolivia, (BOL 22)

Sulla Rural Development Programme: Sylhet District, Bangladesh (BAN 17) *Kaira Milk Producers' Cooperative Union, Anand*, India, (869/23/3045/5111) *Roma Valley Cooperative Society*, Lesotho, (LES 31) *Dry Season Gardening, Ouahigouya, Yatenga*, Burkina Faso (VOL 31)

Plunkett Foundation, Yearbook of Agricultural Cooperation, Oxford, U.K., 1982 (£7.50, £8.30 by post). Annual publication consisting of a collection of articles covering experience of formal cooperatives with a wide range of purposes — some in developed countries, some in developing countries.

Plunkett Foundation, Study Series, useful training manuals which can also be used as general handbooks. Strongly recommended: No.1. B.J. Surridge and F.H. Webster, *Co-operative Thrift, Credit, Marketing and Supply in Developing Countries*, 1978; No.2. Wiseman, *Basic Financial Control*, 1979 (£3.42), and No.3. P. Yeo, *Basic Economic Concepts*, 1979 (£2.81).

B. Roberts, *Organising Strangers: Poor Families in Guatemala City*. University of Texas Press, 1973.

UNRISD, *Rural Institutions as Agents of Planned Change*, 8-volume study, 1969-75, including a *Review of Rural Cooperation in Developing Area*s (vol 1); reports on Latin America (vols 2, 3), Africa (vols 4,5), Asia (vols 6, 7); and *Rural Cooperatives as Agents of Change — a Research Report and Debate* (vol. 8), United Nations Resarch Institute for Social Development, Palais des Nations, 1211 Geneva, Switzerland.

K. Verhagen, *Cooperatives and Rural Poverty: Eight Questions Answered*, Plunkett Foundation, Oxford, U.K., Development Series No.1 (1980) (£1.50).

K. Verhagen, *Guidelines for the Preparation and Appraisal of Cooperative Development Programmes and Projects for use by Project Officers*, Plunkett Foundation, Oxford, U.K., Series No.2 (1981) (£1.00).

World Neighbours, *Cooperatives*, in Action Series No. 10, No. 4E. Based on the experience of the Quetzal Central Marketing cooperative in Guatemala. Gives practical information on managing and organising a cooperative, and on the affiliation of small cooperatives into larger regional organisations.

P. Worsley (ed), *Two Blades of Grass*, Manchester University Press, Manchester, U.K. 1971. Book of readings on cooperatives in developing countries, with especial reference to use of traditional institutions as a basis for modern cooperative development.

A.S. Walford, *Book-keeping: Outline of Double Entry Book-keeping for Small Businesses*. Plunkett Foundation, Oxford, U.K. 1983 (£1.00). A simple system providing a useful guide to small cooperatives and other projects.

P. Yeo, *The Work of a Cooperative Committee: A Programme Learning Text*, I.T. Publications Ltd., London, 1979 (£2.75). Guide for committee members of primary cooperatives, with material for six meetings of study groups.

II   Organisations

The following are particularly concerned with the work of cooperatives in developing countries, and have publications and consultants available:

International Cooperative Alliance (ICA), 35 Rue des Paquis, 1201 Geneva, Switzerland.

Plunkett Foundation for Cooperative Studies, 31 St. Giles, Oxford, OX1 3LF, UK. Organises in-country training courses at appropriate levels for practising managers and project leaders.

Cooperative College, Stanford Hall, Loughborough, Leicestershire, UK. Organises training courses in the UK.

University Centre for Cooperatives, University of Wisconsin, 610 Langdon Street, Madison, Wisconsin 53706, USA.

COPAC, Via Terme di Caracalla, 00100 Rome, Italy. COPAC has a useful *Directory of Agencies Assisting Cooperatives in Developing Countries* (1981). Also publishes a series of country reviews containing information on historical and current cooperative development. Countries include Sudan, Kenya, Egypt, Zambia, Tanzania, Thailand, Indonesia, Sri Lanka, Burkina Faso, Swaziland and Namibia.

PART 6 AGRICULTURE

Oxfam

Part Six AGRICULTURE

6

6

Section 1   Agricultural Guidelines

I   General aims of agricultural projects

Oxfam aims to raise the nutritional level and the general standard of living of the poorest members of rural communities through improved, appropriate and sustainable use of land and related resources. Thus, higher production per unit of area can be obtained and sustained, and a higher proportion of the population can be supported at an adequate nutritional and economic level in rural areas. This would reduce migration to the towns. Where food in excess of normal family needs can be produced it may:

- [ ] act as a 'hedge' against shortages in below-average seasons (a prerequisite for survival in much of the tropics);
- [ ] under appropriate marketing conditions, provide benefits to those urban and rural poor who have to buy some or all of their food. (However, the threat to individual farmers who sell from a position of weakness vis-a-vis the trader cannot be overemphasised.)

To support increased production and productivity is not to ignore the importance of:

- [ ] who controls the land and other productive resources employed in agriculture;
- [ ] how the fruits of effort are distributed.

These are matters about which Oxfam is very much concerned. As regards land tenure, a landlord-tenant system usually operates in poor countries — where most of the land is owned by a relatively small number of people, while a high percentage of the population is dependent upon the land for their livelihood — to the great disadvantage of the rural poor. Tenancy agreements are likely to be verbal only, and security of tenure may not exist. Not only does the tenant or share-cropper get a poor deal but, since he/she has no security, the system effectively prevents him/her from improving the land. Even if the law is on the side of the poor tenant, a combination of ignorance, fear and the prohibitive cost of legal procedures will usually alienate him/her from it.

The poorer farmers usually have marginal lands with poor soils and climate, few resources or facilities, and inadequate access. Where they live in a community which has high social differentiation, they will probably have land in the most disadvantageous location; for example, high on a hillside or far away from the water source. Landless people who cannot even secure a share-cropping arrangement are in the worst situation, because they are totally at the mercy of landlords for work, and hence also for food.

Although the NGO cannot effect fundamental change, it is able to help to mitigate the worst aspects of an inequitable system of land access and resource supply by:

- [ ] funding those local legal aid organisations which assist deprived people both to understand and to obtain their legal rights (see PART FOUR Section 4, Legal Aid);
- [ ] making sure that the projects it supports will be socially equitable as well as technically sound;
- [ ] helping people to see their predicament more clearly, building their confidence and supporting their initiatives — particularly those communal activities mounted for the social and economic benefit of the group; (see PART FOUR SOCIAL DEVELOPMENT)
- [ ] helping landless groups to strengthen their position through various kinds of social organisation or by equipping them so that they can provide a service. For example, in Bangladesh a community of landless people became successful sharecroppers by obtaining a

6

loan to purchase a mobile irrigation pump-set which gave them bargaining power with the landlord (BD 123).

It is important that the NGO's general style and the activities it chooses to support should have the effect neither of distancing people from governmental or other assistance (such as banks) which should be available to them, nor of enabling authorities to shirk their responsibility.

II   Priorities for project designers

Project designers will have a number of objectives which they try to reconcile and rank in order of priority. These are likely to include:

- [ ] the production of an assured supply of food and other materials for a given number of people;
- [ ] the conservation of basic resources (soil, water);
- [ ] the maintenance and repair of any existing capital goods, i.e. seed, tools, houses, grain stores, etc.;
- [ ] the production of a surplus with which to obtain cash for 'necessary' purchases ('necessary' will, in part, be culturally determined);
- [ ] the production of a surplus for 'social' investment for a particular event such as a marriage or pilgrimage, repaying an obligation, or old age;
- [ ] creating a further surplus for investment to improve living standards and security.

Achievement may be limited by lack of a particular resource, by mischance or miscalculation, or because extra physical or mental effort is greater than the expected benefit. Objectives and priorities will differ according to age, family size, culture, current fortunes and a host of other factors relating to the project's target group.

Establishing the true cost of any project plan includes knowing how each resource is regarded by the decision-makers (farmers, group leaders) concerned. If without the project the resource would remain unused (unemployed labour, land or hoarded funds), the true or opportunity cost might be assumed to be nil. Whether this is so depends on the judgement of the people, both men and women. There may be important social or religious reasons for labour inactivity at certain times, land may be regaining natural fertility, and hoards may be regarded as a necessary contingency. In these circumstances, the opportunity costs to the people of employing these resources is not nil. To take another example, a farmer who attends a training course may have to forego production work during that time, a cost which should be taken into account.

Project designers also need to take into account longer-term community considerations which may weigh little in the thoughts of individuals or groups whose poverty concentrates their minds on survival. For example, a farmer or group leader who is short of irrigation water may spread it as far as possible so as to keep plants alive, irrespective of developing soil salinity in the longer term; similarly, although the farmer might like to plant trees, he or she may not be able to afford either the trees or the annual crop land they occupy. Careful project design may be able to take a different line, and so ensure income in the future, and avoid or counter trends caused by individual decisions which, in the long term, may prove detrimental to the community as a whole. Similarly, a project designer should be able to see what effects his/her actions may have on other communities; for example, the taking of water to the inconvenience or serious detriment of communities downstream, or cultivating land hitherto grazed by passing nomads.

Again, the project designer should have a good knowledge of broader issues such as market opportunities, or political instability and impending change.

All *decision-makers* are operating a cyclical system of interrelated production and consumption. (The landless labourer's 'product' takes the form of wages paid for a single resource, work.) It is helpful to see this cycle as a whole.

As a *self-contained economic system*:

a: The production-consumption cycle can continue only as long as a flow of natural resources is available. Over-cultivation or over-grazing can easily damage this flow.

b: Labour productivity, i.e. work, depends on an adequate food supply for the worker (and his/her draught animal).

c: Work is a major contributor to investment goods, such as irrigation channels, grain stores, tools and equipment, which are in turn crucial to raising output in the future.

d: Agricultural products, including stored and/or processed goods, become either directly consumable or capital for investment. Not only the main product (e.g. grain) but also by-products (e.g. straw) are likely to be vital to the system.

e: Wastes of all kinds will have a value for fertility maintenance, fuel and sometimes feed. This is particularly relevant where natural resources are limited and labour abundant.

When the system becomes 'open', so that investment goods like mineral fertilisers, hired labour, and consumer goods such as purchased food or clothing are brought in, it must, of course, be enlarged sufficiently to produce a surplus adequate to cover the cost of the inputs, and insure against the increased risk of indebtedness.

While the objective must be to assist all members of a target group selected for funding, a policy which relies on the people themselves to make and implement suggestions will require the identification of those capable of taking *initiative* and *leadership*. An example of how Oxfam may have an impact by funding and encouraging simple, community-orientated endeavours, is provided by the Oxfam West Orissa Programme (OXWORP — ORS 20). A number of the villages affected by OXWORP are extremely poor, but have been encouraged to help themselves and to form groups which are now accepted by banks as possible recipients of loans.

Social and Political Aspects are elaborated in PART FOUR. Their relevance in terms of land-ownership and access is crucial.

6

| IV | **Group action** |

There is probably no agricultural activity carried on by one individual which does not affect others. Oxfam-supported agricultural projects involve groups of people, and their design may have to take into account the interests of others not directly involved (See Cooperatives, PART FIVE Section 4).

Homogeneous groups made up of small farmers and their families or landless labourers may be in target groups because:

☐ they have common needs and are economical to service with seed, or a training course, for example;

☐ they have, or can be persuaded to develop, a sense of identity and to act as a group in production and marketing activities;

☐ having identified themselves as a group, they will have strength for future group action in new projects, and in dealing with, say, money-lenders and local officials.

It may be impossible to focus agricultural projects on one stratum of a community, except in the initial stages. Inevitably the interests of others will be affected, such as when hired labour suffers by farmers becoming better equipped, and when the growth in crop-yields increases the work-load of those (usually women) responsible for post-harvest processing. (see PART TWO, Section 2).

Ideally, agricultural projects which receive NGO support will go on to grow in scope and depth. At the end of the funding period they will leave a community that is confident and competent to stand on its own feet, and make social and material progress. Stabilisation at least must be achieved so that natural resources are saved from continuous decline.

V	Food aid: effects on food production

(see also PART EIGHT Section 2)

Except when a disaster has destroyed the food supply or the immediate means of producing food, *the cause of hunger is more likely to be poverty than non- availability of food locally*. Therefore, it is the causes of poverty that need to be tackled so that people can produce or purchase food. *Food aid can never be a long-term solution*. Where it is used (or before it is used) as a short-term solution, every effort must be made to ensure that importing free or cheap food will not be detrimental to the local market, or a disincentive to local farmers or the target group itself to produce food.

Where it is used in a food-for-work context there are dangers that the poor will benefit only in terms of the short-term food supply (the food aid) while richer farmers and other members of society will benefit from the development (roads, water supply, buildings) even to the detriment of the poorer (wider economic gap, fewer jobs). A food-for-work project must be designed to avoid or minimise this effect.

A good use for food aid (where the project is properly designed and controlled) is in new land settlements of willing, poor people in areas where there is no local food supply, or where it is inadequate. Food-for-work is then supplied for a strictly limited term to produce infrastructure for the common good, and to enable farming families to be fed until their own land can be productive. (see PART FIVE, Section 2)

Section 2

Land —
Soil types, Conservation, Fertility

I   Principles of land use

Oxfam is concerned to help people recognise the importance of, and the means of protecting, the most basic agricultural resource — land — and to show how to use it for optimal output of food, fuel and other products consistent with:

☐ sustaining this output in the future;
☐ using other inputs appropriately;
☐ providing employment and life support.

Historically, the natural vegetation of the world once consisted of 42% forest, 24% grassland and 34% desert, but within recent times humankind has reduced the forest cover to 32% and allowed deserts to increase. Not only has the pressure of population, associated with unwise cultivation and grazing practices, reduced the area available for production, but it has also caused the quality of much land to deteriorate.

To halt the degradation and erosion of soils and to restore structure and fertility where it has been lost, it is essential to apppreciate not only the physical characteristics of soils in relation to climate and use, but also the social factors affecting long-term productivity.

Possibilities for expanding food production by sensibly extending the area of cultivation now exist in only a few places in the world and are fast disappearing.

Therefore, the way to expansion must be by the fuller and better use of land. Intensification will put even greater emphasis on the need to manage soils correctly in order to maintain their production potential for the future.

1   Physical Factors Affecting Productivity

The major factors are:

☐ the type, structure and fertility of the soil;
☐ the moisture status (enabling plant roots to function well and to take up nutrients);
☐ the temperature of the soil (which at any latitude will be modified by altitude and aspect);
☐ the nature of the terrain, particularly the degree of slope.

These factors will affect decisions about land use; the experience and knowledge of local people will be valuable.

2   Social Factors Affecting Productivity

Poverty and the distribution of land are the main social factors affecting the productivity of land. There are others, notably the size of families, number of labourers, health of the people, cultural attitudes, and economic considerations.

At a given level of technology, land has a human 'carrying capacity'. When this is exceeded — as in Java or the foothills of Nepal, or even where the population is relatively sparse, as in some infertile parts of sub-Saharan Africa — food supplies become increasingly insecure and inadequate as over-cultivation with inadequate management leads to a downward spiral of soil degradation.

The only possible long-term solution is likely to be a political one which redistributes land and alleviates poverty. In the meantime, the agricultural approach must be to improve yields where this can be done without degrading the soil.

If a large proportion of land is in the hands of a few owners, output may be low due to lack of incentive or inclination to farm intensively

6

(though that is not always the case). Conversely, if land is too fragmented, it may not form viable production units. Fragmentation arises partly from the sharing of land of different qualities in subsistence communities, but may be made worse by inheritance rules which cause further subdivision. The amount of land necessary to sustain a family will vary according to climate, soil quality and other resources, and this should be carefully considered when plans for land reform or resettlement are being drawn up. Where small farmers are concerned, the formation of groups or cooperatives can provide a means of putting limited resources to fuller use (see PART FIVE Section 4).

Redistribution of land should increase opportunities for the proper involvement of people in agricultural production and associated activities, with the twin benefits of increased output per unit of land, and a more equal distribution of food and income. However, land redistribution alone is seldom sufficient to achieve an appreciable, long-term rise in production, or large-scale socio-economic improvements.

3 Increasing and Sustaining Output

A rapid increase in agricultural production that cannot be sustained by either the people or the soil is highly undesirable. It is far better for the initiative to come from the people who will be involved, and for progress to be made at their pace. Moreover, the main priority should be to make better and fuller use of local, readily available resources, such as labour and waste materials, rather than to put confidence and cash at risk by the rapid introduction of purchased inputs. Often the structure, fertility, and hence the output, of land can be improved and maintained by putting crop and animal residues back into the soil. In the humid tropics there is a mass of vegetation and litter that is vital for soil stability, fertility and moisture retention.

4 Some Oxfam Experiences

Oxfam puts great emphasis on the importance of maintaining the resources of land. It has had some disappointing experiences when it has funded agricultural projects and has subsequently found that advice regarding the prevention of soil erosion has not been taken sufficiently seriously. Sometimes land and water conservation measures are given direct support, especially when they help small groups or communities. Among many projects assisted in a post-cyclone area of Andhra Pradesh, India, was one involving the planting of coconut and eucalyptus trees to control erosion (AP 24J). Another example is sand dune stabilisation by building vegetative cover and using barriers of brushwood in Somalia (SOM 12). The barriers consist of three rows of brushwood placed 10 yards apart at right angles to the normal wind direction. (See PART EIGHT, Section 2.)

Holders of comprehensive agricultural projects should be encouraged to maintain the organic level and structure of their soil by composting, green manuring and returning animal waste to the soil.

Loss of fertile land due to water erosion can be severe along lake and river banks; this can be extremely serious for poor communities living along those banks. In a project in Malawi, matting made from bamboo was used successfully to protect a lake shore from erosion (MAL 25C).

II   Soil

1 Formation

Topsoil, which is the portion of the earth's crust which supports life, may be either shallow or deep. In many parts of the world it consists of only a few inches overlying subsoil or rock. It is being formed perpetually from parent rocks by physical, chemical and biological processes. It may remain overlying its parent rock, or be transported by water, glaciers or wind to be deposited elsewhere over quite different rocks.

2 Texture and Structure

Soils are made up of mineral particles varying in size from coarse sand down through finer sand and silt to clay. Those with a high proportion of sand are generally referred to as 'light' and those with clay as 'heavy'. A fertile soil will also contain organic material (humus) and sustain a useful population of micro-organisms and soil fauna, such as worms and insects.

The particles will aggregate, and the extent to which they do this determines the soil structure. A good structure means that it is sufficiently open for plant roots to penetrate well, for excess moisture to drain away and for air to be retained in the soil, and yet not so open or loosely held together that it fails to hold sufficient moisture and is vulnerable to erosion by wind or rain. Nothing can be done to alter the basic soil type but a lot can be done to build up the organic matter. This will improve both fertility and structure. Conversely, soil structure can be affected adversely by impoverishment, over-compaction or excessive water-logging.

Under natural vegetation, the supply of organic matter comes from plant litter. Under agricultural systems that do not allow for the return of sufficient residues to the soil, the organic matter level will decline as natural oxidation continues. Traditionally, shifting agriculture allowed sufficient years of rest periods for humus to be restored by natural processes; but with increasing populations, it is now all too common for people to come back to cultivate the same land again too soon. Putting back into the soil as much vegetable and animal waste material as possible will help to counteract the effects of soil exhaustion. Integration of livestock and cropping systems may give important benefits in this way.

3 Acidity and Alkalinity

Soils are termed either acid, neutral or alkaline; the degree of acidity is measured in terms of pH values which run from 0 to 14. Acid soils have a low pH, neutral are around pH7, and alkaline soils have a high pH number. Crops vary in their tolerance of acid or alkaline soils, but almost all prefer a pH of about 6.5.

Acid soils are said to be sour, and sourness can be corrected by adding lime (calcium carbonate) or gypsum (hydrated calcium sulphate). A rough guide is that the application of one ton of lime to one acre will increase the pH by 1. Care should be taken not to apply too much lime, as this can affect the availability in the soil of other minerals that crops require. Where soils are too alkaline, the addition of iron sulphate is recommended. If inorganic (mineral) fertilisers are to be used, they should be chosen in relation to the pH value of the soil.

4 Some Important Soil Types

a: Latosols, sometimes called lateritic or red earths. By far the most extensive group of soils in the humid tropics. Reddish or yellowish, acid and often strongly leached by rain. Agricultural potential ranges from high to very low. Three main sub-types are:

☐ plateau soils of low fertility;
☐ red loams, which are moderately good soils of the savanna zone and suitable for growing annual crops;
☐ basisols, developed on volcanic rock, which are of high potential and able to sustain continuous cultivation, but are rare.

Laterite is a rock-like material that may occur as particles or impenetrable layers in any latosol. It has a use as road gravel.

b: Vertisols, also called black cotton soils or dark clays. Essentially fertile. In the dry season these soils become hard and develop deep cracks, while in the wet season they become exceptionally plastic or sticky. For this reason they can pose problems under irrrigation and are difficult to cultivate by conventional methods, but the natural cracking makes them 'self-cultivating soils', and shallow tillage is generally sufficient to prepare a seedbed.

6

c: Clay and alluvial soils: dark-coloured clays occurring in valley floors, commonly mistaken for vertisols. Highly fertile, particularly valuable under irrigation, but often suffer from poor drainage. In Asia they are used mostly for paddy. In Africa (dambo clays, mbuga soils, vlei soils) they are mainly under pasture, overgrazed and suffering from erosion and compaction.

d: Sands. Where sands are the natural product of underlying rocks, they are frequently inherently poor, subject to leaching and, being low in organic matter, are easily eroded.

e: Saline and alkaline soils (a soil condition rather than type). Salinity and/or alkalinity occurs in arid zone mineral soils where the evaporation rate is very high. It is also a growing problem in many irrigated areas where a combination of inadequate drainage and a high rate of evaporation means that salts in the water are deposited in the soil and crystallise out at, or near, the soil surface.

5 Degradation and Erosion

Population pressure associated with deforestation and unwise farming techniques is the major factor leading to the degradation, or even total loss, of soils. If soils are exploited beyond their carrying capacity, animal and/or human food supplies will become insecure, as attempts to produce more lead to a downward spiral of soil fertility and structure. Climatic characteristics will influence the process, so that as structure becomes poor, vegetative cover decreases and soils become vulnerable to wind and rain erosion, leading to *desertification* in arid zones, and to *water erosion* of slopes in areas of heavy rain.

Protective measures must be taken to reduce these harmful effects (see also Section 5 on Trees). Farmers must be encouraged not only to return organic matter to the soil (living vegetation and crop and animal wastes) but also to:

☐ avoid cultivating in a way which will encourage erosion, i.e. not straight up and down hillsides;

☐ use well-maintained terraces for crop production on slopes;

☐ maintain a cover of vegetation whenever possible, particularly in seasons of heavy rainfall or high winds;

☐ grow crops with a good root system which hold the soil and which improve fertility, e.g. pulses and other legumes, and avoid crops which remove a lot of fertility, such as cassava, unless this is returned in fertilisers or followed by legumes like groundnuts (this policy will, of course, have to be balanced against dietary needs and customs);

☐ plough in, rather than burn, crop residues, except when burning is recommended as a disease control method.

The extension of deserts (desertification) is largely brought about by overgrazing, and by the removal of tree and bush cover for fuel and fodder (see Section 5). It is encouraged by successive droughts or by long periods with low rainfall. Ideally, livestock numbers should be reduced, but in most poor communities where animals are communally grazed, to have large numbers is the best insurance against heavy losses in periodic droughts. Also, the number of animals that a family possesses has a social and cultural significance that takes precedence over their productive ability. Farmers should be encouraged to keep their animals under tight control by herding or fencing so that grazed land and browsed trees are rested; areas around watering points need special protective care (see Section 8). A solution to the overgrazing problem generally requires an integrated regional policy (or at least a community policy) affecting access to land and water.

Both over-grazing and over-cultivation associated with poor management can lead to erosion of soil, and this takes three main forms:

☐ wind erosion, which selectively removes fine particles of soil and leaves sand behind;

□ gully erosion, when small rivulets steadily grow into deep channels;

□ sheet erosion, when the whole layer of top soil is washed away leaving poor subsoil or bare rock.

Wind erosion can be reduced by endeavouring to maintain soil structure, keeping a vegetative cover, by irrigation if practical (damp soils do not blow away), planting wind breaks, and fixing sand dunes by planting trees and/or grasses with a spreading root system (see Tree Planting, Section 5).

Gully erosion begins when heavy rainfall finds its way into a slight channel arising from a natural indentation or a pathway trodden by cattle. It must be quickly spotted before a gully develops; then the water flow can be diverted, the vulnerable area of soil made more permeable to water and more stable by mulching, i.e. covering with crop residue etc., or by giving plant cover.

Where erosion is advanced and extensive, the measures necessary will be beyond the scope and control of farmers, e.g. building of check-dams and large-scale reforestation. However, village communities or farmers' groups may undertake terracing of slopes, contour and tie-ridging of their land, grassing of banks and small-scale tree planting. Oxfam helps projects such as these; for example terracing and land reclamation were subsidised in a highly successful venture in the West Bank area of Jordan (JOR 54) and similarly in Guatemala (GUA 12). The raising of tree seedlings to give to small farmers for planting on terrace edges is part of a project in the hills of Nepal (NP 11).

6 Soil Surveys and Maps

The cost of simple soil surveys is small compared with the benefits of having the knowledge on which to base soil management, cropping and fertiliser plans. Where applicable, project holders should be encouraged to seek out existing reliable soil maps and survey results, or to undertake simple surveys themselves. If facilities are available, soil testing or plant tissue testing should be carried out to discover the plant nutrients needed.

III Plant nutrients (manures and fertilisers)

Crop production depends upon there being a supply of plant nutrients in the soil. These nutrients must be available; that is to say, present in such a form that they can be taken up in solution by plant roots. For this to happen efficiently the soil must have adequate, but not excessive, moisture. Plants will respond best where soils are of good texture and free of weeds which compete for moisture and nutrients.

Organic matter (humus) helps soil to retain nutrients, and reduces the rate at which they are leached by rain. Leached nutrients are usually lost into lower layers of soil or in drainage water, but deep-rooted shrubs and trees can perform the useful function of bringing some minerals back to the surface. It is worth remembering that, while a vigorous crop uses most nutrients, its well-developed root system subsequently adds to the organic matter in the soil. Nutrient requirements vary according to the crop (e.g., cassava is a hungry crop) and to the level of yields. High-yielding crops remove more nutrients from the soil.

Supported projects should always give high priority to maintaining — or preferably improving — soil fertility and soil structure.

1 Essential Elements

There are sixteen elements which are essential to plant growth (quantities and combinations differing for different crops) and these are obtained from:

a: the air — carbon (C) as carbon dioxide (CO_2);

b: water — hydrogen and oxygen (H_2O);

6

c: the soil, including organic materials and applied fertilisers (some may be sprayed on the crop).

Nutrients are categorised as macro and micro, the former being required in quantity and the latter in minute amounts (they are nevertheless very important).

The primary macro (or major) nutrients required for plant growth are: *nitrogen* (N) which is an essential constituent of protein, has a great influence on yield, and is removed in a harvested crop and therefore has to be replaced; *phosphorus* (P) which plays a key role in the use of solar energy through photosynthesis and is important for root growth — it is removed by crops (large amounts by sugar cane) and also gets 'locked up' in the soil; and *potassium* (K) which has many functions, including activation of some 60 enzymes — it is present in stalks and leaves and so returns to the soil if these remain.

The secondary nutrients are: *magnesium* (Mg) which is in chlorophyll and hence aids use of the sun's energy; *sulphur* (S) which is an essential constitutent of protein; *calcium* (Ca) which aids root growth and is in cell walls, and *sodium* (Na) which some crops need, e.g. sugar beet. *Chlorine* (Cl) is also essential, but plants normally have more than they need.

The micro nutrients (or trace elements) are *iron* (Fe), *manganese* (Mn), *zinc* (Zn), *copper* (Cu), *boron* (B) and *molybdenum* (Mo). These are key elements, roughly comparable to vitamins in human nutrition, but plants need them in only minute quantities. If purchasable at all in poor countries, they will be expensive and should not be used unless expert advice and local knowledge suggests that they are necessary and will be cost effective.

② Organic Matter

The organic matter (humus) in cultivated soils rarely exceeds 5% of the total dry matter, and is often much less; however, because it not only contains nutrients, but also affects soil structure and water-holding capacity, it has a big influence on productive capacity. It is formed mainly by the action of soil micro-organisms on dead and decaying plant and animal materials. It is lost by destructive processes of degradation and oxidation, and these can occur at a high rate in tropical conditions.

In the wet tropics the naturally high level of plant production is possible only because nature recycles plant material and converts it into humus. When this cycle is interrupted by cultivation and the removal of crops from the fields, the balance of inorganic and organic soil constituents is altered. With less organic matter, the soil loses texture and becomes not only less fertile, but also more vulnerable to drying out and to erosion. Therefore, under cropping conditions, it is vital to put back into the soil as much organic material as possible in the form of vegetable and animal waste, and to make use of surface vegetation and litter.

In intensive farming with high-yielding crop varieties and inputs of inorganic fertilisers, the situation becomes exaggerated. The fertilisers not only produce an increase in grain yield, but they also stimulate the micro-organisms into greater activity. These organisms use the carbon of organic matter as a source of energy, so degradation of the soil humus is accelerated, and this is only partly compensated for by the greater volume of plant roots of the heavier, more vigorous crop.

As wood fuel has become increasingly scarce, some countries which have not traditionally burnt cattle manure are beginning to use it for fuel instead of fertiliser to the detriment of soils. Encouragement should be given to the use of organic waste materials such as crop residues, animal manures and suitable human waste.

In general, organic materials used as fertiliser have the following *advantages*.

a: They cost little or nothing in cash.

b: They release nutrients slowly so there is less loss by leaching than when inorganic fertilisers are used.

c: They improve soil structure so that plants are able to make better use of any available inorganic fertiliser.

d: They encourage useful soil micro-organisms and fauna.

e: They can be used as surface mulch to reduce moisture evaporation.

f: They can be used as fertiliser; in some cases, obnoxious wastes, rather than being a hazard to health, become valuable after treatment (see PART SEVEN Section 11).

But there are some disadvantages.

a: The nutrient value of organic material varies widely and is generally unknown, except in the case of purchased fertilisers such as hoof, horn and bone meals, dried blood, shoddy, leather wastes, processed fish waste and bird guano, which should be accompanied by an analysis of main ingredients. In intensive, high output farming, organic materials are usually used only as a means of maintaining soil humus content, and the nutrients removed by crop production are supplied by inorganic fertilisers.

b: Being bulky they require more physical labour than inorganic fertilisers.

c: There can be a health risk, particularly where human excreta are used, unless the material is treated to kill any disease organisms.

d: Cellulose material (such as cereal straw) uses nitrogen in the process of break-down, and so will deplete the soil of this nutrient if more is not applied. When there is insufficient nitrogen or moisture, the decomposition of straw in the soil can be very slow. Rice straw has a high content of lignin which is difficult to decompose.

Good *compost* can make an excellent fertiliser. Vegetable and/or animal matter piled into a heap or placed in a shallow pit can be broken down by bacteria and fungi within a few months and provides a fertiliser especially useful for small areas such as vegetable gardens. Almost any waste of animal or vegetable origin can be made into compost and the heat generated during decomposition kills many weed seeds and disease organisms in the material. The top of the compost heap or pit should be covered by a few inches of soil to protect the organic matter from strong rains. The sides and the top should be turned in at least once to ensure that they also decompose. To speed up the process, mix a few pounds of nitrogenous fertiliser with the material to be composted; this is unlikely to be necessary if animal manure, which is rich in nitrogen, is present. Also, since air is necessary for the aerobic (air breathing) bacteria, posts can be buried in the heap and pulled out to form chimneys which enable the air to circulate.

6

3 Nitrogen Fixation and Mycorrhizal Activity

Plants of the legume family (such as beans, peas, clovers) together with several trees (such as leucaena, prosopis and acacia species) have a particular value in enhancing the nitrogen level of soils. Bacteria called rhizobia, living in nodules on the plant roots, trap nitrogen which they obtain from air between the soil particles, and change it into a form that can be used by the plant. These bacteria work well in the warm conditions of the tropics. Different legumes may need their own species of rhizobia; some have more specific needs than others. If it is not present in the soil, seed must be inoculated, i.e. mixed with purchased inoculum of the bacteria (see Section 7).

Legumes are particularly useful for inter-row cropping as they provide some nitrogen for the other crop, while producing useful high-protein food themselves (see Section 7). Legume trees, such as leucaena and acacias, can be grown inter-row in forestry schemes to provide nitrogen, fodder and fuelwood (see Section 5).

Blue-green algae, free-living or in association with other plants, have a similar function in trapping nitrogen for crops. They are best known in the Far East on paddy fields where they live in association with the tiny azolla fern. There are many other living organisms in the soil which help plant growth. Increasing interest is being shown in fungi called mycorrhizae. These seem to make minerals more available to plants. They are particularly significant in relation to phosphorus, which is prone to become locked up in the soil.

All these natural ways in which atmospheric nitrogen is turned into plant food are of great significance to tropical agriculture in general and poor communities in particular. The husbandry methods should encourage them wherever possible.

4 Mineral or Inorganic Fertilisers

The application of inorganic fertilisers, especially nitrogen, will usually increase yield, but is not to be recommended without full consideration of other factors, such as those listed below.

a: Availability of water as rainfall or irrigation (plants cannot use fertilisers unless there is adequate moisture).

b: Cost of the fertiliser in relation to the likely increase in financial return due to a bigger crop.

c: The increased risk which may arise as a result of the increased investment in the crop (probably on borrowed money), should there be disastrous crop loss due to climate or pests or a down-turn in the market.

d: The need to have good weed control so that weeds do not take some of the benefits of the increased nutrient supply.

e: The ability of the farmer to manage the crop and to handle, store and market effectively the increased production.

It is particularly important to remember that a farmer's confidence in the use of fertiliser, or even in improved technology in general, may be destroyed if he/she is encouraged to invest too much in fertilisers or other bought inputs all at once, and the crop is subsequently badly damaged by weather or pests. In practice it is advisable to act with caution.

a: Have soil and/or plants analysed, a service which is usually available from research centres, government services and colleges.

b: Conduct local trials, to confirm the wisdom of the lessons being taught (these should be continued over a series of years)

c: Counsel a step-by-step introduction of fertilisers, starting on part of the crop in question, and applying carefully.

d: Encourage farmers to calculate the value in terms of increased output of the net increment of fertiliser input, leaving plenty of margin for possible adverse effects of rainfall, pest and market imponderables.

Potash. The need for potash varies mainly according to the soil type. If required, it is usually applied during the preparation of the seed bed. It is often a constituent of a compound fertiliser, i.e. one containing two or more of the main nutrients.

Phosphorus is important for root growth and for early maturity. It may be used as a 'straight' fertiliser, such as super-phosphate, or in a compound. It is usually applied as phosphate in the seed-bed or as a dressing in the early part of the season. There is little advantage in applying more phosphate than can immediately be used by the crop; the excess will quickly become almost unavailable to plants. Phosphate fertilisers vary in quality and cost according to the percentage of water-soluble nutrient that they contain. Mineral rock phosphate, such as is mined in North Africa, has a lot of insoluble phosphate, but this is gradually dissolved by soil acids and made available over a period. Solubility is usually increased by fine grinding.

Nitrogen is used by plants to synthesise protein and chlorophyll, the green colouring matter of plants which uses sunlight to create carbohydrates by the process of photosynthesis. It increases stem and leaf area and so has a direct effect on the growth and vigour of a crop.

Crops respond positively to applications of nitrogen fertiliser, provided there is moisture and a reasonable balance of the other major nutrients. A good response can often be obtained from quite a small amount. It is wasteful to apply more than the crop can use because nitrates are readily leached from the soil and lost in drainage water. In large quantities they can be a pollutant.

The type of nitrogenous fertiliser that is most commonly available in developing countries is *urea*. This is a highly concentrated form, but tends to be less efficient than others because it decomposes rapidly and releases ammonia into the air, or, as it is soluble, the nitrogen is lost in water. It is common for as much as 50% or 60% of the nitrogen applied to paddy rice to fail to be recovered by the rice crop. A technique of putting urea in mud balls, and deep-placing them, has been used to reduce losses and the International Fertiliser Development Centre (IFDC) and the International Rice Research Institute (IRRI) have developed a super-granule fertiliser which has urea encased in a sulphur coating. The effect is to slow the rate of release of nitrogen so that the plant is able to take it up; this means less wastage. Research work at the Indian Agricultural Research Institute, Delhi, has shown that the seed of the neem tree (or the residue after oil extraction), if mixed in the soil, will substantially increase the efficiency of utilisation of urea nitrogen.

Other nitrogenous fertilisers are *ammonium nitrate, sulphate of ammonia* (not recommended when soils are acid), *nitrate of soda* (with a lower percentage of nitrogen than the others but useful if sodium is also required) and *nitrate of potash* (only where high rates of potash are also needed). Nitrogen fertilisers tend to be used in compounds for basal or seed bed dressings and as 'straights', i.e. without other elements, to use as top dressing on growing crops.

The amounts of the major nutrients contained in a mixed or compound fertiliser are stated as percentages of N, P_2O_5 and K_2O, always in that order. For example, 20: 10: 10 means that the compound contains 20% N, 10% P_2O_5 and 10% K_2O.

6

Section 3 Water for Crops and Livestock

I The importance of water

Water is a crucial natural resource; in many regions the lack of it may be more limiting to development than problems related to land. Optimising its use is likely to be a central issue in many rural development programmes.

1 Social Organisation

Many aspects of social organisation are related to water availability — the siting and size of villages may be determined by supplies of drinking water, or the need to avoid periodic flooding. The organisation of village life is intimately related to rainfall, not least through the staple crop(s) that it makes possible, and to supplementary irrigation water for which the cooperation of communities is usually essential. Therefore the further development of water supplies, for whatever use, is likely to be an exercise in social development. The implications of the development should be anticipated as far as possible. For instance, in sinking wells for irrigation, one new sinking may have an effect on the water level in neighbouring wells, thus causing widespread increases in extraction costs; in such a case a new social organisation may be necessary to ensure that the whole society dependent upon that groundwater obtains optimal benefit. With irrigation schemes, proper control which ensures equitable distribution of water at all times (particularly at times of water shortage) is essential.

2 Some Oxfam Experiences

Oxfam is heavily involved with projects which are concerned with the conservation and use of water by means appropriate to poor communities and small farmers. Among such projects has been one in Kenya, a clean water and irrigation scheme (KEN 99), and another in Brazil designed to provide community education in appropriate technology, with particular regard to trickle irrigation and windmills (BRZ 141).

Field staff should give priority to projects involving the hand-digging of wells, small dams and small reservoirs. Small dams to provide water to make cereal production possible, and micro-catchment areas associated with tree planting have been constructed in arid areas in Burkina Faso (VOL 41); in contrast, displaced people settled in a large valley in Kenya were vulnerable to flash flooding, so simple check dams and bunds were built for protection (KEN 2 and 31).

A micro-catchment/run-off system was developed in Jordan's semi-desert regions (JO 41 and 44), and assistance has been given to water conservation for nomads in Niger (NGR 5 and 8).

There are numerous examples of Oxfam supplying pumps and materials for building wells (e.g. BIH 10), and for building irrigation reservoirs and wells (WBE 23, GUJ 18 and RAJ 12).

II Water conservation and use

1 Rainfall Pattern

The *pattern* of rainfall is important in two ways; first, in the number of seasons in which it falls and can contribute to crop growth, and second, in the profile of precipitation within any one season. The most fortunate areas are generally those where:
- [ ] the onset of the rains is predictable within a very few days;
- [ ] the rainfall within the season is well distributed;
- [ ] the season is long enough for a food crop to mature;
- [ ] heavy downpours and storms do not occur.

The *amount of water available* for use depends not only on the characteristic of rainfall, but also on the rate of evaporation (or transpiration through plants); this, in turn, is dependent on atmospheric conditions of temperature and humidity and on run-off, which is related to land slope and soil permeability. Much is lost to people's use by these means, and conservation practices are likely to be necessary to avoid these problems.

2 Conservation by Agricultural Methods

Water conservation is closely linked with soil conservation and with good crop and animal husbandry (Sections 2, 7 and 8). The objective must be to reduce water losses caused by excessive evaporation from bare earth, rapid run-off, or the 'mining' of underground stores. Water conservation should be a preoccupation, if not a primary objective, in project planning. Where water is a limiting factor, certain steps should be taken.

a: Optimise the use of the amount of rain which soaks in, for instance, by cultivating to break-up impervious surfaces and/or maintaining a vegetative cover either of trees, grasses or crops, or of mulch or dead vegetation, so as to break the direct impact of rain.

b: Slow down surface run-off, particularly on slopes of loose soil, for instance by ridging and/or ploughing, tie-ridging (see below), hillside terracing, or range-pitting on pasture land.

c: Reduce evaporation of soil moisture by a judicious balance of minimal disturbance of the surface soil and suppression of weeds, in some cases with protective mulching.

3 Water Harvesting and Storage

This is designed for areas which suffer from a shortage of water for crops and livestock, and where some rain falls, but only intermittently. Often so much rain falls in a short period of time that most of it is wasted by run-off and evaporation. The principle of water harvesting is to catch as much of this rainfall as possible and hold it efficiently for later use. Listed below are some ways in which this is done.

a: Run-off farming. This uses a catchment area not suitable for cropping, to collect rainfall that is then channelled either directly onto a smaller cropped area, or into a tank from which water can be subsequently supplied (by gravity or pump) to the crop. A gentle, rocky slope is ideal. Steep slopes will erode and wash silt and stones onto the arable area. Work in the Middle East suggests that a catchment area to cropped area ratio of 20:1 is appropriate; this will vary according to topography and soil type. If the cropped area receives 100mm of rainfall a year, the use of a 20:1 catchment will treble or quadruple the amount received.

b: Catchment dams. This technique is well developed in Australia (information from CSIRD or Department of Agriculture, Perth, Western Australia). A small reservoir or dam is made and the bottom sealed with clay, plastic or cement, and an area around or to one side of it is compacted and angled so that heavy rain will be directed into the storage area. An existing catchment surface, such as a road or concreted holding area for livestock, can be used by siting the reservoir alongside.

c: Micro-catchments. This technique has been used effectively in the Negev, Israel, the Middle East and Burkina Faso. The principle is the same as for catchment dams, but on a smaller scale. In a tree or shrub-planting programme, each sapling may be provided with its own catchment area so that rainfall water flows towards it instead of away.

d: Tie ridging. This involves creating a chequer-board of small basins in which crops grow. First, small ridges are earthed up on the contour to stop soil erosion. Then some ridges are made at right angles to these. In Burkina Faso , trials conducted by the International Institute for Tropical

6

Agriculture with maize have shown that yields are much higher with tie ridging than with ordinary row cropping. Yields were nearly doubled when each maize plant had its own basin, although fairly high fertiliser rates were also used.

e: Gabions and underground tanks (the trammel system) designed to hold water in flash floods. Gabions (loose rocks in strong wire netting made up to form rectangular blocks) are used to form low walls in, but not right across, stream beds. Inside each block is a vertical fin of porous plastic, attached at its base to buried plastic pipe that leads from the stream bed to constructed underground tanks. As the flood comes down, the fin directs some of the water into the tanks. An experiment in Syria is assessing the best layout and angle for the gabions. When adopting this technique it is important to be aware of the total effect on the area and on the supply of water downstream, and to adjust the water take-off accordingly.

f: Constructed catchment tanks. There may be a choice, according to local circumstances, between materials, expensive brick tanks and less expensive (in materials) labour-intensive tanks. Ferro-concrete tanks have proved useful in many areas (COL 54). It is quite possible to make tanks from polythene sheet, or a weak sand-cement mixture (using locally collected sand) and mud, although termites, rats and mice can be a hazard.

Tank size, specifications and construction will vary according to local conditions and needs. They may store roof water or surface collected water, and incorporate simple filters, covers and water lifting devices.

In an area where the evaporation rate is high a cover will be required. Also, the deeper the pond or tank in relation to its surface area, the less water will be lost through evaporation. A cheap cover can be made by straining wires across the surface and laying bundles of sorghum or maize stalks or any suitable stems on the wire.

g: Weirs can check or divert small rivers, or be constructed in the beds of sandy wadis (dry beds) as sub-surface dams. Weirs may result in considerable water storage in the sand beds upstream. Such weirs will usually be made of stone or concrete, but care must be taken to allow for the destructive effects of flash floods along the river.

h: Springs can provide an excellent natural water source. Usually the spring is captured or boxed to enable the water to be directed into a chamber, tank or piping system. Where possible, the water is led downhill by gravity as a piped supply to the community. It is important to ascertain whether the spring flows only seasonally.

III   Wells, tubewells, pumps

Field staff should give priority to self-help well-digging programmes, or the purchase of a simple pump set. For example, in Bangladesh a loan to a landless group enabled members to become effective share-croppers by purchasing a mobile pump set which gave them the means of increasing crop output.

1 Wells

Hand-dug wells are a prime source of safe drinking water and year-round supplies. Few communities exist which do not have a well-digging tradition, and the objectives of well projects should include the upgrading of local well-digging practices, knowledge and training. Hand-dug wells require careful siting, and reputable water diviners may be able to locate sources and avoid expensive trials. The use of pre-cast concrete rings has been a major advance in well linings.

2 Tubewells

These can be of varying depths, and collect water from different subterranean strata, although it is more usual to drill these only as far as the first level where usable quantities of water are found. They can be drilled by hand or machine. The drilled hole usually requires lining and this can be done with metal or plastic pipes. Bamboo tubewells are indigenous to the Indo-Gangetic plain and are appropriate for drawing water from shallow depths. Tubewells are particularly suitable for use with hand-pumps for the provision of drinking water supplies, but present greater difficulties when extensive irrigation use is planned, because of the higher capital and maintenance costs of the extraction equipment.

3 Pumps

There are several common categories:

The handpump is limited in yield and is mainly suitable for domestic use, irrigating small gardens or providing water for stock. Pumps are mounted on tube wells and are of two types depending on the depth of the water.

The shallow well pump has the piston, which is enclosed in a cylinder, inside the pump stand itself. This is the simplest handpump and relies on atmospheric pressure to lift water to a maximum of 6 or 7 metres only.

The deep well pump has the piston located below the water level in the well itself and can lift water from as deep as 50 metres. It is expensive and difficult to repair and, since the piston is connected to the handle by rods, its power requirements are higher, and children may have difficulty in using it.

Animal powered water lifting devices are quite simple in design and can supply up to 26,000 litres of water per day under suitable conditions. A bullock harnessed to a pole-drive connected to the main gear turns a chain wheel. This pump can irrigate small areas cheaply.

Wind pumps are suitable for exposed sites in trade wind zones and can operate over 24 hours a day. They are generally expensive to install and therefore used by larger farmers and cooperatives.

Diesel and electric powered pumps. The most common is the centrifugal pump coupled directly to a diesel or electric motor. Running costs can be relatively high, and villagers must be trained in care and maintenance, but usually these systems are very effective for substantial water lifting needs.

Hydraulic rams provide a simple method of lifting water up to a height of 150 metres from a perennial stream, where a substantial head, or fall, of water can be obtained within a reasonable distance. Rams are driven by flowing water and can operate 24 hours a day for many years with relatively little maintenance.

Solar pumps are not used very widely at this time mainly because of their high capital cost. They are operated by flat-plate or focusing collectors or by using photovoltaic or silica cells. Considerable experience has been gained from their use in Somalia. It is expected that these systems will become more accessible as the costs fall and the technology develops further.

As with all mechanical devices that are introduced, the beneficiaries must be taught how to maintain and repair them, and must establish a clear and agreed structure of responsibility to ensure proper operation. It is necessary to provide suitable tools and spare parts for long-term maintenance.

IV   Irrigation

1 Organisation

It is essential to any irrigation system involving the supply of water to a number of farmers that it should be properly managed under rules drawn up by a group which is representative of all concerned. These rules should cover a number of areas, as listed below.

a: They should ensure that everyone has access to the water on an equitable basis.

b: They must lay down the principles under which water will be distributed when in short supply. In one project near Pune, India, it was agreed that no one should have more than enough to irrigate 2 acres. Therefore farmers with 2 acres or less had a full supply, while farmers with more land were restricted to a supply sufficient for 2 acres.

c: They should also clearly establish the amount and time of payment for water and the procedure for dealing with defaulters.

d: They must make clear who will be responsible for the control of the water on a day-to-day basis and for the management and maintenance of the pumping system, piping and communal channels.

e: They should cover the setting aside of money for payment for electricity or fuel, repairs and spare-parts for equipment, and other contingencies.

f: They should lay down procedures for admitting new members to the group (which should be open to everyone within the area that the system can reasonably supply) and for releasing members who wish to leave (e.g., if they have made an initial investment in the scheme, which they will want repaid).

g: Finally, if possible, they should anticipate any legal liabilities that might be incurred by the officers or members of the group in relation to the irrigation system. If free or cheap legal aid is available, legal advice would be advisable in drawing up the rules.

2 Crop Requirements

These differ according to the crop (some are more drought-resistant than others) and according to the type of soil; ideally the soil should be moist throughout the growing season. Properly timed irrigation should avoid wilting of the plants, which causes a check to growth. Good drainage should avoid waterlogging in most crops (paddy, of course, being an exception).

Crops vary, but as a general rule plants are more subject to stress from water shortage at some times of their life than at others — commonly during germination (seeds need to take up water) and seedling establishment, and again at the time of flowering and seed set.

Water is lost from the system by evaporation from the soil surface and transpiration by the plant. Live mulching (ground cover crop interplanted) or ordinary mulching reduces evaporation, and shade can reduce transpiration.

Salts may collect on or near the soil surface, particularly in conditions of high evaporation, if the water is at all saline and drainage is inadequate. Where this has already happened, flush the soil with a lot of water (if available), improve drainage and, if necessary and feasible, switch to a crop which is more salt-tolerant.

Remember that use of irrigation means more crops/higher yields with all the consequent implications for labour requirements and crop use, storage or marketing.

3 Irrigation Systems

Ideally, when water is to be supplied to a number of small farmers' fields, it should be possible:

☐ for each farmer to control the supply of water to his/her land;

☐ for each field to drain directly into a ditch;

☐ for each farmer to have direct access to his/her land, preferably by a road which can at least take an ox cart.

Unfortunately it is common for land-holdings to be segmented, with portions interspersed with those of neighbouring holdings. Where the farmers can agree to rationalisation and redistribution of the area to be irrigated, this will increase farming efficiency, not only in terms of water use, but often of a net gain of land area that can be cropped. Unfortunately, though, there are often difficulties and legal problems involved in such a procedure.

When an area of land is set up for irrigation, it should, as far as possible, be graded to a steady slope (the slope being determined by the porosity of the soil). Thus the water will move steadily across to the drainage channel at a rate which will achieve optimum uptake of water by the crop, with minimum evaporation and minimum soil and fertility loss into the drainage system. On all but the gentlest slope, terracing (steep slopes) or contour bunding will be necessary.

The actual system used will depend on water availability, fuel availability, financial situation, soil type, area to be irrigated, crop(s) to be grown, and local preference. Obviously, local expert advice will be needed.

For any system to work well, it is essential that the users have security of tenure of the land they are farming. They should also have a system of paying for the water, which encourages them to make the best use both of it and of the land in the long, as well as the short, term.

Whatever the system, beneficiaries must be educated to avoid undesirable side-effects of irrigation, such as bilharzia, caused by bathing in infected canal water, or the increase of plant pests and diseases through excessive crop specialisation and failure in control measures.

V   Aquatic weeds

The vigorous growth of weeds in waterways, lakes and irrigation channels in developing countries is a serious nuisance. It is important that efforts be made to keep them in check. There are various possibilities for attempting control.

1 Removal by Hand

Where the scale of the operation allows, removal by hand will be the cheapest method and the material can be used as a mulch around crops. Some weeds, including the two worst aquatic weeds, water hyacinth and salvinia, can be fed to animals.

2 Mechanical Control

This may be considered for a fairly small-scale operation, but it is unlikely to be feasible for clearing a large expanse unless the harvested material can be put to good use. Harvesting is energy-expensive because the plants, which are 85% to 95% water, are heavy, intertwined and sometimes rooted. There are several possible ways to deal with the harvested weeds.

a: De-watering, using a screw press. This can save energy. A small press designed at Florida University removes 50% of the water content of 4 tons of chopped water hyacinth per hour. The material then becomes useful as animal feed, and the water can be returned to the waterway without causing harm.

b: Water hyacinth and other aquatic weeds have been made into pulp for paper manufacture in some countries, notably India.

c: It has been reported that in the USSR the cellulose from water weeds is used for paper, building materials, and for chemicals for industrial uses.

3 Herbivorous Fish

Some fish, particularly the grass-eating carp, will feed on the aquatic weeds, giving the double advantage of weed control and production of a high-protein human food. This kind of fish farming/weed control is now being developed on a large scale in the Nile Delta streams and canals in Egypt.

6

4 Chemical Control   There are a number of chemicals that will successfully kill aquatic weeds but there are serious disadvantages:

☐ high cost;

☐ toxicity hazards, danger to fish and aquatic animals, possible hazards to adjacent crops or through irrigation using water containing chemicals;

☐ de-oxygenation of the water caused by the decaying mass of killed weed. This is serious for people who are dependent on the source for safe, palatable drinking water, and on fish (which will die) for nutritious food.

VI   Water for livestock

Water supply is a primary consideration when introducing or extending any form of livestock production. Ample water that is not infested by debilitating micro-organisms is needed; milk-producing animals have the greatest need.

There is always a serious danger of over-grazing close to watering points (and possibly under-grazing elsewhere), unless the need for balance is properly understood and close control is kept of animals and their numbers. When developing a water supply system for livestock (e.g. sinking wells in range land), it is important to relate the number of wells to the availability of feed in the same area.

The investment in water must be adequately matched by investment in measures to control insects and infectious diseases, which are intensified by a concentration of animals.

In any project where water was formerly a limiting factor, the improved productivity of the animals may raise problems of use or marketing. Plans and provisions for this must be made which, in turn, may require changes in the attitudes of the owners.

The wider implications of water supply projects must be fully discussed and understood by all the people concerned and their agreement established. Wherever possible, the local community should provide some of the resources, particularly labour.

Section 4   Agriculture and Power

I   Justifying the use of added power

Increased power application in rural areas is justified only if it provides greater net human benefits. This is not necessarily easy to assess. Reduction in drudgery, which may affect health and life expectancy, might be as important a benefit as additional food, materials or cash sales.

Power can be thought of in terms of the output of human, animal or mechanical 'engines' of various sizes. Such engines seldom have single-purpose roles, are rarely used singly, and are seldom used consistently throughout the seasons. Therefore, maximising the productivity of a specified group of engines can present a complex problem.

Power problems need to be considered and solved in both a family context and, more generally, at the community level.

Some specialisation may be both inevitable and economically desirable, both in terms of the type of work done and the control of specific capital items. Excessive inequalities must be avoided, for example, in the distribution of work between men and women, particularly where payment is in the form of food. This is especially important at crucial periods of women's lives, e.g. during pregnancy, and at important times of the year, e.g. before harvest. The advantage to those who hire out animal and mechanical power units must be minimised as this can often lead to debts and an irreversible increase in the discrepancy between rich and poor.

II   Increasing and making optimum use of power

1 Human Power

a: Increasing the human power available. More work will probably lead to an increase in food and clean water consumption. Oxfam's primary concern is to transform a cycle of declining human productivity that leads to a decline in ability to work, into an upward spiral where returns on work exceed the purely physical energy required for that work. This will lead to improved health and productivity. Choosing the most productive activities possible in the circumstances and improving the overall environment is important. For instance:

☐ it is useless to persuade a family to grow an extra crop, if the physical power expended exceeds the energy value of the product;

☐ it may be highly rewarding to introduce a labour-saving device, e.g. for the roof-catchment of drinking water, if this allows time for other productive employment, the value of which exceeds the capital cost of the innovation.

6

b: Improving deployment. There may be a variety of possibilities:

☐ after *very* careful consideration, change from traditional to new ways of doing things;

☐ better work routines and better organisation of the working environment.

c: Improving associated equipment. Simple, well thought-out equipment (preferably home- or village-made from local materials), pedal power (if culturally acceptable) and suchlike can reduce drudgery and increase output.

d: Supplementing with other types of power. Supplementation of human power with power derived from animals, water, wind, sun, or engines, using chiefly fossil fuels, may be desirable. For instance, multiple-cropping made possible by irrigation creates new peak labour requirements at those times in the year when one cropping season ends and another begins; extra power may be required to relieve pressure on post-harvesting labour by employing a power-driven thresher, or to ensure timeliness of the next planting by using better-equipped bullocks or a tractor.

Before such decisions are taken, care is required to ensure:

☐ that the increased productivity is more than enough to pay for all the extra costs (see Cost-Benefit Analysis in PART THREE Section 2);

☐ that individuals do not incur obligations which are to their long-term disadvantage. For instance, incurring a loan which will be difficult to repay because of the lack of immediate benefit.

☐ that the introduction of extra power units and associated equipment, frequently attractive to men, does not, by increasing other work like manual harvesting, put more strain on already over-burdened women.

A successful solution may lie in a group or community project where costs can be spread, provided management is strong and just.

2 Animal Power

a: *Increasing the animal power available.* Where animal power is already employed, increased power per animal will be directly related to better food intake and health (Section 8). Decisions to improve the feeding of work animals must balance the advantages, (e.g. the increase in milk, in dung and/or biogas to enhance soil fertility), against the disadvantages (e.g. the competition which high quality animal feed may present to the growing of food for direct human consumption). In practice, crop by-products and, in less densely populated areas, waste land, will provide a high proportion of animal feed.

Where animals, whether they be oxen, buffaloes, camels, horses, donkeys or llamas, are introduced into farming systems previously powered wholly by human means, the main considerations are:

☐ that the real capital and operating costs are understood and that increases in long-term benefits should exceed the costs;

☐ that there will be adequate food and water for the animals, and that there are no health hazards that cannot reasonably and cheaply be controlled (e.g. tsetse fly);

☐ that good adult or rearing stock is available;

☐ that the substantial costs of training both operators and animals can be met.

b: *Improving deployment.* Work animals should be put to a variety of work to ensure their full employment without over-taxing them.

c: *Improving associated equipment.* Despite recent attention to yoking and the design of ox-drawn equipment, there is still considerable opportunity for new adaptations suited to specific conditions. The National Institute of Agricultural Engineering in Silsoe is collaborating with research centres in developing countries in the design and development of ox-drawn equipment. There is also the Sierra Leone project (see Section 14, Resources).

3 Mechanical Power

a: *Increasing the mechanical power available.* With the exception of small-scale power units, an increase in mechanical power, either stationary or mobile, involves:

☐ substantial capital investment;

☐ use of fossil fuels, which probably have to be imported;

☐ an adequate maintenance and repair service (often involving imports of expertise and materials).

The capital cost of power units increases with size more slowly than power output, so power from small machines is more expensive than from large ones. There may be a few tasks which can be done only by large units. Mobile power units range from 7 to more than 70 brake horse power; with stationary engines, the range is even wider. Small, multi-purpose, easy-to-maintain stationary engines may be feasible, but small, mobile power units present more complex problems.

Mechanical power units may save fodder-growing land, increase the area that can be cultivated at the optimum time, and save manual labour. However, these advantages may easily be overrated. The fodder area saved may be a small proportion of the cultivated area, and in any case, it may not be practicable to replace all animal power. The amount of machine use required to justify its purchase may be such that the amount of cultivation at the appropriate time is not increased. Saving labour in most developing countries generally means, at best, limiting hired labour to casual, unmechanised work, and at worst, to wholesale unemployment if new types of employment do not result.

As a general rule, Oxfam does not fund tractor purchase or support mechanisation, but there are exceptions to this. For example, the use of mechanical power may be justified:

☐ where heavy soils are uncultivable by existing methods;
☐ where a receding flood leaves a very short period of time in which to cultivate;
☐ where supplementary irrigation, together with mechanical power, makes a radical intensification of cropping possible without adverse social effects;
☐ where mechanical power allows land reclamation where other forms of power have been withdrawn, e.g. following salinity and water-logging, cyclones, war, or the slaughtering of work animals.

A tractor to be used by a number of farmers is best driven and maintained by one responsible operator from the group. The organisation of the tractor's use must be under proper management (*not* one of the bigger farmers, who might use it mainly for himself and his friends).

Mechanical power and equipment by themselves are likely to increase yields only in limited ways, for example, by burying weeds better, by deep ploughing to break hard pan, by more efficient cutting and carrying of forage, and by less grain damage in well-designed threshers. Sometimes heavy machines may cause soil compaction. In paddy fields the wheels may need modification to simulate the puddling effect of frequent animal traffic.

b: Improving deployment. If mechanical power units are available, it is essential to use them to maximum advantage to spread the cost. This implies:

☐ looking for non-seasonal tasks (these are relatively few in agriculture, but milling and water pumping may be exceptions);
☐ using power units as flexibly as possible; though tractors are not being used to their precise design capacity when engaged as stationary engines or in transport, the increased return may exceed the additional cost;
☐ concentrating the use of mobile power units, so minimising the time spent in travelling to work;
☐ combining what were hitherto successive operations in one operation with minimum disturbance of the earth, e.g. minimal cultivation plus seeding and fertilising (see Section 7).

c: Opportunities for the joint use of power units. The bigger the power units or pieces of equipment used, the less likely it is that they would be appropriate for small farms. Economical use of big machines is in theory possible by:

☐ the owner hiring out the machine;
☐ shared ownership and costs;
☐ group organisation in which the operation of one or more power units may be the sole reason for cooperation, or one of a number of reasons (see Co-operatives in PART FIVE Section 4).

6

Although success stories about all these alternative types of sharing can be quoted, the owner hiring out the machine requires the participation of at least one substantial farmer, and the other two alternatives require sophisticated organisation and a reasonably sized farm. Unless these exist already and those involved are experienced and confident, mechanisation involves many difficulties and risks.

Hire charges, arrangements for the allocation of the machinery, responsibility for maintenance and repair, and acquisition and repayment terms applicable to the original capital sum, are all potential problem areas.

III   Developing power from other sources

The cost of oil is a particularly serious problem for poor countries. Discovery of alternative power sources, together with the development and renovation of existing ones, is, therefore, important.

1 Wind Power

In open or elevated positions, windmills may be geared to pumping drinking water, irrigating, threshing, winnowing and grinding grain, pressing sugar cane, as well as generating electricity for other power uses.

Many local designs exist, their advantage lying in the use of local materials, labour and skills and their adaptation to local conditions and needs. The main constraints are the unreliability of wind, and the inefficiency of converting wind to power output. The latter may be improved at some cost by the introduction of superior materials and components. Alternatively, designs for constructions using parts from obsolete vehicles are available, the cost depending on local availability.

2 Water Power

Designs of equipment depend on the nature of the site, the size and constancy of water flow, and whether storage of water is feasible, prolonging power supply beyond the natural season. Power offtake ranges from the very simple principles of the hydraulic ram, where large quantities of water are utilised to power the lift of a relatively small quantity, to the sophisticated use of electricity-generating turbines. The *Plata Pump* is a simple water-powered pump consisting of 2½ m x 75cm cylindrical fibre-glass housing which is the turbine. Water passing or falling through this turns a crank which operates a two-piston pump. A *Vertical Axis River Turbine*, connected to a pump, has been developed by ITDG over the past few years. This utilises the energy in the river to pump water for irrigation or other purposes. It is basically a small raft with a turbine suspended from its pump, connected by a crank mechanism on the raft.

3 Priorities in Power Development

In situations in which Oxfam assistance may be appropriate, the order of priority, taking both economic and social considerations into account, is likely to be as below.

a: The introduction of methods and equipment well suited to local circumstances so as to increase the productivity and reduce the drudgery of manual labour.

b: The introduction of animal power and therefore higher technical standards.

c: The introduction of small power units, whether driven by wind, or water, or small stationary engines using fossil fuels, to supplement both human and animal effort.

d: The introduction of small mobile power units designed for a wide variety of work under conditions which pose no serious physical constraints.

e: In exceptional circumstances, large mobile power units may be introduced where opportunities and organisations already exist for their management, and where the back-up service for the machinery is satisfactory in that region.

IV   Energy sources and use

1 Wood

As is discussed in Section 5, shortage of fuelwood and depletion of the forests is increasingly serious and, in consequence:

☐ the poor spend energy and time in searching further afield for fuelwood;

☐ by-products of agriculture, including dung, are burnt instead of being put back into the soil to maintain organic matter (see Section 2).

There is a need to plant trees for fuel, to use fuel more efficiently and to look for other sources of energy.

2 Charcoal

In Africa and Asia, wood charcoal is widely used as cooking fuel. It has five important advantages.

a: It is a concentrated energy source, with 3-4 times the heat value of green wood and twice the heat value of dry wood; thus it is more practical for high temperature work like blacksmithing, although the energy conversion from wood to charcoal is poor.

b: It is easier to transport over long distances than wood.

c: It does not make smoke when it burns.

d: It uses 'waste' wood and can be stored indefinitely in all kinds of weather because it does not rot.

e: It adds no taste to food which is cooked over it, nor is there any residue in the smoke to damage the cooking pots.

Although charcoal can be used in confined spaces since there is no smoke, there must be ventilation to prevent the build up of poisonous carbon monoxide gas.

Charcoal is made by heating wood in a limited supply of air. Small quantities can be manufactured very easily in a 200 litre (44 gallon) drum. One drum will produce about 20 kg of charcoal a day. Improved, small kilns have been developed, for example, the Cusab kiln designed in Kenya to use rangeland shrubs and bushes.

3 Biomass

The use of biomass to produce gas has been developed in some countries, notably China, to the extent that it provides the rural population with much of their energy. The great advantage is that it can be produced from all organic matter, such as animal faeces, human excreta and crop waste, when these are enclosed in a container or tank. The organic matter decomposes by anaerobic bacteria under uniform conditions of humidity and temperature to produce a gas. The decomposed solids are not wasted, but can be used as a good quality fertiliser.

Experience has shown that biogas plants may be more successful when used in a community project or commune, or in an institution such as a school. Individual families may not be able to afford to buy the materials needed to make a gas digester or to produce sufficient organic matter to create gas in useful quantities. However, low-cost designs are now available. The use of night-soil combined with other waste would appear to be a solution to give power *and* a reduction in human disease. However, in some societies there are social, psychological and religious barriers to using human excreta.

In India, a Biomass Research Centre has been set up at Banthra, Lucknow. In the Philippines, there is a government programme to study

6

the Petroleum Nut tree (*Pittosporum resiniferum*), an indigenous species producing nuts which yield oil containing hydrocarbons which burn like petrol. The University of South Pacific is conducting research into the use of coconut oil as a fuel.

4 Alcohol

Two types of alcohol are commonly used as a fuel, either by themselves or blended with petrol:
□ *Methanol* (or methyl alcohol) known as wood alcohol, and
□ *Ethanol* (or ethyl alcohol), grain alcohol.

Some countries, notably Brazil, are converting to the production of Ethanol from sugar cane, sorghum and cassava. The alcohol can be blended with petrol up to 25%. The production costs of such a process are high, using specially grown plants, and can only successfully operate on a large scale.

5 Producer Gas Energy

A combustible gas can be made by gasification of charcoal or uncarbonised agricultural wastes. This gas, if fed into an internal combustion engine, can produce motive power that will drive a pump or agricultural machinery such as grinding or milling machines. TPI in the UK have a small experimental unit on trial that is capable of developing up to 4kw shaft output.

6 Solar Energy

In its broadest sense, 'solar energy' refers to all the energy sources derived from solar radiation. This includes both *indirect* solar energy such as fuelwood, biomass, wind etc., and *direct* solar energy such as thermal and photovoltaic.

Direct solar technologies include solar architecture, solar cookers, solar furnaces, desalination, solar ponds, solar water heaters, solar water pumps, solar dryers and photovoltaics.

The non-commercial solar technology of water heaters, solar cookers, vegetable dryers etc. can be made quite easily from locally available materials. Although solar cookers are technically able to cook food, there have been social barriers to their providing a practical solution to people's needs.

Commercial solar technology, that of photovoltaics for use with water pumps, refrigeration, small-scale power supply (12 volt charging etc.), while apparently offering an independent, decentralised power source, is nevertheless susceptible to many of the problems of monopoly and nationalism found in other spheres of economic activity. This means that items such as low lift solar pumps and solar fridges that could play a useful role in 'cold chain' programmes, or rural dispensaries etc., are still so expensive as to be completely beyond the reach of most development programmes. Widespread use of photovoltaics will only happen if there is a dramatic lowering in price.

7 More Effective Use of Heat

a: Improved drying. The proper drying of fruits, vegetables and other foodstuffs means that they can be stored for out-of-season use. The cheapest effective method of drying is to use a simple solar drier. The techniques have proved to be very successful, and can be used at village level. Solar driers can be used for drying anything from groundnuts to large planks of wood for seasoning.

b: Improved cooking stoves. Traditional methods of open-fire cooking and a lot of local cooking stoves are very inefficient in their use of fuel. Efficiency of wood-burning, for example, can be greatly improved without radically changing domestic habits by using a better design of stove. This may be made of clay or possibly concrete. Oxfam is in favour of supporting work to design and introduce such stoves, and of the training of artisans to make them (e.g. VOL 68, and JD 45). ITDG and VITA have produced publications which give detailed guidance.

ITDG have made the points listed below.

- ☐ It is an essential part of the job of the designer of a stove or oven to determine specific socio-cultural factors that will affect the success of an innovation. No one knows these factors better than the people themselves.
- ☐ Innovations are much more likely to be successful where people have participated actively in their design and introduction. If close involvement of the designer with the community is not possible, the design should be flexible so that it can be adapted to local resources and skills in construction.
- ☐ The most inefficient way of cooking food is to cook it slowly in a lidless pot on an open fire, as is common in Africa.
- ☐ If fuel is scarce, there is evidence that people will change their cooking methods fairly readily, provided that the technological changes are small.

For example, in all urban and semi-urban areas of East Africa, metal charcoal stoves (jikos) exist in one form or another, and are usually constructed by local tinsmiths using scrap tins, metal or old oil drums. These stoves are usually 90% inefficient. In the last few years, new, improved designs have been developed in Nairobi by several groups and they now claim a 35% fuel saving. However there are many people who have seriously criticised the wood stove projects for failing to have found any real answer to the problem of domestic fuel shortages and over-use of timber. Wood stoves have often been found to be technically deficient and also lacking in social acceptability.

6

Section 5   Trees, Shrubs and Forests

I   Introduction

1 Social and Environmental Considerations

In many areas there is a very great need to plant trees and to protect and up-grade forests in order to:
- [ ] protect soils from erosion and degradation;
- [ ] provide much needed fuelwood;
- [ ] improve the environment (trees raise the humidity and provide shade and wind protection);
- [ ] increase food and fodder supply;
- [ ] maintain the traditional needs of people who live in the forest.

An FAO survey of Africa south of the Sahara has found that some 50 million rural people now have no fuel for cooking and that another 130 million are getting some wood, but at the expense of valuable forest. FAO predict that if present trends continue, some 500 million people in that area of the world alone will have no fuel for their basic needs by the year 2000.

Across all continents — and particularly in hilly or arid and semi-arid areas — fuel gathering, uncontrolled commercial tree felling and accelerated slash and burn agriculture associated with increased human population are leading to serious water and wind erosion of soils, leaching of nutrients and loss of soil structure. Erosion not only impoverishes the area immediately affected, but often reduces food production at lower altitudes by silting up water channels and dams.

Where forests are destroyed or cut and replanted unwisely with ill-chosen, exotic species, the effect on poor people dwelling in the forest and dependent upon a range of products found in a natural, mixed forest can be devastating.

2 Objectives

Field staff should encourage initiatives that help people to understand that trees:
- [ ] check erosion (their leaves break the force of heavy rain and their roots, if vigorous and spreading, hold soil);
- [ ] enrich soil (by bringing up nutrients from lower levels and passing them into their leaves — which subsequently fall to the ground and become incorporated in the top soil);
- [ ] capture nitrogen from the air (leguminous trees and some others) and make it available, by the action of bacteria, for other plants;
- [ ] transpire water into the air, thus improving humidity in arid climates;
- [ ] act as wind-breaks;
- [ ] provide fuelwood (some trees and shrubs being capable of being repeatedly cropped for fuel);
- [ ] provide materials for building, for fences (live or dead) and for all manner of useful objects;
- [ ] provide shade;
- [ ] produce food (some species), fodder (particularly the legumes), oil, raw material for pulp for paper, medicines and so on.

Projects can support efforts to plant community or village woodlots and to plant small groups or individual trees near homes, around schools, on bunds, on the edges of terraces, on land not fit for food crops and along roadsides. Some species will grow on very poor, apparently barren land. Shrubs may be nurtured for firewood. Utilization of improved cooking stoves that make more efficient use of fuel should be encouraged (see also Section 4). Where possible, forestry should be regarded as integrating with agriculture — an old tradition in parts of Asia, but not usual in Africa.

264

More often than not, women are the wood collectors so it is important to obtain their views when a tree project is planned. A few years ago in Burkina Faso Oxfam encouraged a survey to discover what species the women found most useful. It was found that they would rather have woody shrubs than large trees, because shrubs are easier to cut and carry. In one programme in Burkina Faso tree and shrub planting was combined with a rain harvesting technique (VOL 93). School plantations have become a feature of the countryside in northern Tanzania (TAN 88) A nursery raises saplings to be planted out at the schools and plans to supply other villages eventually. (See Section 3 for more on water harvesting).

The foothills of the Himalayas, both in India and in Nepal, are the scene of some of the worst erosion in the world due to tree cutting. There exist in these countries various projects to protect trees and to help very poor people to earn a living in ways other than selling firewood. In a major area of Nepal the government has banned all tree cutting; this may be correct from the national point of view, but is quite devastating for the poor who depend on the wood. In a group of villages affected by this ban, a project is trying to help improve food production and to encourage carpet weaving by, for example, supplying lamps to extend the working day (NP 15). In Lebanon, where tree-planting is taking place on the stony, once-forested hills of the south, the Mennonites have developed a three-year programme to raise fig and almond tree seedlings and grape vines and make them available to village people at subsidised prices (LEB 22).

Sometimes there is a need to plant trees after a natural disaster. When a tidal wave and cyclone hit Andhra Pradesh in eastern India in 1977, thousands of coconuts were uprooted. An Oxfam adviser assisted in the replanting and produced a simple guide for local extension workers. The trees planted included not only coconut but also eucalyptus, cashew, leucaena (koo-babul) and acacia. Establishment of trees in the dry conditions was helped by sinking unglazed clay water pots into the ground so that the young seedlings could obtain a steady supply of water.

II   Social forestry and tree planting

The most inhibiting factor in introducing social forestry has possibly been the term 'forestry' itself: its use leads to thoughts of forest officials with their rules and regulations; conventional forestry practices; governmental control; the dissociation of the lay person from the programme and his or her total lack of involvement, often leading to destruction of work done by officials. The main objective in social forestry must be to introduce the tree to the people, and not to institutionalise the concept. It is essential to establish who will own the trees produced by forestry programmes.

'Social forestry' refers to a specific objective; i.e. the benefits should be of a social, rather than of an industrial, nature. It can embrace many types of forest:

☐ village or community woodlots;
☐ farm forestry;
☐ planting on wastelands;
☐ planting on institutional land;
☐ planting in government forest reserves;
☐ urban and recreational planting.

Of these, the first is the most desirable, especially where real community involvement can be developed. Farm forestry is necessarily on private land and may cover the production of wood and fodder for the

6

farmers' own consumption, but frequently involves the sale of forest products to factories, occasionally at the expense of food production. Planting in government forest reserves is frequently for the protection of soil and water supplies, and for industrial wood production, but it is sometimes done with largely social objectives in mind.

1 Village and Community Woodlots

There is great value in a community undertaking its own tree-planting rather than having it done by governmental or other officials assuming that the community will receive some direct benefit from the programme (i.e. percentage of trees cut etc.). This engenders a feeling of responsibility to protect the woodlot from uncontrolled cutting, animals and fire. Prevention of damage by cattle, sheep or goats must be strictly enforced (or the area fenced) and, depending upon size of plantation and climate, fire-breaks and/or fire wardens may be necessary.

a: Fuelwood species. The possibility of using and improving existing local species is always worth considering first. If exotic species are introduced, this should be done with great caution, bearing in mind suitability for soil, climate and use. The most desirable trees for fuel will be species which grow fast and coppice well (grow vigorously from stumps or from root suckers after being cut).

A list of species is given in the National Academy of Sciences' book *Firewood Crops* under three categories — for arid and semi-arid lands, for humid tropics and for tropical highlands — together with useful notes on them. (For a list of the most important fuelwood species see Appendix I).

b: Nurseries. With some species, e.g. neem, *Cassia siamea* and *Gmelina arborea*, it is often possible to sow direct in the field, and this is effective where seed is plentiful and cheap. Usually, however, it is necessary to raise plants in nurseries. In most tropical situations where conditions in the field are severe, plants are best raised in containers, but care has to be taken to ensure that the roots do not coil, and that they are trimmed or pruned while the plants are in the nursery to ensure a vigorous bushy root system. Seed can be sown direct into the pots but is more usually sown first into seedbeds; it is best to use pure sand only. The young plants are then pricked out into the pots when they are about 3-5 cm high. The soil mixture in the pots is important, and local advice on materials should be sought. In general, some organic matter (manure, peat), clay soil, forest soil and gravel may make useful components of a soil mix. It is possible in some situations to issue baskets of germinating seedlings, i.e. at the seedbed stage, to individual farmers or schools for them to pot and raise the plants themselves. Usually trees are planted at the onset of the rains, after a suitable period in the nursery when they have reached 25-30 cm.

c: Planting and management. The site for planting, whether with saplings or seed, must be cleared of vegetation that would compete with the young trees. When transplanting from the nursery, keep as much moist soil as possible around the roots, place in a small hole, replace top soil and heel in by pressing the soil around the plant from several directions. Place a clearly visible stick by each small tree or planted seed so that it will not get damaged in subsequent weeding operations. The young trees must be kept clear of any vegetation until they can withstand competition.

One form of agroforestry (the combination of trees and agricultural crops) which has been used for hundreds of years to create plantations is the '*taungya*' system which originated in Burma but which is now used all over the tropics. In it, farmers agree to clear the forest or bush in return for time to grow their crops, among which forest trees are planted. When the trees have shaded out the crops, new and fertile land is allocated. However, care has to be taken to ensure that, in areas of land hunger, peasants are not exploited. As a general principle, rules should be simple and non-vexatious, and all labour should be paid for. Agroforestry in which farmers raise both trees and crops for themselves is of course a different matter.

d: Shelter belts and windbreaks can be created by planting rows or blocks of trees and shrubs. They can reduce soil erosion, limit the spread of air-borne weed seeds and improve the immediate environment both for crops and for workers. The preferred types of tree, height of the windbreak, width, density and angle of the rows or block will vary according to the particular circumstances, so it is advisable to get on-the-spot expert advice. In general, the higher the shelter belt, the greater the distance over which it has an effect, particularly if the density is only moderate; the greater the density, the more confined is the effect. A carefully chosen mixture of trees and shrubs can not only protect houses and crops from cyclonic winds but also yield food, fodder, fuel and/or resins; where breaks are deep enough, a path can pass through at an oblique angle, which can provide a respite of shade for field workers and also livestock under control.

e: Use. The quantity of fuelwood burnt can be reduced by designing and using more efficient stoves, provided that their cost is within people's means and they are culturally acceptable. This is discussed fully in Section 4.

② Trees for Food and Fodder

The growing of local trees which have edible parts like fruits, nuts, palm hearts, soil and sago should be encouraged, provided they do not compete for land with annual food crops. They can be complementary, drawing on water and nutrients from a lower depth than annual crops. They usually have added advantages such as providing shade, holding soil or yielding other products. One or two can occupy unused ground near the home.

The value of fodder trees is immense where livestock is kept, particularly as they are likely also to provide fuel and to enrich the soil. The leguminous family of trees, such as leucaena and some acacias, is particularly good for fodder because of the high protein content of leaves and pods (but see notes about leucaena). (For a list of suitable species for food and fodder see Appendix II.)

III   Forests and forest people

① Agents of Deforestation

Tropical forests have not only shrunk in area due mainly to encroaching agriculture, but vast areas have also been severely degraded by exploitation and by forest fires. The exploitation is of two kinds. One is by companies who remove top quality trees, but who do not manage the forest so that other qualities are utilised, and do not plant trees or foster natural regeneration. The other is by forest dwellers or villagers who practise shifting agriculture on too intensive a rotation, because of increasing population and/or shrinking land area, and without adequate care to confine their fires. If these people have no involvement in commercial forestry, and if they have not appreciated the value of the trees to their environment, they will feel no responsibility for them.

② Needs of Forest Dwellers

Forest dwellers are unlikely to be consulted, and are sometimes not even considered, when forests are cut, planted or 'improved'. Therefore their traditional needs, which embrace a wide range of forest products essential to their diet and livelihood (honey, fungi, small animals providing protein food etc.) are overlooked. When, for example, an exotic tree species replaces an indigenous one, an ecological chain may be broken.

Oxfam is unlikely to be involved in direct funding of major forestry projects, which are usually run by government forestry departments receiving international aid, but there may be scope for helping local communities to establish small-scale industries associated with forest by-products or waste products. (See also PART TWO, Section 5.)

③ Forest Management

Tropical rain forests naturally contain a great variety of species, only a few of which are saleable or desired by the local people. Management systems in which more of the 'secondary' species are used, thereby clearing the ground for replanting with species more useful to humans, are gaining ground. However, in general it is extremely difficult to manage a natural forest in perpetuity and to maintain its present state. Such successful systems as there are, for example in Ghana or Sarawak, tend to be expensive both in time and skills. There is a good case, however, for conserving large areas of natural rain forests for the sake of the complexity of plant and animal life and the ecosystems that they contain. If possible, it makes more sense to supply people's needs for wood and fodder from nearby fast-growing plantations, and to protect the natural forests which can continue, with little damage, to be used as sources of protein ('bush meat'), fish, honey, medicines etc.

④ Reforestation and the People

Large-scale replanting of trees (reforestation) has become very important on eroding, deforested slopes and in areas, particularly arid ones, where removal of tree cover has led to a run-down of humus content and fertility of the soil. Where the soil is either badly degraded or naturally low in fertility, leguminous trees will help to restore fertility. Where conditions are better, it may be possible to produce food and/or fuel as well as forest trees in two ways.

a: Agroforestry is the name given to mixtures of trees and agricultural crops. The range of possibilities is considerable and complex. Expert advice and local experience should be sought. The choice of trees is important: for instance, leguminous species that fix nitrogen also improve soil fertility. The balance between trees and crops or grass is crucial, however, and there is a risk of one dominating or killing the other. In grazing systems, great care is needed to control the rate at which the sward is consumed, and to prevent damage to the trees.

b: Silvipasture (the combination of trees, pasture and animals) is a particular form of agroforestry. On slopes or in other conditions unsuitable for arable crops, grass can be grown for livestock feed. Unless the trees can be properly protected from the animals, it is better to cut and carry the grass to the livestock. Again, this can be helpful for forest labourers.

Section 6   Grass and Rangeland

I   The state of the rangelands

Rangelands are those areas of arid and semi-arid land where rainfall is insufficient for the purpose of crop production. These may be defined as follows:

0-150 mm: true desert, grazing based on areas of natural rainfall collection. Thick stemmed grasses and scattered browse trees.

250-400 mm: semi-desert, contains areas of shrub with open canopy forest on seasonal flood plains where sand predominates over clay.

400-800 mm: savanna, contains the agronomic dry boundary where annual rainfed cultivation competes with pastoralism and where the cow competes with the camel.

For many years there has been a conviction that conditions in the rangelands are deteriorating, and that although in some areas unwise attempts to cultivate rangelands have caused this, in most cases desertification is the result of overgrazing by domestic animals. This theory also maintains that the process of desertification has increased during the past century.

The main reasons for this belief are that:

☐ overgrazing is due to an increase in the number of livestock;

☐ this itself is due to an increase in the number of pastoralists, and in advances in veterinary medicine;

☐ increase in the livestock numbers is also due to a system of land tenure of communal grazing and private ownership of livestock which provides an incentive to increase the number of privately owned livestock even when this damages the communal grazing land, the so-called 'tragedy of the commons'.

This theory relies on a number of hypotheses that are increasingly being challenged by recent researchers, for the following reasons.

a: pastoral populations in Asia and North Africa appear to be static in size, declining, in some cases (USSR, Egypt) very fast. Outmigration, the severity of pastoral life, and in some African cases, lower fertility (due to social mechanisms designed not to allow human population to grow faster than livestock population), all contribute to a lower growth rate for pastoral populations than for non-pastoralists. This is of course one of the reasons why pastoralists have been progressively marginalised and why farmers whose numbers are growing faster have taken over areas of land previously used only by herders.

b: Livestock populations, especially in Africa, appear to have increased dramatically compared with those at the turn of the century. However, due to diseases and drought, the livestock population at the time may have been unusually low.

c: The deterioration of the rangelands and the speeding up of this process depend likewise on the point of comparison; the long-term trends may not be easy to distinguish. In any case desertification is a complex phenomenon, and to assign it to overgrazing alone is a gross and inaccurate over-simplification. Farmers who cultivate pasture land, the exploitation of resources by charcoal and wood merchants, and other factors contribute to the degradation of the rangelands.

d: The benefits allegedly stemming from the private ownership of grazing land have not occurred in dry rangelands; in Angola, for instance, environmental degradation has been worse on private communal ranches than under traditional communal systems. Similar cases can be found in Botswana, the USA and Australia.

e: The 'tragedy of the commons' argument supposes that herders are unable to organise themselves in order to manage rangeland in an economically rational way. Many traditional pastoral systems around the world once had mechanisms of one sort or another to limit access to rangeland (often through control of water) and were thus able to control grazing pressure to some degree. Now in much of Africa and Asia governments claim that pasture land is government property and that herders may not organise to manage it; this has resulted in a 'free-for-all', where very short-term gains have prevailed over sustainable patterns of land use.

Overgrazing does occur but is only one of a number of factors contributing to desertification.

II   Some causes of overgrazing

What causes herders to begin to revisit the same pastures with increasing frequency, changing the grazing cycles?

a: Increased water availability. When available water for drinking is limited, so is the grazing land in the area, allowing it to recover. When watering points are artificially increased, herd sizes can be increased temporarily near the water-holes to levels which are unsustainable in times of poor rainfall, as the grasslands do not benefit from the provision of artificial water points. It is said that in the Sahel drought animals died of hunger rather than thirst.

b: Farmers who cultivate pasture land put additional grazing pressure on the remaining land.

c: Increased pressures on rangelands is also due to sedentary livestock keeping, either by farmers or by herders, where the animals stay close to the settlement and damage the land more severely than a more mobile nomadic system.

d: Insecurity in land-use rights is a prime cause for the free-for-all attitude observed in some pastoral systems.

III   Methods to improve rangelands

1 Additional Water Supplies

Effective water development can open up for pastoral use new areas or areas of limited use. Consumption of water increases the appetite of livestock, thereby increasing its milk yield and weight gain. Less time is spent looking for water and more for grazing. However, water development and management must take into consideration several points.

a: Water management can only be acceptable when grazing is also managed.

b: Water is often provided free, and attracts sedentary populations, breaking down the traditional disciplines of water management, and destroying rangeland by depletion of forage for a large area around the water point. Such a development neglects the true purpose of the water point, i.e. the support of the nomadic flocks and herds, which are traditionally camped a distance from the water and come to the water point once every two to three days (cattle, goats, sheep), and every eight to twenty days (camels). Such permanent waters are traditionally used in the dry season only; during the rainy season the areas around the water points are rested, and the livestock taken out to seasonal rainponds. Settlers around water points may eventually shut pastoralists out.

c: The essential prerequisite for the introduction of new water points is the imposition of rules and regulations, and the organisation of users of the water into groups capable of taking joint action to manage the water supply. If managed and improved grazing systems are not acceptable to pastoralists, then the provision of extra water should be resisted.

d: Special attention must be given not just to how to *extract* water from the ground, but also how to *distribute* it; otherwise pollution of water and damage to structures may result.

e: Dependance on boreholes reduces uncertainty from climatic causes, but increases it from mechanical failure, shortage of fuel, and social causes — they can be easily sabotaged, or they can give the operator of the borehole a large degree of control.

f: Well planned surface water development may be considered as an alternative to boreholes, as in some Ethiopian cases. Here surface water points with a limited capacity are constructed to provide water for a limited period only. At the end of this period, the livestock is forced to move from the range area served, which is then rested. (See also PART TWO Section 5)

2 Improvement in Rangeland Productivity

Rangelands are characterised by their low and unreliable supply of moisture for plant growth, and so improving the productivity of rangelands is largely a question of effectively managing the water. Watershed management systems to control and retain run-off by water harvesting, water spreading and spate flow diversion to basic irrigation structures are all well-known means of raising productivity. Also, planting of shelters to divert the wind from the water surfaces can reduce water evaporation significantly.

Selection and propagation of plant species that maximise water resources, the use of fertilizers and introduction of legumes for nitrogen fixation are all technically feasible, but can be very expensive. These techniques and the techniques of watershed management require a great deal of technical and economic organisation, demonstration, dialogue and follow-up on the part of the development institutions, and a great deal of self-organisation and discipline on the part of the pastoralists.

In some areas, the real requirement may not be dry forage, which may be available in large amounts outside the immediate neighbourhood of permanent water points, but high quality feed during the dry season to provide protein and in some cases vitamin A. This can often be achieved by protecting or replanting shrubs and/or browse trees. Alternatively, a feed supplement such as cotton seed or cowpeas is sometimes economically feasible.

The right succession of grazing by different animals can increase the total quantity that is harvested rather than trampled or ignored.

In some highland areas there has been success in introducing improved grasses to previously natural pasturage.

3 Improvement in Animal Health and Husbandry

This includes:

☐ measures against diseases, wounds and injuries;

☐ routine operations such as castration and dehorning;

☐ introduction of superior breeds through community animal improvement schemes and the upgrading of existing ones;

☐ application of health measures to neglected species such as goats and camels;

☐ training schemes for community animal health workers;

☐ the importance of using the correct animal species, e.g. the camel for drought-prone, badly watered areas. The Danakil Ethiopian drought disaster of 1983 was largely due to there being far too many cattle and far too few camels;

☐ the use of breeds that are tolerant to Trypanosomiasis infection, particularly when the stockowners are forced in dry spells to use permanent water which carries the disease. (See also Section 7: Fodder Crops, Section 8 Livestock)

6

4 Credit and Other Financial Development

Recent research in Africa has indicated that pastoralists are particularly vulnerable to seasonal price fluctuations of both cereals and animals. This is because high seasonal prices of cereals tend to occur at a time when animal prices are low. Pastoralists are therefore in crucial need of credit in order to buy cereals without selling their animals at a low price, or better still, to buy cereals at a time when their price is low, i.e. at the end of the rainy season, and store them for the dry season. Credit schemes and cereal bank schemes may be particularly indicated not only in this situation, but also for herd reconstitution among poor pastoralists, e.g. after a drought or a serious animal disease.

Another financial development to be considered would be the provision of insurance against seasonal loss or a bad year.

5 Food Security and Weather Warning Systems

In arid areas, dry years and droughts are normal occurrences. Measures to warn pastoralists need not be in the hands of the government only, but pastoralists themselves as a group can operate some form of early warning system. Similarly, measures to counteract the worst effects of such disasters need not depend solely on government action. Food reserves and cereal banks can be operated by pastoralists themselves. (For further details on pastoralists and programme recommendations, see PART TWO Section 5. Also PART EIGHT Section 2.1.)

I Introduction

☐1 Priorities When Choosing Crops and Cropping Systems

Oxfam seeks to improve nutrition chiefly by supporting those endeavours which attack the conditions which cause malnutrition. These include:

☐ poverty caused by inadequate earnings or insufficient land to cover the food requirements of the family;

☐ farming systems which produce unbalanced diets that are often deficient in protein, vitamins and minerals (coupled with an ignorance of dietary needs);

☐ certain farming and post-harvest methods, inadequate storage, and financial pressure affecting the sale of produce, all of which may result in seasonally inadequate food supplies;

☐ ignorance of, or resistance to, suitable improved crops and improved techniques;

☐ over-burdening of those who work with the crop (often women);

☐ pressures to sell food which should be consumed by the family, resulting in a cash income insufficient to guarantee an adequate diet;

☐ increasing population per unit of land;

☐ practices which reduce the productive capacity of land.

Therefore priority should be given to projects which:

☐ increase total food production, or dietary balance where necessary, especially for those most disadvantaged, e.g., aid to school gardens, or to small-farmer and share-cropper groups and cooperatives to support simple irrigation, joint efforts to improve seed supply, savings clubs and revolving loan schemes to increase production;

☐ provide combined agricultural and nutritional advice in rural areas;

☐ encourage cropping systems which reduce risk and, where practical, spread and diversify food supplies through the year;

☐ reduce post-harvest losses of food and protect small producers from unfair market pressures (for example, simple community-owned or cooperative processing facilities and stores not only reduce losses, but also improve diet in traditionally lean parts of the year, or enable sale of surplus when market prices are good);

☐ provide training in crop production and storage, particularly training of a practical nature and preferably with an emphasis on giving on-farm advice and training in the village;

☐ foster the introduction and improvement of animal-power (or, where justified, appropriate small-scale equipment and machinery); taking account of the effect that increased cropping or new systems will have on field workers, with special regard to the burdens laid on women and to their traditional sources of income;

☐ make optimum use of natural resources, but avoid exploitative practices which might result in degradation or erosion of land;

☐ involve processing and handling schemes which give work and income to those who would be otherwise un- or under-employed.

☐2 Some Oxfam Experiences

Most of those projects supported by Oxfam which are concerned in one way or another with crops concentrate on improving either the supply of seed or the techniques of crop storage and marketing. It is at these two ends of the production sequence that the small producer is most vulnerable to exploitation or loss.

Support has also been given to enable farmers to introduce new crops, thereby diversifying and spreading the risk, or to improve their diet, or to grow a cash crop (where land can be spared without jeopardising the food supply). Such projects include the development of oil seed crop production in those villages in the Tabora region of Tanzania which do not benefit from tobacco growing (TAN 106); the introduction of passion fruit in Brazil (BR 225), small-scale cocoa in Bolivia (BOL 31), and coconut production and marketing in Colombia (COL 17). Part of a drought-relief programme in the south of Andhra Pradesh, India, consisted of the introduction of high quality seed appropriate for dry land use (AP 42). Impressive improvements in maize growing have been achieved with small amounts of aid to farmers' groups in Zimbabwe (ZIM 8).

Although farmers may be able to save their own seed satisfactorily for normal purposes, they will need additional, reliable supplies (sometimes very urgently) after a crop disaster, or for a diversification or expansion programme. Drought in an area of Ethiopia where shallots are the basic cash crop, left growers in danger of being without seed at the critical planting time; Oxfam was able to obtain seed for them (ETH 111).

In West Orissa (ORS 20), an exceptionally poor part of India, some 200 families were involved in a project to encourage family vegetable production and consumption. In Burkina Faso (VOL 31, 83 and 84) support was given to a large scheme for establishing market gardening groups by assisting in the purchase of seed and in the purchase and erection of a cold store. This was not very successful however and the cold store was under-utilised. A more typical form of support for programmes involving storage is in helping to improve village storage of staple foods for home consumption or sale. Improved storage facilities can help in two ways: to reduce grain loss and, if they are well managed by a cooperative or group with access to credit, to protect the small producer from the vagaries of the market.

In Burkina Faso, two projects (VOL 53 and 80) are concerned with establishing village grain banks, the grain being stored in simple but effective constructions, under a share-holding system and managed by a village committee. In Haiti (HAI 43) an interest-free loan has helped a grain storage cooperative with facilities for treating grain for protection against pests and storing it. In addition, members were able to sell their crops to the cooperative for immediate cash and to receive a share of the profit after wholesale. In Tanzania (TAN 107), assistance has been given for purchasing materials for village 'go-downs' or stores; the villagers organise and provide the labour for construction.

A number of lessons have been learnt from the experiences of storage groups and cooperatives. Chief among these is the essential need both for adequate management to preserve the crop in good condition, free of pests, and for adequate finance to be able to maintain a reserve of grain for the community's own needs following a poor season, when local production is low and market prices are high.

Mechanised grain milling is becoming the norm everywhere, but there are still remote areas where it may be worth introducing improved hand-mills, or pedal-mills with bicycle mechanisms. While such improvements may seem small, for the people involved they may require substantial adjustments; e.g. the introduction of a pedal mill in the Sudan (SUD 21) failed because men do not grind grain and women do not ride bicycles. Mills have been provided for feed grain at poultry cooperatives (ZAI 31 and 108), and one water mill for grinding grain (MAL 25).

II Crop Types

1 Food Crops

a: Cereals: rice, maize ('corn' in American usage), bread wheat, durum wheat, sorghum, millets, barley, oats, rye, triticale.

As harvested traditionally, cereals generally have an acceptable nutritional balance, but sophisticated methods of de-husking and polishing seriously reduce their nutritional value. Breeders' attempts to improve protein content and quality (high lysine) in maize and wheat have had considerable success, but such varieties would rarely be available in the poorer parts of the Third World. Triticale is a man-made cereal (with wheat and rye parentage) bred for higher protein; its use is now spreading. Each variety of cereal is best suited to a particular environment of soil, rainfall, temperature and day length and to local tastes. Therefore, it is not advisable to introduce a different cereal or type of cereal, except with extreme care and under special circumstances (such as to grow an out-of-season crop quickly when the main crop has been destroyed or when an additional crop is included in an intensified annual cropping programme). Local experience, custom and taste are important. For details of research centres specialising in particular cereals and of national research programmes, see Section 14, Resources.

High Yielding Variety (HYV) refers to varieties (usually of wheat, maize and rice) based on improved plant breeding material originating from two of the international agricultural research centres, IRRI and CIMMYT. The first varieties to come from these centres had weaknesses with regard to disease and pest resistance. However, for some years now research and breeding has been directed towards developing varieties resistant to a wide range of hazards and therefore suitable for poor environments, i.e. for the conditions in which many poor people farm. While having a greatly increased yield capacity they will perform satisfactorily (though not to full potential) with a traditional level of inputs.

Hybrid maize varieties can be produced only in countries where there is a developed seed trade or government seed organisation. A hybrid in this context is a variety produced by crossing two or more specially bred and selected, genetically homogeneous parent lines (maintained by the breeder) to achieve hybrid vigour as well as a desired combination of characteristics in the crop. However, the farmer cannot save hybrid grain for seed. The progeny of the cross will be heterogeneous and therefore, if used as seed, it would produce a mixed collection of plant types. Since the late 1950s, all CIMMYT's work to develop improved maize varieties for the tropics has been with open-pollinated conventional varieties, not hybrids. This has been the general policy of CGIAR centres (recognising the problem that poor people are not able to obtain a crop from their own saved seed); but now highly promising hybrid *rice* varieties have been produced in China and appear to be being introduced into a number of Asian countries.

b: Pulses and other edible legume seeds, including peas, beans, lentils, soya beans, groundnuts, winged bean, the tepany bean for hot dry areas and the tarwi being 'revived' in the High Andes.

All these are rich in protein, and so they are valuable for improving diets based on starchy crops such as cassava, yams and sweet potatoes. *Nitrogen-fixation*: legumes also have the advantage of obtaining nitrogen from the air and making it available in the soil for other crops, so they are useful for inter-cropping. It is usually necessary for the strain of rhizobium specific to the crop to be present in the soil if the crop is to flourish and fix nitrogen effectively (though some are promiscuous). If, however, it is not present, the crop seed can be inoculated (mixed) with the desired bacteria before planting, if an effective inoculum is available in the area. Legumes, like cereals, differ in their tolerance of soil types and climatic conditions, so advice should be sought before introducing a new type.

Recently interest has been stimulated in one particular legume, the

6

winged bean. It is native to Papua New Guinea and South-east Asia, where it has been used for food for a long time (the beans, leaves, flowers and tubers are all edible). It thrives in warm tropical climates with moderate or heavy rainfall (not long dry periods) and is a climber which requires staking. Support has been given to an Indonesian organisation, Yayasan Dian Desa, in a project to develop the processing and marketing of the winged bean. Many trials with winged beans have now been conducted around the world. See Section 14, Resources for further information.

c: Roots and tubers: cassava (manioc, tapioca), yam, taro, potato, sweet potato, winged bean (see above).

The root crops — cassava, yams and taros — are extremely important subsistence crops in many areas and especially in Africa, giving high starch — and therefore energy — yields per acre. Cassava is propagated by sticking portions of the stem of mature plants into the soil. It is important to see that these stakes are free of disease; where possible and practicable, good results can be obtained by dipping the stakes into a solution of fungicide before planting. The storage of stakes intended for planting can be a problem; stakes of one metre length treated with fungicide can be stored for about 6 months. Another well-known problem with cassava is the rapid deterioration of the harvested tuber — pre-harvest pruning of plants reduces susceptibility to deterioration. Excess cropping with these root crops leads to loss of fertility unless steps are taken to replenish the soil.

One of the main tuber crops in tropical lowlands is the sweet potato, but the so-called 'Irish' or 'European' potato is important in high areas of the tropics such as parts of Mexico and Peru, where it originated. Work at the International Potato Centre (CIP) in Lima and at major national potato research centres, such as at Simla in India, aims to produce varieties tolerant of higher temperatures, and hence to extend the geographic limits of this very high-yielding crop. Though essentially a high starch energy crop, it also gives a good protein yield per acre and contains valuable vitamins and minerals (CIP are selecting for higher protein in their breeding experiments).

d: Vegetables. The growing of green vegetables should be encouraged, as they are rich in vitamins and minerals. However, it is important that they do not compete with subsistence crops for scarce land, or for water which, if it is scarce, could add seriously to the burden of the water carriers, usually the women.

School vegetable gardens are particularly useful as an aid both to teaching about the nutritional importance of vegetables and demonstrating their production and use (TAN 80, BHU 03). Oxfam has often helped with the supply of seed. Full use should be made of crop waste by green manuring, i.e. digging in or composting. There is often an opportunity to integrate small pond fish farming and vegetable growing to their mutual advantage (see Section 1). Sometimes yards or kraals where cattle have been penned at night make excellent fertile areas for vegetable growing, as in Zaire (ZAI 78C) where vegetable growing, night-yarding of cattle and small-scale fish farming are all complementary.

e. Food trees: sago, fruits, nuts.

Sago is a starchy food reserve produced when nearing maturity (12 to 15 years) in the trunk of the sago or swamp palm. It is obtained by cutting down the tree and is the staple food for marsh dwellers in parts of Papua New Guinea and Indonesia. It usually grows wild, regenerating from suckers. Where cultivated, the suckers are taken from the base of old trees and planted in fresh water swamps at 5m x 5m. In PNG, yields range from 100kg to 350kg per tree (requiring much labour for processing). A protein constituent must be added to the diet. Leaves are used for thatching. In some areas selected types are now growing in upland, rain-fed conditions (W. Malaysia, E. Sumatra, Sarawak).

Oxfam has funded the planting of coconut seedlings in post-cyclone areas of southern India. Coconut palms are very useful, providing nutritious food, oil, fuel, coir and wood for building, but they are seriously affected by diseases, and scientists are worried at the depleting effect both disease and the change over to hybrid varieties are having on genetic resources of coconut in, for example, the West Indies and the Philippines.

Fruits and nuts are nutritionally valuable and are possible raw materials for cottage industries. They are often available to, but may not be appreciated by, local communities. Their production and use is well worth encouraging where they will not compete seriously for land with the subsistence crops; often they play a valuable part in a mixed garden: their roots occupy a different stratum of the soil from annual crops, and they provide shade.

f: Beverages. Where crops such as coffee, tea and cocoa are grown by small-scale producers, there may be scope for supporting group and cooperative ventures which will encourage and assist in crop hygiene and protection against pests, and improve marketing.

g: Sweeteners, relishes. Sugar cane — see remarks about beverages above. Waste materials from both fields and factories are valuable cattle feeds (see also Section 8). They are also biomass for fuel production. Relishes may be important local crops.

2 Fodder Crops

These include: grass, forage legumes, fodder trees, crop by-products (see also Sections 6 and 8).

Where land suitable for growing crops is limited, food crops must always take priority over fodder. In an intensive, irrigated, multi-crop area it may be possible to sow a quick-growing fodder crop between two food crops. A legume, such as lucerne (alfalfa), a clover, or one of the improved tropical legumes, will all provide fodder rich in protein and minerals, and they will also enrich the soil. A vigorous grass planted on steep slopes, as between trees in a reforestation scheme, can hold the soil and provide fodder for cutting and carrying to livestock. A quick-growing tree, such as *Leucaena leucocephala* (see Appendix II) will provide fodder while at the same time serving many other purposes.

a: Grasses. Where grass is growing on land not suitable for an arable rotation, it is highly unlikely that destruction of the natural grass and replanting will be advisable; one should rather try to improve the old pasture by better management — controlling grazing, applying appropriate fertilisers (where available and economic), and possibly seeding into old turf.

b: Legumes. Some highly-productive improved tropical legumes are now available. Most have been developed from plant material originally collected in Latin America. By over-sowing a natural pasture with a suitable legume, i.e. seeding into the existing sward, and grazing with cattle, the fertility of the land can be improved, and this will encourage the natural development of the best of the indigenous grasses and fodder legumes.

c: Other fodder and browse crops. Many 'weeds', including aquatic weeds such as water hyacinth, can be useful as fodder; for this, local knowledge will be valuable. For multi-purpose trees which can be browsed or cut for fodder, see Section 5.

d: Crop by-products and waste products. Most waste products from crops such as tops, haulms and straw provide valuable fodder, as do trimmings and rejected material from canneries and factories. Digestibility of cereal straw can be improved by chemical treatment; care is needed, but fairly simple techniques have been developed.

6

a: Fibre, including kenaf, jute, cotton, sisal, agave. These are unlikely to be significant crops in many of the areas where Oxfam-funded projects are located, although there will be exceptions to this. Kenaf can be used for paper-making.

b: Oil crops. Apart from edible oils extracted from crop seeds such as rape, sunflower, olive, palm, linseed, cotton, coconut, there are also a number of important industrial uses for edible and non-edible oils. The bean of the jojoba plant, which will grow in dry conditions, has attracted particular interest recently because the quality of its oil is comparable to that of the industrially valuable sperm whale oil.

The residue after extraction of the oil from oil seeds is usually useful either as an animal feed or as a soil conditioner. Research in India has shown that after the removal of oil from the seed of the neem tree the residue will not only improve the structure of soil, but will also increase the effectiveness of applied urea (nitrogen fertiliser).

c: Rubber and guayule. It is anticipated that there will in the future be a shortage of rubber. There may be some scope for village rubber production in suitable circumstances, but first-class advice both on production and marketing is essential. A perennial shrub, guayale, which thrives in semi-arid conditions and is native to Mexico, produces rubber when under moisture stress. This is being tested in low rainfall areas of North Africa and Asia.

d: Straw (and water hyacinth) for paper and board. Cereal straw makes very satisfactory paper and board and is highly desirable in view of the world's growing shortage of wood and the general level of deforestation. Water weeds (water hyacinth and salvina) have been successfully used for manufacturing high quality paper in India. In the USSR, the reed *Phragmites commins* is used for paper and building materials.

e: Crops and by-products for fuel. There is some scope for small-scale biogas (methane) production. Where large areas are cropped primarily for fuel alcohol (as in Brazil), the prime concern should be for the implications of such a development for the poor.

III   Production systems and techniques

1 **Good Husbandry**

Any system which progressively reduces the fertility and/or structure of the soil must be discouraged. Intensive cultivation without proper attention to the maintenance of soil structure and nutrient status, or ploughing straight up and down slopes, will encourage loss of soil by erosion. In some environments, a near-continuous ground cover is necessary to avoid erosion, and mixed annual and perennial cropping may be the most suitable system. Repeated similar cropping, without putting back humus (i.e. organic matter such as green manure, crop waste or animal manure), will make the soil more likely to erode and will reduce its ability to retain moisture and nutrients for crops. Some crops are more 'hungry' than others — they remove more nutrients, which must be replaced (see also Section 2).

a: Rotation/fallowing. It is usually best to grow a range of different types of crop in rotation rather than to grow the same or a very similar crop repeatedly. Continuous growing of one crop is likely to encourage the build-up of weeds, pests and diseases favoured by the seasonality of the one crop or by the micro-climate which it produces. Under shifting systems of agriculture, if the fallow period is of sufficient years, it will restore the soil, but as land has become limited in relation to population, the fallow period has become too short in many areas for natural restoration of soil fertility.

b: Cultivations or 'no till'. The purpose of cultivations is to kill weeds (which would compete with the crop), aerate and improve drainage of soils (most applicable to heavy soils) and provide a good tilth (i.e. tilled land) for a seed bed. In recent years, a practice known as direct drilling in the UK and 'no till' elsewhere has been developed which involves planting into undisturbed land; sometimes after killing vegetation with a herbicide or sometimes planting between the rows of a growing crop or using plants as a live mulch. This technique has been pioneered at IITA, Ibadan, Nigeria and in Sri Lanka for the tropics.

c: Timely planting of crops. This is most important, enabling the farmer to make full use of seasonal supplies of water, ensuring the maximum possible time for the plants to reach maturity, and allowing the crops to reach the required stage to make maximum use of optimum day length. Weeding early before the weeds stifle the crop is also important. Careful planning is necessary to achieve correct timing for each of the crops sown, and, where available labour is inadequate, animal power (or, exceptionally, mechanical power) may be desirable.

Yields can be increased by optimum spacing and depth of planting and the placement of mineral fertilisers close to the roots of plants minimises wastage. For some crops (e.g. rice), transplanting from nursery beds can radically raise yield per acre and speed up crop turn-round if the extra labour required is available.

d: Suitability of crops. Weeds can be good indicators of soil condition, acidity or alkalinity, and long-range climate, and can therefore point to suitable crops where no long-range climatic or rainfall information is available. Of course, before introducing new crops, it is also vital to consider their nutritional value, the amount and seasonal distribution of labour they require, and local tastes and customs.

2 Cropping Systems

a: Mono-cropping, i.e. the planting of large areas with single crops. This makes for the efficient use of energy, but can have disadvantages in terms of land use, crop protection and soil condition, unless good management and inputs are available. Mono-cropping can also be highly risky in areas where rainfall is unreliable.

b: Inter-cropping, relay cropping. For peasant farmers the practice of mixed cropping, which is often traditional, is worth encouraging. It can provide greater stability of production; crops can complement each other both in their growth habit and requirements, and a more balanced diet will be produced for the family.

A good method of mixed cropping is inter-cropping in which two different crops are planted in separate rows (either in alternate rows or alternate blocks of rows). If a legume such as a bean is inter-row cropped with a cereal such as maize, the legume will provide nitrogen for the other crop and the maize may support the bean.

A variation of the system is relay cropping, in which the second crop is sown near to the harvest time of the first, e.g. groundnuts just before the maize harvest, and makes most of its growth after the first crop is removed.

Many crops are suitable for inter-row or relay cropping, but an understanding of the needs of likely partners is necessary. Sugar has been inter-cropped successfully with cow pea. *Sesbania* species, small leguminous trees, are inter-cropped with rice as some varieties will tolerate the water of the paddy fields. Inter-cropping usually suppresses weeds and often one crop will protect another from pests; the spread of disease is likely to be slower than in a mono-cropped field and, because the leaves of two different crops will be at different angles and heights, fuller use will be made of solar energy.

c: Intensification, mechanisation (see also Section 4). One advantage of introducing animal or mechanical power into a cropping system is that more land can be cultivated per family, and/or the time

6

between crops can be reduced so that more crops are grown per year, assuming adequate moisture is available. When increasing the area or intensity of cropping, it is important to be aware of the increase in work which will take place during other seasons. The timing of operations must be maintained, and there should not be too great an increase in manual work. Although mechanisation often benefits men, it frequently adds to the work done by women (see PART TWO Section 2). Handling, storage, use or marketing of the increased crop output from the land must also be considered in advance.

d: Hydroponics and nutrient film technique. Hydroponics is the growing of crops without soil in a solution of nutrients. It is a highly-specialised technique requiring good management and is generally practised under glass or polythene. The newly-developed nutrient film technique (NFT) involves growing crops, usually vegetables, in a continuously-flowing shallow stream or film of nutrient solution. The solution is recycled so that the only loss of water and nutrients from the system is in that which is taken up by the plants or by evaporation. This is a very efficient use of water, but requires a high degree of hygiene (as disease in the system will immediately be carried to all plants), together with constant monitoring and topping up of the solution. While using space intensively, hydroponics is generally capital and technology-intensive. However, a practical adaptation of the NFT technique has been developed by National Organic Chemical Industries of Delhi for use by landless or virtually landless people; the nutrient solution is contained in plastic-lined channels made in flat ground and replenished by syphon from a tank, and the plants grow on capillary matting.

③ Seeds and Cuttings

a: Good seed. It is always worth using the best available seed. That is, seed which is not contaminated with weeds, is free of seed-borne diseases, will germinate well and will be true to type. Unfortunately, in the poorer areas of the Third World sources of such reliable seed tend to be few. When home-grown seed is saved for use, it should be selected from the best plants of a healthy and vigorous crop; any weed seed or other extraneous matter should be removed; and it should be stored in dry conditions, secure from vermin, and in as cool a place as possible. Even when all these precautions have been taken, seed can degenerate so that it is incapable of maintaining yield potential. Many countries now have seed testing facilities. Seed of hybrid varieties such as hybrid maize (see II.1 above) should not be saved. It will not breed true, it will be without hybrid vigour, will show a 10-15% yield reduction and, in the case of maize, the soft grains do not store well.

b: Cuttings. Some plants are best propagated by cuttings or other vegetative means. Plants from which propagating material is taken should be carefully inspected to make sure they are healthy. Where available, fungicidal dips should be used. In sophisticated circumstances, cuttings can be encouraged to root quickly by dipping them in a hormonal rooting powder before planting. A simple alternative is to use a honey solution; pollen in honey contains growth substances.

④ Gardens

Food from small plots around homes may supplement field crops, but for some people, such as widows or landless labourers and their families, it may be the only direct source of food. Best advantage can be achieved in garden plots by intensive use of organic matter, water, labour and purchased inputs where possible. Full use of the land will be made by growing crops of different types and heights together — such as fruit trees and various kinds of nutritionally-rich crops.

IV	**Crop protection**

Although crop hazards can be considerable in tropical conditions, a great deal of food could be saved by careful attention to simple protective measures in the field and the store.

1 Rodents

Destroy rodent habitats. This, of course, will not be practicable where there are woodlands close to cropping areas. The best way to protect stored crops from attack by rats and other vermin is to prevent access to the grain, and this is always more effective than curative methods. However, when preventive methods fail, Warfarin baits can be used *with care* to kill rodents, but *not* in communities where rodents may be used for human food. *Warfarin* is a dangerous poison for humans as well as rats.

2 Insects

Rotations and mixed cropping systems keep many forms of insect attack under control. Chemical insecticides, where necessary, can be applied with knapsack/hand-held sprayers. *Chemical manufacturers' instructions must be followed precisely to avoid harm to the user, to other people, and to the crop.* If insecticides are used in excess, the pests may develop resistance. Biological control methods using natural processes, such as the release of sterile males or the introduction of natural predators, and integrated control methods in which cultural, biological and chemical methods are used to complement each other, are now the subjects of a great deal of research and development around the world, and are to be encouraged where possible.

3 Diseases

Crop diseases are caused by fungi, bacteria or viruses. Disorders due to mineral deficiencies may be called deficiency diseases. Minute organisms in the soil, such as eelworms (*nematodes*), damage plants and give rise to 'disease' symptoms. Airborne insects such as aphids can be vectors of disease, carrying it from field to field or from one type of susceptible plant to another; sometimes disease organisms can remain over winter or during a crop-free period on a host weed. Use disease-free seed, practise crop hygiene to destroy disease-carrying plants, and promote vigorous growth of crop plants. The development of resistant varieties is a major objective in all plant-breeding work.

4 Weeds

Hygiene and good husbandry should always be the first line of defence against weeds. Avoid bringing in weeds in seed and irrigation water, on vehicle wheels etc. Practise timely cultivations to kill weeds before they compete with the crop. Rotations and continuous coverage of the ground by vigorous crops help to keep weed growth down. However, some profitable systems favour certain weeds, and in such cases spraying with a suitable herbicide to manufacturers' precise instructions, using a knapsack sprayer, may be necessary. When any weed is pulled out by hand, care should be taken to pull early or, if that is not possible, to remove seeding weeds completely and destroy them: some weed seeds, such as wild oats, are viable even when still green and soft.

6

Millions of tons of food are lost each year because of poor post-harvest handling, storage and subsequent use. In addition, there is frequently a reduction in nutritional value, particularly during processing by certain modern techniques, e.g. rice hulling and polishing. Therefore, support should be given to projects which:

☐ improve farmers' understanding of good post-harvest handling of crops, e.g. by discouraging the leaving of cut rice on stubbles in damp paddy fields;

☐ help communities or cooperatives to store crops in improved and appropriate storage facilities. This reduces waste, reduces nutritional seasonal 'peaks and troughs' by extending the period of availability of the home-grown food, and permits the sale of cash crops or crops surplus to domestic needs without duress, and when prices are good;

☐ improve local techniques used for processing crops, help to foster community or village processing, and assist in the establishment of local agro-industries;

☐ improve domestic use of available crops, e.g. by drying and cooking.

If crops can be processed within the rural community, rather than being sold for processing elsewhere, a greater proportion of the income arising from the crop will remain in the area and there will be more employment locally. However, if the product is to be sold, the standard of processing and presentation will be important.

☐ General and Grain Storage

Family-scale storage, where a high proportion of the crop is consumed at home, has the advantage of reducing the practice of selling after harvest and buying back later. This can be both a major factor in the build-up of chronic indebtedness and a cause of seasonal malnutrition and physical weakness which can adversely affect work.

Loans and grants have been given to enable groups of farmers or village communities to set up improved, small-scale grain 'banks' or cooperative storage and marketing ventures. An example is in Tanzania (TAN 107 and 108), where finance was given for the construction of village grain stores together with the expenses of seminars on crop protection. The educational aspect (seminars, etc.) is considered a vital part of the funding of grain stores. At Maradi in Niger, 12 cooperative cereal 'banks' are being supported.

a: Important requirements of storage schemes are as follows:

☐ a reliable, trustworthy person to take charge, who has sufficient understanding of the technical aspects of storage;

☐ provision for financing the maintenance of a reserve, even after a bad harvest;

☐ storage facilities that will keep the stored crop dry. Dampness causes sprouting (of grain), rotting and mould. Grain must be thoroughly dry before it is stored and then protected from rain and condensation. Traditional small-scale methods usually provide adequate ventilation, but the need for ventilation must be observed in larger stores;

☐ hygiene and protection against insects (such as weevils and beetles in stored grain). Hygiene means preventing new crops from being infested with insects or contaminated by moulds. Permanent stores should be swept clean, and if necessary fumigated, as soon as they are emptied; standing crops in the vicinity of stores can be infected by pests coming from an old stored crop. Simple storage facilities that cannot be thoroughly de-infested are best destroyed and replaced;

☐ complete protection against rats and mice which not only eat grain, but also cause even greater loss through damage. Rats can jump at least a metre vertically and can climb vertical walls. Beware of using a rodenticide in communities where rodents may be used as human food.

Obviously, it is far better to prevent grain deterioration than to have to deal with it once it has occurred, but vigilance is always necessary.

b: Signs and symptoms of trouble are:

☐ insects visible in grain, hollow or eaten grain, mould, silk spun by moth larvae;

☐ small, black, stringy droppings from rats;

☐ grain caked in lumps due to mould and dampness;

☐ close examination — a pocket magnifying glass is useful — shows minute holes on the grains caused by penetration of insect larvae or insect eggs;

☐ dust and crumbled grain. A good method of testing the larger grains that may be infested is to pass them over a sieve: serious damage is indicated if a great deal falls through.

If bags are re-used, they should be carefully inspected for contamination and turned inside out.

c: Types of grain stores. Well-established traditional methods of storing grain for the family or small community are not to be despised. Mud-brick or wicker-work stores above ground or in sealed pits in the ground are usually suited to the environment, and are cheap and effective in keeping grain in good condition provided that they are properly maintained and that strict hygiene is observed. An example is the Pusa bin developed in India, which is made of plastic, coated inside and outside by mud. Oil drums and old water tanks may also be adapted by farmers for use as grain stores. Bins or silos made from metal must not be exposed to the sun, as this may lead to moisture condensation on the cool side of the container.

Inflatable butyl rubber balloon-type grain stores are now widely used for bulk or in-bag storage for both reserve or emergency grain, and also for more permanent situations. There is an element of physical vulnerability (though this is not great), and these stores often lack the prestigious look of the large concrete silos favoured by some governments. Advantages are:

☐ their fairly easy transportation and speed of erection;

☐ low cost per ton compared with brick or cement structures;

☐ the fact that they are hermetically sealed, so that if there are any insects in the grain when it goes into the silo they will die from lack of oxygen.

6

Expert advice should be sought on construction and planning of storage buildings and permanent silos. Ventilation should be controllable; external paint chosen to reflect light and heat; interior paint of gloss type and not whitewash which could react with insecticides; walls which provide good thermal insulation and roofs which overhang to protect walls from sun. The store must be rat-proof, i.e. walls smooth, doors tight fitting and with metal along the bottom to prevent gnawing. If possible, rat habitats in the vicinity of the store should be destroyed. The building should be easy to clean with no cracks or ledges to harbour insects. The site must be dry and well-drained and located away from fields of growing crops. Grain should be stacked so that inspection from all sides is possible, and bags should be stored on wooden pallets. (See Section 14, Resources for suggested technical references.)

d: Storage of roots and tubers. The general principles which apply are to select disease-free, undamaged produce; to allow good ventilation (particularly important in hot, humid conditions); to shade from very hot sun; to inspect regularly for signs of disease or pest attack, and to remove any tubers that show signs of being affected.

2 | **Preservation and Processing**

Oxfam will be concerned mainly to aid small-scale crop preservation and processing, and to promote the development of simple manual, pedal-powered, animal-powered or mechanical aids to processing. The Tropical Products Institute does a lot of work on developing such equipment, and this includes simple tools and machinery for small-scale processing.

a: Rice hulling and polishing. Machine methods usually remove not only the husk but also an outer, very nutritious, layer of the grain; as a result, people may suffer from beri-beri. Vitamin B additives can restore the nutritional value, but reliance on these is unwise in small-scale processing. It is preferable to encourage parboiling of rice before processing. This practice, traditional in parts of Asia and West Africa, causes the natural vitamin B to move deeper into the grain so that it is not lost. Parboiling also makes grain less susceptible to physical and insect damage in store.

b: Grain milling. Mechanised milling has now become the norm, even in rural areas of Africa (BUR 17 and TAN 120). Field staff should be concerned to see that the technique used minimises the loss of feed value and maximises the recovery and use of waste products, e.g. for animal feed. IDRC, UNICEF and TPI all have research in hand, some of which includes the milling of legumes to make high-protein flour.

c: Cassava drying. Cassava deteriorates rapidly after the root is harvested, so, if it is not to be peeled and eaten immediately, it must be processed. A common method of processing is for the root to be peeled, cut into pieces, dried in the sun, milled or grated, soaked in water to remove the hydrocyanic (prussic) acid (the water then being thrown away), and baked as flat cakes that can be stored. Improved methods of cassava preservation are being investigated at CIAT, Colombia.

d: Oil extraction. Crops giving the highest yield of oil are soya, sunflower, safflower, groundnut, cotton seed, palm, coconut, rape and olive. Engine-driven oil expellers, suitable for small projects, are widely available, but where quantities are not great, a manual press operated by two people may be preferable. The residue left after oil has been pressed from the seeds can be a useful animal feed.

Soya and groundnuts are important as protein foods and not just as oil seeds. However, soya contains a substance which negates the food value of the protein, and can cause dysentery in children unless it is destroyed by processing, such as lengthy cooking. Full fat soya flour is valuable for adding protein to starchy diets. Oxfam has supported village soya flour production: this involves 5 hours soaking, 15 minutes boiling, 24 hours drying, then cracking, winnowing and grinding.

e: Drying fruit and vegetables. Much wastage of fruit and vegetables can be avoided by drying the produce and storing it. The process can be simple, as follows: trim and slice to 7mm slices. Some vegetables, such as green beans, carrots, turnips and cabbage then require blanching in boiling water, and dipping in a bisulphite preservative; but others such as tomatoes, mangoes and bananas are not blanched. Sun drying is then done on wire, gauze or muslin trays or in glass-roofed boxes called solar dryers or cabinet dryers.

f: Leaf protein extraction. The process involves the pressing of protein-rich juice from green leaves. Nutrition trials with the dried extract from the high-protein crop, lucerne, are being done at Coimbatore in South India. Research has not as yet produced all the dietary answers. The protein extraction methods are either capital-intensive or very labour-intensive. Where crops (e.g. lucerne) have been grown especially for leaf protein, it is suggested that the land could be better used for other food crops.

g: Microbial protein production on starch. Starch from grains, cassava or potatoes is one of a number of proven substrates on which microbial protein can be grown. The product results from growing yeasts, bacteria or fungi on the substrate, skimming them off and drying.

3 **Home Use/Cooking**

Two types of pressure have developed in some areas in recent years which have resulted in practices which contribute to malnourishment.

a: The growth of urban populations able to pay attractive prices for food tempts poor farmers to sell food really needed by their own families. Where possible, help should be given to avoid this by supporting nutrition education, increased production, and actions which reduce poverty and hence the need for cash from such sales.

b: Shortage of fuel wood is forcing many families to eat food raw, as they are unable to cook it: this can have disadvantages in terms of palatability, digestibility and hygiene and, in the case of some foods, cooking is necessary to destroy bitter or even toxic substances, e.g. some bitter varieties of cassava must be steeped in water and then cooked to destroy cyanogenetic glucosides and hydrocyanic acid. Oxfam should promote improvement in the efficiency of cooking methods in terms of fuel use and a greater understanding of the need for cooking (see also Power/Energy, Section 4 and Nutrition, PART SEVEN Section 3).

4 **Marketing**

In most situations encountered by field staff, the producers' primary concern should be to achieve an adequate, well-balanced diet for the whole family, all the year round. Virtually all farmers will want to produce a surplus where possible to exchange for necessary commodities and services not produced on the farm, and to ensure adequate supplies in below-average years. It will be in the farmers' interest to market:

☐ unpredicted surpluses, after making adequate provision for year-round consumption, and

☐ crops they can grow more economically, which can be exchanged for others so as to balance family diets and/or increase their overall income, e.g. sell maize and purchase cowpeas or dried fish; sell cotton and purchase food stuffs.

(For Marketing Schemes see PART FIVE Section 2.)

6

Section 8   Animal Production and Processing

including small animals, poultry, bees

I   General principles

1 Why Assist Livestock Development?

It has been suggested that the development of livestock production is an inefficient use of resources. This need not be so, for the following reasons:

a: Much food, and therefore energy, is available in forms not directly consumable by man, such as fibrous roughage and crop by-products, which can be processed into edible foods by livestock. Examples are: leucaena meal for poultry, ensiled banana waste, fish silage, and shrimp waste for poultry. Further information is available from the Tropical Products Institute (TPI).

b: At least 60% of the world's rangelands are unsuitable for cultivation and can be exploited only by ruminant livestock or wild game.

c: Many species ultimately killed for food have a primary role as draught animals or as producers of milk, wool and fuel (in the form of dried dung).

d: Crop and livestock production can be complementary and are often interdependent. Mixed farming systems are usually ecologically more stable and, in some circumstances, can produce more edible food per energy input than separate crop and livestock systems.

e: Animals are the primary source of traction in many parts of the Third World. They are also used extensively as the main form of transport.

2 Priorities

Normally, ranching systems will not come within Oxfam's concern, but there may be exceptions, such as where an effort is being made to develop group, cooperative or communal ranching for the benefit of nomadic or seasonally migrant people. Oxfam is unlikely to become involved in aiding the development of the intensive modern sector of the livestock industry.

Of the other livestock systems, priority should be given to those that will benefit the poorer members of the community, are rational in ecological terms, and are likely to be economically viable (bearing in mind all the advantages and disadvantages to the family).

In the technical field, improvements in animal health should receive priority, followed by improvements in nutrition and management, and, finally, in breeding. Major mistakes are still being made in attempting to improve the breeding of livestock by the importation of exotic animals. This should be advocated only after extensive investigation shows that it would be advantageous, that animal health measures are satisfactory and that the quantity and nutritional quality of the food available will be adequate. Artificial insemination should never be advocated unless appropriate, and only when efficient support services are available.

3 Some Oxfam experiences

In many areas Oxfam has been assisting community or cooperative livestock ventures where either the product, e.g. milk, will contribute to much-needed improvement in nutrition, or where livestock will provide an income while being fed largely on waste or by-products, without competing for human foods or scarce land.

Support has been given by Oxfam to enable poor villagers to come within the State dairy development schemes in India. Assistance has also been given for dairy cattle improvement in a small farmers' dairy scheme in Malawi (MAL 50). Oxfam is keen to encourage animal draught power where appropriate and funds ox-training and ox-cart production (TAN 86, 122 and 95).

Sheep and goat schemes have mainly been associated with training in management (e.g. SUD 29) and in supplying finance for revolving loan schemes to purchase goats, as in India (MP 5). In these goat schemes the adults are first loaned and then purchased by giving back to the project the first two kids to be born for distribution to others. There has not been much involvement with pigs, an exception being support for a pig improvement centre and extension service in Brazil (BRZ 143).

There are many examples of support for poultry projects, e.g. at children's homes in Ecuador and Colombia; a comprehensive project in Tanzania (TAN 80) which included poultry, starting in schools and then developing to involve village communities; in Brazil, where a poultry feed cooperative (BRZ 123) has been supported; in Indonesia, where credit has been provided for chicken and goats (ID 34D, 77 and 73).

Projects involving small animals kept by schools or village communities lend themselves well to Oxfam support. Part of a comprehensive project in Guatemala (GUA 12) is concerned with rabbit breeding, and another in Peru involved bees and the guinea pigs traditionally kept by the Peruvian Indians. Funds have been given for the production of simple manuals in Spanish, e.g. on alpaca.

II Principal farming systems using livestock

1 Extensive Grazing

This includes ranching, and nomadic and transhumant (where pastoralists migrate seasonally from a fixed base) subsistence systems. The major types of livestock are camels, cattle, sheep, goats and, in the Andean countries of South America, llama and alpaca.

In nomadic systems, or those where pastoralists seasonally migrate from a fixed base, the livestock owners' usual objective is to maintain as many animals as possible consistent with the availability of forage (for grazing and browsing) and water. Nomadic systems often function in an environment which is subject to unpredictable seasonal and annual fluctuations in rainfall and thus in the availability of these resources.

In many such systems overgrazing is common, as land is communally owned while livestock are individually owned and female livestock are often the only available outlet for investment. In recent years, overgrazing has worsened with the almost continuous withdrawal of land for settlement, the advent of improved veterinary services, the uncontrolled provision of additional water supplies and, in some circumstances, the influx of refugees and their animals into areas which cannot support them.

The modernisation and development of these systems does not necessarily imply settlement. In fact, most arid areas can be exploited for human food production only by the use of migrant livestock or by irrigated crop production. Development does, however, require:

a: a transition from nomadism to transhumance (seasonal migration from a base) so that the livestock owners can take advantage of medical, educational and other social facilities;

b: that efforts should be made to improve the environment. This requires: attempting to restrict the size of the human population dependent on the system; improving the productivity of livestock by better management, feeding and veterinary attention, reducing total livestock numbers, and possibly changing the animal system to one which sells young stock out of the area to be reared or fattened elsewhere. (See PART TWO Section 5, and Section 6 above.)

2 Shifting Cultivation

Such systems are fast becoming obsolete and a change to sedentary cultivation is common except in some rare circumstances. Shifting cultivators usually raise only small stock such as goats, sheep and

6

poultry. A change to sedentary cultivation requires not only crop rotation, but also a much higher labour input, and may be assisted by introducing work-animals (which, of course, may also supply food and fertiliser or fuel). This is not an easy task, as experience in Africa has shown, but it is a vital one in which Oxfam can help. Some of the difficulties to be expected are pest and disease problems (e.g. tsetse fly), inadequate capital, lack of understanding of animal management, and traditional attitudes to their proper use.

If sheep and/or goats are not already raised by the farmers, it may also be possible to introduce them into the system.

3 Sedentary Cultivation Systems

Sedentary systems of agriculture are the norm in regions of dense rural population. The use of working cattle and/or buffaloes is widespread. In drier areas most farmers keep a few sheep and/or goats and some fowls, while in wetter areas the rearing of pigs and large flocks of ducks is common. Integrated systems with rice, fish and ducks, or pigs, fish and vegetables are also viable in these areas. Their development requires the introduction of improved feeding and management methods and animal health support services.

4 Mixed Farming

In the wetter tropics animal production can and should be combined with tree-crop production. This is particularly so in the coconut-growing areas where ruminant livestock may be allowed to graze beneath the trees, but such systems can also be practised in fruit, rubber and oil-palm plantations. Alternatively, the fodder can be cut and carried to the livestock. (For tree planting, see Section 5.)

5 Livestock in Urban Areas

With the migration of large numbers of rural people to the areas surrounding large cities, there is an opportunity for them to improve their diet by raising small animals. In the less densely populated areas people can keep small numbers of sheep, goats, pigs or poultry and let them scavenge around the houses, while in more densely settled areas poultry, rabbits and guinea pigs may be fed on waste foods and garden forage.

6 Intensive Commercial Farming

With few exceptions, this sort of activity — raising cattle, poultry, pigs or rabbits using purchased feed based on cereals or other food which could be used directly by humans — is not recommended for poor communities.

III   Types of farm livestock

1 Camels

Camels are being replaced as transport animals, but should be assured of a future role as a meat-producing animal in very arid areas, as they are the only domestic animal that can be used to exploit this type of environment. There are two major types, riding camels and baggagers. Well-grown male dromedaries (one-humped camels) weigh 450-600 kg; the females are somewhat smaller. Under desert conditions a female can produce 1000 — 1500 kg of milk per lactation with a 3.8% fat content. Well-fed females may produce 2500 — 3500 kg of milk in a lactation lasting 16 to 18 months. Both male and female camels are sexually mature at 3 years old and are normally used for breeding at 4 years. A male can serve up to 50 females during one mating season. On average, one calf is born to each female every 2 years.

Camels are exceptionally tolerant of heat and of water deprivation. They can lose up to 25% of their body weight and still survive. They prefer browsing to grazing and require adequate quantities of salt in their diet. In arid areas they are not particularly subject to disease or parasitic attack, but if moved into a tsetse infested area, they are very susceptible to trypanosomiasis.

2 Llamoids

The llama and the alpaca are important domestic livestock in the Andean countries of South America. Both animals produce wool, the llama also being used for transport purposes. Their meat is rather similar to that of sheep. Well-grown male alpacas weigh approximately 65kg; the females are somewhat smaller. If milked, the females produce about 2 kg of milk per day. Alpacas and llamas produce 1 — 4 kg of wool each year. The wool is particularly valuable as it is of a fine texture. The wool of vicunas is of even better quality, but the animals are not domesticated and are now relatively rare, so it is expensive.

Male alpacas are first used for breeding when they are about 3 years of age. Females are usually mated at 2 years. Normally one calf is born every two years. Llamas and alpacas can thrive on very fibrous forage and do better than sheep at high altitudes in the tropics. They suffer from most diseases common to other ruminants, but are less susceptible to foot-and-mouth disease.

3 Water Buffaloes

There are two types of buffalo: river and swamp. River-type buffaloes, used for work and for their milk, are found in the Indian sub-continent and as far west as Italy; major breeds in India are Murrah, Nili, Surti and Jaffarabaldi. Swamp-type buffaloes, used primarily for work, are found in South-east Asia and as far east as the Philippines and Guam. Both types may be used for meat.

Female river buffaloes in India produce 2000 kg of milk per lactation with a butterfat content that averages about 7% but may be as high as 15%, and with more protein and lactose than cows' milk. River buffalo bulls mature at 3 years of age, swamp buffalo bulls later. One bull can serve at least 50 cows per annum. Female river buffaloes mature at 3 to 3½ years, swamp buffaloes at a later age. Adult females usually produce two calves every 3 years, having a gestation period about one month longer than that of cows. Oestrus is difficult to detect.

Buffaloes are less tolerant of hot dry conditions than cattle; they are better suited to hot and wet climatic conditions. They thrive best where water is available for swimming or wallowing during the hottest time of the day and are more likely to survive floods and cyclones than cattle. They suffer from exposure to hot sun; if worked or driven for long hours in the sun, they can suffer exhaustion or sudden death (young animals are very susceptible).

Buffaloes can pull heavier loads than cattle and, with their wide flattened hooves, are superior workers in rice paddies. They are generally healthy animals, liable, but on the whole less seriously, to the same types of disease and parasites as cattle (they are more susceptible to rinderpest and less susceptible to foot-and-mouth disease). Greatest losses are among calves.

4 Cattle

The majority of cattle in the tropics are general-purpose, being milked, used for work and, where religion and custom allow, killed for meat. As they are not generally very productive, there is a trend towards the use of more specialised types, particularly for milk production purposes.

a: Dairy-type cattle. The milk production of almost all breeds of cattle in developing countries is low, averaging less than 700 kg per lactation. The most productive native breeds in the tropics are the Sahiwal, Red Sindi, Tharparker and Kenana, and good, more-or-less stabilised crossbred breeds are the Jamaican Hope and the Australian Milking Zebu. The Damascus is a good breed in western Asia. Purebred, high-producing temperate-type cattle are not normally suitable for use in the tropics unless they can be raised at high altitudes or kept under

6

(expensive) controlled conditions. Under wet tropical conditions and on oceanic tropical islands, when the level of management is high, the best milk producers are ¾ temperate-type crossbred with ¼ Zebu cattle. Average production should be of the order of 2000 kg per lactation. In drier areas half-breeds are more suitable, and under poor managerial conditions only a quarter-bred temperate-type animal should be used.

The most suitable and economic managerial system is one where the cattle are grazed at night, in the early morning and late afternoon, and kept under shade or indoors during the hottest part of the day. Indoor feeding and managerial systems where the cattle are fed large quantities of concentrates are common, but they produce expensive milk of interest only to the richer members of the community. This type of system will not normally receive Oxfam support.

Well-grown female cattle can be mated first at 15 to 18 months of age, but it is usual for them to be older. Bulls are used for first service at 3 to 4 years of age.

b: Beef-type cattle. Under extensive grazing conditions (see Section 6), the most productive animals in subsistence systems are likely to be indigenous cattle. When improvements can be made in feeding, management and health control, crossbred cattle may prove to be more productive.

Some of the best indigenous breeds are: the Hariana, Ongole and Gir from the Indian sub-continent, widely used in South America and part-ancestors of American Brahman cattle, but not used for beef production in India; the Bali in South-east Asia that provides very lean meat; the N'dama in West Africa that is tolerant of trypanosomiasis; and such breeds as the Adamawa and White Fulani in West Africa, the Boran in East Africa, the Mashona in Central Africa and the Afrikander in South Africa. Some exotic breeds that may be used, when managerial conditions are good, are the American Brahman and Santa Gertrudis, originally from the United States, and the Droughtmaster from Australia.

Beef cattle on range usually grow more slowly than cattle in the temperate zone, and average age at slaughter is 4 to 5 years. The feed lot system is unlikely to be practical or economical in most developing countries, where grain is scarce and expensive. However, some countries have cattle-feeding enterprises associated with plantations or canneries where there is a good supply of by-products and waste products. Smallholders in some areas feed a small number of animals on crop by-products or by gathering road or field-side forage and browse material from trees. Raising beef cattle on irrigated forage is always uneconomical, though it is often advocated.

c: Working cattle. Work ability in cattle depends upon many factors, including breed, sex, size, training, management, feeding and health. A good pair of bullocks, working 4 to 6 hours per day, should be able to plough 0.4 ha of land in two days to a depth of 20 cm, and carry out all cultivation work during one year on a 6 ha holding.

Sturdy, compact animals with well-developed muscles should be chosen. They can be trained in six to eight weeks, but should not be worked until they are well-grown. In the tropics they should not be used for more than 4 to 6 hours a day and then only in the early morning or late afternoon. They should have access to shade and water when not working, and unless their grazing is of very good quality they should be fed small quantities of additional feed. Females can also be worked, except prior to calving. The calving interval may be affected by working, and the benefits and costs/losses of working females must therefore be weighed carefully.

The need to study factors affecting draught animals' work capacity and to improve and develop equipment suitable for their use is at last being recognised. At the National Research Institute in Sri Lanka an improved bullock cart has been designed, with large wheels and with shafts made from laminated coconut timber.

Some of the *major diseases of cattle* in the tropics and in most other parts of the developing world are as follows:

☐ *Trypanosomiasis*: spread by tsetse fly. This precludes the use in 37 African countries of some seven million square kilometres of land. Prophylactic drugs are available but expensive. The Tsetse Research Laboratory has shown that it is possible to breed tsetse which will not carry the trypanosomes; it is envisaged that these might be used in conjunction with programmes which release sterile male flies (as in Nigeria).

☐ *Rinderpest:* now being brought under control in most parts of Africa. A vaccine is available.

☐ *Contagious Bovine Pleuropneumonia* (CBPP): still common in Africa. A vaccine is available.

☐ *Foot-and-mouth Disease*: still found everywhere in Africa and in some areas of Asia and South America. Vaccines are available.

☐ Disease transmitted by ticks such as *East Coast Fever* (only in East Africa), heartwater, *anaplasmosis* and *piroplasmosis*: avoided by the control of ticks, usually by frequent dipping of the cattle, but this takes time, can reduce animal production, and ticks have begun to become resistant to the acaricides used. The International Laboratory for Research in Animal Diseases (ILRAD), Kenya, is developing a vaccine. In 1982 the Wellcome Foundation launched what appears to be the first drug to provide an effective cure for East Coast Fever (which kills 500,000 cattle a year), and it is thought that it may also be a means of immunising calves.

☐ *Internal parasites*: prevalent everywhere. Control measures are available.

5 Sheep

Sheep kept for meat, wool, sometimes milk, manure or pelts, thrive best in drier climates and when grazed on short pastures, though a few breeds do quite well in wetter climates.

They can be classified as thin-tailed, fat-tailed or fat-rumped, or as hair or wool producers. Some of the most productive meat-type breeds in the drier areas of Africa are the Fulani, the Sudanese Desert, the Somali and the Blackheaded Persian. In Western Asia there are many indigenous breeds: the Awassi is an outstanding milk producer. Two of the more productive indigenous breeds in the Indian sub-continent are the hairy Nellore and the woolled Lohi. Breeds suitable for use in the wet tropics are the Black-bellied Barbados, the West African Dwarf (not exactly a dwarf animal) and the Thai or Malaysian. Nilotic sheep from the Sudan can thrive in swamp areas.

In regions where woolled sheep can be bred and standards of management are good, Merinos can be introduced for crossbreeding. The Dorset Horn and Wiltshire breeds may be used in crossbreeding for meat purposes. The Dorper, a Blackhead Persian crossbred with a Dorset Horn, is a very productive meat breed for use in arid areas.

Females are usually mated at about 14 months. In the tropics they mate in any season but in higher latitudes they come on heat as daylight lessens.

The largest indigenous male sheep in drier areas may weigh up to 70 kg but average mature weight is approximately 40 kg. Sheep acclimatised to wet tropical areas are usually small, often weighing only 18 to 25 kg.

Awassi ewes under good management have yielded up to 350 kg of milk per lactation. The ewes of most breeds, however, produce only about 68 to 80 kg of milk per lactation.

6

Indigenous woolled breeds may produce 1 to 4 kg of coarse wool per animal per annum.

Sheep are subject to a number of diseases such as bluetongue, Nairobi sheep disease (only in Africa) and sheep pox, and are usually particularly susceptible to the ill-effects of internal and external parasites. Dosing regularly with anthelmintics and dipping are standard practices in circumstances where availability and economics allow.

6 Goats

Goats also thrive best in drier climates, but unlike sheep they are primarily browsers. Thus they can be herded together with sheep to advantage. They can thrive on coarser forage than sheep, but they are not so tolerant of water deprivation.

The goat has a bad reputation as being destructive of vegetation and one of the main agents of range degradation. The real culprits, however, are the goat-owners who allow their animals to destroy vegetation. Well-controlled goat herds can suppress unwanted forage species and improve grazing. Alternatively, goats can be reared in pens constructed of indigenous materials and fed on cut forage, as they are in many parts of Indonesia.

The primary products produced by goats are meat, milk, hair and skins. The milk is particularly nutritious. One breed, the angora, produces mohair; others, in Asia, produce pashima (cashmere). Before introducing goats to an area where they are unknown, it is important to make sure their milk and/or meat will be acceptable to the local population.

Among indigenous goats, the Jumnapari from India possibly yields the highest milk production — about 200 kg per lactation when well managed. The majority of indigenous goats produce only 60 to 80 kg of milk per lactation. One exotic breed that thrives quite well in the tropics is the Anglo-Nubian; it is capable of producing up to 900 kg of milk per lactation, but is unlikely to achieve this level of production in a tropical environment. The lower capital cost and shorter breeding cycle of goats (when increased numbers are required) are advantages over other dairy animals.

Female goats mature sexually at 4 to 6 months of age, but are best if not served until 12 months old. On average, females in indigenous flocks will produce 3 kids every two years. A mature male can mate up to 100 females; males not wanted for breeding can be castrated.

7 Pigs

Pigs are an important class of livestock in some regions of the tropics, such as the non-Muslim areas of South-east Asia and the Pacific Islands. They provide high quality protein and energy food and are an important source of manure.

Often they are used as scavengers, running loose around the holding or in the forest. They are also used in very intensive integrated pig/methane production/fish pond/vegetable and fruit systems.

Often it is suggested that no effort should be made to develop pig production as pigs compete to some extent with humans for food. Except in very extreme cases this is a mistaken idea. If small quantities of concentrated feeds are provided, large quantities of inedible by-products can be used and converted into meat and fertiliser. Pigs are particularly efficient converters of food.

In scavenger systems only indigenous or three-quarter-bred indigenous pigs should be used, but in intensive systems there is probably an advantage in using commercial strains of exotic crossbred pigs. These should preferably be Large White (Yorkshire) or Duroc crossbreds. The sows of some South-east Asian indigenous breeds are very prolific and might also be used with advantage for crossbreeding.

Gilts (virgin females) of most breeds can be served when they are about 9 months of age, although there is tremendous variation according to breed and management. Sows come on heat about every 21 days and their gestation period is 112-115 days. Females should be de-wormed before mating.

Boars are sexually mature by 8 months of age, but are usually not used until they are older. When mature they can serve up to 20 females per month but an ideal boar to sow ratio is 1 : 25 if the animals are housed; they should not serve their own daughters and they should be changed from time to time to avoid in-breeding. The aim should be to wean about 7 to 10 piglets per litter, each weighing 14 to 18 kgs. The normal suckling period is 8 weeks, and sows should return to the boar 5 days after weaning.

Pigs grow rapidly, and if well fed and managed they will weigh 50 kg at a slaughter age of 5 months. Some areas prefer heavier weights, but a pig will be uneconomic to feed above 80 to 90 kg. Scavenger pigs will grow much more slowly, mainly due to the effect of internal parasites. It is, therefore, highly desirable that pigs should not be fed on the ground. They should be penned, and each pen should have feeding and drinking troughs. It is important that an animal is slaughtered humanely by a competent person.

The following are some of the *sources of feed* for pigs: groundnut cake, rice bran, chaff from the grinding mills, boiled cassava, cotton seed meal, soya bean oil meal, market wastes (boiled), food leftovers (boiled) from institutional dining halls (schools, hospitals, hotels), leaves and fruits from some of the fodder trees and shrubs, young shoots of pigeon pea, green vegetables and grass, maize, yam peelings, surplus sugar cane (chopped) and bananas.

The type of *housing* commonly used for pigs is usually in need of improvement. Piggeries should provide about 3 square metres for each adult pig. Hard-surfaced (but not so rough as to damage feet) floors of brick, cement or pounded laterite are necessary for good drainage and to avoid an accumulation of mud in the pens. There should be an exercise area, and it is helpful to plant shade trees around the piggery. Pigs are clean animals if looked after properly.

The major *diseases of pigs* in South-east Asia are swine plague and hog cholera. The latter disease can be controlled by vaccination. African swine fever is prevalent in Africa and has spread into adjoining areas. Pigs are very susceptible to the ravages of internal parasites. Kidney worm is probably a major cause of lack of growth and vigour in scavenging pigs. It can be controlled if the animals are kept in clean pens.

6

8 **Rabbits**

Rabbits have many advantages in the kind of circumstances in which Oxfam projects operate. They provide a good protein food, require only a low capital outlay, have high reproductive and growth rates, consume many kinds of readily available materials as food, and have a good conversion rate of food to meat, making them highly efficient producers. They prefer a diet of greenstuff, even plants regarded as weeds, which require much less energy than grain to grow and which are not normally part of human diet. Rabbits require only about one-quarter of the food energy to produce 1 lb of meat that is required by beef cattle.

A female rabbit weighing about 4.5-5 kg (this is perhaps a little heavier than most native-kept stock will currently be) can, under good management, produce 40 or more young each year. A doe can often produce 15 or more times her own body-weight in marketable young each year if the offspring are slaughtered at about 2 kg. The rabbit also needs little living-space — a cage about 9ft square. In general, it is a trouble-free animal if simple rules are observed, and hygienic conditions maintained. However, a number of projects have failed through lack of proper attention, so many view rabbit production with a degree of caution, especially in those areas where it is of recent introduction.

When setting up a unit, it is best to consider indigenous breeds first because they are adapted to local conditions. It will probably be possible to improve their productivity by better management, good feeding, and selecting and upgrading the males to a good meat production type. If improved breeds are to be introduced, good meat breeds are the New Zealand White and the Californian, especially if they come from areas with similar climates. Non-adapted introduced males may be temporarily infertile.

Bucks mature at 6 months and does can be mated at around 6 months, depending on the breed. One buck will be required for up to 10 does. The aim should be to have 4 to 5 litters a year with 6 to 8 rabbits in each litter. The progeny can be killed at 4 to 6 months of age when, depending upon quality of feed and on the breed, live weights will be in the range of 2.2 to 3 kg. The sexes should be kept separate from 12 weeks of age onwards.

Emphasis should be placed on feeding locally available foods that are of little or no value to people. In the wet season this will be easy since broad-leaved weeds, grasses and vegetation are generally abundant. Materials which can be used to advantage include banana leaves, vegetable wastes and agricultural by-products.

Housing is important, but can be simple, provided that it successfully excludes predators such as rats and snakes and gives adequate protection from direct sunlight and heavy rain. A useful design consists of an open section with wire or split bamboo floor through which droppings can fall and an end section with a solid floor that can be closed at night. If available, wire is preferable to bamboo as rabbits can gnaw their way through the latter. If the cages are well raised from the ground, this will aid the collection of the manure for use on growing crops. Cage size is important, and for adults should be about 2-3 sq ft per kg of weight. Rabbits need dry, airy but draught-free conditions; their appetites, and hence their productivity, tend to be lower in high humidity/high temperature conditions. Therefore the positioning of cages and the provision of shade should be given careful attention.

The main disease problems are coccidiosis and pasturella (snuffles). Hygiene is important; where circumstances justify and permit, a coccidiostat (as for poultry) should be given regularly in the drinking water.

Religious and cultural constraints may affect the successful initiation of rabbit-keeping, so this aspect (and the alternatives, such as keeping guinea pigs) should be investigated before a pilot scheme is set up. For example, white rabbits are unacceptable in parts of India.

9 Guinea-pigs

Among the main varieties of guinea-pig are the Peruvian (very long-haired), Rosettes (long-haired) and Abyssinian (smooth-haired). The smooth-haired varieties are rather better adapted to hotter climatic conditions. Much of what has been said above about rabbits is relevant to guinea-pigs, except that guinea-pigs are usually hardier and easier to manage. Females mature at 4 to 5 months of age; litter size ranges from 1 to 6, depending on the breed and the age of the female.

Guinea-pigs are raised in captivity and sold for fresh meat in the Andean region of South America, where they are an important source of good protein in cold and arid areas. In certain areas of Nigeria, domestic guinea-pigs are popular as a source of meat and are either caged or allowed to run around the main dwelling area with an open box for food in one corner. The latter system appears to work well under these conditions, since guinea-pigs are relatively docile creatures and require virtually no capital or labour inputs. However, they are less efficient than rabbits in converting food to meat.

In some parts of the world, there are other small herbivores which are reared for meat. These are usually indigenous, such as the Capybara in South America and the African Giant Rat (*Cricetomys gambianus*) in Nigeria. The establishment of programmes for the domestication of small herbivores, which might be more acceptable to the local people as a source of food than larger animals, should be given consideration in the development of village livestock programmes.

IV   Poultry

Poultry provides one of the most palatable and easily digested meats; eggs are a good source of minerals, vitamins and protein but, unfortunately, are rarely consumed where they are most needed.

Some of Oxfam's most encouraging experience in this field has been in Africa, particularly Zaire (ZAI 52) and Tanzania (TAN 64 and 81). Support has also been given to poultry schemes in Latin America (BRZ 111), and chicken cooperatives (MEX 6).

1 Traditional Poultry Farming

Most breeds of chicken come from a common ancestor, the jungle fowl of South-east Asia. Chickens behave like birds of the forest: they dislike intense sunshine, preferring the shade of bushes, trees and houses, and are not accustomed to low temperatures. The climate in many parts of equatorial Africa, for example, is ideal for poultry farming.

Traditional poultry keepers do not feed the birds but expect them to find their own food. This method results in low egg production and a high mortality rate. Egg production under these conditions will be only approximately 40 eggs per annum, with the hen capable of, perhaps, 100 eggs; whereas a hen from a selected laying breed which is given correct food and management may lay 180 - 200 eggs per annum.

2 Improved Management

Compared with most other livestock projects, poultry farming has the following advantages: it is not alien to most rural cultures; it can be introduced on a family basis; and it requires little capital. The hen has a short incubation period and is potentially a prolific breeder.

The adoption of new stock by peasant farmers to improve egg and/or meat production can be approached in several ways by:

☐ providing improved cockerels to run with local hens (existing cockerels must be killed as they would be aggressive towards the introduced stock);

☐ giving chicks to farmers from a central distributor or hatchery;

☐ hatching eggs for local incubation if husbandry is good.

Normally chicks are distributed at 6-8 weeks old when they no longer require artificial heat and will have been inoculated against disease.

Improved management by means of vermin-proof housing of adequate size to allow each bird sufficient space (around 2 sq ft) to move about, to roost and to lay in, along with a properly balanced diet, will be most effective. If small family groups can be persuaded to pen their birds, a combined house and run, holding up to, say, 20 birds in a locally-made portable unit (giving about 4 sq ft per bird) which can be moved across harvested fields or vegetable plots, providing access to insects, light, plant material and seeds, while distributing manure, will be seen to be a profitable arrangement. Poultry runs are fairly expensive to build, as wire netting is used and, unless covered, they do not provide protection against hawks. Water is essential, and unless a reasonable supply is maintained in a shaded situation, optimum production of eggs or meat will not be obtained.

6

To *incubate* eggs naturally, it is necessary to find good broody hens to sit on them. Such hens are rare among the improved strains of light breeds because broodiness has been virtually eliminated during selective breeding. The farmer may need to introduce a cock from a heavier breed from time to time to improve meat production and provide the occasional broody hen. Artificial insemination is the norm in modern poultry schemes. Breeding stations will need electric or kerosene incubators designed to take about 200 eggs at a time.

All chicks need vaccination against Newcastle disease, fowl typhoid, etc. Hygiene is of great importance in preventing outbreaks of disease. Dead diseased birds should be burned and hen houses thoroughly disinfected.

Feeding. Use grain, extracted oil cakes, crushed pulses or fish meal (in moderate quantities to avoid tainting meat and eggs); grit and oyster or snail shell is needed to help digestion and shell-making. The cost of bought food needs to be balanced against income from eggs and meat. Each bird needs about 4 oz of food per day (in two feeds), but should not be overfed. 100 birds may consume 25 litres of clean water per day. The scheme should not be dependent on water carried for long distances by women.

Marketing. Eggs cannot be sorted in tropical heat, and the packing must be adequate to guard against breakage. Because transport and packing of eggs presents a problem, small farmers may prefer to concentrate on raising chickens for meat.

3 Types of Bird

Improved stock falls into two distinct categories: the light or Mediterranean breeds (Leghorns and Minorcas) which usually lay white eggs and the heavy breeds (Rhode Island Reds, etc.) usually laying brown eggs. The Leghorn-type bird is smaller bodied and a more prolific layer, but has a relatively small and not very useful carcase at the end of its laying life. Where the standard of human nutrition may be marginal, the heavy types of birds are possibly more suitable. The Light Sussex, a largely white bird, does well in tropical situations. Generally speaking, good feeding will achieve more in improving production than breeding plans, which may easily fail. The difficulty with many poultry keepers is their unwillingness to cull or cut down on numbers to comply with availability of food and housing: conditions can soon become overcrowded with the consequent risk of disease.

4 Ducks

Ducks are hardy and not so susceptible to disease as hens. They find most of their food by grubbing in the fields and need only simple shelter. The Muscovy is the most common type in tropical areas and thrives under the toughest conditions. The females will lay up to three clutches of eggs in a year and are excellent brooders. The large size of the Muscovy duck and its naturally slightly aggressive nature enables it to survive under haphazard management. Its egg-laying abilities do not compare favourably with other highly-bred species, but as it is used mainly as a meat producer, this is not of primary importance. Unlike most other breeds, the Muscovy does not require access to open water.

One outstanding highly-bred duck is the Khaki Campbell. Under good conditions of management, difficult to attain in many developing rural areas, they are capable of 300 eggs per year. Under adverse conditions such as high rainfall, high temperature, excessive humidity and poor housing, they can still exceed the best laying strains of chickens. They have the additional advantage of immunity to most disease problems prevalent among fowls and turkeys, and no vaccinations are needed. For small family flocks, they can be ideal. More vulnerable to predators than fowls or Muscovy ducks, they tend to wander in search of water, which they constantly need for washing their bills as well as for other functions.

Indian Runner ducks are almost equal to Khaki Campbells in egg production. The White Pekin duck is a very efficient meat producer. The large Pekin, bred especially for meat production, is able to provide 3.5 kg of duckling in 7-9 weeks under good management.

Ducks can be profitably kept in conjunction with inland fisheries projects, where they provide a lucrative additional income. Their manure stimulates the production of plankton and phytoplankton within the fish pond. Similarly, they can also be combined with paddy rice growing.

5 Geese

A pair of geese kept with access to green feed (grazing) and water at all seasons of the year, such as is common in tropical climates, can produce 40-70 kg of meat for the family yearly, for a period of 20 years or more. A '*set*' of geese is a gander plus four or five females. Adult geese will lay about 30 eggs a year.

Their reproductive life is much longer than that of any domestic fowl and their diet can consist largely of fresh growing greenstuff, such as grass, weeds or legumes. They need plenty of space to wander and forage. They are easy and cheap to keep, and goslings put on flesh quickly.

A by-product of keeping geese and ducks is their soft downy feathers. These fetch a good price if separated from the coarse feathers and dried. Geese are also good house-guards, making a great deal of noise night or day if anyone approaches.

6 Guinea-fowl

Are more resistant to disease than hens. They forage for most of their food, may produce 60 to 120 eggs per year according to their level of nutrition, and reach a mature weight (with a meaty breast) of ½ to 2 kg. They like to lay eggs in the open but can be trained to go into nest boxes. They like to perch fairly high and they are also excellent homestead guards.

V   Bees

Beekeeping has long been practised in tropical Africa and parts of Asia. Traditional techniques of keeping local honey bees are generally appropriate to their environment and are satisfactory, except for the destructive method often used in collecting the honey; their productivity, however, may be low.

Traditional beekeeping provides peasant farmers with:
- [ ] honey — for home consumption, barter or sale, ceremonial use and, in Africa, beer-making;
- [ ] beeswax — this has some important local uses and may be sold as a cash crop, but is often discarded. The world beeswax market depends largely on supplies from traditional hives in tropical countries;
- [ ] pollination — hives of bees near fruit or seed-bearing crops can increase the quantity and quality of the crop.

In a project in Kenya (KEN 2), it was estimated that 50% of farming families in some famine areas derived 50% of their cash income from bees kept by traditional methods. In Bangladesh Oxfam has funded short-term training in beekeeping, and provided simple box hives for landless people and disadvantaged women in association with the Institute of Apiculture.

1 Projects Suitable for Support

Various attempts have been made to improve beekeeping systems in developing countries. Some have been very worthwhile; others have had limited success or have even been harmful. Projects on a small rural

6

development basis are generally more beneficial than national ones: they are in more immediate and direct contact with the beekeepers and can be modest in capital outlay. They only involve teaching improved husbandry methods and giving assistance towards acquisition of hives, etc. (see Section 14, Resources).

National projects are already under way in a number of countries, and at least one government officer is known to be involved in each of the following countries, among others: Kenya, Uganda, Tanzania, Ethiopia, Senegal, Malawi, Rwanda, Burundi, Angola, Sri Lanka and India. National projects are often involved in extension work, research and sometimes in marketing schemes. Marketing can be a real problem in remote agriculturally marginal areas where there are many beekeepers but few and poor market outlets (see PART FIVE Section 2).

A project relating to processing, packing and marketing is not an easy undertaking, and a person knowledgeable in this field must be in charge of it. If marketing collection is to be done on a cooperative basis, establishment of a separate honey cooperative is likely to be more successful than adding honey to an existing multi-purpose cooperative where it may be regarded as of secondary importance.

People can derive great benefit from advice and training in bee management and honey production. This may take the form of short courses on the use of modern, movable-frame hives. Workshops that produce hives locally can be of considerable value.

2 Honey Bees

a: Tropical Africa. The indigenous honey bee is the same species as the European honey bee *Apis mellifera*, but is adapted to the tropics, where it is productive, but also very likely to sting, so that its management can present difficulties.

b: Tropical Asia. Here there is a smaller species, *Apis cerana*, which produces only about one third as much honey as the European bee, but is well adapted to its environment. In no circumstances should the tropical African bee be introduced into Asia. Two Asian species (*Apis dorsata* and *Apis florea*) cannot be kept in hives, but honey is collected from them in the wild, throughout much of South-east Asia.

c: Latin America. There are no indigenous honey bees. European honey bees were introduced 150 years ago, and tropical African bees in 1956. The latter cause much trouble because of their stinging, but they do produce larger amounts of honey. Stingless bees (*Melaponinae*) also produce honey in smaller quantities collected in the wild; colonies are kept in hives in some areas.

3 Hives

The comparatively high cost and complexity of movable-frame hives are often considered to be prohibitive in attempts to improve or to introduce beekeeping as a commercial venture in developing countries where simpler hives are traditionally used. Where frame-hives have been used successfully, they have been thoroughly supervised; they require precision in manufacture and use.

Cheap hives which have been successful in various places are movable-comb hives without frames, originally adapted from a Greek basket hive but built like a trough with sloping sides. One of these types is the Kenya Top-Bar Hive. A queen excluder made from coffee wire (mesh 5 wires per inch) may be used if desired. The movable-comb hive has the advantage over the fixed-comb hive that the combs can be inspected and returned and management manipulations can be carried out without damage to the brood and colony as a whole.

4 Economic return

The economic return from beekeeping can be good, provided there are adequate supplies of nectar and pollen close by. It should be possible for a beekeeper to pay off the capital cost of a manufactured hive within two years, or even after only one year. The recurrent costs are normally small and the labour input only a few hours per hive per year.

In Kenya it has been estimated that a person with 10 good hives of bees, reasonably managed, can earn in one year what a labourer would earn in the cities in a month. World prices for beeswax, of which there is a world shortage, and for honey are good.

The commercial use of honey bees for pollination should not be forgotten, but this is still in its infancy in the tropics.

6

Section 9   Fish Production

I   Oxfam practice

In most countries fish is an accepted food. Where fresh fish is rare, one of the processed forms is usually available. Generally, people would eat more fish if it were available at a price they could afford. Because of its high nutritional value and acceptability, Oxfam is interested in helping to improve the availability of fish where possible and appropriate. One way of improving the supply is to encourage fish farming; another is to make greater and more effective use of natural sources of fish by introducing better methods of capture and/or by improving methods of handling and processing, thus reducing the losses caused by decay. The potential for improvement is great, but specialised knowledge and experience is essential for surveying, planning and implementing new fish projects. Badly planned projects can result in huge wastage of funds and manpower as well as causing serious ecological side-effects. Oxfam policy is to identify and support low-investment projects which will have a positive impact on the communities concerned. Improving fishing and fish farming need not require high capital input.

Oxfam has already supported various types of fish project, including:

☐ building or excavation of ponds for fish farming using animal or vegetable wastes as fertilisers to increase fish productivity. For example, there have been a number of fish farming projects in Zaire (ZAI 78, 230);

☐ stocking new reservoirs, rivers, swamps or paddy with species which can survive even in polluted or saline waters. In Lesotho assistance was given to a village fishing scheme which used mountain streams (LES 19);

☐ introduction of new species of fish from their natural habitat into communal farms or enclosed off-shore areas. There are many examples of types of fish successfully cultured in certain tropical waters which could be grown elsewhere;

☐ fishing boats have been provided for village communal fishing on lakes and dams (TAN 99);

☐ cooperative marine fishing, including locally-made wood and wire cages;

☐ the growing of rice and fish simultaneously in countries where rice is the traditional staple. However, in Thailand, for example, problems have arisen from the use of pesticides on the rice which kill the fish;

☐ provision of facilities for processing fresh fish and their transport, storage, curing and training for marketing. Help has been given to fish marketing schemes, as in Tamil Nadu, India (TN 2);

☐ collection of inedible fish and offal for fish-meal production.

II   Production

1 Fish Farming

Fish farming can be defined as the deliberate introduction of fish into custom-built ponds, shallow water areas created by damming streams, lakes, swamps, reservoirs or other permanent water supplies where they will be cultivated, fed and harvested by approved techniques. This is the main area of potential and of risk, particularly with regard to distribution, marketing and human health as well as that of fish.

In all fish farming projects initial funding should be given for consultants to visit and survey the possibilities. Once demonstration farms have been established, further funding can be given for tools, farmer training or extension in fish culture. Expert advice on fisheries is essential; experience confirms that the planning and implementation of projects without such guidance is not only false economy but carries the risk of harming the very people that the project is intended to assist.

In principle, all artificial waters can be regarded as ponds, provided they are not too deep and are capable of being drained. Some waters are more nutritious than others. Evaporation can be high, and a permanent supply of water is essential to top up and maintain pond water-levels.

Natural depressions can be used if they are capable of being filled with water and drained. In the case of artificial ponds, when soil is permeable, the bottom of the ponds can be improved by plastering with dung before filling.

a: Types of fish. Fish culture in artificial waters requires more skill than in natural waters. The range of suitable fish is large, and includes tilapia, carp, catfish, milk fish, buffalo fish, yellowtails, mullet, eels, salmon, trout, prawns and oysters. Salmon and cod can now be bred and reared in fresh water. Ponds can be stocked with herbivores, plankton-eating and bottom-feeding fish to make full use of all available foods and space.

b: Feeding of fish, or the cultivation of their food source, makes for fast growth rates. Integration of fish farming and local agriculture has evident advantages. Manure should be placed in heaps and not spread around. Sometimes livestock pens are built over ponds. Vegetable material can be thrown in. If grain fodder is used, it should be pre-soaked so that it sinks. Maize, rice and beer wastes make suitable feeds, as do crushed maize cobs. Plants such as green grass, leaves of banana, cassava, papaw, sweet potato, cabbage and lettuce are all favoured; large leaves should be chopped. Feed points should be marked with poles and changed from time to time. Household scraps are usually satisfactory.

Ducks, with nesting accommodation *over* the pond (or pigs kept on platforms over the pond) increase pond fertilisation. The ducks will subsist on material in and around the pond.

c: Pests and diseases. In well-managed ponds, which are regularly drained, there should be little danger of serious disease or parasitic problems developing. All the same, disease is the most serious danger if sufficient care is not taken, once fish farming becomes intensive. Fungus diseases, such as gill-rot, may be introduced inadvertently through water inlets, if these are not well protected. The incidence of gill-rot can be minimised by reducing feed intake, particularly of organic matter in periods of exceptionally hot weather.

Fish can 'catch colds' and suffer from anaemia and sleeping sickness. But most bacterial, fungal and parasitic problems will be avoided by good husbandry and by not allowing overcrowding. The introduction of some saline water, such as the run-off after irrigating saline soils, is a useful prophylactic. Many species, especially carp, can tolerate slightly saline waters.

Routine drying-out of ponds and sluices will check diseases. If the bottom of the pond can be cleared and limed, the parasites and worms which attack fish should be kept under control. One of the greatest hazards comes from predators, including other fish, birds (which can also introduce disease) and small mammals.

d: Types of ponds and systems. The fish used in the Zaire fish programme were *Tilapia nilotica*, and the projects were based on the *derivation pond* system, each pond with its own water supply and drainage, separately controlled. This aims at providing a stagnant water environment and gives complete control over stocking and harvesting the fish; it has worked well.

Disease was rarely a problem in the Zaire programme; selection of clean, healthy stock and drying out ponds between harvest and re-stocking helped to control disease. The advantage of the derivation pond system is that water never runs from one pond to another, whereas in the *barrage pond* system, the sharing of water between ponds serves to transmit disease.

Contour ponds are made on sloping ground with the water entering from a stream or conservation dam at the highest point. Walls are built at the lower end. *Barrage ponds* are preferred where a wall can be built along the side of the stream, each pond having its own inlet and outlet or overflow pipe. *Paddy ponds* are for flat areas and require walls (bunds) to be built on all four sides.

Ponds may be treated with phosphate, but chicken or duck manure should be used if available. Cattle dung is not recommended.

There is some debate about the value of pig manure, even though so many systems seem to do well on it. It is suggested that the effluent nitrogen may inhibit the production of the blue green algae which are very effective nitrogen fixers, and that consequently, only phosphatic fertilisers are necessary to encourage the growth of micro-organisms. in Zaire it was found that well-manured ponds do better than those which are not manured. If pig manure is used, a rate of 1,200 lb to 2,400 lb per hectare per week is ample. Poultry manure at one-fifth of this rate is required.

There are immense opportunities for the culture of fish in rice paddy, provided that there is no danger from chemicals used on the rice. The tilapia group of fish is the most popular for this purpose. Pits or trenches should be excavated in the soil to retain water and fish for the next season while the field is drained for weeding or harvest. More fertiliser, in the form of green manure or night soil, is required for this integrated type of production. Ducks are sometimes used on paddy fields.

2 **Fish from Rivers, Lakes or Sea**

An important danger to guard against is over-exploitation of the fish stocks — this is a complex matter requiring expert knowledge. Oxfam's main involvement is likely to be in the supply of boats and equipment, helping in the organisation of the people in groups or cooperatives, training in the use of modern fishing gear and in the maintenance of boats and engines, etc.

III   **Marketing**

Even if a fish project is intended to provide food only for the local community, marketing opportunities should be studied before proceeding. In areas where fish is not a normal part of the diet, educational programmes should be incorporated into the scheme.

If fish are to be sold fresh, great care must be taken in gutting and washing them as soon as they are removed from the water. They should be kept as cold as possible, using ice if available; or failing that, fish should be held in an airy, sack-covered container and doused with water to benefit from the cooling effect of evaporation. Refrigeration will probably be beyond the local economic and technical resources. Alternatively, fish can be sun-dried, dried in the sun after preliminary heating or smoking, cold-smoked, or sun-dried after salting.

In Zaire, the farmers usually sold all their production at the pond side at harvest; some farmers harvesting larger quantities carried their fish to market. River fish were sold at variable prices; normally demand was greater than supply, so prices remained high. (See Cooperatives, PART FIVE Section 4.)

Unmarketable or inedible fish and offal can be collected for fish-meal production, if such processing facilities already exist in the vicinity and if the quantities are sufficient; alternatively, small quantities can be rendered into fish silage for pig feeding on a simple village basis. The Tropical Products Institute can assist with details.

Section 10 — Agricultural Extension and Training

I   Introduction

'Training' and 'extension' may be crudely differentiated as follows:
Training is aimed at enabling people to acquire specific skills (see Social Development, PART FOUR Section 2 for a definition of 'training'.)
Extension is aimed at encouraging changes in behaviour; in the agricultural context, this usually means encouraging people to change the way in which they practise farming (e.g. to adopt line planting instead of random planting) or to change the farming activities in which they engage (e.g. to adopt irrigated rice instead of — or in addition to — upland rice).

Training and extension are of course mutually reinforcing (there is no point in exhorting farmers to use fertiliser unless they have learnt about storage, placement, and rates of application); and in practice they are likely to overlap a good deal.

For training of farmers and farm-level extension to be effective, it is important that the extension workers and trainers in direct contact with farmers should themselves be trained for their tasks. Table 6-1 shows the relation between farmer training, farmer extension and the training of village extension workers (also known as change agents, development facilitators, community contact workers) and of farmer trainers:

Table 6–1   Relation between Training and Extension

The boxes enclosed by solid lines above represent *objectives* of training, those enclosed by dashes represent training *activities*, each of which is dealt with in turn below.

II   Farmer training

Much of the effort and resources which have gone into farmer training in the past have been devoted to training programmes for farm youth; often these have involved quite lengthy courses (3 months or more) in purpose-built centres, intended for the sons (less frequently the daughters) of farm families, with the objective of transforming them into 'modern' or progressive farmers and of keeping them on the land. Such programmes tend to have the following features.

a: They are attended largely by members of farm families who can be spared from the farm for an extended period and who are not therefore key workers or key decision-makers in the concern (usually because they are too young).

b: They are used by participants as a second-best alternative to academic or vocational schooling or as a stepping stone to further training or off-farm employment.

c: Their objectives are confused: to the extent that (b) is true, they fail to achieve their aim of 'keeping people on the land'; but this is an unreal objective if non-agricultural employment *is* available to their graduates; it tends to be based on an over-optimistic assessment of the livelihoods to be earned in agriculture as compared with non-agriculture.

d: They sometimes promote technologies which have not been validated at farm level, and more often than not promote technologies which have not been validated economically.

Any NGO-supported farmer training programme should avoid these pitfalls and should:

☐ aim to impart specific skills which have been identified as needed and valid (see PART FOUR) and not a battery of 'progressive methods';

☐ be carried out in an environment familiar to farmers, so they can see the relevance of the training to their own farms;

☐ be sufficiently short to be attended by key farm decision-makers.

III   Farmer extension

Many kinds of activity which the NGO supports can be classed as 'extension', since the term describes activities aimed at promoting among rural people an awareness of:

☐ the possibility and feasibility of change;

☐ the likely impact of change on their income and livelihood;

☐ the means by which change can be achieved.

Much agricultural extension has in the past been based on the assumption that the problems of poor productivity and low incomes are technical ones, and the solutions are to be found in agricultural research, the results of which could be transmitted to farmers via an agricultural extension system. Effects of this approach ('top-down' extension) have been:

☐ a generally poor response to extension messages;

☐ uptake mainly by better-off farmers who are able to finance investment and take the risks inherent in adoption of new technology;

☐ subsequent concentration on such farmers ('progressive' farmers) by agricultural extension services, and neglect of poorer farmers.

☐ use of credit programmes to reinforce extension and encourage farmers to adopt new technologies — in which credit programmes themselves become dysfunctional (see PART FIVE, Sections 3 and 4).

Extension programmes should attempt above all to avoid the vicious circle in which the farmers who least urgently need to improve their income and productivity are those who respond to extension most readily and therefore receive most attention from extension workers. They should:

a: seek to identify poorer members of the community and the particular constraints they face in adopting new technologies (e.g. labour or capital shortage, inability to take risks);

b: encourage farmers both to identify their own problems and to develop their own solutions to them;

c: use the indigenous technical knowledge amassed in any rural community as a bank from which much of the basic information needed to devise new technologies can be drawn;

d: always ensure that new technologies are tested and validated on farms, before expecting them to be widely adopted;

e: always allow farmers to change at their own pace; if they perceive change to be beneficial and if there are no serious constraints, their pace will not be slow.

The best-documented case-study of 'bottom-up' agricultural extension is a grain storage project in Bwakiri Chini village, Tanzania. A team of five extension workers visited this village in Morogoro District of Tanzania for a period of eight weeks. They brought with them no set ideas about solutions to the villagers' food problems, but by the end of the eight weeks, the villagers themselves had identified key shortcomings in their food system and had designed and built a number of improved food stores to overcome the most pressing of them. The success of the villagers' own ideas and efforts resulted in a new level of self-confidence in the community.

The difficulties of implementing an extension programme should not be minimised; even where a particularly poor village or region has been selected, extension workers seeking to promote mainly technical change will almost certainly find that they receive the readiest response from the relatively wealthy, articulate, innovative and 'progressive' members of the community who will always be at the forefront in the village meetings; they may find it difficult even to make contact with the poorest groups (landless families, widows, the elderly, etc.). Two approaches to this difficulty can be suggested.

a: To form (or facilitate the formation of) groups of individuals/households with similar problems and needs, identified as specifically as possible.

b: To be prepared always to recognise that problems of poverty may not be amenable to technical solutions but may need social and political change, and to help groups to identify and work towards such solutions.

Agricultural extension has in recent years begun to attract a certain amount of attention from multilateral and bilateral aid donors, most notably the World Bank, which has promoted widely its Training and Visit (T & V) model of extension. Central features of T & V are: frequent briefings of village-level extension workers with specific messages which they are required to carry to their clients; close managerial control of the activities of extension workers; and a network of farmer groups within which it is intended that extension messages should pass from 'contact farmers' to other group members. T & V should probably be regarded as top-down extension in another guise. It appears likely that the impact of T & V is in general inequitable. This is partly because it seeks out progressive farmers as its main contacts and relies on messages reaching others by a 'trickle-down' process.

IV   Training of trainers and extension workers

An important factor in the success of village extension is that the extension workers (change agents/village-level workers/community contact workers) should be well prepared for their tasks. Apart from personal qualities and motivation, trainers should be:

a: technically competent;

b: competent in communication methods and strategies (see PART FOUR Section 3, Communications);

c: competent in social development and social education skills (see PART FOUR Section 1).

The balance of these components will vary with the programme; it is important to recognise that, contrary to the generally held view of agricultural extension workers, (b) and (c) are likely to be at least as important as (a).

There are likely to be important differences between farmer training programmes supported by the NGO and training programmes for village extension workers. While the most effective (and cost-effective) farmer-training programmes are likely to be short, low-cost and highly specific, basic training programmes for extension workers may be longer (up to 3 months in duration) and are likely to carry a relatively high cost per trainee.

Also important for village-level workers may be short refresher courses, which should aim to allow workers to exchange experiences within a structured framework and upgrade specific skills such as the use of drama as a tool for raising awareness of development issues.

Section 11   Agriculture and Labour

I   The demands on labour

1 The Role of Labour in Third World Farming Systems

Labour is the main source of power in many Third World farming systems. Even where animal or mechanical power is used, it is rarely applied to operations other than seedbed preparation, the raising of water, and crop processing. Weeding and harvesting are almost always carried out by hand. Over large areas of the world, most notably in the tropical rainforest zone, draught animals are little used, and farming depends entirely on human labour. Where alternative sources of power *are* used, access to these tends to be easier for the relatively prosperous than for the relatively poor, who have to rely on their own labour power.

Access to labour-power is of great importance in determining levels of farm output and income. This is true even in areas where population density is high and where there is an apparent surplus of labour, because many tasks in cultivation have to be completed within a very short period. This means that a household with several strong sons and/or daughters is likely to be prosperous — a fact which has significance for rural people's attitudes to family size.

2 Seasonality of Labour

The demands placed on labour-time by most farming systems are highly seasonal. Even where population density is high and labour is apparently plentiful in relation to land, labour 'bottlenecks' can occur — in other words, the farm family finds it difficult to yield enough labour-time to complete certain crucial operations (e.g. planting, harvesting) within the short period during which they must be carried out if crop yields are not to suffer. Conversely, seasonal unemployment can exist even where population densities are low and land relatively abundant in relation to labour. The reason for this is that the growing season depends on seasonal rainfall and is therefore short (sometimes only 4-5 months). During this time activity is intense, but when there is no water on the land there is no farm-work to be done.

The seasons during which farm people are required to work hardest tend to coincide with those when they are least fit to do so. Levels of nutrition are best in the period immediately following the harvest, and because this falls in the dry season, it is also the period at which disease incidence is lowest in most areas. Later in the year, during the growing season, people have to work long hours at a time when food stocks are falling and when their general health is poorer. The *loss* of working time through illness can have a direct effect on the size of the harvest and on nutrition and income during the following year. The impact of this is most severe on poor people with little margin over subsistence.

For the poor, a loss of output through illness may lead into a downward spiral of intensifying poverty; one season's low output leads to poor nutrition in the following year, which in turn damages the capacity to work. (For further information on nutrition see PART SEVEN, Section 3.)

II   Performance and economy

1 Returns to Labour and Farm Decision-making

The levels of performance of different farm enterprises are usually expressed in terms of output per acre or per hectare (i.e. return to *land*). However, the return to *labour* (which can be expressed as output per work-day) is often a more important factor in farm decision-making than

6

the return to land. Output per work-day is calculated as follows:

a: Physical

$$\frac{\text{Output per ha (kg)}}{\text{no. of work-days per ha}} = \text{output per work-day (kg)}$$

b: Financial

Output per work-day (kg) × Price per kg (£):
= income per work-day (£).

In devising new crop varieties or improved cultural practices, agricultural scientists have often set themselves the objective of increasing output per hectare, without paying due regard to the labour requirements of the technologies they are attempting to develop. When such new technologies are introduced to farmers, they may be rejected for the following reasons:

☐ that although they may increase yield per hectare, they do not increase return per work-day (e.g. using more fertiliser increases weed growth which means more time has to be spent weeding);

☐ that they require most labour at times when other farming activities also have labour peaks (e.g. in a semi-arid environment where all crops depend on rainfall, an introduced cash crop will probably have to be planted at the same time as traditional food crops);

☐ that they require a heavy input of labour in general, which does not give a short-term benefit (e.g. fertility-enhancing and soil-conserving measures such as composting and terracing are usually laborious and time-consuming, yet do not immediately enhance yield).

In allocating family labour between activities, farmers take into account not only time-cost but also the *energy-cost* of using labour. Especially where nutrition levels are poor, all possible opportunities will be taken to save expenditure of energy.

2 Labour as a Means of Investment

Much of the investment made by farm households is achieved by spending longer hours in work than those necessary to maintain livelihood from year to year. These extra labour hours are used for: planting permanent crops; digging fishponds; soil conservation measures and constructing drainage and irrigation structures. Substantial improvements to livelihood can sometimes be made in this way *without* large cash inputs. These may in any case flood the rural economy with supplies of cash which are out of proportion to the amounts normally in circulation (see Savings and Credit, PART FIVE Section 2). Owing to the seasonality of agricultural production, there is almost always some time available for work not directed at meeting immediate subsistence needs. However, it is not possible to mobilise this time effectively when people are debilitated by seasonal food shortages.

III Division of labour

1 Division of Labour by Sex

In most societies, tradition governs quite strictly which activities are carried out by women and which by men. The main types of labour division are, by *enterprise*, so that women usually have exclusive responsibility for food crops and/or kitchen gardens, and compound livestock, while men have responsibility for cash crops and/or non-homestead plots, cattle-herding, etc., and by *job*, so that of the various types of work to be done on farm crops, some are regarded as belonging to women (usually planting and weeding) and some to men (typically land clearing and harvesting).

Divisions of both these types are usually unequal and unfavourable towards women for the reasons listed below.

a: Divisions by enterprise and by job often occur simultaneously, so that women have the responsibility of providing their families with produce from their kitchen gardens at the same time as they are required to carry out certain jobs on field crops (cash or food); men rarely have a corresponding commitment.

b: As well as sharing farm work with men in this way, women typically have to spend much time and energy in crop processing (almost everywhere done by women).

c: Men's perceptions of the contribution of women to farm work are often distorted; they are likely to regard women's work on, say, weeding cash crops, as fairly insignificant, when in fact it may amount to as much as half the total work input.

d: Newly introduced crops are often burdensome to women; they are almost always brought in *via* men (since extension workers' contact is mainly with men), so men take the main decisions regarding them. However, they are likely to require women still to perform the tasks which are traditionally regarded as 'women's' without diminishing their other, newer responsibilities.

In summary, 'men's' farming activities are very unlikely not to involve women in some way, while women often have farming responsibilities in which men do not take part. These are *in addition to* the responsibilities they bear for collecting water and firewood, cooking and child-care. (For more information on the division of labour by sex, see PART TWO Section 1, Women.)

2 Communal Labour

Many societies have a strong tradition of working together for various purposes. The arrangements for this vary widely and are often quite complex, but the following main types can be distinguished.

a: Exchange arrangements by which members of different households exchange labour-time. The purposes of these are:

☐ to mitigate the boredom of working alone on tedious jobs such as weeding;

☐ to exchange particular skills (e.g. house or barn construction for blacksmithing);

☐ to enable households which are short of certain types of labour-power (e.g. male labour for land-clearing) to gain access to it.

b: Communal work arrangements by which households contribute some of their members' time to projects which benefit the community (i.e. village, hamlet, cooperative, group of households) as a whole. Examples are: the construction and maintenance of irrigation works; building a footpath or road; constructing a school or church.

It may be that labour exchange will become less widely practised as modernisation takes place, and farm households become increasingly cash-orientated, and decision-making becomes more diverse.

Communal labour remains a strong tradition in many societies and can be utilised in community self-help projects. When attempting to mobilise communal labour, it is important to understand the way in which the communities concerned have traditionally organised themselves for communal work.

6

Section 12   Checklist of Questions

I   General guidelines

1. Will the project raise the nutritional level and general standard of living of the poorer people of the area by increasing production without an inappropriate increase in the level of inputs?

2. Do the people own sufficient land or have access to land under fair terms which will be adhered to? If not, will the project strengthen the position of the tenant/share-cropper/worker in this respect?

3. What effect (good or bad) is the project likely to have on the target group — in the short term? — in the long term?

4. What effect (good or bad) will there be on basic resources, particularly soil and water in the long term?

5. Will other people benefit (e.g. richer farmers)? If so, will this be counter-productive for the target group?

6. Will other people be affected to their disadvantage (e.g. farmers downstream, nomads or transmigrants, forest dwellers)?

7. Is there adequate potential leadership in the target farming/landless labourer community, and will this leadership be sensitive to the needs of all, including women?

8. If imported inputs are envisaged, will income be adequate to pay for them and to stand the risk of disasters such as crop loss?

II   Land

1. What information is available about soil type and capability?

2. Are there facilities for testing soils (or plants) for nutrient content and availability?

3. Is sufficient attention being paid to maintaining soil structure and fertility? What organic material is available? Is it being put back into the soil? Compost?

4. Is use being made of nitrogen-fixing legume crops or trees, as appropriate?

5. If the intention is to use inorganic fertilisers: — Has the nutrient need been assessed? — Is the cost justified and has the risk aspect been considered? — Is there enough moisture? Weed control? — Has the farmer the capability, storage, home use or market outlet to handle the increased crop?

III   Water

1. Have the rights to water, its equitable distribution and the system of payment been thoroughly discussed and agreed within the community?

2 Is the organisation and management of the project adequate to maintain the installations?

3. Is the scale of the project such that water may be expected to be available indefinitely?

4. If crop land is to be irrigated, have all the implications for requirement of other inputs, for labour and for use and/or marketing of the additional crop or increased yield been thought through?

5. If for livestock, have the effect on the environment near the water source(s), the amount of feed available, the need to control animals and the use or marketing of any increased product been thoroughly thought through?

IV Power

1. Is this a project where the energy flow through the farming system needs to be considered as a whole to ensure maximum economy and avoid long-term deterioration in natural resources?

2. Does the project improve the annual spread of employment and reduce the strain during the peak labour season? Does the project increase labour productivity and, therefore, individual consumption?

3. Is the design of tools and equipment the best available to meet the project's needs?

4. Is it reasonably certain that the large capital investment associated with mechanical power could not be put to better use in some other way?

5. Will the equipment be used to its full capacity? Is the back-up in fuel, spares, maintenance and repair services, adequate in the area? Is the more sophisticated management required for mechanical operations available?

6. Are there other power sources available which are not being used? Is there a case for using them? Does a local tradition and skill already exist?

V Trees

1. Are there any eroding or degraded soil areas which could be saved or restored by planting trees? Have the most suitable and useful (e.g. multi-purpose) species been chosen for planting?

2. For fuelwood, is there scope for planting a village or community woodlot? Or for planting trees as live fencing or along roadsides, on bunds etc?

3. Is proper provision being made for management of the trees and for ensuring their protection from wood-cutters, roving animals, or fire?

4. Is the right species being planted for the soil, climate and use? If a leguminous tree is to be introduced, will the seed require inoculation with the appropriate strain of nitrogen-fixing bacteria?

5. Are the people trying to use fuelwood economically and efficiently, e.g. using improved stoves?

6. For a nursery, is there an adequate and reliable source of water? Is the site conveniently situated, with road access, for distribution of the saplings?

7. Are all the possibilities for multi-purpose trees understood? Is there scope for inter-row cropping tree plantations with annual food crops? Or grass for livestock?

6

8. Are the needs of the traditional forest dwellers understood and being respected?

9. Are possibilities for developing village industries based on the products of trees or forests being considered? Who owns the trees and the land on which they stand?

VI   Crops

1. Have the consequences of improved, intensified or changed production been thought through? e.g.:

— the managerial and workforce ability to cope with the crop at all times, including storage and marketing;

— the greater financial risk/confidence of increased outlay;

— the effect on the soil: can structure, fertility and low weed/pest infestation be maintained?

2. Is the crop(s) suitable, bearing in mind the climate, soil, traditions of the people, market requirements and demand, nutritional needs, and processing facilities?

3. Is good, reliable seed or vegetative material available?

4. In the case of a fodder crop, is it to be grown on land not suitable or needed for food crops?

5. Is proper provision being made for storage?

6. Are waste and by-products to be put to good use?

7. Are the farmers (share-croppers, cooperative members) free to make wise choices about marketing? Will the policy they plan to adopt be in the long-term interest of the community?

VII   Livestock

1. If livestock is being introduced, what is the justification? To provide power? To utilise waste products and by-products and turn them into human food? To provide additional income?

2. Will there be adequate food and water for the type and number of livestock planned?

3. Will the introduction or expansion of a livestock enterprise compete with food production or consumption, or with some other more marketable or more remunerative type of production?

4. Will management be adequate for the scale of development envisaged? Are the dangers of over-concentration of livestock fully understood, e.g. cattle around water sources, or uncontrolled grazing on eroding hillsides?

5. Is the type of livestock chosen really the best for the circumstances? Have alternative types (or breeds) been considered? Are the animals and their products culturally acceptable?

6. Will the manure be put to good use?

7. What veterinary and extension services are available in the area? Are there any particular pest or disease hazards?

8. If the introduction of exotic breeds or cross-breeding is planned, have the following points been given full consideration: why is the productivity of the existing animals of local breeds poor? Could it be improved by better health, nutrition and management? If animals were introduced from elsewhere, could they adapt to local climatic conditions? Would they be resistant to local diseases? Would available food and management be adequate for their increased productive ability?

9. Have the implications for utilisation, storage or marketing of the livestock products been fully explored and considered?

VIII	**Fish**

1. Is off-shore or inland fishing traditional in the area? If not, is there evidence that it would be readily adopted and that the people would consume the product?

2. What is the availability of local expertise in the country to provide technical advice and guidance. Failing this, will expatriate consultancy be needed?

3. Will there be negative side effects? e.g.

☐ increase in mosquitoes and consequently malaria;

☐ increase in bilharzia;

☐ restriction of water supply to others;

☐ pollution of water supply;

☐ danger of erosion or flooding on other properties;

☐ negative influence on the ecology of the natural water courses.

4. Will the farmer(s) be able to maintain fish production with labour and feed inputs without negatively affecting his/her existing farming system?

5. Is the farmer(s) or fisherman/woman trained and thus capable of implementing the project? Practical on-farm training (e.g. with a successful fish farmer) is best. Does the local fisheries administration provide extension services or will some provision for training have to be built into the project?

6. Will the produce be readily consumed or marketed by the producer(s)? If not, are there possibilities of and is there the knowledge to carry out fish storage or processing? Is ice available in the vicinity, or could it be provided?

7. Is consideration being given to the possible integration of fish farming, e.g. to utilise crop waste or to have an associated livestock enterprise?

8. Is there a government or other reliable fish nursery farm where fingerlings can be obtained, or will the project have to set up its own?

9. In the case of fishing from natural waters, is there a danger of over-exploitation of existing fish stocks?

10. If the project involves the purchase of boats or equipment, is there provision and skill for their maintenance? Will the cost of operating and maintaining these boats enable the fishermen/women to sell their catch at prices competitive with prevailing local prices for fish?

6

1. Are the demands on people's labour made by the project realistic in terms of other seasonal demands, general labour availability, and the existing sexual division of labour?

2. Will the project entail major changes in the existing division of labour within the community or within the household?

3. Are the assumptions made by project personnel about migration in or out of the area reasonable?

4. Would the introduction of new agricultural crops/techniques alter the existing seasonal demand for labour or the existing division of labour? Would a transfer to cash crops reduce the food available to the local population?

5. Specifically, what is the role of women in agriculture in the area, and would any technical or social changes assist or be detrimental to women?

Section 13   Appendices

I   Trees: suitable fuelwood species

A small number of fuelwood species likely to be of interest to Oxfam projects is given below.

1 For Dry Conditions

Acacia tortilis (umbrella thorn, Israeli babool), leguminous tree withstanding extremely arid conditions. Important to have right subspecies for zone. Fast-growing, coppices well, good firewood, excellent sand stabilisation, wood for poles etc., fodder. Likes alkaline soils, lowlands, easily raised from seed. Thorny.

Albizia lebbek (lebbek, karana, kokko etc.). Legume. Used in Tanzanian schools project. Good for reforesting dry, alkaline soils. Good fuel, moderately fast growth, coppices fairly well, good soil binder, fodder, green manure. Tolerates range of conditions. Limitations: not wind firm (roots near surface); seedlings easily damaged by animals.

Azadirachta indica (neem). Traditional fuel in India and Africa; potentially one of the most useful of all arid zone trees, thriving in dry areas and low fertility soils. Good yield of fuelwood, coppices well. Has wide range of other uses including medicinal and insecticidal. Neem seeds lose viability if not planted within 2 or 3 weeks of harvest.

Cassia siamea (yellow cassia, minjri, muong etc.). Legume. Used in Tanzanian schools project (see Oxfam Experiences, Section 5). Will grow in wide range of warm climates from arid to humid, but survives in drier areas only if water available at depth. Cannot withstand cold. Excellent fuel, but smoky. Grows fast, coppices well, useful for reforesting, best in deep relatively rich soils but tolerates soils containing laterite and limestone if drainage good. Grows easily from direct-seeding. Limitations: seeds, pods and leaves highly toxic to pigs (not to cattle and sheep); wood may contain yellow powder that is skin irritant; needs protection from livestock; fuel yield not as good as some other species, such as some eucalyptus.

Eucalyptus camaldulensis (red river gum, red gum). Very good firewood and charcoal wood. Will grow in many tropical and sub-tropical climates, mainly on river plains but can grow at e.g. 1,200m in Zimbabwe, can thrive on poor soils in periods of prolonged drought. Provenances vary considerably in yield ability — some grow fast and coppice well in favourable conditions. Also vary from high to low temperature tolerance. Therefore *very important* to select seed from right origin (similar environment to site to be planted). At present being planted in fuelwood projects in Burkina Faso and Senegal but some villagers do not like fast-burning, smoky wood; kills other plants around it. Good for honey production.

Eucalyptus tereticornis. Similar to above.

Prosopis species. Beware — there are very many prosopis species, some of which are valuable on the poorest soils in low rainfall areas while some are aggressive weeds (e.g. south-west USA/northern Mexico mesquite *P. glandulosa*, Bolivia *P. ruscifolia* and Central America *P. juliflora*). Of interest to the NGO is *Prosopis pallida* (algarroba, huarango, kiawe) with which J. M. Nightingale has experimented in Kenya (P.O. Box 23, Njoro). He can supply seed (his original seed came from Department of Forestry, Hawaii). A legume, it appears to stimulate growth of nearby species and to have great promise in fighting 'desertification'. Thrives under hottest conditions to 4,000ft (although NAS say 300m). Remarkably tolerant of salt. Establishes easily (seed may need rhizobia inoculation), protection necessary, pruning to single stem advised. In Kenya has grown fast. Heavy yield of fleshy pods for food and fodder

6

(also leaves). Fuel use mainly for charcoal. Good shade. Important source of honey. Shallow roots, therefore subject to windblow; like all prosopis can become an invader if not kept under control.

② For Humid Tropics Trees grow well in heat and humidity; unfortunately forests in such areas are being destroyed on a massive scale so protection, management and tree-planting are necessary. Again, a first approach should be to select and encourage local species. If new species are to be planted, a list of suitable ones is given in *Firewood Crops* (see above); the following are good examples:

Calliandra calothyrsus (calliandra). Legume shrub, native to Central America, now planted extensively in Java. Grows fast, in Java has been coppiced at one year old and thereafter annually. Can grow on many soils, even infertile. Needs over 1,000mm rain per year but can tolerate several months of drought. Good household fuel. Erosion control, soil improvement, fodder, honey. Easy to establish, suppresses weeds, needs keeping in check.

Leucaena leucocephala (leucaena, ipil-ipil, koo-babul, lamtoro, etc.). A great deal of interest is being shown in, and a great deal has been written about, this fast-growing legume, every bit of which is useful for some purpose. It holds and improves soil, it is a fuel, fodder and food; it can be planted in plantations, around homes, on bunds, as live fencing etc. However, it must be realised that:

☐ it grows poorly in acidic soils (acidic latasols are common in the tropics);
☐ it is killed by frost;
☐ preferred rainfall range is 600-1,700mm;
☐ it grows best below 500m;
☐ it needs to be free of weeds until it grows a canopy.

Leucaena has a reputation for being toxic for livestock if introduced to them too quickly and as too great a proportion of the diet (usually 10% maximum). This arose from Australian research and experience. Recently Dr R. M. Jones, CSIRO, Brisbane (see below) has conducted feeding experiments in several countries and seems to have shown that it is not, as was thought, mimosene in the leucaena that is toxic but a product produced in the animal by certain gut micro-organisms; these gut organisms are not present in animals everywhere. They are certainly present in Australian cattle, but appear *not* to be present in Indian cattle. So the message is to *proceed with care* when feeding leucaena, but it may well be safe (or at least much safer) with tropical breeds than temperate.

The most vigorous type (for fast growth and repeated harvesting) is Hawaiian giant K8 bred at the University of Hawaii, where the world collection of genetic material is held. The following are sources of information on leucaena:

☐ Professor James Brewbaker, Institute of Tropical Agriculture and Human Resources, University of Hawaii, Honolulu, Hawaii 96822, USA (for information on leucaena and other leguminous trees and sources of seed).
☐ Dr Jake Halliday, Director, NifTAL Project, University of Hawaii, P.O. Box '0', Paia, Maui, Hawaii 96779, USA (centre for information and world-wide trials network, mainly concerned with nitrogen fixation, rhizobia bacteria inoculum for leguminous species).
☐ B.A.I.F., Uruli Kanchan, near Pune, India (practical information and simple training based on their own successful trials growing leucaena on virtually barren land for fodder and soil improvement).
☐ International Council for Research in Agroforestry (ICRAF), P.O. Box 30677, Nairobi, Kenya (growing *Leucaena leucocephala* K8 inter-cropped with cowpea at their Machakos field station).

☐ Dr R. M. Jones, CSIRO Division of Tropical Crops and Pastures, St Lucia, Queensland 4067, Australia (for advice on feeding leucaena to livestock, has cooperated with trials in various areas of the world).

Sesbania grandiflora (agati, bacule, katurai, turi, gallito). Leguminous, multi-purpose, outstandingly fast growth rate, sprouts vigorously after harvesting. Useful for reforesting eroded hills. Grows at up to 800m in frost-free zones. Prefers rainfall exceeding 1,000mm and short dry season only; widely grown in areas with extensive irrigation (e.g. rice paddies) and flooding. Adapted to wide range of soils. Good soil improver. Susceptible to nematode attack.

3 **For Tropical Highlands**

In addition to fuelwood supply, the need for tree-planting to hold and improve soil on eroded hillsides is very great. There is a wide range of species which may be suitable, according to temperature extremes. These include acacias, alders and eucalyptus at lower altitudes and pines higher up. The following are two examples:

Acacia mearnsii (black or tan wattle). Leguminous. Native to Australia but planted elsewhere, e.g. small farm woodlots in cool highlands (above 1,000m) of Java. The higher the temperature, the higher the rainfall required, e.g. above 1,000mm in Java. Restores degraded soils if densely planted and carefully managed. Will not grow on calcareous soils. Must not be allowed to spread out of control. Established by direct seeding.

Alnus acuminata (alder, aliso, ramram). Fixes nitrogen, though not leguminous; good for restoring steep mountain slopes. Gives of its best on humid lower mountain slopes in tropical latitudes. Will grow on wide variety of soils; well suited to agro-forestry or with pasture grown for livestock. Usually seeded in nurseries and transplanted. A limitation is that seeds quickly lose viability.

II Trees: suitable food species

1 **For Arid lands**

Acacia tortilis (see under fuelwood species) has leaves which are dry season fodder in Sahelian Sudan, and has a prolific fall of pods which are eaten by sheep, goats and wildlife;

Albizia lebek (see fuelwood species) has young foliage of high protein content — one tree is said to be able to provide about a quarter of the annual feed needed by a buffalo or a cow. Pods of *prosopis* species (see fuelwood species) are a traditional source of a carbohydrate and protein flour for Indians in southern USA and Mexico. The pods' pulp is largely sucrose. They are relished by domestic livestock and wildlife, having a feed value comparable to that of barley or maize. The pods do not split open and they can be eaten fresh or stored.

2 **For Humid and Frost-free Highland Tropics**

Leucocephala leucaena is an exceptional fodder tree or shrub (see under fuelwood species) but, as explained above, should be used with a measure of caution. In Central America and Indonesia, young leaves and small pods are eaten raw and cooked as a human food.

Moringa oleifera (horseradish tree), grown as a village tree throughout the tropics — roots are a spicy condiment, seeds produce a salad and cooking oil, the growing tips of the foliage can be boiled as a vegetable, very young pods or the peas from more mature pods are edible, and it can be repeatedly cut for fodder (useful notes on experience with the horseradish tree in Haiti have been prepared by Elizabeth Mayhood of Grace Mountain Mission, BP 2268, Port-au-Prince).

Desmodium species, vigorous tall shrubs are so palatable and nutritious as forage that they have been called the 'alfalfas of the tropics'; livestock of all kinds relish the leaves and the young branches.

6

3 Trees with Other Uses

Trees in this category are extremely varied both in character and in the environment they require, but they can very often be fitted into areas that might not be obvious, e.g. neem along roadsides and kapok as live-fencing.

Some examples:

Building and fencing: quick-growing pines, eucalyptus; bamboo; neem is good for posts; for fencing, *live* fencing is often best as it provides fuel and/or fodder as well.

Carpentry: *Tamarix aphylla* (tamarisk) for furniture, ploughs; *Prosopis chilensis* (algarobba) dark brown heartwood resistant to decay.

Fertiliser/soil improvement: Most leguminous trees enrich the soil with nitrogen. Leaf fall provides humus. Neem seeds seem to have a special quality in that when they or their residue after oil extraction are mixed with urea (nitrogen) fertiliser in the soil, they enhance its effectiveness.

Gums and resins: Product of many broad-leaved tropical trees. Traditional local uses should be encouraged. *Acacia senegal* (gum arabic), *A. laeta*, *Ceratonia siliqua* (carob — drought resistant, multi-purpose), *Pinus halepensis* (aleppo pine) resin. The Tropical Products Institute has a wide knowledge of various gums and resins.

Honey: Many species, e.g. eucalyptus, acacia, *Prosopis juliflora*, citrus, Tilia species, *Nephelium litchi*, *Grevillea robusta* (silk oak), *Syzygium cumini* (jambolan), *Gmelina arborea* (gmelina, gamar).

Fibres: *Ceiba pentandra* (silk cotton), *Bombax malabaricum* (cotton tree), kapok. Cotton-like floss produced in seed capsules is used for mattresses and pillows. Lofty shade tree, wood for canoes, living fence posts, leaves edible and medicinal for people and livestock.

Medicinal: Several thousand trees and shrubs have medicinal properties, e.g. neem, *Adhatoda vasica* (malabar nut tree).

Oil for burning: Neem seeds contain up to 40% oil used for fuel for lamps; *Derris indica* (pongam) fuel oil and lubricant.

Paper: Apart from trees used for pulping, such as the eucalyptus, there are many materials that can be used, e.g. paper suitable for school writing-books can be made at village level from bark, bamboo, palm leaves, banana stalks, reed, straw and grasses.

Silk: Sericulture can be a good village community activity. The best silk is produced by the caterpillar larvae of *Bombys mori* which feeds exclusively on leaves of white or black mulberry trees (*Morus*), which can also be used for soil-holding to counteract erosion. A coarser thread is produced by tasar insects (genus *Antheraea*) which will feed on a number of tree species.

Soap: Sources of oil for soap-making include coconut, oil palm, castor and neem.

Tannin: Extracts of tannin from the bark, wood and seeds of some trees can be used for village-level tanning of leather, e.g. *Acacia mearnsii* (black wattle), *A. nilotica* (Egyptian thorn), *Syzygium cumini* (jambolan), neem, and bark of *Pilhecellobium dulce* (Manila tamarind).

Section 14   Resources

I   Bibliography

1 Agriculture, general

M.E. Adams, *Agricultural Extension in Developing Countries*, ITAS, Longman,Harlow, Essex, (£2.50)

M. Gaudy, *Manuel d'Agriculture Tropicale*, Paris, 1959. Excellent for French- speaking Africa.

S. George, *How the Other Half Dies*, Pelican Books, Harmondsworth, Middlesex, 1976. Provocative analysis of the inequalities of agricultural production and distribution.

ILACO, *Agricultural Compendium, for Rural Development in the Tropics and Sub- tropics*, Elsevier, Amsterdam, 1981. (US$ 46.50) An extremely detailed and comprehensive reference book.

ILO, *Poverty and Landlessness in Rural Asia*, International Labour Office, Geneva, 1977. (£8.95) A study, headed by Keith Griffin, which examines seven Asian countries to see what factors contribute to poverty and landlessness.

T. Jackson, *Against the Grain*, Oxfam, Oxford, 1982. (£4.50) A study of project food aid in practice.

D. Joy and E.J. Wibberley, *A Tropical Agricultural Handbook*, Cassell, London, 1979. (£2.95) Comprehensive, well illustrated. At standard of O-level and City and Guilds Tropical Agriculture courses.

S. Lindquist, *Land and Power in South America*, Penguin Books, Harmondsworth, Middlesex, 1979. (£2.25) Hard-hitting account of land problems in S. America and the obstacles to land reform.

F. Moore Lappe and J. Collins, *Food First: Beyond the Scarcity Myth*, Ballantine, London, 1978. Lucid and comprehensive, strongly recommended.

Memento de l'Agronome: Techniques Rurales en Afrique, Libraire Eyrolles, 61 Boulevard Saint-Germain, 75005 Paris, 2nd edn. (F60) Concise manual of technical information.

Andrew Pearse, *Seeds of Plenty, Seeds of Want: Social and Economic Implications of the Green Revolution*. Clarendon Press, Oxford, 1980. (£8.50)

T.A. Phillips, *An Agricultural Notebook*, Longman, Harlow, Essex. (£1.95) Not a descriptive book, but full of useful facts and figures, and relevant to many parts of West Africa.

Hans Rutherberg, *Farming Systems in the Tropics*, Oxford University Press. (£13.50 paper cover) A technical book, particularly useful for its orderly treatment. (A farming 'system' consists of interrelated production enterprises and activities.)

Claire Whittemore, *Land for People: Land Tenure and the Very Poor*, Oxfam, Oxford, 1981. (£1.50) Broad discussion of some of the major issues connected with land and agrarian reform, utilising Oxfam's experience. Prepared for the 1979 World Conference on Agrarian Reform and Rural Development.

C.C. Webster and P. Wilson, *Agriculture in the Tropics*, Longman, Harlow, Essex, new edition, 1980. (£14.00)

G. Wrigley, *Tropical Agriculture, the Development of Production*,

Longman, Harlow, Essex, 1982. 4th edition. (£24.95) Probably the most generally useful book on the subject, very comprehensive.

P. Buringh, *Introduction to the Study of Soil in Tropical and Subtropical Regions*, Centre for Agricultural Publishing, Wageningen, Holland, 2nd edition, 1970.

2 Land — Soil Types, Conservation, Fertility

R. Dudal (ed), *Dark Clay Soils of Tropical and Subtropical Regions*, FAO, Rome, 1965. Very clear summary of vertisols.

FAO, *Soil and Plant Testing as a Basis of Fertiliser Recommendations*, Rome, 38/2, 1981. (£2.50)

J. Russell, *Soil Factors in Crop Production in a Semi-arid Environment*, University of Queensland, J.M. Dent, Letchworth, Herts., 1977. (£17.95)

NAS, *Soils of the Humid Tropics*, National Academy of Sciences, Washington, USA, 1980. (£5.35)

Soil Conservation

E. Eckholm and L. Brown, *Spreading the Deserts — the Hand of Man*, Worldwatch paper No. 13, Worldwatch Institute, Washington D.C. A general introduction to the subject.

UN Desertification Conference — Leaflet No. 4, UNEP, includes a list of measures involved in anti-desertification.

R. & M. Williams and others, *Dry Lands, Man and Plants*, Architectural Press, London, 1979.

Fertilisers

V. Ignatieff and H. J. Page, *The Efficient Use of Fertilisers*, FAO Agricultural Studies, 43, revised edition 1968. Good and comprehensive.

3 Water for Crops and Livestock

L.J. Booher, *Surface Irrigation*, FAO Land and Water Development Series No. 3, Rome, 1974.

I. Caruthers and C. Clark, *The Economics of Irrigation*, Liverpool University Press, 3rd edn. 1980.

FAO Irrigation and Drainage Series, particularly *Drainage of Salty Soils*, No. 16, 1973 and *Crop Water Requirements*, No. 24, Rome, 1974.

Goldberg et al, *Drip Irrigation : Principles, Design and Agricultural Practices*, Drip Irrigation Scientific Publications, Shmaruahu, Israel, 1976.

N.W. Hudson, *Field Engineering for Agricultural Development*, Clarendon Press, Oxford, 1975. (Paperback £6.95, English Language Book Society edn. £2.95)

IT Publications Ltd., *Rainwater Harvesting*, King Street, London (forthcoming 1985).

UNEP/WHO, *Handpumps*, July 1977, International Reference Centre for Community Water Supply, The Hague, Holland. Social considerations of putting in handpumps.

P. Stern, *Small Scale Irrigation*, IT Publications Ltd, King Street, London, 1979. (£3.95)

Using Water Resources, VITA, 3706 Rhode Island Avenue, Mt. Rainier, Maryland, USA 20822. ($5.50) Tubewells, dug wells, water lifting, pumps, water storage and purification; easy-to-use construction details.

4 Agriculture and Power

Agricultural Technology for Developing Nations

Appropriate Technology Development Association, Lucknow, India, *Handbook/Directory of Machines, Tools, Equipment Processes Available in India*.

J. Boyd, *Tools for Agriculture*, ITDG, King Street, London, 1976. A buyer's guide to low-cost agricultural implements. (1985 edition forthcoming).

Commonwealth Secretariat, *Guide to Technology Transfer in East, Central and Southern Africa*, 1981, London. A catalogue of agricultural equipment manufactured in the region.

FAO, *Farm Implements for Arid and Tropical Regions*, 1979.

R. Wijewardene, *Conservation Farming Techniques and Tools — for Small Farmers in the Humid Tropics*, IITA Sri Lanka Programme, 133 Dharnapala Mawatha, Colombo 7, Sri Lanka, IT Publications, 1982.

Energy

P. Fraenkel, *The Power Guide: Catalogue of Small Power Equipment*, ITDG, King Street, London, 1979.

A. Makhijani, *Energy Policy for the Rural Third World*, International Institute for Environment and Development, London, 1976 (£0.80). Brings together many of the ideas and sources of the use of energy, and treats it as an aspect of village planning.

Animal Power (see also ANIMAL PRODUCTION AND PROCESSING)

FAO, *The Employment of Draught Animals in Agriculture*, FAO, Rome, English edition, 1972. *Manuel de Culture Avec Traction Animale*, Republique Francaise Secretariat d'Etat aux Affaires Etrangeres, Paris, 1971. A valuable book on animal powered cultivation in Africa; one of a series, 'Technique rural en Afrique'.

P.H. Starkey, *Farming with Work Oxen in Sierra Leone,* Njala University College Ministry of Agriculture and Forestry, University of Sierra Leone, 1981.

Wind and Solar Energy

C. Flavin, *Wind Energy, A Turning Point*, Worldwatch Paper No 45, Worldwatch Institute, Washington D.C., 1981.

P. Fraenkel, *Food from Windmills*, Intermediate Technology, London, 1975. Report of a windmill irrigation project in Ethiopia.

Hayes, *Energy: The Solar Prospect*, Worldwatch Paper No 11, Worldwatch Institute, Washington DC. Covers solar heating, cooling and generation of electricity, use of wind, waves and falling water, energy for plants, and the storage of energy from solar sources.

Charcoal

ILO, *Charcoal-making for Small Scale Enterprises*, Geneva, 1975.

Water Energy

H. Hamm, *Low Cost Development of Small Water Power Sites*, Volunteers in Technical Assistance (VITA), Arlington, Virginia, U.S.A., 1967.

W.G. Ovens, *Design Manual for Water Wheels*, VITA, 1975.

V. Meier, *Local Experience with Micro-hydro Technology*, Swiss Centre for Appropriate Technology, University of Saint-Gall, Switzerland, 1981.

S.B. Watts, *Manual on the Automatic Hydraulic Ram Pump*, Intermediate Technology, London.

Biogas

A. Barnett et al, *Biogas Technology in the Third World: a Multidisciplinary Review*, International Development Research Centre, Canada, 1978. Summarises existing studies.

P.J. Meynell, *Methane: Planning a Digester*, Prism Press, Second edition, 1982.

NAS, *Methane Generation from Human, Animal and Agricultural Wastes*, National Academy of Sciences, Washington, 1977.

P. Rogers, *Producer Gas Systems and Agricultural Applications*, Ind. Nagar New Delhi, 110008, 1982.

Stoves

Eindhoven University, *A Wood Stove Compendium*, Eindhoven University, Holland, 1981.

I. Evans, *Lorena Owner-built Stoves*, revised 1981. VITA Publications.

H. Gould and Joseph, *Designing Stoves for Third World countries*, Intermediate Technology, London.

ITDG, *Boiling Point*, Intermediate Technology, London, Stoves Project magazine, 4 times a year.

Y. Shanahan, S. Joseph and Trussell, *Compendium of Tested Stoves*, Intermediate Technology, London for FAO, 1980.

Volunteers in Technical Assistance, Arlington, Virginia, U.S.A. *Wood Conserving Stoves*. A design guide. From Intermediate Technology, London.

5 **Trees, Shrubs and Forests**

D. D'Abreo, *People and Forests*, Indian Social Institute, Delhi, 1982. The author, Director of Development Education Service, Mangalore, attacks the Indian Forest Bill, and suggests a new policy which considers the poor forest dwellers.

J.G. Berie, H.W. Beall and A. Cote, *Trees, Food and People: Land Management in the Tropics*, IDRC, Ottawa, Canada, 1977.

L. Bremness, *Growing Trees from Seeds*, 'The Green Papers' No. 4 from Green Deserts Ltd, Rougham, Bury St. Edmunds, Suffolk.

E.P. Eckholm, *The Other Energy Crisis: Firewood*, Worldwatch Paper no. 1, Worldwatch Institute, Washington D.C., 1975. (£1.00)

E. P. Eckholm, *Planting for the Future: Forestry for Human Needs*, Worldwatch Paper no. 26, Worldwatch Institute, Washington DC. (£1.00)

FAO, *Forestry for Local Community Development*, FAO Forestry Development Paper no. 7, 1978. Also Spanish and French editions.

N.K.A. Iyer, *A Guide to Extension Workers on Tree Planting* — English translation of booklet in Telugu (Indian language spoken in Andhra Pradesh) produced by local Oxfam consultant. Includes list of trees, their uses and where to plant them.

NAS, *Leucaena: Promising Forage and Tree Crop for the Tropics*, National Academy of Sciences, Washington DC, USA, 1977. ($10.00)

NAS, *Firewood Crops: Shrub and Tree Species for Energy Production*, National Academy of Sciences, Washington, 1980. Groups fuelwood trees according to the conditions in which they grow (humid tropics, tropical highlands and arid and semi-arid regions) with notes on growing and using them.

NAS, *Tropical Legumes, Resources for the Future*, National Academy of Sciences, Washington D.C., 1979. ($18.00) Has a section on trees and tree products.

D. Webb, P. Wood and J. Smith, *A Guide to Species Selection for Tropical and Sub-tropical Plantations*, Department of Forestry, Commonwealth Forestry Institute, University of Oxford, 1980.

F.R. Webber, *Reforestation in Arid Lands*, Peace Corps/VITA publication, 1977.

William, Chew and Rajaratnam, *Tree and Field Crops of the Wetter Tropics*, 1981, Longman, Harlow, Essex. (£3.50)

6 Grass and Rangeland

CIAT, *Tropical Pasture Programme*, annual reports. Cali, Colombia.

G. Dall and A. Hjort, *Pastoral Change and the Role of Drought*. SAREC Report, R2, Stockholm, Sweden, 1979.

FAO *Tropical Forest Legumes:* FAO Plant Production and Protection series No. 2 (ed. P.J. Skerman), Rome 1977.

R.J. McIlroy, *Introduction to Tropical Grassland Husbandry*, Overseas Development Institute, London, Pastoral Development Network, 1972. Various papers.

S. Sandford, *Management of Pastoral Development in the Third World*, John Wiley, Chichester, U.K., 1983.

7 Crop Production

All the *CGIAR Centres* (Consultative Group on International Agricultural Research) produce useful training material and a variety of publications. These include some practical, illustrated handbooks, such as *Field Problems in Cassava* from CIAT, and *Field Manual of Common Wheat Diseases and Pests*, Information Bulletin 29 from CIMMYT — both pocket size. For a list of the Centres and their main subjects of work see *ORGANISATIONS* at the end of this section.

General (see also General Agriculture)

J.D. Acland, *East African Crops*, FAO/Longman, Harlow, Essex, 1972. (£2.85) A useful account of a range of annual and perennial crops for the humid and seasonally arid tropics.

I. Arnon, *Crop Protection in Dry Regions* (2 vols), Leonard Hill, 1972.

FAO, *Better Farming Series*, elementary handbooks, published by FAO with the Institut Africain pour le Developpement Economique et Social, Abidjan, Ivory Coast.

A.A. and K.A. Gomez, *Multiple Cropping in the Tropics*, International Development Research Council, Canada. ($12.00) Focus on Asia. Stresses importance and technology of multiple cropping, with sections on production technology and research methodology.

R. Irvine, *West African Crops*, Clarendon Press, Oxford, 1974. (£7.50)

D.E. Kay, *Food Legumes*, Tropical Products Institute, London, Crop and Product Digest no 3, 1979. (£6.50) Comprehensive coverage of many beans, peas, lentils, lupin and groundnut including local names, description of plant, planting procedures, pests and diseases,

6

harvesting, processing, use, storage, trade and bibliography. Very useful.

D.E. Kay, *Root Crops*, TPI Crop and Product Digest no 2, 1973. (Tropical Products Institute, 56/62 Grays Inn Rd, London WC1X 8LU.) Useful comprehensive coverage.

C.L.A. Leakey and J. Wills, *Food Crops of the Lowland Tropics*, Clarendon Press, Oxford, 1977. (£24)

National Academy of Sciences (NAS), *Underexploited Tropical Plants with Promising Economic Value*, 1975. Summaries in French and Spanish. Information on a wide range of cereals, roots and tubers, vegetables, fruits, forage and other useful plants known to indigenous peoples.

NAS, *The Winged Bean, A High-protein crop for the Tropics*, 1975. Summaries in French and Spanish. Report of an international panel on the potential for this multi-purpose tropical legume indigenous to PNG and SE Asia and now in trial use in many countries.

NAS, *Tropical Legumes: Resources for the Future*, 1979. Discusses the potential and use of a wide range of food crops, fodder and trees of the nitrogen-fixing legume family.

J.W. Purseglove, *Tropical Crops, Dicotyledons* Vol I and II (2 vols combined £9.95) also, *Tropical Crops, Monocotyledons* Vol I and II (2 vols combined £10.95) Longman, Harlow, Essex. Valuable reference work on all the most important crops.

H.F. Schwartz and G.E. Galvez, *Bean Production Problems, Disease, Soil and Climate Constraints of Phaseolus Vulgaris*, CIAT, 1980 (also in Spanish). Brings together the expertise of many research workers; produced by the International Centre for Tropical Agriculture, Ap. Aereo 6713, Cali, Colombia.

M.L. Vickery & B. Vickery, *Plant Products of Tropical Africa*, Macmillan, London, 1979. (£10.95)

C.N. Williams and W.Y. Chew, *Tree and Field Crops of the Wetter Regions of the Tropics*, (ITAS) Longman, Harlow, Essex, 1979.

C.A.C. Herklots, *Vegetables in South-East Asia*, Allen and Unwin, London, 1972. (£5.95) Useful reference book, relevance much wider than South-east Asia.

F.W. Martin & R.M. Ruberbe, *Edible Leaves of the Tropics*, Antillian College Press, Mayaguez, Puerto Rico, 1975.

H.A.P.C. Oomen & G.J.H. Grubben, *Tropical Leaf Vegetables in Human Nutrition*, Royal Tropical Institute, Amsterdam, The Netherlands, 1977. Practical and well illustrated guide on the production and nutritional value of leaf vegetables.

A. Pacey, *Gardening for Better Nutrition*, booklet prepared by Oxfam published by Intermediate Technology Publications Ltd., London, 1978.

J. Samson, *Tropical Fruits*, Longman, Harlow, Essex, 1981. (£15) A large part devoted to citrus but also covers bananas, plantains, mangoes, pineapple, avocado and paw paw.

N.W. Simmonds, *Bananas*, Longman, Harlow, Essex, 2nd edit. 1981. (£12.95)

H.D. Tindall, *Commercial Vegetable Growing*, Clarendon Press, Oxford, 1968. (£6.50) (English Language Book Society ed, 1978, £1.00) Despite title, this is a simple level book covering small-scale gardening in the tropics; includes short descriptions of a wide variety of vegetables.

Seeds

J.G. Douglas, *Successful Seed Programmes: A Planning and Management Guide*, Western Press, Boulder, Colorado, USA, 1980.

FAO, *Agricultural and Horticultural Seeds*, FAO (Rome) reprinted 1961. A substantial book, includes chapters on production.

FAO, *Improved Seed Production*, FAO (Rome) Series no 15, 1978. Chapters include 'Preparation of seed programmes', 'Formulation of seed projects', implementation and evaluation.

O.L. Justice and L.N. Bass, *Principles and Practices of Seed Storage*, US Department of Agriculture Handbook no 506, Washington, 1978.

National Seed Corporation, New Delhi, India, *Seed Processing*, 1970.

National Seed Corporation, New Delhi, India, *Plant Operators Handbook*, 1972.

J.R. Thompson, *An Introduction to Seed Technology*, Blackie (Glasgow), 1979.

Crop Protection

F. Bischof, *Common Weeds from Iran, Turkey, the Near East and N. Africa*, GTZ German Agency for Technical Cooperation, Dag Hammarskjold Weg 1, Eschborn, FRG. Detailed illustrated guide in colour.

D. Bull, *A Growing Problem*, Oxfam, Oxford, 1982. (£4.95) Publication based on PAU/Overseas Division study of problems and misuse of pesticides in poor areas.

J.A.F. Compton, *Small Farm Weed Control*, Intermediate Technology Publications Ltd. London and International Plant Protection Centre, Corvallis, Oregon, U.S.A., 1982. (£6) An annotated bibliography.

D.S. Hill and J.M. Waller, *Pests and Diseases of Tropical Crops* Vol 1. ITAS, Longman, Harlow, Essex, 1981. (£3)

ICRISAT, *Sorghum and Pearl Millet Disease Identification Handbook*, Information Bulletin no 2, 1978, from the International Crops Research Institute for the Semi-Arid Tropics, Hyderabad, India. Pocket size; full colour illustrations of disease symptoms.

G. Matthews, *Pesticide Application Methods*, Longman, Harlow, Essex, 1981. (£7.97)

WARDA, *Weeding of Paddies in W. Africa and Catalogue of the Main Weeds*. Prepared by the French Research Institute on Tropical Agronomy and Food Crops (IRAT), published by W. Africa Rice Development Association (WARDA), P.O. Box 1019, Monrovia, Liberia, 1978. Control methods, herbicides, and colour guide to weeds.

Storage and Processing

CDTF, *Appropriate Technology for Grain Storage*, Community Develoment Trust Fund of Tanzania, 1977, Box 9421, Dar es Salaam. A clear and informative manual on different storage techniques used in Tanzania. Also describes the successful methodology whereby the researchers worked with government officials and villagers to draw up this information. Recommended on both counts.

P.A. Clarke, *Rice Processing: a Checklist of Commercially Available Machinery,* TPI, Slough, Berks., 1978.

Commonwealth Secretariat, London, *Careful Storage of Yams.* (£0.35) also, *How to Build a Low-cost Maize Crib.* (£0.35)

D. Dichter, *Manual on Improved Farm and Village-level Grain Storage Methods*, GTZ, German Agency for Technical Cooperation, Dag Hammarskjold Weg 1, Eschborn, FRG., 1978. Based mainly on African experience; comprehensive field training manual.

L. Druben & C. Lindblad, *Small Grain Storage*, Peace Corps/VITA Manual, USA, 1977. Practical and comprehensive. Strongly recommended. Also in Spanish.

FAO, *Rural Home Techniques: Food Preservation*, Rome, 1979. Collection of 20 pamphlets, including simple instructions in French and Spanish.

GTZ, *Potato Storage in Panama*, 1980. Useful, practical guide to potato storage in tropical and sub-tropical conditions based on the work of GTZ, German Agency for Technical Cooperation, Dag Hammarskjold Weg 1, Eschborn, FRG.

D.W. Hall, *Handling and Storage of Food Grains in Tropical and Sub-tropical Areas*, FAO, Rome, 1970.

T.H. Jackson & B.B. Mohammed, *Sun Drying of Fruits and Vegetables*, FAO, Rome, 1969.

E. Orr & D. Adair, *The Production of Protein Foods and Concentrates from Oilseeds*, TPI, Slough, Berks, U.K., 1967.

R. Wills, T. Lee, D. Graham, B. McGlasson & E Hall, *Post Harvest, an Introduction to the Physiology and Handling of Fruit and Vegetables*, Granada Publishing, St. Albans, Herts, 1981. (£12)

8 Animal Production and Processing

(see also under Agriculture, General)

Livestock — General

FAO Better Farming Series on *Animal Husbandry*, Nos. 11-15.

H.T.B. Hall, *Disease and Parasites of Livestock in the Tropics*, Intermediate Tropical Agriculture Series (ITAS), Longman, Harlow, Essex, 1977.

D. Hallam, *Livestock Development Planning; A Quantitative Framework*, CAS Paper 12. (£10.00 post free) Available from the Centre for Agricultural Strategy, University of Reading, 1 Earley Gate, Reading RG6 2AT, UK.

INADES — formation, *Cours d'apprentissage agricole*, series of practical booklets in French on topics such as 'La nourriture et le logement', and 'L'elevage des moutons et des chevres'. INADES — formation, BP 8008, Abidjan, Ivory Coast.

R.W. Matthewman, *A Survey of Small Livestock Production at the Village Level in the Deprived Savanna and Lowland Forest Zones of South West Nigeria*, Dept. Agric. and Hortic. Study No. 24, University of Reading, UK, 1977.

G.S. Ositelu, *Animal Science*, Cassell, 1981. (£2.95) Written for O level and City and Guild students; the author is at the University of Ife, Ibadan, Nigeria.

R.D. Park and others, *Animal Husbandry*, Clarendon Press, Oxford, 1970. (£6.50)

Ranjhan, *Animal Nutrition in the Tropics*, Vikas (Delhi), 2nd ed, 1982. (£10.95)

G. Williamson and W.J.A. Payne, *An Introduction to Animal Husbandry in the Tropics*, Longman, Harlow, Essex, 3rd ed, 1978. Completely revised and rewritten.

Cattle and Buffaloes

M.A. Barrett and others, *Milk and Beef Production in the Tropics*, Clarendon Press, Oxford, 1974. (£6.50) (English Language Book Society edition £2)

W. Ross Cockrill, *The Husbandry and Health of the Domestic Buffalo*, FAO, Rome, 1974. A large book recognised as 'the bible' of water buffalo knowledge.

National Academy of Sciences, *The Water Buffalo: New Prospects for our Under-utilized Animal*, National Academy Press, Washington, 1981. 1981.

W.J.A. Payne, *Cattle Production in the Tropics*, Vol. 1, Longman, London, 1974.

Peace Corps manual, *Animal Traction*, P.D. Press, 4419 39th St, NW, Washington DC 20016, USA, 1982. ($US19) Comprehensive, practical.

M. Ristic and I. McIntyre (editors), *Diseases of Cattle in the Tropics*, Martinus Nijhoff, 1 Harlequin Ave, Great West Road, Brentford, Middlesex, 1981. (paperback $US39.50) Brings together the knowledge of international experts on a wide range of cattle diseases and pests.

Sheep and Goats

A.B. Carles, *Sheep Production in the Tropics*, Clarendon Press, Oxford, 1983.

C. Devendra and G. McLeroy, *Goat and Sheep Production in the Tropics* (ITAS), Longman, Harlow, Essex, 1982. (£4.25) Good and practical.

C. Gall, *Goat Production*, Academic Press, London, 1981. Comprehensive worldwide review.

J. Halliday, *Practical Goat Keeping*, Ward Lock, London, 1982. (£5.95) Intended for UK but has basic practical information.

World Neighbours/Food Ministry of Peru, *The Control of Parasites in Sheep* (El control de Parasitos en las Ovejas), 1978. A motivational and instructional *filmstrip* in colour. Script English and Spanish. ($US5)

Pigs

6

C. Devendra and M.F. Fuller, *Pig Production in the Tropics*, Clarendon Press, Oxford, 1979. (Paperback £4.95)

J.A. Eusebio, *Pig Production in the Tropics*, (ITAS), Longman, Harlow, Essex, 1981.

Rabbits

H.D. Attfield, *Raising Rabbits*, VITA manual, Arlington, Virginia, U.S.A., 1977. (£0.90) A complete and practical guide to keeping rabbits. Also available in French.

Extension and Research Liaison Services (ERLS), *Guide to Successful Rabbit Raising,* Extension Guide No. 54, Livestock Series No. 7. ERLS Ahmadu Bello University, Zaria, Nigeria.

Poultry

INADES — formation, *L'Elevage Familial des Poules*, booklet in 'Cours d'apprentissage agricole' series, INADES — formation, B.P. 8008, Abidjan, Ivory Coast.

Oluyemi and Roberts, *Poultry in Warm Wet Climates*, Macmillan, London, 1979. (£4.95)

Pascal de Pury, *Comment Elever les Poules*, Editions Cte, Yaounde, Cameroun. Essential handbook for the French-speaking African countries.

D. Sainsbury, *Poultry Health and Management*, Granada Publishing, St. Albans, Herts, 1980. (£8.95, pb £5.95)

W. Thomann, *Poultry Keeping in Tropical Areas*, FAO, Rome, 1968.

Bees

W. Drescher and E. Crane, *Technical Cooperation Activities: Beekeeping. A Directory and Guide*, International Bee Research Association, 1982. This can be purchased from IBRA, or a request can be sent from a developing country to GTZ, Dag-Hammarskjold-Weg 1, 6236 Eschborn 1, FRG. The Guidelines are written for people who may be involved with aid programmes, and the Directory gives details and assessments of all known past and present beekeeping programmes, and of feasibility studies and reports produced, in 85 countries.

9 Fish Production

Annual Reports, *Tropical Fish Culture Research Station*, Malacca, Malaysia.

Marilyn Chakroff, *Freshwater Fish Pond Culture and Management*, Peace Corps/VITA Manual, U.S.A., 1976.

I.J. Clucas and others, *Fish Handling, Preservation and Processing in the Tropics*, TPI Report GI44 (French and Spanish summaries), Slough, Berks, 1981. (£4.05)

C.F. Hickling, *Fish Culture*, Faber, 1971 and *The Farming of Fish*, Pergamon Press, Oxford.

M. Huet, *Textbook of Fish Culture*, Fishing News Books, London.

K. Jauncey, *A Guide to Tilapia Feeds and Feeding*, Institute of Aquaculture, Stirling University, Scotland, 1982.

J. Lemasson and P. Lessent, *Manuel de Pisciculture pour l'Afrique Tropicale*, CTFT, France, 1971.

S.Y. Lin, *Pond Culture of Warm Water Fishes*, UNESCO, New York, U.S.A.

R.S.V. Pullin and Z.H. Shehadeh, *Integrated Aquaculture Farming Systems* Conference proceedings 1981, Philippines, ICLARM, MCC, P.O. Box 1501, Makati Metro Manila. Asian experience.

A. Maara et al, *Fish Culture in Central East Africa*, FAO, Rome, 1966. Very practical for wider application than Africa.

SEAFDEC, *Fish Farming Handbook*, SEAFDEC Institute of Aquaculture, Philippines, 1981. Mainly about fish farming with sea water at the sea's edge.

C.V. Seshadri and others, *Engineering of Photosynthetic Systems*, Sri AMM Murugappa Chettiar Research Institute, Madras, Monograph Series, 1980.

10 Agricultural Extension and Training

'Top-down' Extension

Everett Rogers, *The Communication of Innovations*, Collier-Macmillan, London, 3rd ed. 1983.

Saville, *Agricultural Extension*. Oxford Tropical Agriculture Handbooks. 1965.

D. Benor & Harrison, *Agricultural Extension. The Training and Visit System*, World Bank, Washington D.C. 1977.

'Bottom-up' Extension

Paulo Freire, 'Extension or Communication' — essay in *Education: the Practice of Freedom*. Writers & Readers Publishing Cooperative, London, 1974.

R. Chambers (ed) 'Rural Development: Whose Knowledge Counts?' Vol 10, no. 2 of *IDS Bulletin*, devoted to articles on Indigenous Technical Knowledge (ITK). Institute of Development Studies at the University of Sussex, January 1979.

Richard Schwenk. 'Sarawak: Working Backwards to Development' Interview in Reading Rural Development Communications *Bulletin* No. 8, Reading University, November 1979.

Community Development Trust Fund of Tanzania. *Appropriate Technology for Grain Storage*. 1975. See also article in Reading Rural Development Communications *Bulletin* No. 3, Reading University, September 1977.

[11] **Agriculture and Labour**	R. Chambers, R. Longhurst, and Arnold Pacey (eds), *Seasonal Dimensions to Rural Poverty*, Francis Pinter, London, 1981.

R. Chambers, *Health, Agriculture and Rural Poverty: Why Seasons Matter*, I.D.S. Discussion paper No. 148, Institute of Development Studies, University of Sussex, 1979.

R. Jolly et al (eds.), *Third World Employment*, Penguin, Harmondsworth, Middlesex, 1973.

ILO, *Employment, Growth and Basic Needs: A One-World Problem*, ILO, Geneva, 1977. A succinct report of the World Employment Conference, 1976.

II Organisations

[1] Land

FAO (Rome) and CGIAR Centres (Consultative Group on International Agricultural Research) such as IRRI (Los Banos, Philippines) publish many bulletins, handbooks and pamphlets covering soils, conservation and fertilisers.

The International Centre for Soil Conservation Information (ICSCI) can be contacted at the National College of Agricultural Engineering at Silsoe, Bedfordshire, U.K.

The International Fertiliser Development Centre (IFDC) has a library and information service covering all aspects of fertilisers and fertiliser use, with special reference to the tropics. Address: P.O. Box 2040, Muscle Shoals, Alabama 35660, USA.

[2] Water for Crops and Livestock

A good source of information on irrigation is the Institute of Irrigation Studies at Southampton University. The training given at the Institute is strongly orientated towards the small farmer and his/her crops; it covers all systems from simple flooding to sophisticated, mechanised sprinkle, trickle and sub-irrigation.

CSIR regional research laboratory, Jorhat, India has caried out work on the conversion of aquatic weeds into pulp for paper manufacture.

Sources of information on aquatic weeds are:

6

☐ Weed Research Organisation, Long Ashton Research Station, Bristol.

☐ Aquatic Weed Programme, University of Florida, Gainesville, Florida, USA (who produce a newsletter called 'Aquaphyte').

☐ International Plant Protection Centre, Oregon State University, Corvallis, Oregon, USA.

3 Trees, Shrubs and Forests

Commonwealth Forestry Institute, Oxford, U.K.

The International Council for Research on Agroforestry (ICRAF), Box 30677, Nairobi, Kenya, has prepared many useful publications on agroforestry.

The Tropical Development Research Institute, London. See below.

4 Crop Production

CGIAR Centres (Consultative Group on International Agricultural Research) International Rice Research Institute (*IRRI*) Los Banos, Philippines (P.O. Box 933, Manila). Rice; resistance to disease, insects, pests, drought, tolerance to adverse soils, deep water and floods, extreme temperatures; improved nutritional quality.

International Maize and Wheat Improvement Centre (*CIMMYT*), Mexico (Apartado Postal 6-641, Mexico 6, D.F.). Wheat, maize, triticale; emphasis on pest and disease resistance, nutritional quality of grain, high yield potential.

International Centre for Tropical Agriculture (*CIAT*), Colombia (Apartado Aereo 6713, Cali). Cassava, beans (*Phaseolus vulgaris*), improvement of acid infertile savanna for beef production, feeding of crop wastes and unconventional diets to pigs, collaborative maize work with CIMMYT and rice work with IRRI.

International Institute of Tropical Agriculture (*IITA*), Nigeria (P.M.B. 5320, Ibadan). Cassava, yams, grain legumes (cowpea), collaborative rice and maize work with IRRI and CIMMYT; farming systems to assist transition from shifting to settled agriculture, and on-farm crop storage. The FAO/African Rural Storage Centre is based at IITA.

International Centre for Agricultural Research in the Dry Areas (*ICARDA*), P.O Box 114/5055, Beirut, Lebanon and P.O. Box 5466, Aleppo, Syria. Barley, lentils, broad beans (*Vicia faba*), durum wheat; dry land systems.

International Potato Centre (*CIP*), Apartado Postale 5969, Lima, Peru. Improved potato varieties and potato growing and storage in developing countries and extended range of adaptation to include lowland tropics.

International Crop Research Institute for the Semi-Arid Tropics (*ICRISAT*), India (1-11-256 Begumpet, Hyderabad 500016, Andhra Pradesh). Improved grain yield and nutritional quality of sorghum, pearl millet, pigeonpea, chickpea and groundnut and developing farming systems which make better use of natural and human resources in the seasonally dry semi-arid tropics.

West African Rice Development Association (WARDA), P.O. Box 1019, Monrovia, Liberia. Aims to promote self-sufficiency in rice for a 15-country region where rice is a staple food.

Other centres under CGIAR but not crop orientated are given under ANIMAL PRODUCTION.

Another source of information on arid land plants with food, fuel and other domestic uses is The Royal Botanic Gardens, Kew, UK.

International Board for Plant Genetic Resources (IBPGR), F.A.O., Via delle Terme di Caracalla, 00100 Rome, Italy.

Information or publications on *vegetables* can be obtained from:

The Asian Vegetable Research and Development Centre (AVRDC), Taiwan (P.O. Box 42, Shanhua, Taiwan 741). Chinese cabbage, tomato, white potato, sweet potato, mungbean, soyabean, rice. Dr. George Selleck, Director, has offered advice to Oxfam on the purchase of vegetable seeds.

The National Vegetable Research Station (*NVRS*), UK (Wellesbourne, Warwickshire). Temperate vegetables. Oxfam-supported vegetable gene bank.

The Centre for Tropical Agriculture, University of Florida, USA (Gainesville, Florida 32611).

A range of publications on production and storage of crops is produced by the Tropical Development Research Institute (TDRI), 56/62 Gray's Inn Road, London WC1X 8LU, U.K. and the Storage Department of the TDRI, London Road, Slough, SL3 7HL, Berks, U.K. TDRI are also involved in research into the preservation and processing of a wide range of products and in the development of suitable equipment, e.g. a manually-operated hydraulic oil seed press, a pedal-operated grain mill, and a small scale hand-operated decorticator for sunflower seed. Information available for these includes detailed drawings and advice for local manufacture.

Further information on appropriate tools and equipment for crop processing and production can be obtained from:

☐ The National College of Agricultural Engineering, Silsoe, Bedfordshire, U.K. who have also developed a simple sprayer for chemical insecticides.

☐ ITDG, Agricultural Research Station, Shinfield, University of Reading, Berkshire, U.K.

Legumes — some major centres of expertise are:

☐ Indian Agricultural Research Institute, Delhi, for chick peas, cow peas, pigeon peas, grams and mung beans;

☐ Central Research Institute for Food Crops, Bogor, Indonesia, for soya beans, ground nuts and mung beans;

☐ Plant Breeding Institute, Cambridge, U.K., for *vicia faba* (broad) bean;

☐ TDRI, London, for all legumes.

The CSIRO Division of Tropical Pastures and Crops, Brisbane, Australia has developed some improved tropical pasture legumes for fodder.

The Central Plantation Crops Research Institute (CPCRI) at Kasaragod, Kerala, is collecting material for a world coconut germplasm centre planned for the Andaman Islands and the Indian Ocean.

A world-wide advisory service on the suitability of crops based on the concept of weeds as good indicators is operated by Dr James Duke at the US Department of Agriculture's Research Centre, Beltsville, Maryland, USA.

For details of worldwide seed testing facilities contact the National Institute of Agricultural Botany, Seed Testing Station, Cambridge, UK.

6

Nutrition

The League for International Food Education (LIFE), an American non-commercial consortium of scientific societies at 915 15th Street NW, Washington DC, produces information leaflets and a newsletter on food and nutrition, simple storage and processing methods.

International Food Policy Research Institute (IFPRI), 1776 Massachusettes Avenue NW, Washington DC 20035, USA. Focuses on sensitive economic and political issues surrounding food production, distribution and trade with Third World countries.

5 Animal Production and Processing

International Livestock Centre for Africa (ILCA), P.O. Box 5689, Addis Ababa, Ethiopia. Improved livestock production and marketing systems for tropical Africa; pastoralism, water harvesting, fodder legumes, feeding crop residues.

International Laboratory for Research on Animal Diseases (ILRAD), P.O. Box 30709, Nairobi, Kenya. Seeks to control trypanosomiasis and theileriosis that limit livestock production in many areas world-wide.

The International Buffalo Information Centre is a research centre set up at Kasesart University, Bangkok.

Factors affecting the efficiency of animals such as feeding and types of harnesses are being studied at the following:

☐ The Centre for Tropical Veterinary Medicine, Edinburgh;

☐ The National Institute of Agricultural Engineering, Silsoe, Bedfordshire, U.K. has a collaborative programme with the International Research Insitute for the Semi-Arid Tropics, Hyderabad, India.

☐ Njala University College, Freetown, Sierra Leone, where there is a comprehensive work oxen project.

Scientists associated with the Tsetse Research Laboratory, University of Bristol, Langford, Bristol, and the TPI, London, have carried out research into disease in cattle spread by the tsetse fly.

In 1978 the International Bee Research Association prepared an extensive Bibliography of Tropical Apiculture, the work being funded by IDRC, Ottawa, Canada. A free leaflet listing the titles of the 24 parts, which provide access to virtually all the information available, can be obtained from the International Bee Research Association (IBRA), Hill House, Gerrards Cross, Bucks, SL9 0NR, UK. Institutions in developing countries can apply on form BOTA/3a to obtain the parts they require, free of charge.

ODA funds an Information Officer for Tropical Apiculture, whose help and advice may be sought, at IBRA.

Also available from IBRA are 1981 'Source Materials for Apiculture' leaflets at £1 each including postage, £8 for 10, or individual leaflets free to developing countries. These provide information often needed and sought by beekeepers in developing countries.

6 Fish Production

The International Centre for Living Aquatic Resources Management, MCC, P.O. Box 1501, Makati, Metro Manila, Philippines, was set up in 1975 specifically to study combinations of agriculture/livestock and fish. A good source of information on combining ducks and fish is the Dor Fish and Aquaculture Research Station, Israel.

R & D work to produce suitable feeds for fish farms in developing countries is being carried out by Tropical Development Research Institute and the Universities of Aston, Reading and Stirling, U.K.

Useful reference material on methods of preserving, and preventing contamination of, fish products are available from TDRI and FAO.

7 **General Research Assistance**

International Service for National Agricultural Research (ISNAR), P.O. Box 93375, 2509 AT, The Hague, Netherlands. Provides assistance to developing countries to plan, organise and manage their own research more effectively.

6

PART 7 HEALTH

Oxfam

Part Seven HEALTH

7

7

Section 1   Health Guidelines

I   Introduction

The guidelines laid down in this section are consistent with WHO's general aim, to help in the provision of health for all (and especially the poorest sectors of society) by the year 2000. In practice, it is recognised that Oxfam's input in this field takes place in the context of an increase in the scale of health problems worldwide. Population growth, the decline in health services in many countries, migration and urban congestion, natural disasters and wars all serve to undermine health initiatives and make health targets seem unattainable.

In the past most NGOs have worked mainly with health programmes operating in isolated rural areas. However, the growth in urban populations means that in the future there will be a greater need for programmes in inner city slums and the poorer suburbs. Although the typical urban programme will have much in common with its rural counterpart, it will present new challenges. For example, communications may be easier in the city, but water, sanitation, and congestion, will all present severe problems.

The guidelines given here are intentionally only general, and only where appropriate is specific technical advice given, since to have practical applicability, health intervention must be flexible and must adapt to local socio-economic and geographical circumstances. The implementation of health care should, though, always reflect Oxfam's main priorities and objectives, and these rest on certain fundamental principles.

a: Wherever possible the health measure should be *preventive rather than curative*. Expensive curative services never match up to prevention in terms of improving the general health of a population as a whole.

b: Simple *primary health* care takes priority over sophisticated clinical medicine provided in expensive hospital complexes.

c: While it is important to continue funding individual, small-scale projects, field staff should increasingly be looking at the *overall needs of a geographic or administrative area*. The goal should be to raise the general level of health care in a whole region rather than to create pockets of excellence.

d: It must be recognised that in poor countries it is particularly *difficult to set up and sustain a health programme*, especially where there is little support from the State. In this context health provision must amount to a great deal more than the simple delivery of services and equipment or the construction of buildings such as clinics and hospitals. The programme will be far more effective if there is a high level of community participation. The interaction between service and community is a key area in health care. No one model can be advocated: the relative contributions made by service or community will depend on a number of factors, such as the wealth of the community, the efficiency and motivation of the service and the nature of the health intervention. Health care does require the infrastructure of a service, or a service component; provision must always be made for referral, and health planning itself needs skills which are normally beyond the reach of a small community. However, individual interventions will vary a great deal when it comes to the degree of central control required. Immunisation, for example, requires far more support and supervision than the administration of first aid.

e: The success of all health interventions (and especially those depending on a high level of community participation in planning,

7

administration and, possibly, funding) depends on a *strong element of health education* being incorporated into the programme. Encouragement should be given to the use of imaginative, although inexpensive, techniques for the dissemination of information relating to key areas of health care (see PART FOUR Section 3).

f: Taking this approach to health care means that the various components of the system will not be very distinct. Thus, maternal and child health, the control of communicable diseases, nutrition and environmental sanitation will ideally all be integrated into the programme. However, even though this approach has many advantages, care must be taken that *no one area of health provision should be neglected* as a result of being incorporated into a general programme. It is important that all programmes be both balanced and comprehensive.

g: The greater the community involvement in health provision, the harder it becomes to separate the health measures from other areas of development. In fact, *it is vital that the relationship between health and any other intervention be carefully examined*. It must not be taken for granted that improvements in the socio-economic status of a community will of necessity improve its state of health. For example, an irrigation project may increase the risk of malaria and schistosomiasis infections. On the other hand, health care may be intricately tied to other programmes. For example, a health clinic may also act as a community centre, or an income-generating project may help fund a health programme.

II Definitions

1 Primary Health Care

(PHC) is the delivery of basic curative and preventive health care at the community level, and refers to the work of dispensaries and other static units, mobile clinics and small health centres. Out-patients and community health programmes, etc., can also be included under the primary health care heading even if run from hospitals whose other functions would normally be classed as secondary health care. The services included in PHC are: simple curative treatment; mother and child health; nutrition; health education; environmental sanitation, such as water supplies and sewage disposal; birth control; immunisation, and the training of paramedical health workers. The cost per head of the total population is usually low.

2 Secondary Health Care

— rural or urban hospitals and some large health centres offering in-patient treatment of a curative nature or specialised facilities for handling referral cases provide secondary health care. Cost per patient treated is usually high.

3 Community Health

— in the context of Oxfam's work is the delivery of primary health care to the community with the active training and participation of that community both in the planning and practical execution of the programme. In reality many so-called community health programmes are merely public health extension services with little community involvement. Community health is normally the cheapest method of providing primary health care. The community can make material contributions through:

☐ provision of local accommodation facilities and semi- or unskilled assistance;

☐ payment for services;

☐ provision of local health insurance schemes.

Committees should be formed at the community level to administer, monitor and evaluate the programme and to coordinate health interventions with other development initiatives (crafts, agriculture, etc.).

a: Primary health care
- [ ] Nutrition
- [ ] Emergency nutrition
- [ ] Immunisation
- [ ] Sanitation and water supply
- [ ] Emergency sanitation
- [ ] Control of communicable diseases
- [ ] Mother and child health
- [ ] Birth control
- [ ] Training health workers
- [ ] Simple curative services and first aid

b: Secondary health care
- [ ] Hospitals
- [ ] Care of the disabled

III   Priorities

When considering a project both the overall health needs of the area covered and the input of other agencies must be assessed. However in times of disaster some of these criteria may have to be modified to meet immediate needs. (See PART EIGHT, Section 2 for information on emergency health provision.)

1 High Priorities

a: Primary health care for rural areas and urban slums.

b: The promotion of preventive health measures (with emphasis on the prevention of malnutrition especially in children and other vulnerable groups). This should be linked with efforts to increase food production.

c: Birth control.

d: The training of auxiliary medical personnel.

e: Health education.

f: Immunisation.

g: Rural outreach — including mobile units, but increasingly emphasising static units, possibly with mobile supervision.

h: The provision of adequate water supplies, preferably linked to preventive health programmes (See Section 11.)

i: Sanitation and other environmental health measures. (See Section 11.)

j: Control of communicable diseases.

k: Prevention and rehabilitation of physical and mental handicap.

2 Lower Priorities

a: Mental health (curative services), because it is costly and difficult to implement and is often well catered for by traditional healers.

b: Dentistry, though important, is not as vital as other life-saving inputs.

c: Geriatrics. Senility, and the problems associated with this condition, is far less common in the Third World than in industrialised countries. Retirement is also uncommon. It is therefore more appropriate to address the health problems of the aged through a general PHC programme than to devise a specific programme to cater for their special needs.

d: Tertiary and postgraduate education. This is both outside the range of Oxfam's work and very expensive.

e: Hospital administration, salaries of hospital personnel and expensive hospital equipment.

7

There is enormous variety in the quality and type of health provision being offered by governments in different parts of the Third World. In some areas health care is provided for by the voluntary sector, in others the State has established an extensive health system. But it is all too common to find that large sectors of the population remain without any health provision whatsoever. There are three main variables which serve to categorise individual health programmes.

*a: **The purpose*** of the project — in other words, whether it is based on a generalised (integrated) or a specialised (vertical) intervention.

*b: **The scale*** of the project — in terms not only of the coverage, but also of staffing, capital assets, the technological input and the level of financing.

*c: **The control*** of the project — on the one hand, whether it falls within the primary or secondary health care system and, on the other, to what extent either the health service or the community controls the programme.

1 Independent Community Programmes

The defining characteristic of this category is that the group concerned (which for simplicity is termed community but may in fact be a trade union, cooperative, social club, etc.) has no special skill in health care. In this kind of programme the degree of community control will be high and the focus will be on auxiliary health workers recruited from the community. Local people will contribute their labour for the building of health posts and other structures; they will be in charge of administration and, in addition to paying for the drugs they consume, may even help fund the programme by paying a small tax. However, since supervision by professional services is likely to be minimal, especially at the outset, it is important that the programme should be planned on sound health principles.

This type of project will of necessity have a number of limitations. For example, without some form of laboratory, TB control will be impossible and immunisation will not be feasible without a cold chain. And yet even where there is little external support, a considerable reduction in the level of disease, malnutrition and mortality can be achieved with a number of simple measures. Examples of the activities most appropriate to this type of programme are: the improvement of water sources; simple first aid; general health education; training of village midwives; promotion and use of indigenous remedies and rehydration schemes. The main advantage of programmes run by the community is that they will probably cater more directly for the expressed needs of community members than programmes controlled by outsiders.

*a: **The input*** provided by the funding agency may take a number of forms:

☐ the supplying of an initial stock of drugs and equipment;
☐ the provision of materials for building a health post, channels for clean water supply, etc.;
☐ the funding of a skilled health worker for supervision of the programme;
☐ the provision of a system for referrals;
☐ general health education and training of health workers.

But it is important that provision be made from the outset for the community to assume the running costs of the programme. Thus, the scale of the programme must be commensurate with the economic resources available to, and the organisational capacity of, the community. Although in the short-term the funding agency's main function may be to help the community help itself, the long-term goal should be to encourage the community to demand of the government the basic health service to which it has a right.

b: The planning of community-based primary health care is a critical stage in the development of the programme. It is advisable to seek the help of a skilled health worker to undertake the community diagnosis and assess which measures would be most appropriate given the needs/circumstances of the community. While a PHC programme may function well using primarily health workers who have received a minimum of training, the planning will usually require someone with epidemiological skills.

c: Regular supervision and support will also help sustain the initiative and control the quality of the programme. Even where skilled health workers may not be involved in the daily routine of the programme, it should be possible to find a local health adviser who will make regular visits. Preferably, and for the sake of continuity, the adviser should be the person who undertook the initial community diagnosis. Only in exceptional circumstances, should field staff consider funding community primary health care in areas where there is no local health adviser.

d: The choice of a local health adviser can prove extremely difficult. He/she should have first hand experience of working in poor areas and an understanding of local socio-cultural conditions. In addition, a knowledge of epidemiological principles is vital, as is a sympathetic and understanding approach towards the people with whom he/she will be working. The skilled health worker should be seen as, and indeed should see himself/herself as, a servant rather than a master of the project. Further, it is important to understand that there is a wide variation in the training of skilled health workers and some will be totally unprepared for the task of health planning.

e: It should be borne in mind that *cost per patient* in such programmes is relatively high as compared to large programmes and therefore funding should only be considered in those cases where this is the only viable form of health provision.

2 Independent Health Units

This category includes a wide range of health units — from large hospitals running a traditional curative service to small groups of skilled health workers. These units may or may not have the support of the government, but they have in common the intention of establishing a properly structured and supervised community outreach system in countries where large sectors of the population are neglected by the state health service. The skilled practitioner undertakes to train and work with auxiliaries, frequently offering his or her services for little or no remuneration. While such ambitions are commendable and should be encouraged, it is important to bear in mind that there may be a number of problems with this kind of approach.

a: Frequently there is *little consultation* with the people the project is intended to benefit. For such a project to succeed it is vital that representatives of the community be incorporated at a very early stage in the planning.

b: Often it is the *curative aspect of medicine* which takes priority in these programmes — especially where a charge is levied for services rendered. In these circumstances the preventive/health education aspects of the programme may be very weak. This problem is often compounded by the expectations of the people the project is working with: the health practitioner will usually gain credibility more rapidly by curing patients than by promoting preventive health measures. A balance has to be achieved between curative and preventive measures so that the project will both gain the confidence of the people and improve their general health status.

7

c: If community outreach is to form only one aspect of the work undertaken by the health unit, then there is a risk that the programme will suffer because of the *competing demands on the personnel*. Therefore, a fairly senior person (a doctor, experienced nurse or health administrator) should be elected to take sole responsibility for the project.

d: Because this type of project depends on a strong element of supervision, and cannot necessarily expect state support, the *costs cannot be borne by the community*. Field staff have on occasion found that they are forced to continue funding for much longer than planned at the outset.

e: Ideally, independent health units should be able to use resources controlled by the state health service when necessary. However, in many countries the state system is under severe pressure and its employees are uncooperative, and thus the independent unit may be *forced to rely entirely on its own resources*. It may in fact be in a position to be able to assist or support local, state-run initiatives (dispensaries, training programmes, etc.).

In both the programmes run by the independent health unit and those run by communities, cooperatives and other groups which are not specialists in the area of health, the need for the exchange of information and support is considerable. Encouragement should be given to individual programmes to meet with others as frequently as possible, to exchange ideas and information regularly and, perhaps, even to plan temporary exchanges of personnel or to establish more formal links by founding an umbrella organisation (see PART THREE, Section 5) Programmes operating in fairly close proximity to others should be encouraged to combine their resources to undertake specific health interventions, such as tuberculosis control. The members of the Voluntary Health Association of India (DEL 008) have been collaborating with vertical programmes for a number of years.

3 Regional Health Programmes

This type of programme offers a primary health care system which normally functions as a part of the State health service. Certain governments, such as those of Tanzania, India and Ethiopia, are very much in favour of developing regional health programmes. In more centralised States such as Kampuchea and Mozambique this is often the only acceptable form of health intervention.

The regional health programme is usually organised into a geographic or administrative unit with fairly wide coverage. The service provided is comprehensive and incorporates both vertical and integrated programmes. The primary health care worker has access to a well-established system of referral and receives constant supervision and support. Indeed, one of the long-term goals of such a programme is to institutionalise the system of health care offered at the community level, although only in very few countries is this service free of charge.

Clearly, the regional health programme offers a number of advantages.

a: Even though the rate of progress may not appear quite as dramatic as with small-scale independent projects, the total number of people benefiting will probably be much greater.

b: The commitment to health provision is very long-term (or in most cases permanent), and therefore any improvements that the funding agency is able to facilitate are likely to continue having an impact long after funding has ceased.

c: Because of economies of scale the cost per patient will be much lower than in smaller projects.

d: The quality of health care should be fairly consistent from one region to another.

e: By combining the resources of the secondary and primary health care systems the programme is able to tackle a wide range of health problems, including those which would be too complex or costly for a smaller project.

However, there are also a number of problems with this type of programme.

a: The contribution made by the NGO is likely to form only a very small part of the total funding and it will therefore be hard to evaluate the effectiveness of that contribution.

b: The community served by the programme will have no control over the planning, administration or implementation. This may be a particularly serious problem in areas where there is extreme social and economic differentiation and perhaps a number of contrasting cultures and different languages. It would be hard for those who plan and implement the programme either to understand, or to cater for, the special needs of the more marginal groups (see PART TWO). The risk is, then, that many people will feel that the programme does not serve their real interests, and thus the attendance may well be low.

c: Although the NGO may play an important role, giving advice and support, ultimately its impact will be limited. This means that in many areas of health provision the priorities of the programme may not coincide with those held generally by the funding agency. For example, preventive measures may be given far less importance than curative medicine.

d: The overall scale and cost is such that in the end it should be the responsibility of governments and international agencies such as WHO rather than NGOs to underwrite such a programme. Therefore it is particularly important that before undertaking funding field staff should question whether there really is no alternative source of financing and whether involvement really can be justified.

4 Vertical Programmes

This category includes all those health initiatives which involve a specialised rather than generalised health intervention. In principle it is always preferable to support integrated rather than vertical programmes since they are more cost-effective. Clearly it would prove extremely expensive to set up a vertical administrative system for each and every disease endemic to an area and it could present a serious problem of coordination which at the very least might lead to duplication, but is more likely to result in serious discrepancies. However, there is a place for vertical programmes within a system of general health provision where specific health problems require specialist attention.

There are a number of practical issues field staff must consider when assessing a vertical health intervention.

a: Whether or not this is the most effective way of combating the health problem in question.

b: Can this intervention be combined with another similar one? For example, leprosy control can be combined effectively with a tuberculosis programme. Similarly, once a cold chain is established it is more cost-effective to immunise against a number of diseases rather than just one.

c: Should the specialised intervention be incorporated into a broader programme? For example, an appropriate context for immunisation would be a mother and child health programme.

d: It should be borne in mind that the vertical programme may be especially effective for dealing with those diseases commonly neglected by primary health care systems. The auxiliary health worker will tend to concentrate on the more urgent, immediate problems and may neglect those people suffering from more long-term ailments. If his or her work

7

can be coordinated with that of a specialist operating as a part of a vertical intervention, then a great deal of unnecessary suffering can be relieved.

⑤ Specialist Service Programmes

It is hard to be specific about the type of programme falling into this category, although they all have in common the fact that they service the health sector. The NGO's contribution in this area is bound to be fairly piecemeal and may often be very short-term since it may amount to no more than a one-off donation. There are no specific guidelines on the funding of projects servicing health care, although any such initiative should be consistent with the funding agency's general priorities in health provision.

Examples of this type of programme are: the provision of workshops for disabled people and Samaritan funds for hospitals; the facilitation of inter-project visiting; the provision of books, technical manuals, and teaching scholarships; the funding of seminars for, and training of, health workers and the manufacture of low-cost drugs. (For more information see PART THREE, Section 5.)

V Funding of health care

Health care can often be an expensive undertaking, more so, possibly, than any other development intervention. As a rule, the small independent project is less cost-effective than the larger programme receiving back-up and support from the State. Any programme costing more than £3 (in 1984) per annum per head of the total population covered will need special justification for funding.

Perhaps one of the main reasons why, despite the problem of cost, NGOs should continue funding the smaller independent programmes is that they often form the only source of health provision in an area and, precisely because they are so small and possibly so isolated, do not have access to the larger funding agencies or the state sector.

① The Alternatives

Ultimately, it is the duty of the State to provide health services, although for a number of reasons — not the least economic — in many countries the national system is hopelessly inadequate. The gap is filled sometimes by NGOs (including church-based groups) and more often by large international agencies such as those affiliated to the United Nations.

Long-term funding presents a problem for many agencies and most are reluctant to enter into an open-ended commitment. Whereas it may be feasible for the NGO to pay initial capital costs, or the costs of setting up a project (including training, etc.), or to provide funding for a vertical intervention, it is not practicable to become involved with the payment of salaries and other running costs in the long term.

Unfortunately, in practice many health programmes flourish in the early years of their life, but flounder at a later stage when on the one hand the NGO begins to withdraw its support, and on the other it transpires that the State is either unwilling or unable to take over the burden. Careful consideration should be given to the project's likely long-term aims and financial prospects. An expensive and elaborate programme which is unlikely ever to be adopted by the State and which is beyond the means of the local community can easily become a white elephant. The risk is particularly great with certain health interventions, such as vertical TB programmes, which require considerable external inputs (in this case vaccines, cold chain, continuing drug supplies, etc.).

2 Self-financing

It is argued by many that fees should always be charged (except in emergency programmes) for health services. This is not only to assist in the funding of the programme, but also to give it more credibility in the eyes of the community. However, in practice it is hard to apply this principle: people may be only too happy to pay for curative services — which are seen to be beneficial — but are less convinced of the need to pay for a preventive service, especially where this is delivered to the individual rather than to the group.

Payment for services rendered may be absolutely essential in small-scale programmes which are independent of the state system. However, it is not necessary to place all the burden on individual patients. There are various ways in which the programme can raise money.

a: Communal contributions. There are numerous ways of raising funds communally. For example, a village with access to sufficient land may work specific plots on a communal basis — cultivating cash crops. Alternatively, the community may undertake to maintain its health workers by cultivating their land or raising their animals free of charge. Clearly these options do not exist in urban areas, where other fund-raising techniques must be devised, such as organising social events (dances, raffles, etc.) for which a charge can be levied. In both rural and urban areas communal labour can be organised to build any structures required for the programme or a tax, payable (in cash or in kind) by all the members of the community, could be levied.

b: Sponsorship. In some areas it has proved possible to gain sponsorship for health posts and other capital assets from local businesses, etc. Contributions have also been made by private industrial concerns and agricultural cooperatives towards salaries and other running costs of local health services.

c: Contributions from patients. Many projects levy a small charge for each consultation (whether the patients are attended by a professional practitioner or by the local health workers). However, it must be accepted that with this system the burden will fall unevenly on the population; patients suffering from long-term illnesses, or from illnesses for which the treatment is expensive, will have to pay a great deal more than those who are relatively healthy. Besides, this kind of system will serve to promote curative and not preventive measures. Further, there will always be patients who are so poor that even the most modest charges will be beyond their reach: unless the community can mount some kind of special fund, they will be unable to seek consultation.

3 Payment for Drugs

On the whole, drugs should only be supplied free of charge in very special circumstances (such as emergencies, or with certain diseases like leprosy or tuberculosis). It is extremely difficult to start levying charges if people have become accustomed to being supplied with drugs without payment. Even in emergencies it may not be really necessary to supply drugs free of charge: NGOs can on occasion be guilty of creating among a people a dependence on free supplies simply by neglecting to examine carefully their true financial circumstances. Where drugs are donated, they can be used by the programme as a basis for establishing a revolving loan fund (see PART FIVE Section 4).

VI Health education

As a concept, health education refers to the enlightenment of the community as a whole rather than the training of health workers in preventive and curative medicine. There are a number of ways in which people may learn about health care. Even without the intervention of the health service many societies will have certain customs and practices which have a positive impact on their health status. Although it is important always to encourage health education, it has to be recognised that there are often long delays before results begin to show. Before they are willing to accept a change in practice, most people not only have to learn the facts, but also have to become convinced of the relevance of such innovations to their lives.

The speed of change will be influenced by many factors. Customary practice may be a direct cause of ill-health and yet be firmly entrenched in a culture. Health education to combat the practice of female circumcision, for example, is unlikely to show any impact for some years to come.

On the other hand, where the time-lag between a health intervention and its effect is short, and especially where that effect is fairly dramatic, the impact on people's understanding of health care may be fairly rapid. The time lag between treatment and effect with oral rehydration, for example, is very short. A good immunisation programme which demonstrably reduces the rate of mortality caused by communicable disease will be accepted rapidly by the population.

Traditionally health education has been a fairly restricted concept and in practice consisted of supplying posters and giving talks at health clinics and similar locations. However, the approach to health education has changed radically in recent years. Emphasis should be given to:

☐ counselling, involving the community in a dialogue rather than providing a one-way flow of information;

☐ the use of local forms of communication (such as songs and puppets or street theatre);

☐ where more sophisticated technologies (slides, video cassettes, etc.) are employed, they must be used in a manner appropriate to local conditions and the culture (see PART FOUR Section 3).

VII   Urban Health

One of the most important phenomena of the post-World War II period has been the massive expansion of the urban populations of the Third World (see PART ONE Section 3). In the past Oxfam has placed greater emphasis on health provision in rural areas than in urban zones on the grounds that in the latter the state health service is likely to have greater presence. However, as the cities of the Third World grow, it is becoming more and more obvious that despite the proximity of often fairly sophisticated health services in urban areas, access is denied to the poor who live in the slums. While the provision for casualties such as accidents and maternity emergencies may be fairly adequate, there is little reason to suppose that either the private or state health system is willing or able to tackle the less dramatic health problems of the poor.

1 Factors Facilitating Urban Health Provision

Doctors on the whole prefer to work in the city rather than in the countryside; the vast majority of highly trained medical personnel in the Third World live and work in capital cities. This is partly because the capital provides better career prospects, as well as a higher standard of living and more 'social perks' such as good education for the children and general entertainment.

The problem of facilities and supplies is less acute in urban than in rural areas; there are usually more hospital beds, and drugs and vaccines are more widely available. The whole question of availability of supplies can change the very nature of a programme. For example, while in urban areas there may be an adequate supply of rehydration salts in the treatment of diarrhoea, in rural areas people will most likely have to make up their own solutions. Thus, in the rural context the outcome of the treatment is far more dependent on the patient understanding and following instructions than in the urban context.

Another constraint which has far greater impact in rural areas than in the city is the problem of communications. It is much easier to establish a reliable and efficient cold chain in the urban environment for example. If necessary, supplies of vaccine may be distributed daily from a central store to health posts, and the wastage rate of vaccines will be much lower than in the remoter rural zones.

2 The Problems of Urban Health Provision

The presence of large hospitals and clinics with out-patient departments has, if anything, depressed the service as a whole in cities. While in many rural areas a complete vacuum has forced NGOs, voluntary practitioners, independent health units and communities to create their own health infrastructure, there have been very few initiatives of this sort in urban zones. Further, the sophisticated secondary health care facilities of the city tend to create expectations among the population which, given the limited resources, cannot be met. The poor, seeking to provide their own service, tend to select as their model the elaborate hospital out-patient complex or the private clinic.

Since in most cities there are as yet few community initiatives in health care, most people go to the pharmacist for advice and attention, only visiting the hospital in an emergency. This inevitably means that the curative rather than preventive approach to health will dominate.

The very feasibility of a community approach to health may seem questionable in many urban areas: in some sectors population turnover is high, in others family structures appear to have broken down. The complexity of the social structure may be such that it is hard to imagine that the community will ever be able to run its own health programme. The danger is, then, that all the effort will come from trained practitioners, the voluntary sector, etc., and the community will be no more than the passive recipient of a service.

Crowding, especially in inner-city areas, will usually lead to an increase in the incidence of communicable diseases. Where crowding combines with poor water supplies and inadequate sewage disposal, then mosquito-borne diseases like elephantiasis and certain haemorrhagic fevers are likely to increase. Whereas the NGO can do nothing about crowding, the provision of clean water and the safe disposal of sewage should be a priority. However, sewage disposal is extremely costly in urban areas, since the sewage has to be carried out of the city by cartage or some form of water-borne system. The problem of sewage is far less serious in the countryside, where the main aim is to encourage people to use purpose-built facilities (such as pit-latrines).

The urban dweller is less likely than his or her rural counterpart to have close ties with the land. This can be significant for health care in two ways:

a: Both adults and children are more likely to be dependent on cash earnings than in the countryside. The poor in cities tend to work in conditions which are both inhumane and dangerous. There are numerous health hazards arising from poor working conditions, often due to chemical pollution, noise, dangerous machinery, etc.

b: The evidence on levels of malnutrition in the urban environment is conflicting. It is often claimed that food prices in Third World cities are maintained at an artificially low level by politicians fearful of discontent among urban populations and therefore that food in the city is not a serious problem. However, the more remote from the land the people are, the less control they will have over their own diet; the more they will depend on purchased foodstuffs, and the more vulnerable they will be to increases in food prices and shortages in supply. Indeed, foodstuffs which have no nutritional value whatsoever will be far more widespread

7

in the city. At times of stress, during periods of high inflation,or when harvests fail, the urban poor who are locked into the cash economy are extremely vulnerable.

Especially in those cities which have developed more as a result of migration from the countryside than by a natural increase in population, the nuclear, as opposed to the extended, family predominates. The wider group of kin may lose contact with individual members, and group responsibility for weaker members is eroded. Therefore it is possible for individual families to suffer extreme hardship in a community which is generally quite affluent. Equally, large numbers of people may become detached from their families altogether, living off their own initiative in the streets.

Thus there are large numbers of people in urban areas who have special need of a health service; they are isolated from their families and often from the community as a whole, and are usually far poorer and more vulnerable than the majority. However, these people present an enormous logistic problem from the point of view of setting up a health programme because they are usually extremely dispersed geographically and socially. They may in addition be very mobile — working and sleeping in different parts of the city from one week to the next. Perhaps it has to be faced that even a solid community-based health intervention may not reach these people who are the ones most in need.

There is still much to be understood and much to be learnt about health provision in the urban environment. Urban health care has not been given the same attention as health provision in rural areas: the medical profession has as yet to develop a coherent methodology for dealing with the health problems of the urban poor.

| VIII | **Baseline study and assessment** |

Two important prerequisites for a successful programme are the formulation of clearly defined objectives and the provision of sufficient data to check whether these objectives are being reached. For objectives to be set realistically some form of *baseline study* is needed: this study is also useful for later evaluations, in order that changes can be observed.

The baseline study must examine the overall health problems of the area under consideration. The information required is summarised in Section 12, Checklist of Questions.

By establishing the basic statistical data on health it is possible to make the *community diagnosis* or in other words to discover what is wrong with the community as a whole (in the same way that a health worker diagnoses the problems of the individual and assesses their cause).

The main factors responsible for ill health are:
☐ poverty
☐ poor health provision
☐ lack of health education.

Having determined what are the most important health needs in an area, the next step is to assess already existing health services. Normally these will consist of formal state or voluntary sector services or the inputs of practitioners of alternative medicine. From this analysis it can be assessed which health problems are not being dealt with by existing services, and plans can be made to confront these problems. It is very important that such plans should involve *all* the existing health-related services as well as the community and its healers. Improvement of agriculture and health education should also be included.

In larger programmes this process is made much easier if the area covered is an administrative unit (block, district, region etc.). Wherever possible, a *regional approach* should be the aim: considerable time, effort and money needs to be spent on the coordination of health services in a region but will not be wasted.

The ideal programme should be partly the responsibility of trained health workers and partly that of the community. All those people involved in the programme should have clearly defined roles.

7

Section 2   Primary Health Care

I   The infrastructure of Primary Health Care

1 Support and Supervision

It cannot be overstressed that any PHC system must have adequate supervision and support. *However well trained the medical auxiliaries or health workers are they will achieve nothing if they are not regularly supervised and provided with essential drugs and equipment.* Any project (however small) seeking funding should therefore be seen in the context of the health service and infrastructure of the region or country as a whole.

Ideally, the basic unit of health care should consist of a person with some medical skill (midwife, trained health promoter, etc.) sited in each and every community. In the absence of such people, an aid post can be visited weekly by a medical assistant and/or nurse. If neither of these is possible, mobile clinics may be operated as an intermediary step. However, it has been found that mobile clinics are about four times more expensive per patient than static clinics using auxiliaries.

Dispensaries and *health centres* are larger units sited in population centres, from which smaller communities may be visited. They should be staffed at least by a medical assistant and a nurse. The health centre in particular should offer a fairly complete health service (apart from the more specialised functions); it should also carry out health education and perhaps do agro-nutritional work, in addition to servicing all the health workers and aid posts in its catchment area.

There is overwhelming evidence that it is a complete waste of both time and money to train and deploy health workers in a PHC system if provision is not made for community support. The community must help in the implementation of water, sanitation and other environmental programmes. Equally, the health service must do all in its power to ensure that the health worker is seen to be of value to the community and has an effective referral system for cases which are beyond his/her competence. It is therefore necessary to train nurses and other medical personnel in the supervision of health workers before training the health workers themselves. The medical specialists should take part in the training of health workers. Specialist health personnel must be encouraged to leave the clinic to supervise health workers in the community, and conversely health workers should assist with the health post work. A major role for the auxiliary is the *home visiting* of patients who have been seen at the clinic.

The planning and implementation of a primary health care programme takes much more time than one might imagine. Moreover, the pace at which it develops must be determined by the community rather than by any deadlines set by health personnel. This inevitably makes for problems in budgeting and targeting, but the potential long-term advantages outweigh the difficulties of this approach.

2 Transport and Mobile Units

There are two types of mobile service:
☐ those that supply a curative and some preventive service, e.g. for under-5s;
☐ those that supply a supervisory service for static units at the local level.

Obviously it is necessary to have transport linking the various parts of this infrastructure for the purpose of supervision and referral; but quite distinct from this are the mobile clinics which carry staff, supplies and equipment and function as self-contained units for treatment or vaccination. Oxfam has often funded this type of mobile service, and it usually consists of a single vehicle carrying a team of medical auxiliaries (nurse, midwife and a birth control, nutrition, or health education worker). There should be regular stopping places, often dispensaries or aid posts, where the team works for a day at a time.

A survey of several mobile medical teams in Tanzania showed that typically they would make three day-trips per week, working at the base hospital for the rest of the time. About half the stops made by these teams were at dispensaries, each one being visited weekly or fortnightly. Other stops would be made at villages with no static infrastructure, although local women were sometimes recruited to meet the teams at such stops and help with the unskilled work.

3 Problems of Mobile Clinics

While mobile clinics provide a valuable interim service, their limitations have become increasingly apparent since their inception in the 1960s . Too much time is spent in arranging routine servicing and the replacement of vehicles; too often vehicles break down, the schedule is disrupted and eventually frequently abandoned. Too much staff time is spent in travelling, costs are high and obviously there is no provision for permanent health cover. *The ultimate aim should therefore be to establish a static infrastructure providing the basic services continuously.*

4 Vehicles for Primary Health Programmes

Oxfam support for primary health programmes often takes the form of providing vehicles. The questions below should be considered.
a: What is the cheapest vehicle that will do the job? (Land Rovers are expensive and their 4-wheel drive facility is rarely necessary.)
b: Can the vehicles be *quickly* repaired and serviced when necessary?
c: What provision is being made for ultimately replacing the vehicle? Is there a depreciation fund? (Such a fund may be difficult to arrange because there is usually no income from the programme.)

In view of rising fuel costs and the cost of vehicles, NGOs cannot now supply conventional motor vehicles on the scale that was once possible. Wherever possible, *motorcycles* or *scooters*, *pedal bicycles* or even *animals* should be used. Health centres and other bases should be less widely scattered, so that use of these simpler forms of transport becomes feasible. A better service may result — excessive mobility can lead to a superficial approach.

5 Record Keeping

Normally two types of record keeping are useful.
a: Patient-retained cards. The advantage of these is that:
☐ the cards are available to any visitor to the home;
☐ the patient may attend another clinic or health centre, taking his/her card: the health personnel will be able to see the health situation immediately;
☐ self-retained cards assist the administration of under-5 clinics: the child can be weighed immediately on arrival and the result marked on his/her chart. If the record is only kept in the hospital, it has first to be traced and then removed. Frequently many people have the same name, so patients' records get confused and lost;
☐ it is easier to survey a community if the families are covered by home-retained cards;
☐ as responsibility for health is increasingly placed on the community, it becomes more important that records are available for use and monitoring by health workers.
b: Hospital-retained records. Hospital-retained records are useful in giving an overall view of the pattern of disease within the community, coverage of the service, etc. Data accumulated should be used as the basis for planning and evaluating the programme and for setting new objectives. This information is also useful for following up defaulters when possible. Special emphasis should be given to groups at risk, and

7

records of these people should be kept separately and revised regularly: these records can then be used to guide home visiting.

| II | **Training and function of auxiliaries** |

Auxiliaries can handle 80-90% of common ailments, so it is a waste of money, time and effort for either doctors or experienced nurses to be in full-time attendance at all clinics. However, there must be regular supervision, adequate back-up for referrals, and clear, simple standing orders.

1 Types of Auxiliary

The terminology regarding different types of medical auxiliary is confusing and is rarely used consistently. In some cases an auxiliary may be a technical worker with less than full professional qualifications. Thus, when a professional nurse has to undertake some of the responsibilities of a doctor, he or she is functioning as an auxiliary.

At the other end of the scale, there are primary health care workers, village health workers, health promoters or 'barefoot doctors', and single-purpose auxiliaries (who may have had almost no education, but have been taught a particular skill enabling them to help in specific situations). Some auxiliaries are paid a salary, others not; those that are may be paid by the community or by the central government. The complete spectrum of skills and functions may be classified roughly as below.

Professional medical workers:
☐ senior grade: doctors.
☐ para-medical grade: nurses, pharmacists, physiotherapists, laboratory technicians.
Auxiliary medical workers:
☐ medical assistants and enrolled nurses;
☐ dressers.
Auxiliary community health workers:
☐ health aides, village and community health workers, barefoot doctors, primary health care workers, health promoters;
☐ village midwives;
☐ single-purpose auxiliaries.

2 Community Health Workers

Oxfam has had experience recently of a number of health projects which have included the training and deployment of community health workers, and traditional midwives. While field staff should continue to give this kind of work a high priority, since it remains the only way to achieve a reasonable coverage of the population in many areas, nevertheless there have been more failures than successes in this field. The problems are almost always a result of poor preparation and the failure to plan in detail the long-term commitments of the programme.

In general, much more time and effort must be spent in discussing and planning *with the community*. It is pointless to start a project without the community's support and commitment. Key areas for discussion are:

a: Selection of health workers:
☐ What qualifications are looked for, e.g. literacy?
☐ Age and sex. In practice mature people who have had families tend to be more successful; a mixture of the sexes is probably ideal.
☐ Full-time or part-time?
☐ How will the views of the community be canvassed? The danger is that the health worker will be appointed as a vassal of the local hierarchy.
☐ Whether the health worker has to work independently in several communities, or to form part of a team, the ability to work with

others will be a major factor to look for. Problems may occur where the health worker fails to fulfil his/her role efficiently and the hospital or central authority wishes to dismiss him/her. Should this situation arise, the best way of overcoming the problem is to make the community aware of the shortcomings of the person of their choice, and to leave it to them to dismiss him/her and hire another worker.

b: Reward

☐ It is unrealistic to expect health workers to work without some financial reward. The status that the job carries is in itself unlikely to be sufficient reward, especially if more than a few hours' commitment is needed each week.

☐ The payment can either be in cash or in kind (see Section 1, V. for suggested forms of payment). Perhaps the ideal is a small salary paid by the community, which may be financed by the sale of drugs. The system employed must prove appropriate to local conditions and be approved and preferably initiated by the community.

③ Training

Courses need to be problem-orientated and, if possible, carried out in an in-service situation. They should not be theoretical or contain too much scientific background material.

a: Length of courses. Courses vary considerably in length, say from 10 days, backed up by frequent 7-day refresher courses (BUR 22), to 21 days (KEN 19) or as much as six weeks. In the Sudan Primary Health Care programme (SUD 60) a 9-month training is planned. Refresher courses, sometimes as often as one day per week, are a key element in the system.

b: Content of courses. Most courses should cover the main subjects in the primary health field, but they may vary according to local problems and the exact function of the health worker. For instance they may include: child-care, running pre-school clinics (KEN 19, IS 144); ante-natal care, birth control; nutrition; agro-nutrition and gardening (BD 20, BUR 22); TB, leprosy, malaria; vaccinations; hygiene and sanitation; first aid.

7

Section 3   Nutrition and Diarrhoea

I   Introduction

Malnourishment and diarrhoea are two of the most important causes of sickness and death in poor countries. Malnutrition and consequent infection lead to the deaths of 40,000 children every day. One child in four in developing countries is seriously undernourished. Poor nutrition also means that any illness is likely to last longer, be more profound and carry a greater risk of death than in a well nourished individual. Malnutrition is rarely the actual cause of death; usually it is one of the simple infectious diseases, such as measles, malaria or tuberculosis, which gives the final blow. There are five million deaths every year from dehydration, caused by diarrhoea. Most of these deaths could be prevented by Oral Rehydration Therapy.

Both malnourishment and diarrhoea are multifactorial and in different circumstances different causes may predominate. Unfortunately, the solution is rarely a straightforward technical one and usually depends on bringing about broad socio-economic changes. The problems of poor nutrition and diarrhoea have been further exacerbated in recent decades by the growth of population in the Third World, polarisation of wealth, migration towards the cities, overcrowding, inadequate food supplies, poor sanitation, inadequate safe water supplies and lack of access to health care facilities (see PART ONE, Section 3).

II   Diarrhoea

1 Prevention and Treatment

Someone who has four or more loose stools a day has diarrhoea. This can most effectively be tackled by introducing a number of general environmental measures, rather than simply mounting a programme of treatment. An 'integrated' approach to the problem would include ensuring access to adequate supplies of safe water, immunisation, sanitation measures and refuse control as well as the more traditional approach through health education. (See also Section 11.)

Diagnosis of the specific cause of diarrhoea may prove extremely difficult, both because many factors are involved and because the technology required for diagnosis is fairly sophisticated. Diarrhoea can be related to problems such as poor sanitation and hygiene and to overcrowding. It can be caused by parasites, bacteria or viruses which are often difficult to treat — especially where there is no change in the economic status of the community as a whole. Less commonly, diarrhoea may be caused by the presence of toxins or the inability to digest lactose (milk). Diarrhoea can result in dehydration and the loss of body salts, which is very dangerous, especially for children.

The best form of treatment is to prevent dehydration by replacing the salts and fluids which have been lost, using oral rehydration. This is fairly straightforward, inexpensive and also very effective. In Bangladesh oral rehydration has not only reduced infant mortality significantly, but also appears to have reduced the level of malnutrition, making the children well more quickly, especially by bringing about rapid weight gain. Other methods of treatment include intra-peritoneal infusions. These can be used if the child is unable to tolerate Oral Rehydration Salts (ORS) e.g. through persistent vomiting, or if the child is deteriorating despite treatment with ORS.

Oral rehydration consists of giving a simple mixture of water and salt by mouth to restore lost fluids and salts. This mixture is made either by dissolving a ready-prepared sachet of salts in water or, more cheaply, by adding finger pinches of salt and sugar to sterile (boiled) water. Sachets cost between 10 and 20 cents (U.S.).

The solution. Glucose (or sugar) is needed to promote the absorption by the body of salt solutions given by mouth. Oral rehydration fluids thus consist of water, salts and a sugar, preferably glucose.

Solutions vary from a simple salt and sugar solution to the more scientific, pre-packaged WHO mixture, which is then dissolved in one litre of safe drinking water.

Ingredients per 1 litre (standard formula)

Sodium chloride (table salt)	3.5 grams
Sodium bicarbonate (baking soda)	2.5 grams
Potassium chloride	1.5 grams
Glucose (anhydrous)	20.0 grams

It is important that the concentration of the salt is not too high. Table 7.1 summarises different OR methods.

Table 7–1   Methods of Oral Rehydration compared

Method	Ingredients	Comments	Person administering
1. Finger pinch of and four finger scoops of One litre of	Salt Sugar Water	Inaccurate but useful as it only needs a fluid measure	Parent or sibling
2. TALC spoon * (type 1) One litre of	Salt and sugar Water	More accurate: ? introducing a new technology? Currently being evaluated.	Parent or sibling + health promoter
3. TALC spoon * (type 2) One litre of	Salts – NaCl (Salt) $NaHCO_3$ (Baking powder) KCl (Potassium chloride) Glucose Water	A better salt solution but more difficult to prepare	Health Worker at a clinic

* available from TALC, Institute of Child Health, 30 Guilford Street, London WC1N 1EH, UK.

Besides measuring the amount of salt and sugar it is important to have a standard volume of water. This is usually 1 litre and it is necessary to identify a suitable 'local' measuring unit, e.g., soft drink bottles or cooking oil tins, etc.

Sachets are easy to use but at present the cheap brands are not sufficiently widely distributed.

7

Normal ranges for quantities of ORS. The amount of solution given depends on the weight of the child and the degree of dehydration: thus a child weighing 25 kg with mild dehydration will probably have lost 5% of body weight and so will need $5/100 \times 25$ kg (or litres) of solution $= 1\frac{1}{4}$ litres. This amount must be added to the 'normal' daily needs. Alternatively, Table 7.2 can be used as a guide;

Table 7 − 2   Quantities of ORS by age

Age	Volume in 24 hours
below 6 months	¼ − ½ litre
6 months − 2 years	½ − 1 litre
2 − 5 years	¾ − 1½ litres

The volume should equal the amount of stool passed, or be sufficient to satisfy thirst, whichever is the greater. It is important to continue feeding, especially breastfeeding, once rehydration is complete.

Progress in oral rehydration is most easily measured by weighing the child, an increase in weight indicating an improved condition.

Oxfam follows WHO guidelines in giving oral rehydration a high priority, but it must be realised that *a large input of health education is needed if OR is to be both safe and beneficial*.

In a minority of cases the child may not respond to oral rehydration. The comatose child must be referred for more specialist attention. Where epidemics are known to occur − particularly in urban areas − it may be necessary to establish rehydration centres which are well stocked with oral rehydration salts and have adequate back-up supplies of intravenous solutions and equipment. In this situation it is essential to look closely at the possibilities of adopting wider preventive measures.

III   Nutrition

Malnourishment is more commonly a result of a total lack of food (or in other words *undernourishment*) rather than a relative lack of one or more ingredients in the diet (*malnutrition*). Thus, the main aim of programmes involving severely undernourished populations should be to increase the overall calorie (or joule) intake. However, where malnutrition results from fairly specific causes, such as early weaning of infants or ignorance of dietary needs, health and nutrition education may take priority.

1 The Causes of Malnourishment

a: Poverty. In the Third World people spend a very high proportion of their real income on food and are therefore extremely vulnerable to fluctuations in food prices. Also, in poor areas incomes tend to be very unstable − employment often being seasonal − and large numbers of people exist for long periods with insufficient money to purchase basic foods. The low-income groups are unable to generate sufficient demand to allow normal market forces to come into effect and bring about an increase in supply. It is also much harder for the poor − who rarely own the land they work and live from − to obtain loans, advice or other inputs for improving food production (see PART ONE).

b: Food supplies. A number of problems can occur with the supplies of food in poor countries. In both rural and urban areas large numbers of people have no access to land, and the levels of food production are insufficient to meet their needs. In the agricultural sector there is often a lack of appropriate technologies, and crops may be grown for export rather than for local consumption. Certain of the staples consumed in poor countries are very bulky (i.e., high in volume relative to the number of calories) and thus present a particular problem for children who, because of their small stomachs, cannot consume sufficient quantities to meet their needs. This occurs particularly in areas where cassava (manioc) and matoke (cooking bananas) are the main staples, and most commonly appears as kwashiorkor. The problem is exacerbated by the fact that in many families meals are widely spaced, because of the demands of work and the level of poverty.

Supplies of food in many areas are severely reduced through wastage. As much as 25% of the food produced in the Third World is lost through poor storage facilities; it is damaged by moisture, insects and rodents. Because storage presents a problem in those years when the harvest is good, the surplus is sold to outside traders rather than being held locally as a reserve for use when there are shortages.

c: Ignorance. Malnutrition can often be caused by not knowing the dietary needs of the individual. For example, mothers are being encouraged by the propaganda of multinational corporations to bottle-feed their babies rather than give them breast milk. Further, many people do not understand the need to supplement breast milk in the infant's diet from the age of six months onwards.

2 Types of Malnutrition

The most common and important types of malnutrition are described below.

a: Protein energy malnutrition i.e., marasmus and kwashiorkor.

Marasmus is the more common and is due to insufficient food (not enough calories). In its early stages, marasmus is recognised only as a loss of weight. As it progresses, severe wasting occurs until the skin hangs loosely from the bones. This especially occurs around the buttocks, where loss of fat and muscle causes the skin to hang like 'baggy pants'. The eyes become sunken, causing the face to appear far older.

Marasmus is often seen in very young bottle-fed babies and in breast-fed babies over six months who are receiving no supplementary foods. Marasmus can be confused with dehydration. Very often children suffer from both.

Kwashiorkor. Swelling or oedema is the earliest and most important sign of kwashiorkor. The swelling usually starts in the feet and lower limbs but can occur all over the body, resulting in a swollen abdomen and typical moon-shaped face. The child appears miserable and apathetic and often has hair and skin changes.

Kwashiorkor is most frequently found in children recently displaced from the mother's breast by the arrival of a new baby. It can also follow acute infections such as measles and is endemic in certain areas where cassava and matoke (cooking bananas) are the main staples.

Marasmic-Kwashiorkor is a combination of the above two conditions.

b: Mineral and vitamin deficiencies, i.e. Anaemia.

The prevalence of anaemia in tropical climates is well known. It may well be the single most serious complication of pregnancy and certainly causes both a failure to thrive and increased vulnerability to disease in young children.

The most common causes are:

☐ nutritional, notably where iron and folic acid are absent in the diet, thus reducing the formation of blood;

☐ diseases which destroy the blood in the body (e.g. malaria), or cause loss of blood (e.g. hookworm);

☐ genetic. This is less important, except in certain parts of West Africa where sickle-cell anaemia is common.

c: Vitamin deficiencies. The most common are:

□ Vitamin A deficiency which can lead to nutritional blindness;
□ Vitamin B deficiency (beriberi) which if untreated can lead to paralysis or heart failure and death;
□ Vitamin B2 deficiency (pellagra);
□ Vitamin C deficiency (scurvy).

These specific deficiencies may need to be treated both at individual level and through improved public health. The best way of preventing such deficiencies is through an adequate diet.

The distribution of multivitamins is usually a waste of time and money as most of them contain only small quantities of individual vitamins.

3 Vulnerable Groups

a: Babies (4 months to 1 year) and *young children* (1-4 years) are especially vulnerable, common problems being early weaning on to unsuitable foods and unnecessary recourse to bottle feeding.

b: Pregnant and lactating mothers, who have special dietary needs. Those women who have frequent, closely-spaced pregnancies are subject to a repeated drain on their reserves.

c: Large, closely-spaced families have been shown to have a low calorie intake and to be more vulnerable to malnutrition.

d: Certain socio-economic groups are particularly vulnerable — for example, landless rural labourers and the urban unemployed. Handicapped people, the chronically sick and disabled are also at risk.

4 Methods of Combating Malnutrition

First the principal *causes* of malnutrition and *at-risk groups* must be identified. These groups may be the poorest section of the community or a certain age group within that community: often they are a combination of both.

Having identified the vulnerable groups, an *appropriate response* must be worked out with the community. It is important to realise that there is no panacea for malnutrition. Bear in mind the points below.

Traditional methods of tackling malnutrition have often proved unsuccessful, so it is especially important to keep an open mind to new ideas and techniques. Wherever possible the responsibility for maintaining nutrition should be laid on the community itself. The components of nutrition programmes can be broken down roughly as below.

a: Education: to improve marketing and buying practices; to achieve a balanced diet, especially for children, and to improve cooking, storage facilities, etc.; to encourage breast-feeding and discourage bottle-feeding.

Education can be offered at antenatal clinics, under-5s clinics, farmers' clubs, schools (via nutrition scouts or primary health workers) and at nutrition rehabilitation units, and women's groups.

b: Encouraging gardening: provision of seeds, tools and advice, especially with women's groups, and small-scale agricultural extension work. Wherever possible *local foods* should be used and people encouraged to grow more of their own food. The use of imported foods should be minimised because they may act as a disincentive to local agricultural production.

c: Emphasis should always be given to the *prevention* of malnutrition.

d: A high level of *community participation* in nutrition programmes is essential. *Home visiting* is vital for follow-up because any programme should be aimed at having an impact on the diet, methods of food storage and preparation, etc. in every household with malnourished people.

e: Income-generating programmes may be the most effective method for helping certain groups — especially those without land or stable employment. Those groups who are unable to produce their own food should be given help in identifying and purchasing cheap and nutritious foods.

f: Breast-feeding should be promoted both for the nutritional benefits to the child and the contraceptive benefits for the mother. Other methods of birth control and child-spacing should also be encouraged. *Supplementation* of breast-feeding from 6 months of age onwards should be encouraged, using local foods traditionally prepared.

g: Agricultural techniques Every effort should be made to *improve* these and especially the storage of crops. It is usually better to encourage people to grow more of their staple foods rather than to introduce new crops, although food enrichment may be necessary when the staple is deficient in protein (as with cassava) or the diet lacks certain vitamins. *Livestock projects* may be justified on nutritional grounds but in some circumstances the same resources would produce more food if used to grow cereals, legumes or vegetables. Converting plant food to meat or eggs can be inefficient. (See PART SIX Sections 6 and 8). Production of local vegetable protein (groundnuts, beans etc.) should be encouraged.

h: Welfare. There will probably always be some cases of severe malnourishment needing food supplements or hand-outs and this responsibility should not be shirked. Preferably any hand-outs should be in the form of food for work programmes, while looking at the possibilities of alternative community interventions. Supplementary foods can be supplied at clinics, schools, creches and Nutrition Rehabilitation Units. In *emergency feeding* programmes(see PART EIGHT, Section 3) the emphasis should be on the provision of the staple food rather than high protein supplements (except for the very young).

i: Inadequate food intake may result from a lack of food supplies, but other factors may be more important. For example, children — who predominate among those suffering from malnutrition — require only small quantities of food, and it is often not understood that their main need may not be for more food as such, but more frequent small meals of higher quality. In many families a 'pecking order' exists in which the newly-weaned and/or female children are given a lower priority than others. Malnutrition tends to peak in the second year of life, when weaning occurs, and special emphasis should be given to this age group in any nutrition programme.

j: Anaemia should be preventable where the diet includes reasonable quantities of meat and green vegetables and the intake of iron and folic acid is adequate. It should be remembered, though, that the iron content of breast-milk is inadequate after the child reaches 6 months. Where the diet is poor, iron can be given orally in tablet form — usually with the addition of folic acid. Prevention of anaemia can be ensured by certain measures such as the treatment of worm infections and the practice of malaria prophylaxis during pregnancy and in early childhood.

5 Practical Examples

a: Under-5s Clinics. Early warning of malnutrition can be obtained from under-5s clinics where each child is weighed and has a 'road to health' growth chart: *weight gain is more important than the absolute weight*. A decline in the rate of weight increase, or an actual loss of weight, are early signs of malnutrition. When this occurs the cause can be identified and the mother advised and helped. This may entail education, the distribution of food supplements, garden projects or a combination of all three techniques. Malnourished children can be encouraged to attend weekly or fortnightly until progress is found to be satisfactory. Wherever possible, home visiting should be included, because it enables the factors causing malnutrition to be better understood and thus prevented, or tackled earlier.

b: Nutrition scouts. These were tried by UNICEF in East Africa and at the Silveira Hospital, Zimbabwe in a variation of the Village Health Worker scheme. The scouts are local people trained to recognise malnutrition in its early stages. They are equipped with tape measures and QUAC sticks designed to check on the nutritional status of children by body measurement. They use bicycles to visit all homes within a 5-mile radius; they give advice to mothers of malnourished children and carry leaflets to reinforce their teaching. They can also refer serious cases to other workers and clinics. The low cost, home visiting and high coverage elements of this approach make it very attractive, and NGOs should certainly investigate it further.

c: Nutrition Rehabilitation Units (NRU). The term NRU has been used to describe several different types of institution, but in general it implies a residential facility attended by mothers and their malnourished children for a period of 3-4 weeks for the purpose of rehabilitating the child and educating the mother. The mothers of the rehabilitated child are expected on their return home to spread the message about malnutrition and its treatment. The key to success of an NRU is not whether children are discharged from the unit in good health and nutrition but whether they remain in this state. Unfortunately this has not proved to be the case in many NRUs. Constant return to the NRU by children can sometimes be avoided by ensuring that on discharge from the unit the child is referred to a local community health worker to ensure that he/she is not allowed to revert to a malnourished state. The difficulties of this should not be under-estimated in poor areas.

If an NRU is set up, baseline data on levels of malnutrition should be established first, in order to know whether the NRU has any real impact on nutritional levels in the community. The NRU should be established in simple surroundings as close to the home environment as possible, seek to use local foodstuffs, ORT, promotion of vegetable gardens where appropriate and should not attempt to care for the really sick child who should be referred to a hospital.

For a number of reasons NRUs are no longer high on Oxfam's priorities. Some of the criticisms levelled against NRUs are that the follow-up of children in the home is inadequate leading to a recurrence of malnutrition, that they remove the mother from the home or farm for a long time leading to disruption for the other children and the domestic economy, many vulnerable children are not identified by the NRU, and the NRU is relatively expensive per capita compared to general community health programmes which include nutrition as one of their concerns. Finally, the NRU may labour under the misapprehension that malnutrition is the result of ignorance rather than poverty.

d: Creches and daycare centres. These are provided for working mothers, especially in urban areas, and often both well-nourished and malnourished children are admitted. The children are both fed and given some form of education in nutrition . The advantage of this system is that it allows the mothers to earn and so afford more food. The disadvantage is that because the mothers are absent they do not benefit from the teaching, although older brothers and sisters may well benefit. Well-run day creches should be encouraged.

e: Mothercraft and mother clubs. Both in urban and in rural areas these form part of the community self-help approach and can be a good base for assisting vulnerable groups, if the group genuinely includes all mothers in the community.

f: Nutrition work among older children. While the first five years of a child's life carry the heaviest toll of disease and death, the older child should not be neglected. School programmes will only pick up those who attend, these usually being children from the wealthier sectors of the community. At present there is no established system for reaching those over-5s who do not attend school. This is short-sighted: *good nutrition is especially important for girls in the 8-12 age group* if short stature and subsequent obstetrical problems are to be avoided. The use of *Nutrition Scouts* would be one means of identifying these cases, especially since the Scouts will have contact with a 'family unit' rather than the 'age group' approach used at clinics.

Nutrition work in schools should include the points below.

☐ Education: nutrition and domestic sciences; agriculture; gardening and other food production; child-spacing and birth control where possible, through family life education programmes.

☐ School Meals: this type of scheme should have a *low priority* except in grave circumstances and should always be phased out over a reasonably short period to avoid dependence on outside funding.

Government and education authorities must be included in the planning and implementation of the scheme, and so must parents. Supervision and control (where possible, this should be local) is essential to prevent pilfering and unjust distribution.

☐ Screening by weighing, measuring and clinical assessment can usefully be carried out in schools, remembering the limitations — that probably those most at risk will not be attending.

☐ School gardens may be developed, usually for educational purposes rather than for providing food for the school as a whole.

g: Adult nutrition. This usually has a low priority except in severe, near-famine conditions when it is normally linked to food-for-work programmes. The exceptions are:

☐ lactating and pregnant mothers, who need 750 extra calories per day;

☐ anaemia in pregnancy. Anaemia is probably responsible for more problems in pregnancy than any other single factor. Treatment is therefore a priority, i.e. iron and folic acid, anti-malarials, anti-worm and adequate diet.

h: Domestic science & nutrition education for adults. At clinics, health centres, women's clubs, schools, etc., teaching should be as practical as possible using, for example, locally-produced visual aid material. Subjects to be stressed are:

☐ shopping, i.e. the best buys to provide a balanced diet with a limited budget: this is especially useful in areas where food is mainly purchased;

☐ growing vegetables for various minerals as well as protein and energy (see PART SIX Section 7);

☐ cooking to make fuller use of familiar foods and to demonstrate the acceptability of unusual foods; cooking stoves may be introduced with improved fuel consumption (see PART SIX, Section 4);

☐ food technology (PART SIX, Section 10) may be taught partly to demonstrate how foodstuffs are produced, partly to introduce new techniques which are usable in the home and sometimes to produce convenience foods which the project can sell to the mothers attending, e.g., dried vegetables (ZAM 22), cooking oil (ZAM 23) or macaroni (BRZ 102).

7

Section 4   Immunisation

I   Introduction

It follows that since Oxfam gives priority to preventive medicine, support should be given to immunisation programmes wherever they are found to be the most effective way of controlling or eradicating disease. Immunisation has an advantage over most other preventive interventions in that its impact can be measured. However, it should always be borne in mind that while the procedures involved in an immunisation programme — for example, the administration of vaccines — may be fairly straightforward, the logistics can be extremely complex. It is all too easy to spoil vaccines, especially by storing them at inappropriate temperatures.

Immunisation must not, of course, be viewed in isolation but as a part of the general preventive armament. WHO has launched a world-wide immunisation campaign (EPI = Expanded Programme of Immunisation), and before commencing with a programme the local WHO or UNICEF representative should be consulted. Under the EPI, loans are made available to national governments to assist funding, teaching material is distributed locally and advice is given on the setting-up and maintaining of a cold chain. National programmes have been mounted in various countries as a part of the EPI scheme, for example in Malawi, Zimbabwe, Lesotho, Pakistan and parts of India.

II   Programme analysis

1 Types of Programme

There are three main types of immunisation programme.

a: Emergency. Where the level of vaccination in a population is very low, health provision is inadequate and disease is endemic, there may be a place for a one-off mass immunisation (with repeats where necessary, at the appropriate intervals). The main priority in this type of programme is to achieve the maximum possible coverage, often using mobile facilities with portable refrigerators, etc. Since emergency programmes of this kind concentrate entirely on immunisation and are only short-term, they usually result in a high level of community participation and are relatively easy to organise. However, this approach will only prove effective in the long-term control of the disease if, once a fairly extensive coverage has been reached, immunisation is then incorporated into an integrated programme of health care.

b: Integrated. In areas where there is a viable health structure, immunisation should be incorporated as an integral part of the programme. This approach aims at the long-term reduction of disease but presents certain logistic problems and can be less efficient than an emergency programme. The main difficulties are sustaining the supplies of vaccines and maintaining the cold chain over a long period. Human error can be critical in this type of programme. The coverage will tend to fall off with time, and therefore it is important to check the project regularly.

c: Refugees/displaced people. In regions with large concentrations of refugees or displaced people, mass immunisation can be integrated as a part of the registration procedure.

2 Programme Structure

Immunisation lends itself to large-scale schemes, and an essential feature of any immunisation programme must be a high degree of centralisation. It is virtually impossible to maintain a cold chain in the absence of clear procedures and effective control at all levels. This is particularly true of long-term programmes. Where immunisation is integrated with other health initiatives in a well established infrastructure, health workers can meet regularly for reporting and discussion. In Gujarat, for example, a large-scale voluntary sector programme successfully combines primary health care with immunisation in addition to manufacturing low-cost high-quality basic drugs for local distribution (GUJ 220).

Clearly, since the delivery of vaccines is expensive and extremely time consuming, it is far more cost-effective to provide immunisation against a number of diseases rather than to concentrate specifically on one.

With the exception of very few countries (for example, India) all the vaccines used in the Third World have to be imported. Field staff should only attempt to import vaccines directly when all other possible channels have been exhausted, because it is difficult to avoid spoilage, the bureaucracy tends to be very complex and the whole procedure is very expensive. It is preferable to use existing supply systems maintained by the State or by UNICEF/WHO. UNICEF, in addition to providing vaccines free of charge, also supplies refrigerators. A third alternative is to make an arrangement with a local private hospital.

3 Programme Assessment

Before starting a programme, the questions below need to be asked.

a: Is immunisation being used in conjunction with other efforts to remove the disease reservoir (for example, the provision of clean water, health education, TB control, under-5s clinics, etc.)?

b: Are vaccines available locally?

c: Do facilities exist for the safe transportation, storage and refrigeration of vaccines? It is essential to establish a cold chain in order to transfer the vaccine intact from the manufacturer to the patient, and all the links should be checked thoroughly. Do the health workers understand the different storage requirements of the different vaccines at each stage in the chain?

d: Is the equipment and staffing adequate for effective and easy administration of the vaccines? Is the sterilisation of instruments adequate?

e: Has a reliable system of recording the immunisation been worked out? For example, 'road to health' charts can be used effectively for the under-5s.

III   Immunisation and vaccines

1 Priorities

Measles. Immunisation against measles should take the highest priority in most poor countries because this disease has a particularly high mortality rate. Until recently dilution was recommended because this was in the past one of the most expensive of vaccines. However, since the price has now fallen considerably, dilution is no longer considered necessary. A single dose can give complete immunity. Care should be taken to avoid metal syringes, as they will spoil the vaccines.

Polio. Polio is a major cause of disability and mortality in poor countries and thus should also be a high priority in any immunisation programme. The vaccine is very cheap. While a single dose can give immunity, it is usual to give three or more to ensure success.

Tetanus Toxoid. In the case of this vaccine, immunity is achieved through more than one dose. The timing of inoculation is critical in the gradual building up of immunity, and this can present certain logistic problems. It is particularly important to vaccinate pregnant women to protect the child against neo-natal tetanus.

TB and BCG. BCG vaccination should normally be integrated into a

7

wider health intervention, for example MCH, a general immunisation programme or a TB control scheme.

DPT (diphtheria, tetanus, pertussis or whooping-cough). DPT alone will not justify an immunisation programme. It requires more than one dose to achieve immunisation and the spacing between inoculations is critical.

Priority should be given to programmes which incorporate all the basic vaccines rather than just one. The possible exception to this rule is measles, which can be sufficiently beneficial to justify a specialised vaccination programme.

With certain diseases, such as cholera and typhoid, immunisation is less important than other preventive measures. The improvement of sanitation and water supplies is more effective and economical in the long-term control of these diseases than is immunisation. Thus cholera, yellow fever, typhus and TAB inoculations should only be funded in exceptional circumstances, such as an emergency.

2 **Individual Vaccines for General Use**

Table 7 – 3   Vaccines

Vaccine	Storage	*Route & dose	Type of jet injector to be used
BCG for TB	Refrigerator (3-5°C)	intra. 0.1 ml	Pan-jet
Measles	Refrigerator (2-5°C)	subcut. 0.5 ml	Port-o-jet
Polio live Sabin	Refrigerator (2-5°C)	oral 3 drops	n.a.
Triple antigen (DPT) for diphtheria whooping cough tetanus	Room temperature or better in refrigerator	Intramusc. or subcut. 0.5 ml	Port-o-jet

*subcut. (subcutaneous) means that the vaccine is injected *under* the skin; intra. (intradermal) is *into* the skin. Polio vaccine is dropped on to the tongue – or on to a sugar lump. Some types of DPT vaccine are given by intramuscular injection to avoid local reaction.

N.B. (a) A new, more heat-resistant measles vaccine, which can survive 37°C for up to a week, is now available.

(b) The 'dead' Salk Polio vaccine may also be used in the future.

3 **Methods of Administering Vaccine**

Syringes and needles are still the methods most commonly used, but under certain circumstances, the improved jet injector or gun may be appropriate. The jet penetrates the skin without the use of a needle.

The intradermal technique has an advantage over the intramuscular or subcutaneous route because only one-fifth of the normal dosage need be used. This could be especially helpful in situations where supply is a problem. However, intradermal injection requires considerable skill.

a: Advantages of jet injectors:

☐ rapid administration, especially by the Port-o-jet;

☐ low cross-infection, reducing the risk of hepatitis.

b: Disadvantages:

☐ need for good maintenance and regular checking for accuracy of dose;

☐ limited range of applications, especially the Pan-jet;

☐ cost, especially of the large types (e.g. the Port-o-jet).

c: Types of jet injector or gun

☐ the Pan-jet (an updated version of the Dermo-jet) is about the size of a pen torch and can only put 0.1 ml intradermally.
Cost in 1984: £129;

☐ the Port-o-jet (the American version is the Ped-o-jet) is versatile and about the size of a suitcase. It can put up to 1.0 ml subcutaneously and vaccinate up to a thousand people an hour. Ideal in crowded communities and camps, but not in small clinics.
Cost in 1984: £937.

4 Target Populations

Priority should be given to the following groups in all immunisation programmes:

a: children under one year — especially for BCG, DPT, polio and measles;

b: pregnant women and women entering child-bearing age for tetanus toxoid.

5 Typical Schedule of Immunisations for Children

a: Neo-natal period. BCG vaccination. Babies seen with an unhealed umbilicus in unhygienic circumstances should have anti-tetanus serum. Issue record cards and inform parents of the need for subsequent immunisations.

b: 1-2 months. First DPT and polio vaccinations. BCG if not already given — otherwise check whether the neonatal BCG was successful.

c: 2-3 months. Second instalments of DPT and polio. Check previous BCG vaccination.

d: 3-4 months. Polio. Ensure that all immunisations have been recorded.

e: 8-9 months. Measles vaccine. Warn parents about reaction (7-10 days) later.

f: Finally, a booster dose of DPT and polio may be given when the child is 18 months to 2 years old. All immunisations should have been recorded on a suitable card at each stage.

6 Benefits of Measles Vaccination

The severity of measles is well known to health workers in those areas where protein-calorie malnutrition is a problem. Not only is it a killing disease with a case fatality rate of between 5-20% for hospitalised children, but it also predisposes children to many months of ill-health. It is often the precipitating factor in the development of kwashiorkor; a third or half of all children with this disease have had measles in the preceding few months. Measured in cost/benefit terms, measles vaccination is usually regarded as the single most effective and worthwhile health measure.

At present, two types of measles vaccine are available; the old type which was highly sensitive to temperature and the new one which is more tolerant of heat. Always check which type you are ordering and make sure that you are familiar with the handling instructions. If in doubt, follow the instructions given below and presume that the vaccine is of the sensitive strain.

a: Age of immunisation. It is generally agreed that immunisation should not take place earlier than 9 months because before that age, maternal antibodies would interfere with the immune response. On the other hand, immunisation should occur before 18 months as this is the peak period of incidence.

b: Price of vaccine. The price of measles vaccine has in the past been very high but recently has begun to fall. One must expect to pay about 20p for a full dose.

7

The cold chain is the system used for the storage and distribution of vaccines. The main purpose of setting up a cold chain is to ensure that the vaccine is administered in a potent state. Once spoiled, the potency of a vaccine cannot be restored, and live vaccines, such as polio, BCG and measles, are especially vulnerable to heat.

The cold chain depends on certain key elements.

a: Vaccine supplies. The quantity of vaccine held in stock will depend on a number of factors, primarily:

☐ the expected interval between delivery of supplies;

☐ anticipated coverage of the programme;

☐ calculated wastage rate (the acceptable wastage rate varies enormously and depends on geographical factors, population density, number of health workers etc. The wastage rate is usually higher in isolated rural areas than in densely populated urban zones).

It is essential that supplies should be kept up and careful records taken of the stock.

b: Personnel. The cold chain will be broken if there are insufficient people to ensure the safe distribution, storage and handling of supplies, and if those involved are not well organised and fully informed as to the requirements for each vaccine at each stage in the chain. The health workers involved must know how to:

☐ assess requirements (in terms of equipment, transport and vaccines);

☐ distribute, store and administer vaccines;

☐ care for vaccines and equipment;

☐ monitor the programme (keep records of the number of immunisations and stock levels and check expiry dates of vaccines).

c: Equipment. Appropriate equipment is required for the safe storage and transport of vaccines. Different items of equipment may be needed at different stages in the cold chain.

Important points to consider when choosing equipment:

☐ refrigerator or freezer. A choice has to be made between an upright (front-loading), or a chest (top-loading), model. The former gives easy access to vaccines and takes up little space, but has a short cold-life when power fails and is not very economical to run. The latter has a long cold-life, can be tightly packed, and is relatively cheap to run, but occupies more floor space and does not give easy access to vaccines. The equipment must be of sufficient capacity to enable the storage of vaccines in trays or boxes, and in addition to allow space both for the diluent and for the circulation of air. It is important to choose a make which is inexpensive and to check that spare parts and fuel are available locally. The ice-making capacity must be sufficient to meet requirements.

☐ cold box. The cold box must have effective insulation, be well sealed and resilient. Cold boxes are used for collecting and transporting large quantities of vaccines and will vary in terms of their capacity and cold-life. The choice of size and type should depend on the expected use and on the number of doses required. Always use a cold box which is larger than the estimated requirement to allow for delays in distribution.

☐ vaccine carrier. A smaller container or possibly a plastic bag may be needed while dispensing vaccines, and this should have the same characteristics as a cold box. If the container is too small to hold ice-packs, then loose ice should be placed around the vaccines.

d: Storage temperatures. The cold chain must be maintained up to the point where the vaccines are administered. Therefore, it is important to choose the quickest and safest route for distribution and to select an immunisation site which is both cool and shady. The vials should never be left outside the container for any length of time and should always be protected from direct sunlight. Vaccines must be used quickly and therefore should not be diluted until all patients are present.

Each vaccine has different requirements when it comes to the safe storage temperature. Make sure that the containers are packed correctly, to ensure the vaccines are stored at the correct temperature. *DPT and tetanus toxoid must never be frozen*. Freezing can be prevented by placing an insulating material between the ice and the vials. On the other hand, polio, measles and BCG will not be damaged by repeated freezing.

If at the immunisation point the ice melts, then measles and polio vaccines should automatically be thrown away. Other vaccines will still be potent so long as the ice has not been melted for more than a day.

Potency testing is an extremely expensive procedure and is only appropriate where the supplies are very large. However, all WHO and UNICEF vaccines are supplied with a cold chain monitor which is sensitive to temperature and will facilitate temperature control at all stages in the cold chain.

e: Wastage. It is impossible to eliminate wastage entirely, but faults in the supply system or problems of organisation and administration can increase the loss rate considerably.

The main causes of wastage are:
☐ the failure of the cold chain (breakdown of equipment, problems of transport, etc.);
☐ damage to vials;
☐ expiry of vaccines;
☐ failure to use all the vaccine from an open vial.

In order to avoid high wastage rates the programme must be well planned from the outset, health workers should meet regularly to exchange information and monitor their work, and the equipment should always be handled with great care.

7

Communicable diseases combined with malnutrition are the main causes of death and ill health in poor countries.

This section is not intended to cover all the major communicable diseases but outlines those which are of particular importance or present particular problems to the NGO. As a general guide we recommend Benenson's *Control of Communicable Diseases in Man*. (See Section 14, Resources)

In any area the community diagnosis should establish the main communicable diseases and in each case look at the feasibility of:

☐ treatment
☐ prevention
☐ control

Ideally there should be input into all three aspects: this is possible and highly desirable for diseases such as tuberculosis (which is the reason for the degree of detail on that disease in this section). However the control of some of the diseases such as onchocerciasis (river blindness) and schistosomiasis (bilharzia) is usually outside the range of the NGO.

Certain diseases should, for a number of reasons, be given priority in primary health care. *Tuberculosis*, for example, is important both because of the incidence and the severity of the disease. Each year 10 million new cases of tuberculosis develop world-wide, and as many as 3 million people die. If these people had access to a health service they could be cured and prevented from passing the disease to others.

Leprosy has also been identified as taking priority in primary health care, not so much because of the incidence of the disease, but more because it is extremely debilitating and in many areas carries a social stigma. WHO estimates indicate that between 11 and 15 million people in the world suffer from leprosy. Yet only 50% of these people are registered, and of this 50%, few attend clinics regularly or for sufficient time to receive adequate treatment. The incidence of leprosy varies considerably from country to country, and possibly the highest concentrations are in India, where there are an estimated 3.2 million cases.

Leprosy and tuberculosis are distinguished from many other serious diseases because they can be controlled with a well-organised health intervention, even without effecting a change in the economic status of the sufferers. Despite this, little financial provision is made for tuberculosis control in most health programmes, and very few health workers are trained to tackle leprosy, even in areas where the disease occurs fairly frequently.

Another disease which can cause high levels of mortality in areas where it is endemic is *malaria*. It is especially prevalent in the lowland, humid tropics. Children tend to be particularly susceptible to the disease, and in malaria endemic areas the mortality rate among children may be as much as 10%.

Eye infections are exacerbated by dry, dusty conditions and are particularly common where water supplies are polluted. In Afghanistan 30-70% of the total population is affected by trachoma — the incidence varying from region to region. Epidemics of infective conjunctivitis can occur, as in the refugee camps of West Bengal in the years 1971 and 1972. Problems with eyes are often associated with malnutrition and with other communicable diseases such as measles. The more severe infections frequently cause blindness. (See also Section 9.)

Intestinal parasites are also common in areas with a poor water supply and where the level of personal hygiene is low. Both parasites and eye infections can be prevented by undertaking certain fairly straightforward health measures such as providing a clean water supply.

Many of the serious communicable diseases require special inputs in the primary health care system. The control of leprosy and tuberculosis, for example, necessitates the use of laboratory microscopy. The primary health worker will need special training to enable him or her to identify and treat communicable diseases, and the community as a whole should be taught about the most important methods of prevention and control. Where the primary health care structure is weak or where the proposed intervention is fairly costly, it may be necessary to organise a specialised *vertical* programme to tackle a specific disease.

II The goals

[1] Control of Disease

Preventive health care aims to control or eradicate disease in a population. To control a disease is to reduce its incidence (i.e. the number of new cases) in a population. This form of intervention normally entails a high unit-cost, but is usually more effective than treatment. Unfortunately — except in the case of smallpox — a comprehensive immunisation policy is unlikely to eradicate most diseases. Therefore, in the majority of cases one is limited to maintaining the incidence at the minimum possible level (i.e. controlling it). Where immunisation is not possible, the thrust of the health intervention should be the identification and treatment of infectious cases and their carriers so that they cannot pass the bacteria on to others.

In many cases treatment is both long-term and extremely expensive and drug resistance is common. Therefore the health service must make a long-term commitment to the work. Field staff must be sure before undertaking to fund a programme of this sort that the indigenous health service will eventually be able to take over the financial burden.

The incidence or prevalence of a disease can be controlled by:
- [ ] identifying and protecting individuals;
- [ ] reducing the reservoir of the disease (in the animal, insect and human populations);
- [ ] educating people as to where, when and how diseases are transmitted so that they may take preventive action;
- [ ] initiating other environmental measures (e.g. reducing the breeding ground of mosquitoes).

[2] Disease Groups

WHO has outlined six *disease groups* which need special attention: they are
- [ ] malaria
- [ ] trypanosomiasis
- [ ] leishmaniasis
- [ ] schistosomiasis (bilharzia)
- [ ] filariasis (e.g. onchocerciasis and filarial elephantiasis)
- [ ] leprosy.

This is *not* a comprehensive list, nor does it necessarily include the most important infectious diseases in any group: notable omissions are all the causes of respiratory disease, both bacterial and viral. Causes of diarrhoea are also excluded because WHO has a separate programme to cover this.

Trypanosomiasis is a group of parasites causing several diseases, the most important being sleeping sickness in Africa and Latin America. Sleeping sickness is transmitted by the tsetse fly and exists in two forms, *Rhodesiense* and *Gambiense*: *Rhodesiense* can be carried by other

7

vertebrates like the wild buck (and is therefore a zoonosis). In Latin America, *Trypanosoma cruzii* (Chagas' disease) is carried by the reduvid bug which lives and breeds in the cracks of poor homes. In general the treatment of the disease is clear cut and is aimed at the parasite. Control emphasises the elimination of the insect vector. Principles of vector controls are changing rapidly (e.g. the new conical traps for tsetse fly), and field staff are advised to check for up to date advice.

Leishmaniasis (Aleppo, Delhi boil, Uta, Espundia, Chiclero) exists in many forms, either affecting internal organs (visceral) like the liver (e.g. Kala Azar) and the skin (e.g. *Cutaneous Leishmaniasis* or *Muco-Cutaneous Espundia*). Once again the diseases are caused by a parasite transmitted by an insect (e.g. the sandfly or *Phlebotomus*) and control measures consist of destroying this vector. Treatment is not very effective and often involves dangerous drugs. Prevention usually involves measures to stop the fly biting: to this can be added inoculation of certain cutaneous forms. Some forms of leishmaniasis are self limiting and heal themselves. Leishmaniasis is carried by many wild creatures, usually rodents.

Schistosomiasis (bilharziasis) exists in three forms, *Japonicum* and *Haematobium*, which attack the bladder, causing blood in the urine, and *Mansoni*, which affects the bowel, causing bloody diarrhoea. *Japonicum* is carried by other vertebrates (e.g. pigs), while *Haematobium* and *Mansoni* are not. *Mansoni* is common in irrigated areas and should be considered in all agricultural projects which involve flooding of any sort. The life cycle of all forms of schistosomiasis is complicated and involves various types of snail intermediate hosts.

Adult worms develop in the blood vessels of the abdomen and excrete eggs either into the bladder or bowel, which then are passed in the urine or faeces. Curative drugs exist. However, prevention takes higher priority and involves avoiding water likely to be infected (usually indicated by the presence of the relevant type of snail). Control involves breaking the cycle (e.g. administering moluscicides to kill snails, or constructing latrines to prevent egg-laden urine and faeces reaching snails). Attempts have been made to reduce the egg load in the human population by mass treatment. Probably a continuation of all these methods is needed to reduce the prevalence of this disease.

Filiarases are a form of worm carried by various flies. The most notorious diseases which they cause are onchocerciasis and elephantiasis. Onchocerciasis is carried by a minute black fly, *Simulium damnosum*; microfilarial tend to affect the retina of the eye causing the characteristic features of river blindness. Treatment is difficult, especially in the advanced stages, as drugs which kill the microfilarial may result in further damage to the retina. Controls involve eliminating the black fly which likes to breed in fast-flowing streams.

Infectious elephantiasis exists in two common forms, *Wuchereria bancrofti* and *Bruggia malayi*: both types are carried by mosquitoes (e.g. *Culex fatigans*) which breed in contaminated water. For this reason the disease is especially common in polluted urban areas. As urbanisation increases the disease is likely to become more common. Treatment is adequate if taken in time, but prevention and control depend mainly on improving the environment and destroying the breeding places of the mosquitoes.

III

The policy and practice of tuberculosis control

1 Tackling the Disease

Tuberculosis is caused by the tubercule bacillus. Where it is common, one person in every hundred may have active disease. Tuberculosis is a disease of *poverty*, its prevalence being related to malnutrition, the frequency of other debilitating infectious diseases and of course regular exposure to infectious people.

Pulmonary (chest) TB, which is transmitted by the infectious ('open') case to others who are vulnerable, is extremely common in poor countries. Bovine and, to a lesser extent, avian TB is common mainly in areas where people drink a lot of milk and raise livestock. Bovine TB is controlled by pasteurising milk, slaughtering infective animals and undertaking continuous surveillance of the animal population. However, the cost of this procedure is enormous, and it is therefore not possible to give it a high priority in the work of NGOs.

The emphasis in any TB control programme should be on tackling the disease in the human population. *Pulmonary tuberculosis is best controlled by identifying and treating infectious cases, thereby ensuring that they cannot pass the disease on to others*. It has been estimated that each infectious, or 'open', case will, if left untreated, infect between 12 and 15 others.

The diagnosis of pulmonary tuberculosis is best made by looking for the bacillus with a microscope. The use of the Xray in diagnosis is less reliable and more expensive. In all tuberculosis programmes it is essential to provide simple but effective laboratory services. Sputum analysis is necessary to:

☐ make the initial diagnosis;
☐ follow up patients being treated (to check that they have ceased to be infectious);
☐ ensure that the patient is cured at the end of treatment.

2 Treatment

Treatment involves administering a relatively long course of drugs (between 6 and 18 months, depending on which of the various drugs available is used). In order to prevent the bacillus from becoming resistant, the treatment must always involve more than one drug.

The shorter courses of treatment (i.e. 6 months) are about ten times more expensive than those which are more long-term. However, the main advantage of the short-term treatment is that compliance on the part of patients is likely to be high (in other words, few will drop out before the course is completed).

Another problem with the shorter courses is that they involve the routine use of back-up drugs such as rifampicin which are normally reserved for treating cases which are resistant to the first-line drugs. If the programme is not well organised, there is a risk that resistance will also develop to these back-up drugs, leaving the population with no protection against the disease, which may then spread unchecked. The details of which drugs should be used, and for how long, vary according to local circumstances. Where a government policy exists on the control of tuberculosis and/or a standardised regime of drugs, this should be followed. In the absence of such guidelines, a local health adviser should be consulted.

It must always be made clear that the drugs used in the control of tuberculosis should never be used for other infections. Rifampicin in particular should only be used under close medical supervision in conjunction with another drug.

The treatment of tuberculosis should be on an out-patient basis, since for a number of reasons it is no longer felt to be necessary to place sufferers in sanatoria. Most patients cease to be infectious after only

7

three weeks of treatment. Besides, long-term segregation in sanatoria frequently destroys the patient's family life and, where the patient works for an income, may also seriously affect the domestic economy. Internment will also tend to increase the stigma associated with the disease. Treatment for tuberculosis can be either on a daily basis or intermittent (e.g. twice weekly). Where patients have to travel for long distances to receive treatment or where the health provision depends on mobile facilities it may be more convenient to arrange to see patients intermittently.

3 BCG Immunisation

It has been questioned whether or not BCG immunisation really is effective in protecting people against pulmonary tuberculosis. Most medical experts accept that it does protect children against the lethal blood-spread (miliary) varieties which commonly cause meningitis and bone disease. BCG clearly, however, is not as effective as case-finding and treatment in the *control* of the disease. Ultimately it is the open cases in the adult population rather than the sufferers in the child population who transmit the disease.

To be cost-effective, a BCG immunisation programme should be combined with a general immunisation programme and/or identification and treatment of infectious TB cases.

For maximum protection, BCG should be given to all children at, or shortly after, birth. In these circumstances previous tuberculin testing is unnecessary.

4 Control Programmes

The *diagnosis* of pulmonary tuberculosis can be quite simple: it involves identifying the bacilli in a sample of sputum using a microscope. Non-pulmonary forms are harder to diagnose and usually require a specialist, but the treatment can easily be carried out by a primary health worker. In larger programmes a more sophisticated *laboratory* is needed to act as a centre for the training of staff, and to provide facilities for culturing tubercule bacilli and testing their resistance to drugs. It will be necessary to form a team to implement the control programme which will work to a rigidly defined schedule of vaccination, diagnosis and drug therapy.

Tuberculosis control lends itself ideally to the concept of primary health care. The primary health worker has an especially important role to play, both in identifying cases and in ensuring the continuity of treatment.

Health education plays an important part in the control of tuberculosis, and the representatives of local groups (women's organisations, community leaders, welfare committees, etc.) should be encouraged to participate in the programme. It must be recognised that, even though for maximum impact the programme should concentrate on treating open cases in the adult population, to achieve a high level of community participation it may be necessary to put a great deal of effort also into the immunisation of children. *The failure of tuberculosis control programmes can usually be attributed to poor planning, a lack of motivation and compliance rather than drug resistance*.

The programme should aim to achieve a maximum possible coverage in as wide an area as possible.

Effective communications and extensive record keeping are vital elements in the successful programme. There must be a streamlined system for the collection of sputums, for testing them in the laboratory and then passing the results back to the health worker. It is essential to maintain records for each patient, especially since treatment is long-term and patients may move from one area to another, becoming the responsibility of a different health service.

5   Funding

The community should not be expected to pay for a tuberculosis control programme. This is especially true in areas where the incidence is not very high, because it becomes fairly costly to identify and treat infectious cases.

Treatment should be free, or at least subsidised, because it is long-term and because continued attendance is crucial. A patient who has to attend a health post or clinic daily for up to a month is less likely to do so if he or she has to pay for the service.

Case identification and treatment is more cost-effective than BCG immunisation.

Hospital treatment may cost as much as twenty times more than treatment provided on an out-patient basis.

6   Other Respiratory Diseases

As in rich countries, simple respiratory diseases are one of the commonest causes of death and ill health, especially in children. Pneumonia, bronchitis and sore throat caused by bacteria and viruses are extremely common. Unfortunately (with the exception of whooping cough — see Section 4) these diseases, cannot be prevented except to some degree by commonsense measures (such as keeping babies away from people who are ill). As with other infectious diseases, attention should be given to good nutrition and maintaining high fluid intake. Antibiotics are useful in certain bacterial infections (in pneumonococcal lobar pneumonia) but the bulk of respiratory diseases can only be treated symptomatically.

IV   The control of leprosy

1   The Cause

The organism causing leprosy (bacillus bacterium) is similar to that which causes tuberculosis. With the lepromatous (the more serious and extensive form of the disease) sufferer the bacillus can be obtained from the skin or nasal mucus. The organisms are excreted into the environment mainly (and some believe almost exclusively) from the nasal mucus. As with tuberculosis, only a very small proportion of those exposed develop the disease.

The form the disease takes is very variable. Some patients will only have one lesion on the skin which contains no leprosy bacilli, whereas those suffering from lepromatous leprosy may have numerous lesions and millions of bacilli in the nerves of the skin, the nasal cavity and many other parts of the body. Approximately 80% of those who suffer from leprosy fall into the former category and are described medically as non- or pauci-bacillary cases. These cases virtually never transmit the disease to others. Their level of resistance to the disease is high and they respond well to relatively short courses of drug treatment. By contrast, the lepromatous sufferer has a poor resistance to the disease and will require much longer period of treatment. If left untreated, the lepromatous case will excrete vast numbers of bacilli into the environment.

2   Diagnosis

Unlike tuberculosis, leprosy is fairly easy to diagnose. The diagnosis can be established in the majority of cases by a clinical examination carried out by a trained health worker. The skin of the patient may be scraped, and the pulp obtained can be spread on a slide for examination. The presence of leprosy bacilli in the smear will confirm the diagnosis. However, it should be stressed that such a positive diagnosis will only be possible with lepromatous or near-lepromatous patients. In all other cases, the diagnosis of leprosy will be entirely reliant on physical examination (which should include testing the skin for loss of sensation — a vital symptom of the disease). Xrays play no part in the diagnosis of leprosy.

7

3 Prevention

At the present time there is no vaccine for leprosy because it is not possible to grow the organism *in vitro* in the laboratory, and therefore, as with TB, the only preventive measure is to find and treat infectious cases. WHO is conducting research which may lead to the development of a vaccine, but it may take as long as ten years before this becomes available for general use. Drugs such as dapsone have been used for prophylaxis (given to relatives and other close contacts of known leprosy cases in order to prevent them developing the disease) in some parts of the world. However, this practice is generally impracticable and unrealistic and is dangerous if overdosage occurs.

4 Treatment

The treatment of leprosy is with drugs (all of which are given by mouth) — dapsone, rifampicin, clofazimine (also called lamprene or B663), and either prothionamide or ethionamide. One of the most important developments in leprosy control in recent years has been the set of recommendations presented by WHO, *Chemotherapy of Leprosy for Control Programmes*. The study recommends the *multiple-drug approach* and gives guidelines on the way in which the drugs should be used for all cases of leprosy, from those with high resistance and localised lesions, to those with little or no resistance and generalised lesions. The periods of treatment range from 6 months to 2 years, after which time it is advised that treatment should be stopped and the patient kept under observation. The regimens consist of a minimum of two drugs; patients with the type of leprosy with few bacilli (pauci-bacillary) receive daily dapsone and monthly rifampicin; patients with the type which has many bacilli (multi-bacillary) have daily dapsone, daily clofazimine and monthly rifampicin. In certain circumstances, ethionamide or protionamide may replace clofazimine in the latter regimen.

Whilst these regimens are more expensive and difficult to administer and supervise, they are of much shorter duration than those recommended previously. Furthermore it must be recalled that approximately 80% of all leprosy cases are in the pauci-bacillary group, so that they could be 'cleared' from the treatment programme in a remarkably short period of time. This would leave time and money for the intensive detection of new cases and the full treatment of those with multi-bacillary leprosy.

5 Constraints and Difficulties — the continuing failure to control leprosy

Many of the points described above for TB apply also to the leprosy problem. However, for leprosy there are additional factors such as fear, superstition, prejudice, social stigma, and the traditional isolation of cases, often in remote and medically inefficient leprosaria. Many of these problems have still to be recognised and overcome. Health education is of particular importance in the case of leprosy and should help patients and the community as a whole to understand the need for sufferers to present themselves for diagnosis and treatment. Additionally, because the leprosy bacillus attacks the nerves, many cases become paralysed or blind and lose sensation in vital areas such as the hands and feet. Disability and deformity requiring institutional care, physiotherapy, reconstructive surgery and rehabilitation are further constraints on the control of this difficult disease.

6 Oxfam Policy

Patients should be detected, diagnosed, classified and treated as out-patients. In-patient treatment (especially in remote leprosaria, without adequate medical advice) is not recommended, except in exceptional cases with treatment reactions, disability or grave social problems.

Multiple drug therapy is the mainstay of the attack on leprosy and should be implemented as soon as possible.

Adequate standards of medical care are essential if treatment is to be effective and safe; in many countries training or retraining may be necessary before the most expensive and potentially toxic of the drugs listed above are used.

In the absence of a vaccine for leprosy, BCG campaigns should be given every encouragement, since there is some evidence that, apart from its undoubted benefit in the protection of individuals against TB, the vaccine may also protect some communities from certain forms of leprosy.

Whilst support must clearly be given to humanitarian activities concerned with deformity, disability and blindness and the care of the rejected outcast, it bears repetition that the chain of infection will be broken most effectively by the proper use of multiple drugs for the treatment of the bacillary infection.

In this context, the NGO can probably do most good by encouraging the dissemination of practical information about leprosy, and by ensuring as far as possible that all grades of medical staff, including medical students, are properly instructed about leprosy control. (Since 1981, Oxfam has been actively involved in the assembly and distribution of a pack of 24 documents for leprosy teaching and training.)

On the question of funding: multiple drug treatment should be administered free of charge because it is expensive and, in most cases, fairly intensive and long-term.

V	**Eye diseases**

1 Causes

There are five main causes of preventable blindness:
☐ trachoma — usually with secondary infections;
☐ vitamin A deficiency — xerophthalmia (see Section 3);
☐ river blindness;
☐ leprosy;
☐ other infections.

Conjunctivitis (pink eye) is also common. There are several kinds, some caused by bacteria or virus infections, some by irritation due to dust or a broken eye lash, and some occurring as one of the symptoms of trachoma. In the population at large, poor personal hygiene and infrequent washing of the face are major contributory factors.

2 Oxfam Practice and Policy

In addition to routine involvement in PHC programmes, Oxfam has supported mobile clinics which work mainly in rural areas — Afghanistan, Kenya, Pakistan and Uganda. The eye conditions encountered most often by these clinics and the usual priorities in treatment or prevention are:

a: trachoma — prevention, or early curative treatment to avoid risk of blindness; the prevalence of trachoma is closely related to the availability of water for washing: the disease is most common in arid areas during the dry and dusty seasons. Once the infection is present, early treatment with an antibiotic eye ointment can prevent damage to eyelids. Once the eyelid is damaged, it may cause abrasions on the eye itself which lead to further infection and ultimately to blindness: a simple operation performed by paramedics can repair the eyelid;

b: conjunctivitis — treatment and prevention;

c: opaque cornea — prevention by attacking infections;

d: senile cataract — restoration of sight by surgery;

e: glaucoma — relief of condition by surgery (iridectomy).

River blindness (onchocerciasis) is common in West Africa, but is usually outside the scope of mobile eye clinics. Effective prevention depends on treating the patient before the eyes are affected, or on eliminating the fly which carries the disease.

7

There are several possible approaches to problems of blindness.

a: Medical treatment of trachoma and conjunctivitis.

b: Medical and surgical work; suitable for cataract, glaucoma and entropion operations in the field (entropion is distortion of the eye-lid, usually due to trachoma.)

c: Preventive work; mass treatment and health education in anti-trachoma campaigns.

d: Eye camps are held in countries such as India, where doctors sometimes give their services free for one year.

e: The health education component and simple treatment of trachoma are well within the capability of the primary health worker.

VI   Malaria

---

1 The Cause

Malaria is caused by a parasite: there are four types of malaria, of which the most lethal is *plasmodium falciparum* or malignant malaria, which often attacks the brain.

Malaria is transmitted to humans by the bite of various types of anophelene mosquito. The mosquito is also necessary for the development of the parasite.

2 Treatment

Malaria is usually treated using one of the quinine analogues, chloroquine (nivaquin) being the most common. There is as yet no vaccine against the disease.

In many parts of the world the parasite is becoming resistant to chloroquine and therefore alternative drugs have to be used. Sometimes it is, unfortunately, necessary to fall back on quinine: this has many more adverse side effects than chloroquine.

3 Prevention

Protecting the body from mosquito bites, using nets on beds and the screening of houses as well as simple practices such as smoke rings and other insect repellants. Breeding sites near houses should also be destroyed (see below).

Chemo-prophylaxis or the taking of preventative drugs: this is usually recommended weekly for expatriates, but a monthly dose can be very beneficial in preventing the development of anaemia and giving some protection to young children and pregnant women. Pregnant women are particularly vulnerable, and it has been shown that, in addition to causing abortions, malaria greatly reduces the birthweight of children, which in turn increases the risk of sickness and death.

4 Malaria Control

Control programmes should attempt to:

☐ eliminate the presence of the parasite in man by mass treatment (chemotherapy);

☐ attack the anophelene mosquito by house spraying with insecticides and by the destruction of breeding sites with the use of larvacides and the filling in of puddles where practical;

☐ protect houses (screens, nets, etc.);

☐ use biological techniques such as the breeding of sterile mosquitoes, encouraging mosquitoes to bite animals rather than humans, and breeding fish to eat the larvae, etc.

The problem with using sprays is that the mosquitoes tend to develop a resistance; they are also toxic and expensive. Finally, many buildings are not very suitable for spraying, for example, mud-brick dwellings.

The destruction of breeding sites by the community (for example with dynamite) is an interesting alternative, especially where the mosquito is not too vigorous, as in countries like Sri Lanka. This also has the advantage that communities can do the work themselves with less technical support.

Simple treatment for malaria should have a high priority in health programmes. This treatment can be given by paramedics and even primary health workers.

Partial protection of young children and pregnant women should be part of any MCH programme in malaria endemic areas.

Field staff should be prepared to experiment with community based environmental control measures such as those carried out in Sri Lanka. Anyone with a swingeing fever (in areas where malaria is endemic) should be treated as for malaria.

VII Intestinal parasites

Worms and other intestinal parasites are common throughout the Third World. They act as a constant drain on the body's resources, causing diarrhoea, preventing the proper utilisation of food in the body, and making the individual susceptible to other infectious diseases. Hookworm in particular can cause severe anaemia and has a very high incidence in certain populations. For example, it was estimated during the recent crisis in Kampuchea that as much as 90% of the population suffered from hookworm. The most important of the intestinal parasites are *roundworm, hookworm, schistosomiasis (bilharzia; see above),* *amoebas* and *giardaiasis.* All of these, with the exception of hookworm and bilharzia, are spread by the eggs which are lodged in the faeces of one individual being eaten by others. Bilharzia and hookworm have more complicated life-cycles and penetrate the human body through the skin.

Intestinal parasites are not only a problem of poor countries, since most children in the wealthier industrialised countries also suffer from worms of one sort or another at some stage in their lives.

It is important to treat individual cases of worms with the appropriate drugs especially where heavy infestations are causing other problems (for example, where hookworm is causing anaemia). However, the only way of having any real impact on the level of disease caused by intestinal parasites in an area is by improving water supplies, sanitation and personal hygiene (see Section 11). With certain parasites, other measures can also be effective; the incidence of hookworm, for example, can be reduced by the wearing of shoes, since this particular parasite enters the body by penetrating the feet.

VIII Sexually transmitted diseases

Sexually transmitted diseases are rampant throughout the Third World. The diseases most commonly encountered are lymphogranuloma, syphilis and gonorrhoea. Also prevalent are serum hepatitis (type B), scabies and pubic lice. Unfortunately, some of the more common sexually transmitted diseases are becoming resistant to the standard treatments.

Although programmes aimed at combating sexually transmitted diseases should fit well into the general primary health care system, certain special demands may arise.

a: The degree of training and the level of supervision of health workers must be high if treatment is to be effective and if resistance to the disease is to be avoided. It is vital that all health workers should be aware of this risk.

b: Those health workers who are involved with the prevention and treatment of sexually transmitted diseases must understand the need for extreme tact and sensitivity in dealing with patients. They must be well

7

prepared for the task of explaining to people how the disease is spread and how to avoid giving it to others. It may be useful, when planning health education, to emphasise the link between venereal disease and infertility.

1 Prevention

Health education with special emphasis on teaching people about the need for responsibility in sexual relationships is an important preventive measure.

The community can be protected by the control of prostitution and the promotion of barrier methods of birth-control.

Facilities should be provided for early diagnosis and treatment — services should be both accessible to all members of the population and *discreet*.

Contact tracing and follow-up are important, including possibly the screening of women in early pregnancy.

2 Treatment

The main form of treatment is with large doses of long-acting penicillin. Alternatively, tetracycline may be used. Extended post-treatment observation and follow-up is necessary.

Compliance with treatment is a frequent problem since some injections have to be given daily.

Treatment can be integrated with other curative measures within the primary health care system. However, in areas where incidence is particularly high, it may be more appropriate to establish a vertical programme, with specialised clinics and follow-up and outreach work.

The need for a vertical programme controlling sexually transmitted diseases may be great in an area where there has been a major social upheaval (such as a war, occupation of territory, the displacement of large numbers of people). The prevalence of sexually transmitted diseases will most likely be very high where there are undisciplined troops and where rape is common. However, a control programme should not even be considered until some degree of normality returns to the area. Where there is a high level of promiscuity in the population, there is little point in introducing a curative service, because the treatment will probably result in the bacteria becoming resistant to drugs. In these cases, curative measures should form part of the rehabilitation process.

Under no circumstances should programmes dealing with sexually transmitted diseases be linked with birth-control centres. It is common in health programmes throughout the world to tie the two services together: this can be extremely detrimental to both services. Especially where the concept of birth-control is new, if it is linked in people's minds with sexually transmitted diseases, it may be seen as totally unacceptable.

I   Mother/child health

1 Background

The health care of young children is the joint responsibility of the health service and the community. Except perhaps in the case of immunisation, the health service is unlikely to alter markedly the pattern of disease in young children without the help of the community.

Children under five and pre-school form approximately 20% of the total population in the Third World. Infant and child mortality often account for 50% to 70% of total deaths. Causes of death among the new-born are associated with the techniques of delivery, the levels of nutrition and health care of pregnant women, inadequate birth hygiene and care of the new-born baby. Many mothers also die as a result of these problems. Deaths among older infants and young children are the result of *common infections* (e.g. measles), mostly simple to treat or prevent, combined with *poor nutrition*, weaning practices and other problems such as dehydration and diarrhoea.

Programmes aimed at improved child health should be linked to other efforts at improving the general living conditions. It is instructive to note that in New York the halving of infant mortality between 1900 and 1930 had little to do with improved medical services, but was due mainly to improved environmental conditions in the home (water and sanitation, improved housing, etc.), better hygiene and better nutrition. (See Sections 3 and 11.)

2 Essential Components of MCH Programmes

The MCH programme is essentially a monitoring or screening service aimed at the section of the population normally most at risk. Although MCH programmes are involved primarily with the individual, there should be a strong emphasis on health and nutrition education at the group or community level. MCH services should be strongly linked to social developmental programmes involving women in particular and also to environmental sanitation projects. MCH programmes should include the following elements:

☐ antenatal care;
☐ maternity services;
☐ post-natal care;
☐ comprehensive child-care (preventive and curative, road to health charts, promotion of breast feeding, use of ORS, etc.);
☐ nutrition programmes;
☐ immunisation;
☐ training of auxiliaries (especially traditional birth attendants);
☐ home visiting;
☐ planned parenthood, birth control.

Some of these services may be organised separately, e.g. immunisation, but as far as possible, any vertical programme involving mothers and children should be integrated with the general MCH service.

a: Clear objectives. An MCH programme should have stated aims in terms of the number of people it will hope to benefit and what it hopes to achieve in those populations. Objectives must be based on those priority needs which are amenable to treatment. The target for achievement can be defined in terms of indices of childhood nutrition, incidence of preventable disease, mortality rates, etc. Some baseline information must be available from which to measure the achievements,

7

and the programme should include a means of evaluation — preferably a system using statistics gathered daily. Collection of data on births and deaths should be undertaken where possible. It must be recognised, however, that some achievements are difficult to measure: e.g. degree of community participation, involvement of women, etc. (See PART THREE, Section 4, IV. and PART FOUR).

b: Target levels of care. These are the stated objectives, which should be clearly identified. For example, a target of 60% coverage of under-five children might be attempted, with the aim of seeing these children on average six times a year. For maternal care, 80% of mothers might be the aim for ante-natal care, with a minimum of three visits per mother.

c: Wide ante-natal outreach. It will be many years before it is possible to provide supervision by qualified personnel at all births. However, a wide outreach in terms of ante-natal care can help with the identification and referral of the mothers at risk, and it can also prevent certain common medical problems associated with pregnancy and childbirth, such as anaemia, malaria and tetanus of the new-born.

d: Health and birth control instruction. Education in health, nutrition and birth control is an important element in all ante-natal services. It is the ante-natal clinic which provides the most useful environment for the introduction of concept/techniques etc. of birth control within the MCH framework.

e: Traditional midwives. In many countries, these women will for many years to come continue to perform the majority of the deliveries. An integrated, well-run maternity service can ensure that only normal deliveries are carried out in the home, that the local midwives' knowledge is supplemented and that they are provided with basic, simple equipment. They should also be assisted by training to identify unforeseen complications early. They should be encouraged to have confidence in, and a close relationship with, the qualified staff, rather than trying to work in competition with them. These women can also act as valuable agents in promoting birth control and identifying families in need of planned parenthood services, but only when they are themselves convinced of the advantages of such a programme.

f: Comprehensive care of children and Under-5s clinics. This is such an important service that it is discussed more fully below. The four main activities involved are:
- ☐ weighing;
- ☐ teaching;
- ☐ immunising;
- ☐ treating and advising the sick.

g: Mobile clinics, rural clinics, auxiliary personnel. Studies have shown that to be utilised effectively, MCH services must be within 12 miles of the home. For outpatient care people are usually willing to travel a longer distance — up to five miles — but they are rarely willing to travel far for preventive care. It is especially difficult for pregnant women with small infants to walk long distances. It is therefore essential to provide services close to people's homes. This can be best achieved by a health worker living in the village and receiving occasional visits from supervisors. To reach those most at risk, often the poorest who are the last to seek treatment, it is advisable that routine contacts be made in the home. If records are kept of all families under care, it will prove relatively simple for even a part-time health worker caring for, say, 100 families, to visit each one every 2-3 months (2 families per day). Auxiliary MCH workers can be used to train, supervise and support the health worker.

h: Community involvement. Without the participation of the community, a MCH project will have little real impact. Health committees may be formed, and through them, MCH can be integrated with other activities including food production and those cottage industries which involve women (although it is important not to increase the workload of women, see PART TWO, Section 2). Participation may involve training local people as part-or full-time health workers, stimulating the community to undertake cooperative activities such as providing children's midday feeding, or day-care centres, or providing volunteers to assist with monitoring or the distribution of antimalarials, birth control supplies or TB drugs, etc. Part of the purpose of an MCH programme is to enable the community and its leaders to understand the underlying causes of ill-health and to take action to improve the health of the population.

3 Pre-school or Under-5s Clinics

The activities which usually take place in such clinics may be categorised as below.

a: Weighing and the use of growth charts. The early detection of poor nutrition is achieved by the regular weighing of each child and by recording weights on a calendar-type growth chart. When growth starts to falter — an early sign of malnutrition — the mother should be advised on feeding and child care by individual counselling and if necessary by the provision of food supplements. It is preferable that the mother herself should keep her child's chart. Visits to the clinic should be backed up by regular *home visits* by the health worker.

b: Immunisation. See Section 4. Also include malaria prophylaxis if malaria is endemic in the area.

c: Health and nutrition education. This may include cooking demonstrations, group teaching/discussions, the promotion of vegetable gardens and the raising of small animals, and individual counselling of mothers. Home visits may also be made to individual mothers. Information may be gathered on currently available produce/local food shortages and fluctuations in food prices.

d: Treatment of the sick child. At health units, out-patient care for children should include both preventive and curative measures and should be integrated with the under-5s clinics. This is facilitated by using the weight card for all curative and preventive visits. A child needs *comprehensive* care, and it is misleading to distinguish between the 'well' and the 'sick' in poor countries. Many mothers only present themselves at a clinic with their sick children, but *all* children need the preventive services of the under-5s clinic as well. Such clinics should therefore *not* be run just as 'well-baby clinics' — an illusory concept — but should provide a truly comprehensive service for children of this age group. Many of the activities of the under-5s clinic can be carried out effectively by primary health workers. One of the most important and simple techniques is to undertake *oral rehydration* of those sick children with diarrhoea. Oral rehydration should be available through all clinics. It allows a potentially serious condition to be treated early, cheaply, and at home.

e: Nutrition Rehabilitation Centres and Malnutrition Clinics (Section 3). These can be regarded as an extension of the under-5s clinic, to which children found to be seriously malnourished can be sent, usually accompanied by their mothers. Discharge from a specialised unit of this kind should involve referral back to the under-5s clinic.

4 School Health Services

Because children in the school age group are healthier and less at risk than those under five, this service is less of a priority than other MCH services, but should become more important as education and health services develop. However, one of the main problems is that in many countries, the children who do not attend school tend to come from the less prosperous families, and this should be borne in mind when developing a school health service. School health services could include:

☐ health and nutrition education, plus broadly based sex education (when allowed by parents);

☐ school feeding (see Section 3);

☐ immunisation;

☐ screening children for parasitic infections, tuberculosis, leprosy, etc.;

☐ weighing and measuring to detect malnutrition or other disease.

5 Payment for MCH Services

Accurate costing of MCH programmes is extremely difficult and seldom takes into account the costs to the patient such as travel, time consumed etc. However, as a rough guide, an outpatient visit should not cost more than 50p. Generally, the use of mobile clinics and trained personnel is more expensive than static units with voluntary or low-paid primary health workers.

Some experiments have been carried out with charges for MCH which may in certain circumstances prove effective. In one example, the pregnant mother is expected to pay a lump sum entitling her to both ante-natal and maternity care. When she has completed a specified number of visits, the bulk of the fee is returned. Alternatively, a higher fee is demanded for delivery without antenatal care than for delivery following antenatal care. For under-5s clinics a registration fee for the weight chart may be charged, and any subsequent visits be free. Sometimes fees are charged for treatment but not for services. Primary health workers may be voluntary or supported by the community in various ways; initially, assistance may be required especially with training and ongoing supervision, but emphasis must always be on community responsibility.

6 Child to Child Programmes

The 'child to child' technique was first put into practice during the International Year of the Child (1979). It is customary in many cultures for older brothers and sisters to take responsibility for their younger siblings. The child to child programme aims to build on this practice and use it to encourage older children to learn about, and concern themselves with, the care and well-being of the younger ones. Since the concept was first developed, more than 1,500,000 children have been introduced to child to child activities of one sort or another.

Oxfam funds have been used to support child to child 'health scout' activities, and this type of approach can prove extremely useful in promoting the idea of primary health care or supplementing traditional health care services. Children can be particularly effective in identifying cases of malnutrition, recognising the symptoms of major communicable diseases and measuring eyesight and hearing.

II · Birth control

1 Introduction

No other area of health care has proved as controversial as that of birth control. Particularly in recent years, the issue has become the subject of conflict between competing political and religious interests. For the most part, traditional family planning programmes have been instigated by governments at a national level — primarily as a measure designed to combat the drain on limited resources which has resulted from massive population growth in the post-World War II period. Indeed, the statistics on the population of the Third World can be extremely daunting; if the present trend continues, many of the countries in which NGOs work will double their populations in the next 30 years.

However, many national programmes have encountered strong opposition. Most have failed to reach their targets; some have been abandoned altogether.

One of the main criticisms has been that programmes designed to control population growth do not address what is probably the main cause of poverty in the Third World today — the unequal distribution of wealth and income (see PART ONE). It has been argued that, by claiming poverty is caused by the sheer size of populations, governments, international agencies and the wealthy elites of the world are abnegating their responsibilities and deflecting the debate away from the main issue, which has more to do with the distribution than the quantity of resources. In many countries, for example, the problem is mainly a political one; the land may be fertile enough to feed the whole population, but because it is controlled by the few and because export crops take precedence over food crops, the majority of people remain hungry.

National population programmes are likely to prove insensitive to the conditions obtaining at the local level and to ignore the real needs and wishes of the people they claim to serve. *Most people will choose to have small, well-spaced families only when they have sufficient economic and emotional security not to have to depend on their offspring.* The large family is part of a sound economic strategy, and those children who survive beyond the age of eight become crucial members of the workforce controlled by the household. Smaller families can suffer extreme hardship, and a family planning programme is ill-considered which is directed at people whose income is unstable, who have no control over their land, their home or their job and cannot expect to receive a pension or social security benefits.

It is all too common to find that the prime objective of a family planning programme is to attend to the maximum possible number of couples or to distribute a given number of contraceptive devices within a specified period of time, the stress being on technologies and statistics rather than on the quality of the service. This means that there can be very little time for consultation. Certain methods of birth control may be promoted without full explanation being given of their effectiveness or, equally important, possible side-effects. Thus, couples are advised to use methods which are totally unsuitable or may in certain circumstances prove harmful. Above all, few of the standard family planning programmes have taken full account of local cultures or social structures. *Extension workers and development planners usually fail to recognise that only in a very limited number of societies do women have control over their own fertility: while advice and research on birth control is by convention directed at women, the actual decisions about family size are usually taken by men.*

It is becoming increasingly clear that the very term 'family planning' is inappropriate. It implies that the aim is to reproduce worldwide a western, idealised model — the nuclear family based on two children. Clearly, single people also have need of birth control, and contraception is used by many people not so much to plan the number of pregnancies or to space their offspring, but more to permit a degree of freedom in sexual relationships. Besides, the 'family' may consist of an extended group of kin, all of whom have a vested interest in the fertility of individual members, or it may exclude men altogether, since a growing number of households in the world are headed by women who are raising children without support or assistance from a man (see PART TWO, Section 2). Many people are not so much interested in planning their family as putting an end to conception once the family has reached a certain size.

Birth control, planned parenthood and child-spacing should all be seen as health issues. While it may be true that a more equitable distribution of resources is more critical to the alleviation of poverty than the control of population growth, this should not detract from the fact that the control by the individual of her or his fertility should be a fundamental right. *It is wrong to assume that, just because in many parts of the Third World people have not articulated a demand for birth control, there is no felt need for this service.* In many areas where people are denied access to birth control services, abortions (carried out in unhygienic conditions, at great risk to the woman) and infanticide are common. *The voluntary adoption of birth control is one of the most effective ways of improving the health of women and children in poor countries, of increasing both the quality of their lives and their life expectancy.* Birth control, therefore, should take priority in the work of NGOs, where appropriate.

Nevertheless, there remains one serious problem: it has already been indicated that child-spacing can be a causal factor in the reduction of maternal and child mortality and in levels of malnutrition, and yet, paradoxically, without changes in mortality rates and in the nutritional status of either children or adults, few people will wish to practise birth control. Thus, the real demand for contraception often only arises when parents already have a large, closely-spaced family, or in other words when the future is to some extent secure. Indeed, it often seems that people are only too ready to adopt practices which exacerbate further the problem of fertility and malnutrition. For example, in recent years there has been a dramatic increase in the bottle-feeding of infants (especially in urban areas), which reduces the intervals between births by eliminating the contraceptive effect of breast-feeding.

It may well be, then, that *in areas of extreme poverty, where food shortages and infant mortality are common, more general programmes designed to improve the quality of life* (such as income-generation, food production, etc.) *should precede birth control initiatives.* Indeed, there is good evidence to show that policies which promote birth control should concentrate simultaneously on increasing literacy and improving health conditions generally. The hope is that the improvement in material conditions and the reduction of infant mortality rates will lead to a demand for a long-term reduction in family size. However, it must be stressed that the one process does not automatically follow from the other. Field staff should be aware that there are still many unanswered questions surrounding the whole subject of fertility control, and in most countries the issue is an extremely sensitive one.

It may seem especially hard to deal with so personal a subject — and one which may have a remarkable impact on individual lives — when the initiative is being made at a community or group level. This is perhaps one of the main reasons that birth control programmes are so often directed exclusively at women. Men are seen as being at best passively acquiescent to fertility control initiatives and at worst openly hostile. Indeed, there is a growing demand among women in poor countries for contraception. But they face fierce opposition, not only from husbands but from older relatives as well. The social and religious pressures can be enormous, especially where women have a low status.

However, *whatever the attitudes of men and whatever the role of women in society, the successful birth control programme should try to establish a dialogue involving both sexes.* Sometimes it may be necessary to discuss the more sensitive issues with each of the sexes separately, but wherever possible they should be brought together regularly in order to establish their feelings on topics such as fertility and child-rearing, and to challenge discriminatory or prejudicial attitudes.

Since medical research is a long way from finding a perfect method of contraception — all known methods have disadvantages — *it is vital that people should always be offered as wide a choice as possible and that the programme should not promote any one birth control technique.* People must be provided with clear information about the different methods available, their advantages and disadvantages and their likely impact on fertility. Any methods of birth control or patterns of fertility traditional to an area should be examined prior to funding a programme, and customary practices — where they are found to be effective — should always be incorporated rather than discarded. The approach to fertility control has to be both imaginative and flexible.

The one aspect in this area of health care which unfortunately has to remain a low priority in Oxfam's work is the problem of infertility. The investigations and treatments associated with infertility are both complex and expensive and involve a high level of expertise and a large technological input. Work in this field, in the Third World at least, tends to be concentrated in hospitals in urban areas, and there is little likelihood of the service being extended to rural areas in the foreseeable future. Moreover, problems of fertility are rarely dealt with outside the private health system.

While having sympathy and understanding for individuals or couples experiencing fertility problems, Oxfam can support work in this field only indirectly by promoting health education within the primary health care structure. Thus, information can be made available on causal factors in infertility such as infection (for example, following circumcision) or sexually transmitted diseases and malnutrition. Preventive measures such as barrier methods of contraception can be promoted. Natural methods of birth control (rhythm method, Billings, etc.) can be encouraged, as they help couples to recognise the 'fertile' period in a woman's menstrual cycle.

2 Programme Structure and Organisation

Birth control programmes may be *structured* either as:
- ☐ a specialised (vertical) intervention, with specialised clinics and/or specially trained personnel;
- ☐ an integral part of the general health care system.

It may be necessary to provide a specialised service where the primary health care structure is not well developed. Equally, a vertical intervention may be most appropriate where the concept of birth control is new, or the intention is to mount a pilot project. There may be a special need for a vertical intervention in areas where men and women differ radically in their attitudes towards birth control and where the prime objective is to give women greater control over their own fertility. Where local or national women's organisations exist, they may be the most suitable agencies to respond to women's demands for contraception.

However, except in rare circumstances, the integrated approach to birth control is always preferable. Where birth control is seen as an integral part of the general health care system, a sense of both individual and community responsibility can be developed. The community can become aware of the relevance of birth control in all aspects of health — and especially in preventive health measures — and at the same time those people who are sexually active can be helped to understand the value of assuming responsibility for their own fertility.

Perhaps the greatest problem when integrating birth control with other health measures is that primary health workers normally have many demands on their time; they may not have the specific training nor the time for counselling. It is in practice hard for them to understand just how much impact child-spacing can have on mortality rates. *Thus, birth control rarely receives the attention it deserves — all the more so because the 'patient' is not ill, or visibly at risk.*

Supportive health care must be available for women suffering from the side effects of contraceptives or the effects of badly-performed abortions.

One of the best ways to integrate birth control is by linking it to a mother and child programme, since ultimately the outcome of the one is so dependent on the outcome of the other. Birth control programmes

7

should never be tied to vertical interventions aimed at controlling sexually transmitted diseases. Although it may seem logical to run these two measures together, the risk is that by doing so birth control may become unacceptable to the population at large.

It is essential that traditional midwives and healers, where they exist, are involved in fertility control education and application.

The *goal* of the programme should be to address the health problems of the family, especially of women and children, and to reduce malnutrition and maternal and infant mortality through child-spacing. However, even though health should take priority, statistics cannot be ignored; the aims should as far as possible be measurable and the programme should be cost-effective. The intervention must to some extent be planned in terms of the numbers of people needing or wanting birth control. Clearly though, not all objectives are easily measured or evaluated. For example, *one of the main goals of a programme may be to change attitudes towards fertility rather than promote the use of contraception* (see PART FOUR, Section 1).

Above all, the goals should not be dictated by health experts or specialists in the field of birth control. The main priority of the people served by the programme might well be to control fertility only after a large number of children have been born to each woman. This pattern may not fit with contemporary Western thinking on health or population control, but must be respected all the same. The more remote the programme from the felt needs of the community, the less likely it is to have a successful outcome.

All social, religious and cultural constraints must be taken into account when assessing projects which deal with birth control, and, where appropriate, discussions should be held with those sectors of the community who might oppose women's desire for contraception.

The provision of *information* and *health education* should always take priority over the delivery of contraceptive devices. The element of education should be fundamental to any programme, all the more so if the concept of fertility control is alien to a culture. No contraceptive method is perfect, and people must understand the problems and risks involved. Family life education may be undertaken with young people in schools. This often proves the most effective way of altering views on child-rearing, fertility, the status of women etc. However, it is essential to obtain parental consent and approval in such initiatives. The participation of men in all debates on birth control should be actively encouraged. In many societies ignorance about the body is equated with virtue, and general sex education is essential for men as well as women.

The *technological input* of a programme is to a great extent dictated by the source of contraceptive supplies and the availability of those supplies. In some countries either the government or international agencies distribute supplies of contraceptives — in some cases free of charge. In other countries it may be necessary to use commercial sources. Although the programme should offer as wide a variety of birth control methods as possible, certain methods will always be preferred to others. In the Philippines, for example, oral contraceptives are very popular, while in India the official emphasis is on vasectomy. Indeed, sterilization is now one of the most common forms of contraception worldwide.

Obviously cost is also a critical factor influencing choice. Certain methods may be preferred simply because the contraceptive device has to be replaced less frequently than with others. For example, devices such as the IUD or the diaphragm may in the long run prove cheaper than condoms.

Training on child-spacing, planned parenthood and birth control should be provided in all health programmes, at all levels, whether it be aimed at skilled practitioners or primary health workers. This policy should apply even in those countries where the State does not promote or support birth control initiatives. The training required for this kind of programme is to some extent fairly specialised since the health worker must learn to treat the subject of birth control with great tact and sensitivity. In addition, he or she must have a comprehensive understanding of all the methods available — even if the more complex cases (for example those requiring vasectomy or IUD insertion) are referred to a specialist.

The promotion of birth control should never be aggressive; a demand may exist but be poorly expressed, or people may be fearful of being frank on the subject. *The motives for adopting birth control are numerous, and the demand can be developed more effectively by promoting health education than by campaigning on the issue*. The more common motives are:

☐ sufficient number of children (family completed);
☐ desire to provide more adequately for each individual child (in terms of health, shelter, education, food etc.);
☐ desire to improve the economic status of the family as a whole;
☐ plans for both or one of the parents to further their education or to take employment outside the home;
☐ the wish to avoid unwanted pregnancies.

There are also a number of indications more directly related to health which provide a basis for adopting birth control:

☐ chronic ill-health of the mother (anaemia, heart or lung disease, etc.);
☐ problems associated with childbirth (difficult deliveries, bleeding, etc.);
☐ diseases transmitted to the baby (abnormalities, sickle cell anaemia, VD, etc.);
☐ numerous previous pregnancies — complications of pregnancy and labour rise sharply after the fourth pregnancy, increasing the risk to both mother and child;
☐ birth-spacing. Closely spaced births increase the likelihood of maternal and child mortality. The minimum gap betwen pregnancies should be 18 months, although 3 years is better from the health point of view.

Birth control programmes present certain *funding* problems. Except in rare cases, contraceptive devices are imported: they cannot normally be produced locally. Further, the technological input is very long-term — all individuals, or all couples, will need to use contraception so long as they are sexually active, for as long as they are fertile and so long as they do not wish to have children. The NGO cannot assume this cost. In communities where there is sufficient wealth, individuals should pay for contraceptive devices. In some areas, governments and international agencies distribute contraceptives free of charge, and in some cases it is possible for the community to subsidise individuals. Otherwise, if the project is functioning in a very poor area, it may be best to encourage the use of birth control techniques which do not involve any devices.

3 Contraceptive Methods

a: No one method of contraception can be universally recommended. All programmes should offer as wide a choice as possible, and local religious, cultural, social and economic factors should always be taken into account. Existing patterns of health and traditional birth control practice must also be considered when recommending contraceptive methods.

b: Programmes which recommend only the *temperature method* of contraception (i.e., most of the programmes sponsored by the Catholic church) should be supported only where there are no alternatives and

Table 7 – 4   Contraceptive Methods

A) TRADITIONAL METHODS AND THOSE INVOLVING PERIODIC ABSTINENCE

Method	Effectiveness	Acceptability	Availability	Side Effects	Reversibility	Comments
1. Rhythm Method	Fair	Fair	Education required – Calendar helpful	None	Complete	Requires compliance by both partners and needs long periods of abstinence. Problem can be erratic nature of most women's 'cycles'.
2. Billings Method	Good if well taught	Good	Education required	None	Complete	As above but shorter abstinence periods required. Local infections may disrupt results and increase failure rate. Needs to be well taught.
3. Withdrawal	Very few studies on failure rates. Seems to be moderately reliable	Variable	Nothing required	None	Complete	Needs mutual confidence. Can lead to tension – which affects acceptability – but is a traditional method which has been used happily for years by many couples.
4. Traditional abstinence	as No. 3					Traditional in many parts of Africa. 'Taboo' periods, e.g. post-delivery, during lactation, etc., lead to beneficial birth intervals.

Method	Reliability	Acceptability	Availability	Side effects / dangers	Reversibility	Comments
5. Breast-Feeding	Not very reliable	Good	Nothing required	None	Complete	Longer gaps between babies. Benefits too for the feeding infant.
6. Home-made tampons & pessaries	Variable – generally fairly high rate of pregnancy but useful in 'emergency'.	May be attractive for some	Described all over the world using many local materials	Variable – some local inflammation can occur. but in some instances overwhelming infection can lead to death e.g. use of cow dung.	Normally complete	Some methods can lead to infection and damage and can be positively dangerous. Other methods effective and harmless providing clean and 'hygienically' prepared.
7. Douching	Unreliable. Usually too late to prevent pregnancy.	–	As No. 6	Can cause local irritation: danger that corrosive liquids may be used causing permanent damage and scarring.	Normally complete.	Not recommended as a method.

B) BARRIER METHODS

Method	Reliability	Acceptability	Availability	Side effects / dangers	Reversibility	Comments
8. Condoms	Good (depends on expertise of user and quality of condom).	Good – variable	Through hospitals, health centres, dispensaries, pharmacies and shops.	None (very rarely allergy)	Complete	Help prevent sexually transmitted diseases (STD). May be culturally unacceptable in some societies.

7

Method	Effectiveness	Acceptability	Availability	Side Effects	Reversibility	Comments
9. Diaphragm	Effective providing it is used regularly and carefully.	Fair – Good	Commercial products through hospitals, health centres, clinics, dispensaries. Some traditional methods available.	None	Complete	Some protection against STD. Needs to be fitted by a trained person. Needs to be regularly checked or replaced.
10. Cervical caps	"	"	"	"		As above but suitable only for some women. Slightly more difficult to use.
11. Spermicides	Fair	Fair	Shops, hospital, health centre.	Occasional allergy	Complete	Not always freely available. Recurrent costs can be high. Better if used in conjunction with methods 8, 9, 10.
C) THE IUD						
IUDs generally	Very good	Good/variable	Hospital, health centre.	Variable risk of heavier bleeding. Risk of infection or perforation and ectopic pregnancy. Best for women who have had one child – to minimise side effects.	Complete, providing no complications experienced (Infection can lead to infertility).	Needs to be expertly fitted in hygienic conditions. May cause some immediate discomfort. If used in women who have had many pregnancies, there is a risk that it may be expelled. Not advisable if history of infection or pelvic disease.
12. Plastic IUDs (Lippes loop)	"	" depends on side effects experienced	"			

Dalkon Shield —————— *No longer recommended* ——————

	Effectiveness	Provider	Side effects	Reversibility	Comments
Dalkon Shield					*No longer recommended*
13. Copper IUDs **Copper 7** **Copper T** **Multiload**	Very good if taken correctly and regularly	As above	Can be used for those not yet child-bearing, but must be aware of infection risk.	"	Considered suitable for those who have not been pregnant. Same comments as 12 above. May reduce some of the side effects. May not be advisable for the childless with frequent sexual partners because of increased risk of STD.

D) HORMONES

	Effectiveness	Provider	Side effects	Reversibility	Comments
14. Combined pills	Very good if taken correctly and regularly	Hospital, health centre, dispensary, birth-control workers, nurses.	Variable – fewer on low dose pills. Risk of thrombosis. May also cause nausea, raised blood pressure, weight gain.	Complete after 1 – 3 months. Occasionally Can take up to 2 years.	Needs to be taken under supervision. User needs to be examined and have history taken by trained person. Women with certain medical conditions should not take the pill. Advantage that it regulates menstrual cycle.
15. Depo-Provera Injections	Good (second only to the pill)	Hospital, health centre, birth-control workers, nurses.	Moderate, irregular bleeding.	Usually complete within 18 months – but not immediately reversible.	Use is discreet, must be used only after counselling of possible problems/side effects. 'Informed' consent essential. Risk of being misused. Controversial.

7

E) PERMANENT METHODS AND POST COITAL METHODS *(see also Section 13, Appendices)*

Method	Effectiveness	Acceptability	Availability	Side Effects	Reversibility	Comments
16. Tubal ligation	Very good. Immediate	Fair	Hospital or special 'camps'.	Risks of operation. some slight risk of heavier periods afterwards.	Usually irreversible	Both partners need good counselling to understand implications. Some women do bitterly regret the operation – others permanently relieved of worries of child bearing.
17. Mini-laparoscopy	"	"	"	"	"	Quickest and simplest method, but needs a skilled operator and special equipment.
18. Vasectomy	Very good but not immediate (may take 2 – 3 months to become fully effective)	Fair	Hospital, health centre, 'special camps'	During procedure only. Psychological worry re sex life (minimised by good counselling).	Irreversible normally.	Simple operation under local anaesthetic by specially trained doctor or technician.
19. Abortion (legal)	*Not* a method of contraception, but effective when contraception has failed.	?	Hospital	The earlier in pregnancy it is done, the safer it is. Risk of operation and infection.	May experience some problems with later pregnancies, i.e. early labour or miscarriage.	Illegal abortion has been used for centuries when contraception has failed. Safer to use modern efficient method of contraception.
(illegal)			Home	High risk of infection and death.		

where it is believed that the initiative will at least have a long-term educational value.

c: Oxfam is prepared to support both male and female *sterilisation*, provided that it forms part of a well-supervised programme run by trained and responsible health workers. In some countries (particularly in Asia) far too much emphasis is given to sterilisation and too little to other, more temporary, methods of contraception. Participation in such a scheme must be voluntary and must require the informed consent of both partners. Oxfam funds must *not* be used to provide inducements to potential patients, and no pressure should be exerted on hospitals or clinics to initiate such schemes. Out-patient methods not requiring general anaesthesia or bed occupancy for more than 24 hours are recommended; these now include methods of female sterilisation (e.g. trans-cervical methods) as well as vasectomy for men under local anaesthesia. (For further details see Section 13 Appendices.)

d: Breast-feeding has an appreciable contraceptive effect and should be encouraged both for this reason and because breast milk is more nutritious for the infant than other forms of milk. Breast-feeding is of course no substitute for the more effective methods of birth control.

e: Depo-Provera (injectable contraceptive). This method could solve many of the logistic and social problems often encountered with birth control programmes. However, Depo-Provera is a relatively new drug (developed within the last 20 years) and therefore should only be supported where constant supervision is possible. Proposals relating to the use of injectable contraceptives should be submitted to the Oxfam Health Unit for assessment.

f: Abortion is not a contraceptive method, and its practice therefore indicates a failure in a birth control programme. Even though abortion remains outside the sphere of Oxfam's work, financial support for other medical work should not be denied to those hospitals where abortion is carried out legally.

| III | **Female circumcision** |

1 Introduction

The term female circumcision is used to describe the practice of changing the appearance of the female genitals by surgical intervention. As a term it is fairly imprecise, since true female circumcision (i.e. the removal of the hood of the clitoris) is but one type of operation. The operations most commonly performed are clitoridectomy and excision. These involve the complete removal of the clitoris and most of the labia minora. A geographically less widespread practice is infibulation. This involves not only the removal of the labia minora and the clitoris but also of the labia majora. What remains of the genitalia is then pinned together to make an almost complete closure through the formation of scar tissue.

Many people have assumed incorrectly that the practice of female circumcision is directly associated with Islam. In fact, many of those who believe in the practice are not Muslim, and the majority of Muslims do not undertake female circumcision. Some women in Europe and America have had clitoridectomy performed upon them, but female genital surgery as a cultural phenomenon is peculiar to the southern end of the Arabian peninsula, to some of the areas of Indonesia settled by South Arabian traders and, above all, to Africa. Female circumcision is an ancient practice of Arab or Semitic origin which began long before the dawn of Islam and was subsequently abandoned in much of the Middle East. Clitoridectomy and excision began as South Arabian religious rites in the early centuries A.D. Infibulation has a more recent

7

origin and occurs today in the Nile valley, in Somalia, Ethiopia and along the caravan routes to West Africa.

Surgery is normally performed on young girls between the ages of 7 and 12. Since it is usually carried out in unhygienic conditions without anaesthesia, it can have a damaging effect on health and poses a very real threat to the physical and emotional welfare of the child.

However, because female circumcision is so ancient a custom and, where it occurs, is so deeply embedded in the social structure, any attempt to end it is likely to have an impact only in the long term.

2 Policy

Oxfam's approach is to promote a better understanding of the custom and to support work being undertaken at the local level which stresses the risks to health involved. Special emphasis must be given to the practice of infibulation, which has a far more devastating effect on health than the more modest operations. Clitoridectomy may pose certain health risks at the time of operation and clearly may have long-term emotional consequences, but is less likely to produce lifelong medical problems than infibulation.

Thus, it is inappropriate to support campaigns against these practices organised at an international level, because the practice varies considerably from one country to another, as does the context in which it occurs. It is beyond the power of international agencies to bring the practice to a halt, and only by working with local groups who understand the social and cultural context will campaigning be effective. Some of the pressure groups operating from the West have misinterpreted the situation, assuming too glibly that men must be blamed for the practice of female circumcision. In fact, in some countries, such as the Sudan, it is the older women who are most anxious to continue the practice, because they see it as enhancing the status of their lineage.

Among Cushites in East Africa, female circumcision has a very different significance; both sexes are circumcised in rituals which mark their advancement to elderhood. Societies that practice infibulation tend to stress female virginity at marriage more strongly than those practising only clitoridectomy and excision.

Within each country it is likely that there will be a wide variation in response to any moves to halt female circumcision. In Kenya for example, even though President Moi has banned female circumcision, the issue is debated daily in the press. The practice is still widespread among pastoralists and farmers in remote areas where control over girls is strong and bridewealth demands high. Perhaps in the case of Kenya, efforts to encourage change should be concentrated in urban areas, or at least in those rural areas not far from urban labour markets, where the introduction of cash crops and other innovations has brought about a number of changes in the social structure, weakening the authority of older people. Equally, in the Sudan, the last people to cease circumcising will be the very poor, who will continue to feel that they must go on endowing their daughters with the traditional qualities for marriageability, since they can settle no other dowry upon their heads.

Many women in Africa have been angered by the actions of western feminists wishing to make a *cause celebre* out of female circumcision and ignoring problems which are more fundamental, such as chronic shortages of water and food. Instead of making an issue out of female circumcision, a more appropriate policy would be to provide information concerning health risks, etc. for those requiring it in areas where the practice is already being called in question, or is on the wane, as in peri-urban Kenya. Support can be given for adult education or for programmes run by women's groups. Assistance may be given for the production of information packs or teaching materials which deal with the whole question of mother and child health, and more specifically with the dangers of female circumcision. In some countries women's groups have already prepared their own educational material which is now available through the mother and child health structures. Support for education on female circumcision can be combined with income-generation or any similar programme which enables the revaluation of the status of women in society.

7

Section 7   Secondary Health Care

Hospitals and health centres

1 Oxfam Policy

While Oxfam's main emphasis will remain on community health at the primary health care level, it has in exceptional circumstances supported and will continue to support the work of hospitals in the Third World for the reasons below.

a: If one is looking at the health care of an administrative unit, the district or regional hospital cannot be ignored. It should act as a referral and training centre in addition to giving supervision for health centres and undertaking outreach work.

b: Many of the hospitals in poor countries are in urgent need of funds to carry out their work. Medical services often receive a low priority in such countries and in certain neglected areas within those countries. It is, therefore, a legitimate use of Oxfam's money to provide funds for medical services in those cases where local funds would not be forthcoming.

c: Much of the disease with which these hospitals deal results from, or is associated with, malnutrition. This applies especially to children under five.

d: To date, a large proportion of those funds given by Oxfam to hospitals has been allocated to institutions run by missions. It should be noted, however, that Oxfam funds are available to all hospitals run in accord with the medical priorities outlined in the Health section, and offering a reasonable assurance of continuity. Health Centres, where the cost of treatment per patient is usually lower than in a hospital, are particularly worthy of support.

2 Types of Aid to Hospitals

Below are the suggested types of aid to hospitals and health centres to be supported (this list does not represent an order of priorities).

a: Building and reconstruction. In normal circumstances, requests for the building of an entirely new hospital should not be considered. It is preferable to assist existing hospitals by funding reconstruction, repair, or additions. The work proposed should be as inexpensive as is compatible with efficiency. New buildings should conform with sound local standards of construction, and be approved by the appropriate surveyor or architect. They should be designed to fulfil satisfactorily the purposes for which they are intended. Preference should be given to building or reconstruction work which will provide increased facilities for child care (e.g. children's wards and out-patients' clinics). Accommodation for doctors should be considered only in exceptional circumstances. Wherever new buildings or alterations are planned, possible future expansion must be allowed for.

b: Transport. Adequate transport is essential for rural outreach services. Wherever possible, inexpensive forms of transport such as motor bikes and bicycles should be encouraged. However requests for supplying and maintaining vehicles (mobile dispensaries, etc.) should be sympathetically considered, provided the cost involved is reasonable and the vehicles are suitable for the purpose stated and for the terrain in which they are to be used. Oxfam does not normally fund ambulances. It is important to analyse the real benefits obtained from transport which is often very costly in relation to its use and benefit to patients.

c: Equipment. All requests for equipment should be considered in terms of the likely demands on its use and the extent of the benefits accruing. Costly equipment serving only a small number of patients should not normally be considered. It is essential that adequately trained staff are always available to operate and maintain the equipment.

Requests for small electricity generators, which would make a substantial improvement to an important medical service, may be supported in the interests of up-grading hospitals.

The question of Xray equipment is dealt with separately (see II. 1(e) of this section).

It is recommended that the fullest use should be made of the facilities offered by ECHO.

d: Staff and running costs In providing assistance of any kind to hospitals and hospital services, it is important to make sure that adequate staff and funds will be available to undertake the work involved. However, where staff and funds are insufficient, Oxfam may consider providing short-term assistance for the payment of salaries, although a long-term commitment should be avoided. Careful account must be taken of the running, maintenance and depreciation costs of equipment and vehicles. Experience has shown that the costs of such items are often underestimated.

e: Drugs. The generic name of the drug should be given and its intended use clearly explained. The drug list should be short, inexpensive and simple (see Section 10). Oxfam should be prepared to advise hospitals of the opportunities for obtaining drugs free of charge or at a discount from WHO and other suppliers. Offers of sample drugs should be accepted only after professional advice has been given.

f: Vaccines. Where money is to be given for the purchase of vaccines, it must first be established that the project personnel understand the principle of the cold chain and the needs of the individual vaccines for refrigeration.

3 Cost and Design

Field staff will find it useful to monitor building costs on a country-wide basis, as this will provide guidance in assessing requests to fund buildings. To obtain the cost per square foot or square metre, calculate the floor area of the building and divide this figure into the total construction cost. Elaborate buildings are expensive to construct and seldom justifiable, and costs can be reduced by a careful study of the building plans. Points to look for:

☐ are there an excessive number of small rooms? Could the larger rooms be more effectively divided by removable, light-weight partitions?

☐ is there a recognisable flow pattern in planning for the movement of ambulatory patients?

II   Hospital and medical equipment

1 Oxfam Practice and Policy

Oxfam frequently helps hospitals and health programmes with the supply of vaccine, drugs and smaller items of equipment for treatment, laboratories, etc. Where the more costly types of equipment are concerned, the practice has been to give grants only for the equipment below.

a: Autoclaves (sterilisers). Wherever appropriate, support should be given for the provision of sterilisers. Large autoclaves, costing around £1,500 in 1984, are needed to sterilise surgeons' gowns, bandages, dressings, etc. Smaller sterilisers for instruments are also needed and supplies from ECHO range from £90 to £135 (in 1984). A field steriliser for use over a fire costs £120.

b: Operating tables. Again, help is often given with these items, but a clear distinction must be drawn between sophisticated models with a variety of movements and attachments, and simple models which would be adequate (e.g. for maternity work) in a context where major operations are likely to be rare.

The more sophisticated models may cost up to £3,750 (in 1984) and are suitable only for well-staffed and properly equipped hospitals. In contrast, the simpler models can be obtained from ECHO for approximately £425 (in 1984). Unfortunately, there appears to be a dearth of tables provided at prices in between these two extremes.

c: Jet injectors. See Immunisation, Section 4, for details of pan-jets and port-o-jets.

d: Blood banks. These have occasionally been provided by Oxfam but not as a part of normal practice.

e: Xray equipment. Oxfam does not support the use of Xrays by mass radiography for TB diagnosis because the sputum smear tests are very much cheaper and more reliable.

While Xray equipment for general hospital use is important, Oxfam in general gives it a low priority because of the high cost and risk of breakdown.

Many enquiries for Xray equipment are received, but it often transpires that the type of equipment sought does not match the hospital's needs, being far too expensive to run. Running costs tend to be high because of the need to pay for film, to replace tubes about once in three years, to run a dark room, and to pay the salary of a specially trained technician.

In a few cases, Oxfam has helped a hospital to install an electricity supply as a step towards acquiring Xray facilities at some later date, for example, as in Kenya (KEN 58).

The following points should be taken into account when considering a request for Xray equipment:

☐ it is essential that trained staff be available in the hospital and that their skills should be appropriate, given the type and expected function of the Xray unit requested;

☐ the proper protection of hospital personnel handling the units should be ensured; the potential risks of radiation, etc. need to be fully recognised;

☐ it is essential that there be an electricity supply capable of providing the current required by the unit.

Types of Xray equipment. A small portable unit (15-30 mA) would serve the many needs of a small hospital, even if the doctor has no special training in radiology. It would be very useful in orthopaedics and in the examination of fractures, etc., in a casualty department. However, this kind of equipment is not definitive enough for detailed investigation of lungs and other organs.

A medium unit (100 mA) would be useful to a doctor with some knowledge of radiography. A technician must be available for maintenance. A strengthened floor would be necessary to hold a unit of this weight. Given these conditions, this type of unit would probably be most useful for a small hospital.

A large unit (300 mA) would require a doctor with considerable radiological training to interpret the various films, a skilled radiographer to work it, and a technician trained in maintenance. These units are generally beyond the scope of Oxfam.

Hospitals must ensure that the unit requested is the one most suited to their needs. Xray units cannot be returned to the country of origin. Some European manufacturers have agencies overseas which will undertake to install and maintain Xray units.

f: Electricity generators. Diesel generators for hospital lighting, refrigeration and for running Xray equipment have been provided fairly frequently. Sets should be supplied only where maintenance is available. Sizes range from 2kW (for a very small lighting system) to 70kW.

The voltage and AC frequency of the generator must be compatible with the electrical equipment which will be run from the supply. When ordering a generator, state:

- [ ] output required in kilowatts (kW)
- [ ] AC or DC (nearly all equipment is now AC)
- [ ] voltage
- [ ] AC frequency in cycles or Hz (the choice is usually 50Hz or 60Hz)
- [ ] Single phase or 3-phase AC

2 Locally-made Hospital Equipment

Many types of equipment can be made in a hospital work-shop at very low cost. Pioneer work by S. W. Eaves at Zaria in Nigeria has demonstrated the practicability of making trolleys, wheelchairs etc. out of simple materials, such as standard metal piping or bicycle parts.

ITDG have published drawings of hospital equipment as follows:

- [ ] hospital blood transfusion drip stand;
- [ ] hospital bedside table and locker;
- [ ] folding bed;
- [ ] hospital ward screens;
- [ ] dressing/instrument trolley;
- [ ] bush wheelchair;
- [ ] hospital wheelchair;
- [ ] bush ambulance.

7

Section 8 — Alternative Health Systems

I — Introduction

It has been estimated that up to 70-90% of all self-recognised episodes of ill-health are managed without recourse to allopathic ('western') health care, either through home treatment or the use of traditional/alternative healers. Hence field staff should find out about these alternatives, and wherever possible integrate them into primary health care and social development programmes.

In developing countries, particularly, the allopathic system will tend to suffer from three disadvantages:

☐ high cost;
☐ serious misuse, especially by over-prescribing;
☐ cultural alienation caused by high technology.

Given the cost of allopathic treatment and the need for highly qualified personnel, the entire population cannot be covered by this system alone, nor is it possible for poor communities to sustain such health projects from their own resources. Hence it is very important to mobilise the traditional systems and make use of their advantages.

One of the major advantages is that alternative systems are generally much cheaper than allopathy and therefore offer real possibilities for self-sufficiency. The possibilities for community participation may also be much greater.

Traditional systems are generally accredited with greater success in some types of treatment than in others. It is important for the practitioners to be able to recognise these deficiencies and to refer cases or adapt to allopathic treatment where it shows a clear superiority. This is particularly so in the case of killer diseases. Tuberculosis, for example, will be better treated under the allopathic system.

Traditional practitioners should also conform to the general policies for primary health care (see PART SEVEN, Section 2) and in particular should conduct immunisations. Teaching of oral rehydration should be included in any such programme.

II — Assessment of alternative health sysems

In the absence of scientific analysis, field staff may have to rely to some extent on local acceptability as an indication of effectiveness. But they should also take the advice of consultants, preferably medically qualified and sensitive to alternative systems.

The systems of acupuncture, ayurved and homeopathy are well-established but other systems may require a careful assessment, and if necessary a small survey.

Oxfam does not fund detailed medical research, but limited studies of local systems may be useful. Any scheme involving alternative systems should be carefully monitored.

In practice a judicious selection from alternative medicine with some borrowings from the allopathic system will combine the advantages of both systems, providing health care at a lower cost and more within the control of the community. This approach has been used effectively by small non-medical groups in Bihar State of India using homeopathic treatment, referrals to Government hospitals and an emphasis on community health.

| III | **Oxfam involvement** |

Field staff contemplating schemes based on alternative approaches should ensure that:
- [ ] they have good local advice;
- [ ] they have the capacity for close assessment and monitoring;
- [ ] they can ensure the training of traditional practitioners in primary health, immunisation, oral rehydration, etc.

| IV | **Traditional practitioners' role in conventional health projects** |

Within a conventional (allopathic) health programme it is important to ensure that all traditional healers have been identified and their potential has been used as far as possible. This particularly applies in the following cases

Traditional Birth Attendants (TBAs)

TBAs usually offer a relatively cheap service which is accepted by the local community. They are readily available and can follow up cases closely. Every effort should be made to understand their methods and offer re-training to those who are prepared to improve their techniques as necessary.

The initial reaction of TBAs may be one of hostility or suspicion but this may be overcome by a sensitive approach and an understanding of their economic and social status.

As far as possible TBAs should be integrated into the health system and should never be ignored. (See also PART SEVEN, Section 6.)

Traditional Mental Healers

Psychosomatic and psychiatric illness may be as common in developing countries as elsewhere. Many allopathic doctors have little training or experience in these matters, and psychiatrists will be few and far between. Where available, medical treatment for mental illness may be of a drastic and harmful nature. In these circumstances efforts should be made to integrate traditional healers within a community health approach, to offer them training and to refer cases for which they are the best available alternative. Traditional healers should in return be encouraged to identify cases beyond their competence and to refer for medical or psychiatric treatment as necessary. (PART SEVEN Section 9.)

| V | **Conclusion** |

The problem of health is so vast that all resources must be mobilised. A sensitive approach will enable traditional practitioners to contribute to the overall effort. Alternative health systems may offer a cheaper approach, more easily controlled by the community.

Field staff should make every effort to identify these possibilities but must accept the need for a considerable amount of time to be spent in assessment and monitoring.

7

Section 9   Handicap

I   Mental health

1 Introduction

The degree of illness in the Third World which has been caused by mental disease has been greatly underestimated by many policy-makers, health administrators, medical practitioners and laypeople. Neuroses are particularly common in poor areas where the stress caused by poverty and social and political marginalisation can be severe. And yet, mental health is poorly serviced in the Third World because priority has been given mainly to fatal diseases (especially endemic communicable diseases) and accurate diagnosis is difficult as there are few psychiatrists in these countries. The treatment of psychiatric patients, however, tends to be relatively inexpensive.

It has also been presumed that the community in the Third World is better able to care for the mentally ill than communities in the more differentiated, complex societies of the industrialised world. It is felt that the mentally ill are well catered for by the attentions of their family, traditional healers and religious leaders. In reality, since they are not usually able to be useful working members of society, they are often neglected by their families and isolated by their community. They receive even less support in congested urban areas or areas where there has been social upheaval or natural disaster. The mentally ill are frequently forced to live off their own resources by begging in the streets or by theft.

Mental health presents a challenge to the NGO and can be a particularly problematic area of health intervention in countries where there is no national psychiatric service. Perhaps the most effective way an NGO can help reduce mental illness is through the support it gives to general development programmes. A number of mental disorders — especially neuroses — can be prevented by alleviating poverty and reinforcing family and community ties. Also, there is no doubt that a number of the patients who present themselves at general PHC units are suffering from minor neurotic illnesses, and these can be treated by the health worker on an out-patient basis without too much cost.

However, the more serious cases requiring specialist treatment and possibly hospitalisation must be referred to the appropriate practitioner in the State psychiatric system. It is beyond the capacity of Oxfam and most other NGOs to fund specialist units for the treatment of psychiatric patients, primarily because the staff/patient ratio is normally so high. It is more feasible to support projects involved in the rehabilitation, rather than the cure, of psychiatric patients. For example, assistance has been given to the occupational therapy unit of a rural Peruvian hospital (PRU 283). In this programme psychiatric patients are rehabilitated and returned to the community. They become involved in activities such as weaving, carpentry, animal-rearing, brick-making and farming — all skills which have a strong practical application in the local economy.

2 Types of Mental Illness

The most common forms are:
- ☐ psychoses, e.g., schizophrenia — related to demonstrable lesions in the brain;
- ☐ neuroses — depression and anxiety — with no obvious lesions, usually with multiple causes (including stress);
- ☐ behavioural disorders, e.g., maladjustment in children, juvenile delinquency, absenteeism etc.; and
- ☐ psychopathic disorders, e.g., constant irresponsible, aggressive or antisocial behaviour.

3 **Causes of Mental Illness**	***a:*** Genetic: probably these have been overestimated, but Down's Syndrome (mongolism) and possibly some of the psychoses do have genetic linkages.

b: Organic brain damage due to:

☐ trauma at birth;

☐ infection, e.g., meningitis, trypanosomiasis, syphilis, encephalitis and cerebral malaria;

☐ malnutrition such as vitamin deficiency, pellagra or severe protein calorie malnutrition;

☐ toxins: alcohol, opiates and other drugs;

☐ degenerative lesions (old age).

c: Socio-cultural. Troubled families tend to produce children with behavioural problems. It is important to differentiate between beliefs in spirits, etc. which may be part of the local culture, and illusions which are not part of a culture and are therefore pathological.

d: Idiopathic (exact cause unknown). Many neuroses fall into this group. It is likely that there are several causative factors, including a predisposition in the patient and a difficult environment which may precipitate the disease.

4 **Community Mental Health Programmes**

The community must care about the problem if it is to help the patients. Often communities have a system for handling the problem of ill health, and it is important to find out about this before planning any intervention.

Certain communities are afraid of mental illness, and in such cases efforts must be made to change attitudes and remove the stigma. It is also important for people to understand that the individual who is badly behaved may need treatment rather than punishment.

The aim should be to teach the patient and the community rather than send him or her to an institution. If the patient needs to attend hospital it should be as an outpatient wherever possible.

II   Physical disability

1 **Introduction**

It has been estimated that one person in ten in the Third World is physically disabled in some way. Severe handicap is obviously much less common (1% or less), but nevertheless poses an enormous problem in poor countries where assistance to the disabled is usually given a low priority. Poverty causes much of the physical disability in the Third World, and therefore the alleviation of poverty is one of the main ways in which disability can be prevented. (Also see PART TWO, Section 4.)

2 **Types and Causes of Disability**

When working in poor areas it is necessary to concentrate on the most common types of disability plus those which can be either prevented or countered by rehabilitation. The minutiae of genetic and congenital causes that fill Western textbooks and conferences are of less immediate concern. The principal concerns are blindness (see Section 5) and locomotor impairment (i.e., of the limbs). It is most useful to consider causes under these two headings, rather than to use a more scientific aetiological classification. It is also useful to consider causes and prevention together where possible. Causes and prevention of disability are listed in Table 7-5.

3 **Prevention**

Most serious disabilities in the Third World are preventable and it is far more cost-effective to prevent disability than to rehabilitate the disabled. (See Table 7-5.) Moreover, in the case of severe disability, rehabilitation can usually only be partial.

a: Methods of *primary prevention* obviously depend upon the cause of

7

Table 7 – 5   Causes and Prevention of Disability

Type	Cause	Prevention
Blindness	**Trachoma.** A viral infection causing vision impairment and blindness. Highly infectious in crowded conditions.	a) Good water supplies to improve general hygiene. b) Mass chemoprophylaxis with tetracycline eye ointment. c) Early detection of infection and treatment with tetracycline eye ointment. d) Surgical intervention to prevent ingrowing eyelashes.
	Vitamin A deficiency. Cause of childhood blindness − initially xerophthalmia, 'dry eye' − then keratomalcia, destruction of the cornea. Effects of Vit.A deficiency can be worsened by measles.	a) Growing of vegetables/fruits containing Vit. A. b) Use of Vit.A capsules in emergency situations.
	Onchocerciasis (river blindness). Infection by filaria parasite transmitted by black-flies. Geographical limitation primarily West and Central Africa, parts of Central America.	In cooperation with WHO/national programme: a) Vector control b) Mass treatment programmes
	Cataract. Mainly natural due to ageing. May also be caused by infection, injury,	Cataract cannot be prevented, but can be 'simply' treated by cataract removal, in hospital or mass eye camps. Glasses are necessary afterwards.
	Accidents. Particularly 'stick' injuries in bush areas.	Early treatment to prevent secondary infection to scratched eyes etc.
Locomotor impairment	**Polio.** Viral infection causing damage to nerves supplying principally limb muscles. Transmitted by faecal/oral route.	a) Improved water and sanitation. b) Polio immunisation campaigns and regular immunisations by MCH programmes.
	Leprosy. Bacterial infection causing principally nerve damage. Consequent loss of sensation leads to easy damage and mutilation of limbs.	a) Early diagnosis, education and treatment on non-institutional lines (see Oxfam leprosy pack). b) BCH immunisation. c) Effective management of damaged limbs to prevent further damage; improved footwear, etc.

Type	Cause	Prevention
Locomotor impairment	**Accidents/Burns.** Important cause of limb impairment in children.	**a)** Attempts to prevent burns by stove/pot design improvements. **b)** Early treatment of burns — first aid to reduce burned area, early medical care to prevent contractions. **c)** Early treatment of accident damage.
	Meningitis. Bacterial infection — occurs in epidemics in Sahel region — widespread low prevalence elsewhere. By residual damage to brain, leads to a wide pattern of physical and mental handicap.	**a)** Early diagnosis and treatment of cases. **b)** Vaccination in epidemic situations.
	Other infections of the nervous system. Any infection of the brain, e.g., cerebral malaria, can lead to handicap. Rare *epidemics* of viral encephalitis may also occur. Transmission by mosquitoes.	Mosquito control
	Cerebral Palsy. A wide variety of physical handicaps resulting from birth injury — which may range from physical damage to hypoxia.	General improvements in antenatal and obstetric care. This means training at the periphery, *not* high technology incubators, perinatal care etc.

the various disabilities, but among the most effective are:
- [ ] immunisation against polio, TB, measles and DPT;
- [ ] improvement of maternal services;
- [ ] early detection and treatment of infectious diseases such as cerebral malaria, TB, leprosy and trachoma;
- [ ] good water supplies;
- [ ] good diet, adequate in Vitamin A and iodine;
- [ ] disease control programmes, e.g. onchocerciasis (river blindness);
- [ ] legislation to avoid industrial and other injuries (including traffic accidents);
- [ ] public education programmes to publicise health hazards.

b: Secondary prevention: either stopping a disability from worsening or reducing it; methods include physiotherapy, provision of appropriate calipers for polio patients, prevention of secondary infection, etc.

4 **Rehabilitation**

Even though prevention should have priority over rehabilitation, those who are already disabled still need assistance. Rehabilitation has both physical and social elements (see also PART TWO, section 4). The physical element includes the provision of physiotherapy, calipers, wheelchairs, and artificial limbs to facilitate mobility. Basic physical aids can considerably increase the patient's morale and physical and mental independence. Social rehabilitation involves the acceptance of the

7

disabled person by the community and the involvement of the community in caring for the individual. It also includes the provision of employment if necessary in both workshops and agriculture.

As much of the physical rehabilitation as possible should also be done close to the community: this may involve the provision of mobile services which can carry supplies of calipers, etc. This is particularly important in child polio cases where the size of the caliper needed changes rapidly and where constant admissions to a central hospital are difficult to arrange. Repeated admission of children to hospitals can lead to their being abandoned in the institution by the family.

5 Recommendations

Emphasis should be given to the prevention of disability: this has the highest priority.

Secondary prevention and physical rehabilitation are also vital. Appliances should be as simple as possible and should be available and accessible to a large number of people, especially the poor.

Preference should be given to projects that involve not only the family but also, wherever possible, the community as a whole.

Social and economic rehabilitation should be stressed. Wherever possible, steps should be taken to help the disabled attain economic independence (see PART TWO, section 4).

Section 10   Other Health Problems

I   Drugs for poor countries

1 Introduction

Over 90% of all the illnesses in the world can be cured by a small number of inexpensive drugs. Unfortunately, though, the tendency in many countries — especially in some of the poorer ones — is for drugs to be prescribed which are both highly inappropriate and extremely costly. For example, expensive antibiotics are often recommended where sulphur-based drugs would be much cheaper and equally effective. When costly drugs are prescribed, many people — possibly unaware of the risks involved — fail to complete a course of treatment because of lack of funds. The prevalence of certain drugs over others on the market is usually a result of the effective publicity campaigns of multinational pharmaceutical companies rather than the relative attributes of the drugs concerned.

Another problem — particularly pressing in isolated rural areas — is that of supply. In many poor countries the supply of basic drugs is hopelessly inadequate. Moreover, drugs are often prescribed without sufficient follow-up by the medical practitioner, or without providing the patient with sufficient information on possible side-effects, etc. This situation is further complicated by the use of proprietary names for many drugs. For example, the generic drug tetracycline carries more than five different names and may have as many prices, even though the basic ingredient is the same. This is confusing for both patient and health worker. Thus, it is recommended that all projects should work with a standardised supply of essential drugs, identified by their generic title and, where possible, approved by the Ministry of Health for the country concerned.

2 Essential Drugs

Table 7-6 lists the generic drugs which should cover the needs of most primary health care programmes.

Table 7-7 lists the additional drugs for use by trained health workers (requirements for 10,000 people over 3 months).

ECHO supplies all the basic drugs at reasonable prices, and field staff should hold a copy of its catalogue. ECHO is preparing a 'basic drug pack' which will include the essential items in balanced proportions.

3 Funding Policy

Except perhaps in disaster situations, drugs should not generally be distributed free of charge. The costs of drugs should be borne by the people using them; where this is not possible, the finance should come from the government or some other source, rather than Oxfam. However, in those cases where Oxfam is involved in funding drugs, the commitment should never be open-ended. Priority should be given to those projects which need finance only for an initial supply of drugs and which will thereafter assume the costs themselves. For instance, Oxfam may help set up a revolving loan, whereby supplies are maintained through the sale of drugs to patients — the cash being held in a central fund used exclusively for replenishing stock.

Wherever possible, encouragement should be given to the local manufacture or wholesale of generic drugs, providing it can be assured that the prices will compare favourably with those of drugs supplied by ECHO or UNICEF. This approach has proved highly successful in a project in Bangladesh (BGD 20), where despite the complexity of organisation and management a local factory is producing basic drugs at low cost, with the additional benefit of providing employment for a number of people.

7

Table 7 – 6   Basic Drug requirements for 10,000 persons for 3 months

Drug	Pharmaceutical form and strength	Total required for 3 months (rounded up)
Analgesics		
acetylsalicylic acid	tab 300mg	17,000 tabs
paracetamol	tab 500mg	4,500 tabs
Anthelmintic		
mebendazole	tab 100mg	2,100 tabs
piperazine	syrup 500mg/5ml (30ml bot.)	5.1 litres
Antibacterial		
ampicillin	suspension 125mg/5ml	420 bots.60ml
benzylpenicillin	inj 0.6g (1 million IU)	500 vials
phenoxymethylpenicillin	tab 250mg	9,500 tabs
procaine benzylpenicillin	inj 3.0g (3 million IU)	375 vials
sulfamethoxazole + trimethroprim	tab 400mg + 80mg	7,500 tabs
tetracycline	tab 250mg	9,000 tabs
Antimalarial		
chloroquine	tab 150mg	8,000 tabs
chloroquine	syrup 50mg/5ml	3 litres
Antianaemia		
ferrous salt + folic acid	tab 60mg + 0.2mg	30,000 tabs
Dermatological		
benzoic acid + salicylic acid	oint.6% + 3%, 25g tube	100 tubes
neomycin + bacitracin	oint.5mg + 500 IU/g, 25g tube	50 tubes
calamine lotion	lotion	5 litres
benzyl benzoate	lotion 25%	35 litres
gentian violet	crystals	200g (8 bots.)
Disinfectants		
chlorhexidine	solution 20%	5 litres
Antacid		
aluminium hydroxide	tab 500mg	5,000 tabs
Cathartic		
senna	tab 7.5mg	400 tabs
Anti-diarrhoea		
oral rehydration salts	sachet 27.5g/litre	6,000 sachets
Ophthalmological		
tetracycline	eye oint. 1%, 5g tube	750 tubes
Solutions		
water for injection	2ml	500 amps
water for injection	10ml	500 amps
Vitamins		
retinol (Vitamin A)	caps 60mg (200,000 IU)	500 caps
retinol (Vitamin A)	caps 7.5mg (25,000 IU)	400 caps

Table 7-7   Drugs for use by Doctors and Senior Health Workers

Drug	Pharmaceutical form and strength	Total amount
Local anaesthetic		
lidocaine	inj 1% vial/50ml	10 vials
Analgesic		
pethidine	inj 50mg in 1ml amp	10 amps
Antiallergic		
chlorpheniramine	tab 4mg	100 tabs
Antiepileptic		
diazepam	inj 5mg/ml, 2ml amp	10 amps
Anti-infective		
metronidazole	tab 250mg	1,500 tabs (2 tds* 5/7 for 50 patients)
benzylpenicillin	inj 3.0g (5 million IU)	100 vials
chloramphenicol	caps 250mg	2,000 caps (2 qds* 5/7 for 50 patients)
cloxacillin	caps 250mg	3,000 caps (2 qds* 7/7 for 35 adults) (1 qds* 7/7 for 30 children)
Antimalarial		
quinine	inj 300mg/ml	20 amps (2ml) (avg of 4ml per patient)
sulphadoxine + pyrimethamine	tab 500mg + 25mg	150 tabs (2 − 3 stat* for 50 patients)
Plasma substitute		
dextran	inj sol 6%/500 ml, with 10 giving sets	5 litres
Cardiovascular		
glyceryl trinitrate	tab 0.5mg	100 tabs
propranolol	tab 40mg	100 tabs
digoxin	tab 0.25mg	100 tabs
digoxin	inj 0.25mg/ml, 2ml amp	10 amps
epinephrine	inj 1mg/ml, 1ml amp	10 amps
Dermatological		
nystatin	cream 100,00 IU/g, 30g tube	10 tubes
hydrocortisone	1% cream, 30g tube	10 tubes
Diuretics		
frusemide	tab 40mg	100 tabs
frusemide	inj 10mg/ml, 2ml amp	10 amps
Gastrointestinal		
promethazine	tab 25mg	100 tabs
promethazine	syrup 5mg/ml, bottle of 250ml	10 bottles
codeine	tab 30mg	100 tabs
Hormones		
hydrocortisone	inj 100mg	10 vials
Opthalmological		
sulphacetamide	eye oint. 10%, 5g tube	250 tubes

continued overleaf

7

Drug	Pharmaceutical form and strength	Total amount
Oxytocics		
ergometrine	tab 0.2mg	100 tabs
ergometrine	inj 0.2mg/ml, 1ml amp	10 amps
Psychotherapeutic		
diazepam	tab 5mg	100 tabs
Respiratory		
aminophylline	inj 25mg/ml, 10ml amp	10 amps
salbutamol	oral inhalation, 0.1mg per dose	5 aerosols
beclomethasone	oral inhalation, 0.05mg per dose	5 aerosols
Solutions		
compound solution of sodium lactate	solution/500ml	10 litres
glucose	inj sol, 50% hypertonic, 10ml amp	10 amps
sodium chloride	inj sol, 0.9% isotonic, 500ml, with 10 giving sets	5 litres
water for injection	10ml amp	100 amps

*Key: Stat = at once
tds = 3 times a day
qds = 4 times a day
x/7 = x number of days treatment

N.B. These lists are based on UNHCR/WHO specifications. Details can be found in WHO Technical Report Series No. 685, 1983.

4 The Storage and Transport of Drugs

Care should always be taken when transporting or storing drugs to ensure that they are held in the optimum conditions to avoid spoilage. The need to protect supplies from damage may present a number of practical problems — especially in humid tropical regions. All those involved in handling and storing supplies should be aware that damaged drugs may become ineffective and in some instances may even prove dangerous to the user.

The main causes of spoilage are listed below.

a: Moisture. Certain medicines absorb water from the atmosphere and are therefore very vulnerable to humidity. Capsules and tablets, if kept in a moist environment, may become sticky or discoloured or may disintegrate. Drugs should always be stored in properly sealed containers in a dry place. If possible, a few grains of rice should be put in the container to absorb at least some of the moisture from the atmosphere.

b: Heat. All drugs should be kept as cool as possible. However, with the exception of the polio vaccine, drugs in a liquid form should never be frozen. The store should be well ventilated to allow the free circulation of air.

c: Light. Many drugs — especially those in liquid form — will spoil very quickly when exposed to direct sunlight. It is therefore important to ensure that supplies are kept in the shade at all times. They should preferably be contained in coloured glass bottles and stored in a box or cupboard.

When dispensing, open the container for as short a time as possible and only dispense in small quantities. Supplies should never be stored on the ground, and care must be taken to protect them from rodents and insects. All drugs must be stored in a given order, regular inventories must be taken, and those drugs with the earliest expiry date must be used first.

II   Alcoholism

1 Introduction

The excessive consumption of alcohol is a rapidly increasing problem in poor countries. As well as undermining the health of the individual, the addiction harms the whole family physically, mentally, financially and socially. Alcoholism increases thieving and violence both within the family and in the community as a whole. It is common to find that the root cause of a child's malnutrition is a poor home where the father is either absent or spends a large proportion of his money on drink.

There are of course other addictions having similar effects such as the chewing of chat in Yemen and East Africa and the opium smoking of the Far East: however, alcoholism is more widespread and increasing faster than any other addiction, with the possible exception of tobacco smoking.

2 The Cause

The cause is a combination of availability and the susceptibility of the individual.

a: Availability. A high proportion of alcohol consumed in poor countries is brewed locally, e.g., palm wine, tedj or tella. Although local brews have always been available, there is a growing tendency to distil these products in the home to yield high alcoholic spirits. Proprietary brands of beer and spirits are becoming increasingly available. With the exception of some Islamic countries, few, if any, governments enforce legislation on the sale of alcohol, for example, to young people, licensing hours, etc. With the growth of urban centres there has been a marked increase in the number of bars and similar places which sell alcoholic beverages.

b: Susceptibility of the individual. With the breakdown of religious and other taboos and traditional family structures, many of the natural controls which communities previously exerted on individuals have diminished. Urban drift in particular accelerates the trend towards a nuclear rather than an extended family structure. Associated with this development is the growth of the cash economy in which the husband commonly controls the family income and is therefore more able to squander the earnings on alcohol. Social pressures such as those resulting from extreme poverty, unemployment, frustrated ambition and an unhappy home life cause people to take refuge in drink. The transition from light drinking to heavy drinking and then to alcoholism can be quite gradual.

3 Prevention

There are various ways in which the problem of alcoholism may be reduced:
- ☐ tighter legislation on the sale of alcohol would help, but in most countries would be hard to enforce;
- ☐ the banning of local distillation;
- ☐ socio-economic improvements such as increased employment opportunities;
- ☐ education in schools and elsewhere about the dangers of alcoholism.

4 Cure

A real alcoholic is difficult to cure and, unless the underlying cause can be dealt with, will usually relapse. Probably one of the best ways to help is through out-patient treatment, with the assistance of the alcoholic's family: this will require regular home visiting.

7

Where alcoholism is a major problem in the community, discussions need to be held with community leaders as to what measures should be taken.

In the past, Oxfam has not been involved directly in the cure of alcoholism: however, field staff should be prepared to explore and experiment in a small way in its treatment and rehabilitation, especially where projects are community-based and alcoholics are treated on an out-patient basis.

Section 11   Environmental Public Health

I   Water supplies

1 Introduction

Humans are the reservoir of most diseases that destroy or incapacitate them. A great number of these diseases are caused by infected human excreta and other body wastes constantly contaminating food and water. When consumed, these cause illness and sometimes death. This is known as the anal-oral link, and it is a major health hazard faced by all communities, but particularly those where living standards are low. For improvements in health to be made or maintained, it is vital to break this cycle. This is done by ensuring that basic human needs are met, such as uncontaminated, wholesome food, adequate supplies of safe water for drinking, cooking and personal hygiene, and proper sanitary disposal of human excreta and other wastes.

2 Water for Human Usage

Water, essential to life and health, is required by humans in considerable quantities for drinking, cooking, personal hygiene and for washing (about 15–30 litres per head per day), quite apart from the demands of any livestock or irrigation requirements. To fulfil this fundamental requirement, an adequate supply of safe water should be available to all human beings.

A *safe water* supply is one that cannot harm the consumer. Polluted drinking water is a constant and major cause of enteric (i.e. intestinal) diseases; the possibility of contamination increases enormously under disaster conditions.

In any project undertaken to improve water supplies, the local people must be encouraged to participate in identifying and expressing their needs and preferences. This extends to promoting an awareness of hygiene standards and precautions, as well as motivating them to achieve these by communal self-help. The supply of safe water should, wherever possible, be integrated with the improvement of other social needs such as agriculture, maternal and child welfare, women's training, child education, etc.

If water is found to be too *brackish or saline*, it will be unsuitable for drinking or cooking purposes. It can however be used for personal hygiene or the washing of clothes. Where safe water is in short supply, it is important to enquire as to the availability of non-potable supplies for these purposes.

Transportation from source to user can be a problem. Since one litre of water weighs one kilogram, and the average daily requirement per person is between 15 and 30 litres, piped supplies are clearly the only feasible long-term solution. Transportation by other methods, such as buckets and mobile tanks, may have to be used in the short term for disaster situations. Pipes make for ease of access and reduce the chances of contamination.

Even the simplest of water supply schemes requires a great deal of technical expertise and administrative follow-up to guarantee its long-term viability. Training may be necessary to ensure that maintenance and care of tools and equipment is continued, and education must be kept up.

The need for water is such that human beings tend to use any water that is readily available to them, whether it is polluted or not. The source must therefore supply a quantity of safe water that is adequate for the needs of the community. In general, the main source of water is rain, which runs off the land in either streams or rivers. Alternatively, it will penetrate the ground where it will be stored, or eventually seep to rivers or the sea. *Research into local conditions*, some of it of a highly technical

Table 7 – 8   Choosing a Source of Water

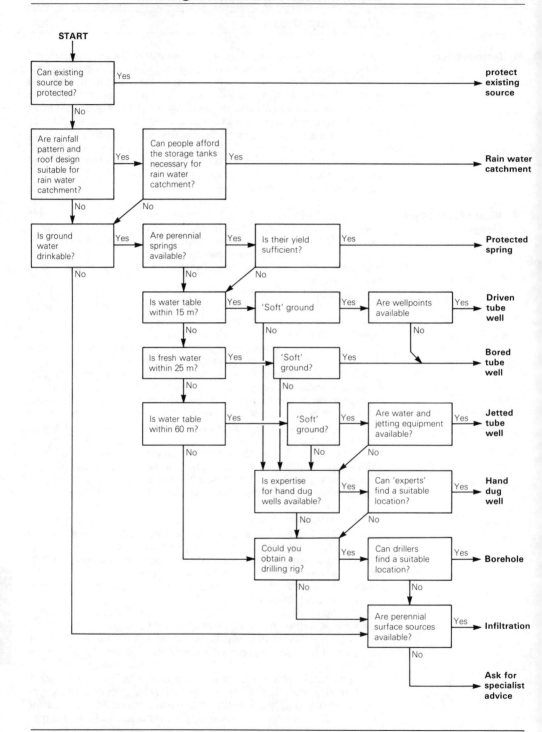

Source: *Small Water Supplies*, Richard Feachem, Sandy Cairncross (Ross Bulletin 10, January 1978).

nature, is necessary before any water supply scheme, of whatever magnitude, can go ahead.

It is useful to *measure the quantity* of rainwater and record the results. The extent of its penetration through the ground should be ascertained and, if necessary, this should be improved by slowing the run-off.

Storage of rainwater can be arranged by constructing small dams to create ponds or reservoirs, by land-terracing or by providing tanks.

Some rainwater may be collected and stored before it reaches streams or passes into the ground. It may be feasible to improve this at either the household, the institution or the community level. However, rainwater or soft upland waters should not be stored in metal containers, as soft waters can dissolve metals such as zinc, lead and copper into the supply, causing a health hazard.

Further, such waters may *lack essential elements such as iodine and calcium*. Iodine deficiency causes goitre, and can only be rectified through the large-scale provision of iodised table salt. Calcium is the basic building block for sound bones and healthy teeth, and a vital dietary ingredient. Calcium deficiencies in the water supply can be offset by making sure that foods high in calcium are available. The best sources of calcium are milk and milk products, e.g. dried milk, sour milk, yoghurt, tinned milk. If these are not available, finger millet and sesame (sim-sim) are also good sources, as are any small fish eaten whole, such as tinned pilchards and sardines. Any varieties of *amaranthus* (a dark green vegetable), sweet potato leaves and cassava leaves also provide a useful source.

③ Treatment of Water

Water in streams or rivers is called *surface water*, and that flowing underground or emerging as springs is called *ground water*. Where available, the latter is the preferred source for human consumption, as, having filtered through the ground, it is usually fairly safe, whereas surface supplies are liable to be polluted. If surface waters *must* be used for drinking water, the points below should be taken into account to minimise the risks to health.

a: Always select a surface water that is *upstream of any pollution source*. Every effort must be made to avoid polluting the source with human excreta or other wastes. Water extraction points must be protected and fenced.

b: Excessively *turbid (i.e. murky)* waters should be avoided if at all possible, as they carry the highest possibility of pollution.

c: *Storage reservoirs* can reduce bacterial populations by 50%-90% depending on the storage time. They should have a minimum capacity of two days' water, and ideally of up to two weeks'. If not available at the community level, adequate household storage containers can upgrade water quality. Without storage, it is difficult to provide a safe supply from surface water. There are at present no simple water testing or treatment techniques using chemical disinfection that are suitable for long-term use by poor communities.

d: Other methods of water improvement are:

☐ *infiltration galleries*. This is an old technique which involves digging a well above the flood line that connects to a perforated pipe on the river bed. It allows for a fixed pumping point, regardless of the conditions of the river;

☐ *river bed filters*. A screen is placed inside a hole in the river bed which is then covered with clean sand or gravel from the river. The screen is connected to the suction pipe of a hand- or power-operated pump. The pump draws water through the screen so that dirt and sand gather and are held on the river bed above the screen. This increases the amount of filtration. The surrounding river bed itself becomes a biological filter that destroys bacteria and reduces the level of ammonia and iron that may be present.

Both these techniques provide an effective primary filtering process.

e: One of the best ways of purifying water is by *slow sand filtration*. Probably no other single treatment process can simultaneously improve the microbiological, chemical and physical quality of water to the same extent. It is simple, inexpensive, and reliable, and remains the most universally appropriate treatment available. Properly operated, a sand filter can remove 90% or more of all bacteria caused by the presence of human sewage. It is suitable for treating water of reasonable quality, possibly contaminated, but low in turbidity (i.e. fairly clear). If the water is turbid, then some preliminary *sedimentation* is recommended. This is a process whereby water is stored so that the heavy matter in suspension settles out.

Essentially, the slow sand filter is an open-topped box, partly filled with a filtering medium (usually clean sand) and a layer of stones or gravel. Raw, untreated water is introduced into the space above the sand and then works its way down through it. The treated water is discharged through the under-drains. The filter will run for weeks or even months before cleaning is necessary. It is essential that the sand should not be allowed to dry out as this destroys the microbiological organisms which make it so effective.

This method has numerous advantages for developing countries:

- [ ] low construction costs, particularly when manual labour is used;
- [ ] the simplicity of the design and operation means that a filter can be built and used with very limited technical supervision;
- [ ] very few materials and little equipment need to be imported, and no chemicals are required;
- [ ] if water can be delivered to the filter, no power is required, since there are no moving parts;
- [ ] variations in incoming water quality need not be a problem, provided that its turbidity is not excessive;
- [ ] since large quantities of wash water are not required, water is used economically;
- [ ] there are fewer problems associated with sludge than with more sophisticated methods.

4 Extraction of Sub-surface Water

Three particularly useful and well proven drilling techniques exist for the extraction of sub-surface water, and these are outlined below:

a: Hand augers can be used in 'soft' soils, e.g. clays, alluvials (river deposits or silts), soft rocks. They are available in sizes ranging from 2 — 12 inches (10 — 30 cm) and can drill in suitable soils to a depth of 10 — 15 metres. Small diameter augers are being used successfully as survey kits for determining where water tables can be located and either drilled or hand dug to obtain water. Costs of such augers are around £3,000 (1984).

b: Jetted well points can be made by using a tubular device down through which water is forced by a powered pump. The water emerges through the 'jetting shoe' and washes away the surrounding soil, thus allowing the well point to be forced further into the earth.

c: The driven tube well uses a specially perforated tube which is driven down into the ground by means of a small weight until water is reached. The water rises up the pipe and is available for pumping, the tube itself acting as the suction pipe. Tube wells of this type can be driven into all classes of strata, provided that the ground is not too hard. One advantage of this method is that if no water is found, the tube and drive point can be withdrawn and used elsewhere. However, wells of this type can only be used when water is less than 25 feet from the surface, as this is the maximum depth for use with standard pumps.

Table 7 – 9   **Three basic methods of groundwater extraction**

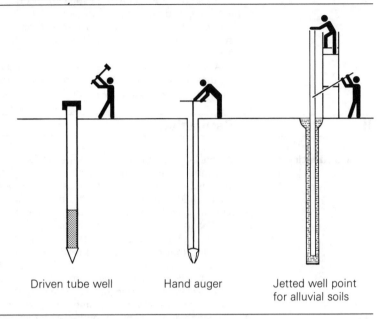

| Driven tube well | Hand auger | Jetted well point for alluvial soils |

Source: *Small Water Supplies*, Richard Feachem, Sandy Cairncross, (Ross Bulletin 10, January 1978)

The kind of help that the NGO can give ranges from providing funds for small-scale improvements such as well-digging, to fully integrated community supply schemes. In Western Kenya for example, Oxfam has provided a submersible dewatering pump to remove the water while the diggers complete the bottom section of 50ft wells for small communities (KEN 168).

5 Oxfam Water Packs

These are used in disasters and refugee or long-term projects and are designed with a population of 5,000 people in mind. However, the packs and their accompanying equipment can readily be modified or multiplied, and much of the technology used is as relevant for long-term community development as for short-term disasters. They all share certain characteristics which make them appropriate to a variety of circumstances:

☐ clear, easy to follow diagrams detailing all stages of each particular operation;

☐ very low construction costs — a complete supply of tools and equipment accompanies each pack;

☐ they are capable of rapid installation by semi-skilled workers within hours;

☐ they are reusable and have low maintenance and energy requirements;

☐ they are applicable to very many situations, regardless of soil conditions or slopes.

There are four packs available.

Pumping: this contains a pump with its suction and delivery hoses, together with ancillary fittings to suit the particular source.

Storage: this consists of three tanks, and may be used to provide

7

storage for an existing distribution system or for other purposes such as fire-fighting.

Distribution: this contains a supply main kit, a water collection point kit, a venting and draining kit, and a tool kit.

Treatment: this pack is basically an easily assembled slow sand filtration system using two separable tanks to allow for isolation and cleaning of each.

A further pack on hand augers and hand-dug wells is in preparation. Further information can be obtained from the Oxfam Technical Department, 274 Banbury Road, Oxford, OX2 7DZ, U.K.

II Sanitation

1 Introduction

Where large numbers of people are without adequate latrines, disease is inevitable, often fatal, at times reaching epidemic proportions in crowded conditions. This is caused by pathogens present in excreta or urine which then reinfect the population. Effective environmental sanitation and sewage disposal are a basic requirement for improving the health and well-being of all communities.

The worst effects of inadequate sanitation can be alleviated by ensuring that human excreta is *always* discharged into a chosen *safe* location, so as to keep food and drinking water free from contamination.

This discharge may be via a family latrine connected to a sewer, septic tank, pit latrine, or even just a simple purpose-dug hole in a field; any of these solutions requires an integrated programme of local participation, education and research, together with technical knowledge and advice to ensure that the most appropriate scheme is selected.

2 Education and Participation

The complexity of the problem is far greater than may at first appear. The installation of amenities may be a relatively simple matter, but care must be taken to ensure that adequate maintenance and educational facilities are available. Motivation has to be stimulated and encouraged, since for the majority of people in the Third World hygiene of this nature is simply not a priority.

No sanitation programme will succeed without the understanding, support and active participation of the people concerned. Technical improvements that fail to take into account local customs, traditions and beliefs have usually proved to be of little lasting value.

The priority target groups for health and hygiene education are schoolchildren and mothers with children.

Oxfam is at present involved in a scheme run by the Catholic Church in Ethiopia (ETH 74) to provide 200 hand-dug pit latrines as a follow-up to a gravity water supply for a village. Although it was difficult to promote awareness of hygiene, the project did succeed in encouraging villagers to apply for materials for their own latrine. The labour and much of the raw material was provided by the people themselves, with Oxfam supplying concrete slabs and metal sheets for lining.

3 Soil Conditions

The choice of latrine facilities depends upon local geology. It is vital that pollution of groundwater supplies should be avoided; no latrine should be placed within 50 metres of a water extraction point.

When considering the choice of latrine facility, the following points should be kept in mind:

☐ sanitation problems are much greater in wet, swampy areas than in dry conditions;

☐ it is more difficult to provide adequate sanitation facilities on flat, undrainable land than in hilly regions;

□ if the area is susceptible to flooding, above-ground latrines should be used;

□ good percolation soils are preferable to rocky, impervious soils, but where the ground is excessively hard, a suitable 'sanitation without water' technique may provide a solution.

4 Appropriate Sanitation Technologies

Any sanitation system breaks down into four basic stages: deposition, collection, transportation and treatment.

Deposition. This is the act of defecation itself, and there are three ways of dealing with it: the water seal with cistern flush which uses ten litres or more from the mains water supply (the system used almost universally among high-income groups in all parts of the world); the water seal with pour flush (using between one and three litres of water); and the squatting slab that uses gravity to remove the excreta into a hole. The last is probably the simplest and most efficient method, because — except in areas where anal cleansing with water is customary — it requires no water.

Collection. After deposition, the excreta are normally collected in a pit, tank or bucket. This is usually when decomposition takes place, which occurs best under the following conditions:

□ the environment should be as dry as possible;
□ plenty of ventilation is required;
□ the area should be free of rodents, insects and smells;
□ the excreta should be left undisturbed.

The length of time required for decomposition varies, ranging from two weeks, in optimum conditions, to six months. During this 'retention period' the majority of bacteria will be eliminated without the necessity for water, resulting in a fairly harmless compost material. For composting to work well, some organic waste such as food left-overs and ashes should be added to the excreta.

There are several collecting methods that are in use in poor areas. (See also PART EIGHT, Section 4 III).

*a: **The pit latrine***, where the excreta are collected in a hole in the ground, usually located directly beneath a squatting slab, and generally dug by hand. When two-thirds full, it is filled-in with earth and a new pit is dug nearby. If two pits are dug side by side, the contents of one can be composted while the other is used, and vice versa. The Vietnamese double pit latrine works on this principle. Pit latrines are not generally suitable in highly populated areas.

*b: **The aqua privy*** is a water-filled tank located directly beneath the defecation point. The solid material settles and forms a sludge that is digested anaerobically (i.e. without oxygen), while the liquid effluent flows out through an outlet pipe. The tank has to be kept full of water to provide a seal against odours. The effluent usually flows to a soakaway. The sludge must be removed periodically.

Aqua privies have worked much better in parts of the world where water is used for anal cleansing (as in Muslim countries) than where sticks, stones or heavy paper may be used (as in parts of Africa). No refuse of any kind should be placed in an aqua privy as it causes premature filling of the tank and then failure. The water used for anal cleansing has the effect of keeping the squatting slab and chute clean and of providing the necessary water to flush through the tank and maintain the seal. Aqua privies are more permanent than pit latrines and offer the most cost-effective solution to the problem of sanitation.

*c: **The bucket latrine*** is one of the oldest and generally least hygienic systems. A squatting slab is placed directly above a bucket which is accessible from the street or lane, and is emptied by a sweeper - preferably once a day, but usually once or twice a week.

*d: **Communal latrines*** are another useful alternative. They must be

7

Table 7–10   Generic Classification of Sanitation Systems

On site Dry	On site Wet	On site or off site Wet	Off site Wet	Off site Dry
1. Overhung latrine	8. Pour-flush latrine, soakaway	14. Low volume cistern flush, soakaway, or sewer	17. Conventional sewerage	18. Vault and vacuum tank
2. Trench latrine	9. Pour-flush latrine, aquaprivy, soakaway	15. Low volume cistern flush, aquaprivy, soakaway, or sewer		19. Vault, manual removal, truck, or cart
3. Pit latrine	10. Pour-flush, septic tank, vault	16. Low-volume cistern flush, septic tank, soakaway, or sewer		20. Bucket latrine
4. Reed Odorless Earth Closet	11. Sullage flush, aquaprivy, soakaway			21. Mechanical bucket latrine
5. Ventilated improved pit latrine	12. Sullage flush, septic tank, soakaway			
6. Batch-composting latrine	13. Conventional septic tank			
7. Continuous composting latrine				

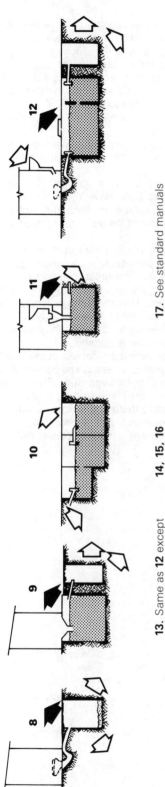

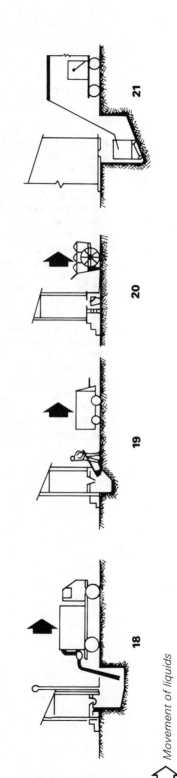

13. Same as **12** except conventional cistern-flush.

14, 15, 16 Same as corresponding configuration in **8** to **12**, except for elevated cistern with low volume-flush.

17. See standard manuals and texts

Movement of liquids

Movement of solids

Source: The World Bank, *Water Supply and Waste Disposal*, Poverty and Basic Needs Series, Washington D.C., September, 1980.

7

within easy access of the population served and should have separate entrances for each sex. Due attention should be given to privacy as well as the provision of water for cleansing and washing. These latrines can often supply the needs of quite substantial numbers of people. In Nicaragua for example (NIC 40), 250 latrines are providing sanitation facilities for a community of some 4,000 people. Local materials are being used to reduce costs and to involve the local people. A full-time competent attendant is required to keep the latrine in the best possible condition.

Transportation. Waste material is either piped in a flow of water through a sewerage system (a very expensive alternative) or is carted without added water (nightsoil collection).

Water-borne sewage systems must be built on self-cleansing gradients otherwise costly pumping stations have to be provided. Obviously, sufficient water must be available to flush the system. Ventilation must be provided, so as to prevent the build-up of poisonous and explosive gases.

Nightsoil collection is still used in many parts of the world, and some of the collection systems use quite sophisticated motorised vehicles equipped with suction facilities. However, in a great many areas, the collection is carried out by poor, ill-equipped, badly paid workers, often working in the most degrading conditions. This method is generally one that should be discontinued or replaced by other schemes. It would be appropriate to supply communal latrines where other options are not possible, and where good local management is assured.

Treatment. The best and most appropriate system for use in the Third World is that using waste stabilisation ponds. These can be constructed for small communities (such as a school) or for large cities. The ponds can receive sewage from a conventional sewerage system, or from an aqua privy, or nightsoil which is carted to them and sluiced into them down a ramp. They stimulate natural biological and physical processes and require no mechanical parts or equipment. They are economical and easy to construct.

Table 7–11 Comparison of Several Sanitation Technologies

Sanitation System	Rural Application	Urban Application	Construction Cost	Operation Cost	Ease of Construction	Water Requirement	Hygiene
Pit latrines	Suitable in all areas	Not in high density suburbs	Low	Low	Very easy except in wet or rocky ground	None	Moderate
Bucket and cartage	Suitable	Suitable	Low	High	Easy	None	Bad
Vault and vacuum truck	Not suitable	Suitable where vehicle maintenance available	Medium	High	Requires skilled builder	None	Moderate
Aqua privies	Suitable	Suitable	High	Low	Requires skilled builder	Water piped to privy	Good
Septic tanks	Suitable	Suitable for low-density suburbs	Very high	Low	Requires skilled builder	Water piped to privy	Excellent
Pour flush and soakaway	Suitable	Not suitable	High	Low	Requires skilled builder	Water source near privy	Good
Sewerage	Not suitable	Suitable where it can be afforded	Very high	Medium	Requires experienced engineer	Water piped to privy	Excellent

Source: *Small Excreta Disposal Systems*, Richard Feachem, Sandy Cairncross, (Ross Bulletin 8, January 1978).

7

Table 7 – 12.1   First-stage algorithm for Selection of Sanitation Technology

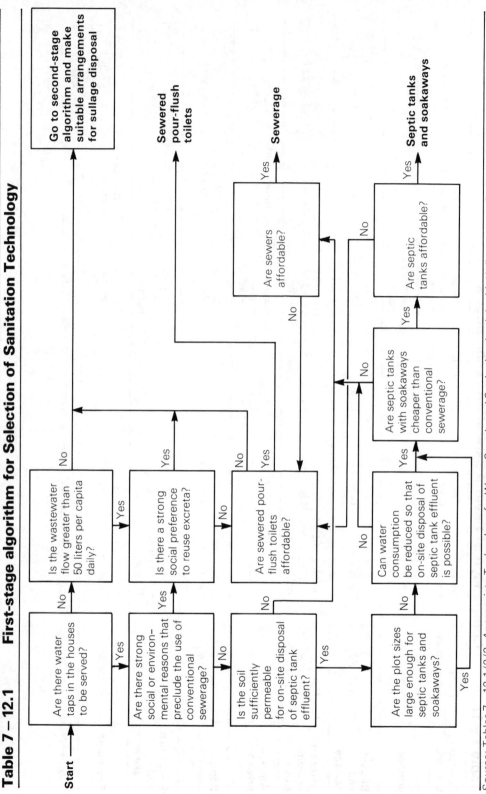

Source: Tables 7 – 12.1/2/3. *Appropriate Technology for Water Supply and Sanitation*, by John M. Kalbermatten, DeAnne S. Julins, and Charles G. Gunherson. (World Bank, December 1980).

Table 7 – 12.2   Second-stage Algorithm for Selection of Sanitation Technology

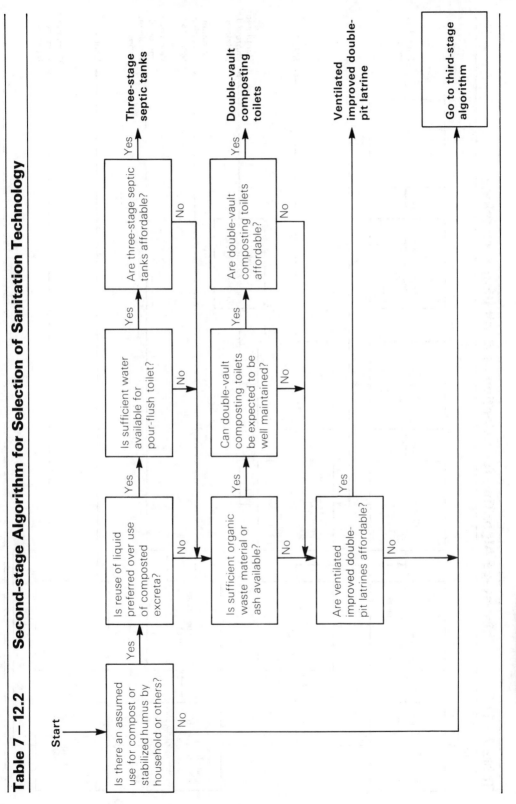

7

Table 7 – 12.3   Third-stage Algorithm for Selection of Sanitation Technology

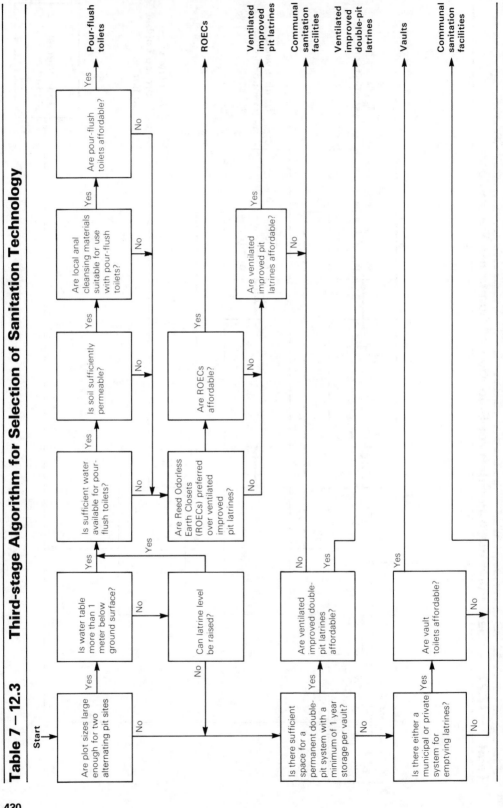

I   Primary health care

1. Population of catchment area; description (e.g. socio-economic or tribal groups).

2. Brief details of geographical, climatic, agricultural and sociological background.

3. Map indicating terrain, communications and all health facilities of the area — both voluntary and governmental (this need only be roughly to scale). Indicate distance from nearest alternative health facility.

4. What are the main health problems in the area and how many people are affected? Differentiate by gender and age, distinguishing particularly between adults and children (where possible, give reasons for distinctions). What proportion of the total population is affected.

5. What are the underlying causes of these problems? (e.g. ignorance, poverty, large families, lack of immunisation, poor water management, lack of sanitation, etc.) At what times of the year are the problems greatest (e.g. pre-harvest famine)?

6. How does the community see its needs?

7. What is the best method to cure the problems and remove their underlying causes? How does this approach relate to the government programme? Be specific about the use of mobile services, if any. How will they be related to static clinics?

8. How do you intend to involve the community both in the planning and implementation and financing stages?

9. How will you ensure adequate *coverage* of the population in your catchment area?

10. What staff do you intend to use? What form of training of para-medical personnel is planned? Will health promoters be used? What will be the sex distribution of the staff?

11. Trends in attendance over the past year. Will the project have an impact on attendance?

12. How will patients be referred?

13. How will this work be financed? e.g. by charges or insurance schemes. If no charges are made, why not?

14. If primary health workers are used, how will they be selected, rewarded, supported and supervised? To whom will they be responsible? Are they full-time or part-time?

15. What is the system of record-keeping and accounting? What is the approximate cost per patient?

16. What will be the birth control content of your programme?

17. How do you intend to assess your programme?

☐ *Infrastructure:* e.g. how many clinics opened — personnel trained — meeting held — drug supply system — transport, etc?

☐ *Coverage:* i.e. what proportion of the target population in each category is to be reached? (antenatals, under-5s, TB patients, proportion of children immunised, proportion of families practising birth control).

□ *Quality:* Good supervision, correct diagnosis and treatment.

□ *Health statistics:* e.g. reduction of under-5s mortality, reduction of TB or measles incidence.

Please state therefore what measurable improvement you hope to achieve by the end of each year.

II   Nutrition and diarrhoea

---

1. Is the child's weight adequately monitored during the rehydration process? When rehydration is complete, is feeding, especially breast-feeding, continued? Is there need for a rehydration centre? If so, are back up supplies of intravenous solutions and equipment available? Should wider preventive measures be adopted?

2. Have the principal causes of malnutrition been identified? Are these due to ignorance, e.g. improper diet practices, or due to external factors, e.g. lack of food supplies?

3. Have the most vulnerable groups been identified? Is special attention given to babies and young children, especially those just weaned, pregnant and lactating mothers, certain socio-economic groups?

4. In combating malnutrition, is an appropriate response being worked out *with* the community?

5. Is enough being done to use local foods and are appropriate groups encouraged to grow more of their own? Are seeds, tools and advice available? Are agricultural techniques being improved and is there a good reason for the introduction of new crops?

Is there sufficient assistance in marketing? Are those without access to land or employment given sufficient help in procuring cheap, nutritious foods? Are there sufficient income generating schemes?

6. Food supplements, if necessary, should be in the form of food-for-work programmes. Have appropriate alternatives to this been considered? Where food distribution is unavoidable, have the best distribution centres been identified and is there provision for follow-up visits?

7. Is health and nutrition education the priority? Are education programmes run at relevant places, e.g. ante-natal clinics, schools, women's groups, etc.

8. Is breast-feeding adequately promoted together with the supplementation of local foods? Is there sufficient emphasis on the nutrition requirements of children just weaned?

9. Is home visiting being followed? Is there an impact on the diet, food storage and food preparation of each affected household? Are children receiving the correct diet?

10. Are under-5s clinics being supported for sufficient health monitoring, education, food supplement programmes? Is there opportunity to organise nutrition scouts? Has some baseline study been carried out before the establishment of an NRU and what provision is there to evaluate the impact on the community? Is there sufficient follow up and home-visiting? What are the results of other NRUs in the area?

III	**Water**

1. What research has been made into local ecology?

2. Is rainwater storage possible?

3. If surface water is used, are appropriate methods of water improvement available?

Has provision been made for education in basic hygiene?

4. Has slow sand filtration been considered?

5. If ground water is chosen, which method of extraction is feasible and appropriate?

6. Have the Oxfam water packs been considered?

See also Table 7-8: Choosing a Source of Water

IV	**Sanitation**

1. Has adequate provision been made for involving and educating the local population?

2. Have the local geological conditions been investigated? What latrine facilities are therefore appropriate?

3. Are nearby water extraction points free from the risk of contamination?

4. Have all the possibilities been considered as regards deposition, collection, transportation and treatment?

7

Section 13   Appendices

I   Basic nutrition

1 Types of Nutrients

a: Carbohydrates (starches and sugars) provide energy, for physical activity, body heat and body functions.

b: Fats and Oils are a concentrated source of energy, stored as a reserve mainly under the skin.

c: Protein: needed for body building and maintenance; can also be used as an energy source if carbohydrates and oils do not meet the body's caloric needs.

d: Minerals: e.g., iron, calcium, iodine and salt: small amounts of these are needed in the formation and maintenance of various tissues, for example, bones (calcium), teeth (calcium fluoride), hormones (iodine), body fluids (salt).

e: Vitamins: 'vital-amines', small quantities of which are needed as catalysts in the production of body chemicals essential to growth and health:
- [ ] Vitamin A for preventing blindness;
- [ ] Vitamin B for preventing pellagra, beri beri, etc;
- [ ] Vitamin C for preventing scurvy;
- [ ] Vitamin D for preventing rickets.

N.B. A balanced diet provides all these ingredients without any need for supplements or vitamin tablets.

Table 7 – 13   Contents of Common Foodstuffs

Foodstuffs	% Protein	% Fat	% Carbohydrate	Vitamins/Minerals
Cereals	7-14	negligible	75	B vitamins, iron
Roots & tubers	1-2	negligible	20	some B & C vitamins
Soya & groundnuts	25-40	rich	20	iron,
Other legumes	20-30	negligible	20-35	calcium
Oilseeds, nuts	20	40-50	20-25	
Vegetables, fruits and immature pulses	2-5	negligible	little	carotene for Vit.A, Vitamin C, iron, calcium
Animal foods	10-25	0-30	nil, except milk & liver	Vitamins A, D

Table 7 – 14   Energy requirements per day

in KCalories (Mega Joules shown in brackets)

Age	KCals/Kg	Ave.Total KCals.
0-1 year	112 (0.470)	400-1200* depends on weight
5 years	91 (0.381)	1870 (7.8)
Adult Male	46 (0.192)	2000-3000 depending on weight,
Female	40	activity and sex.

*Note (1) For first 3 months, fluid requirement 150mls per kg body weight in 24 hours.

(2) 1000KCal = 4.184 Mega Joules

1 Mega Joule (MJ) = 239 KCals.

2 Protein requirements	These vary with age, being proportionately higher in young children than in adults. The situation is complicated by the fact that different proteins have different nutritional values. This is because protein is made up of various essential amino acids and, the more closely the relative quantities of these amino acids match the relative needs of the human, the higher will be the value (net protein utilisation or NPU) of the protein. The NPU is highest for animal proteins, but certain vegetable foods complement each other, e.g., cereals and legumes. In practice this means that a mixed diet should be encouraged.

Protein needs can either be expressed as:

a: grammes per kilogramme live weight: 3-3.5 gm for infants and 1.2 gm for adults; or

b: the percentage of energy or calories in the diet which is provided by protein (NDP Cal %). For instance, this is about 8% for older children and 5-6% for adults . Most staple foods have an NDP Cal of 5% or over (cassava and cooking bananas being exceptions). Thus, high protein supplements are only necessary in the very young and the severely malnourished.

It is important to stress that a proper balanced diet may still be insufficient in total calorie or energy content: if this is so, the protein will be burnt off to provide energy, and growth will be retarded.

3 Minerals and Vitamins	These are usually contained in adequate quantities in a normal diet. The routine use of vitamin pills should be discouraged. Much has been written about the lack of Vitamin A leading to blindness: it is certainly a risk among people living on diets deficient in milk products, eggs and green vegetables. It is important to realise that Vitamin A is contained in the fat portion of milk and is thus lacking in dried skimmed milk (DSM). Therefore, when DSM is used as a relief food, alternative sources of Vitamin A must be provided. As Vitamin A is stored in the liver, deficiencies may take some time to develop, early signs being roughening of the skin and a reddening and itching of the eyes proceeding to an opaque cloudiness. Where the diet is deficient, Vitamin A can be given in large doses at 3-monthly intervals (100,000 units: larger doses give longer protection but can be toxic).

Folic acid and iron are also frequently inadequate in the diet, resulting in anaemia (especially where other factors such as malaria and hookworm predispose the population to this condition).

II   Assessment of malnutrition

1 PEM (Protein Energy Malnutrition)	*a: anthropometry*: this measures the state of nutrition of the individual and is an indication of what has happened in the past.

☐ *slowing of weight gain or weight loss*: the evidence of inadequate growth rate is a more useful indication of malnutrition than a single low measurement.

☐ *weight for age*: this can be demonstrated graphically on the 'road to health' charts and is a cornerstone of the under-5 clinic. However, in certain circumstances it is difficult to estimate age.

☐ *arm circumference*: this is a less accurate measure, but is extremely useful for screening. The normal arm circumference of children between the age of 1 and 5 years is fairly constant at about 17 cms. Low arm circumference indicates a loss of subcutaneous fat and muscle which in turn indicates malnutrition. A child with an arm circumference under 13.5 cm is malnourished.

☐ *weight for height* is probably the most useful single measurement as it distinguishes between short, fat children and tall, thin children, both of which types might fall into the same grouping on a weight for age basis. Weight for height is the most useful measurement in surveys and famines. Any child below 80% weight for height is malnourished.

☐ the QUAC stick involves the use of the arm circumference in relation to height and is similar in its purpose to weight for height, but not as accurate. It is useful for screening and has the advantage of not requiring scales.

b: present diet: it is important to have some idea of what people are eating at the time of assessment.

c: reserves of food and cash/storage/harvest prospects: these all give some index of future trends. But it is important not to take too simplistic a view; many other factors, like market prices, food imports, hoarding tendencies, etc., also play a part. It may be useful to list a few basic food substances and compare market prices and availability over a period of time.

② Anaemia

This can be measured:

a: roughly, by looking into the inside of the lower eyelid for paleness;
b: by pinprick, using special paper strips and comparing the colour of the blood with a standard;
c: by standard laboratory testing.

III   Essential information for those contemplating sterilisation

---

1. The operation is irreversible and permanent.
2. Details of operation, anaesthetic, days off work, etc.
3. Post-operative check-up.
4. Effects of operation.
5. Post-operative fertility/tests for sterility.
6. Post-operative counselling.
7. Disadvantages of vasectomy/tubal ligation.
8. Informed consent.

IV	Comparison of vasectomy and female sterilisation	
	VASECTOMY	**FEMALE STERILISATION**
Effectiveness	Very effective, but slightly higher rate of spontaneous recanalisation and sperm production.	Very effective; slightly lower failure rate.
	Effective 6–10 weeks after surgery.	Effective immediately.
Complications	Procedure involves almost no risk of internal injury or other life-threatening complications.	Procedure involves slight risk of serious internal injuries and other life-threatening complications.
	Very slight possibility of serious infection	Slight possibility of serious infection.
	No anaesthesia-related deaths.	Few anaesthesia-related deaths.
Acceptability	Minute scar.	Scar can be small but still visible.
	Slightly more reversible.	Slightly less reversible.
	Less expensive	More acceptable in many cultures.
Personnel	Can be performed by one trained person with or without an assistant.	Team needed, including one doctor, one trained anaesthetist, and at least two assistants with more training than is needed for vasectomy assistant.
	Safely performed by trained paramedics.	Not suitable for parademics to perform.
	Can usually be performed in half the time of female sterilisation.	Usually only physicians with training in gynaecology can perform laparoscopy and laparotomy. Minilaparotomy is simpler.
Equipment	Requires no specialised equipment. Equipment readily available.	Laparoscopy requires expensive, complex equipment, which needs to be carefully maintained. Minilaparotomy requires only simple standard surgical instruments.
	Can usually be performed under local anaesthesia.	Systemic sedation necessary as well as local anaesthesia.
Back-up Facilities	No back-up facilities needed for immediate complications.	Back-up facilities needed in case of damage to abdominal organs and blood vessels or other complications that may occur in laparotomy.
Possible Long-Term Side Effects	None demonstrated. Uncertainty about effect of increase in sperm antibodies.	Slight risk of ectopic pregnancy.

7

L. Liskin, J.M. Pile, and W.F. Quillin, *Vasectomy — Safe and Simple*, **Population Reports**, Series D, Number 4, November-December 1983, p. D-69. Population Information Program, The John Hopkins University, Baltimore, Maryland 21205, USA.

Section 14   RESOURCES

I   Bibliography

1 Health Guidelines

H.H. Abramson, *Survey Methods in Community Medicine*, Churchill Livingstone, Longmans, Harlow, Essex, 1984.

V. Djukanovic and E.P. Mach, *Alternative Approaches to Meeting Basic Health Needs in Developing Countries*, WHO 1975.

H.E. Hilleboe and others, *Approaches to National Health Planning*, WHO 1972.

M. King, *Medical Care in Developing Countries*, Oxford University Press, Oxford, 1970.

Prof. D. Morley, *Pediatric Priorities in the Developing World*, Butterworths 1973. (Also available in French and Spanish.)

Prof. Morley, Rhodes and Williams, *Practising Health for All*, Oxford University Press, 1983.

Prof. F.J. Bennett, *Community Diagnosis and Health Action*, Macmillan Tropical Community Health Manuals, 1979.

D. Melrose, *Bitter Pills*, Oxfam, 1982.

M. Muller, Tobacco and the Third World *Tomorrow's Epidemic*, War on Want, London, 1978.

R. Feacham (ed.), et al, *Water, Washes and Health in Hot Climates*, Wiley, Chichester, U.K., 1980.

O. Shirley (ed.), *A Cry for Health* — poverty and disability in the Third World, The Third World Group for Disabled People, 16 Bath Street, Frome, Somerset, BA11 1DN, 1983.

D. Melrose, *The Great Health Robbery*. Baby milk and medicines in Yemen. OXFAM, Oxford, 1981.

2 Primary Health Care

D. Werner, *Where There is No Doctor*, Hesperian Foundation, 1977. £3.00. Highly practical, many illustrations. Essential for those villages developing programmes. Also available in Spanish, French, Arabic.

D. Werner, *Helping Health Workers Learn*, Macmillan, London.

M. and F. King, *Primary Child Care*, Oxford University Press, Book One: Comprehensive child care in simple language, well illustrated. £2.00. 1983. Book Two: Guide for the community leader. £3.95. 1979.

E. Barton et al, *On being in Charge*, WHO, Geneva, 1980.

A.S. Benenson, *Control of Communicable Diseases in Man*, American Public Health Association, Washington D.C., 1975. £4.50. Also available in French and Spanish.

J. Elford, *How to look after a refrigerator*, AHRTAG, London, 1980.

A. Battersby, *How to manage a Health Centre Store*, AHRTAG, London, 1983.

M. King, *A Medical Laboratory for Developing Countries*, Oxford University Press, 1973. £4.00. Also available in Spanish, from Editorial Pax-Mexico.

M. Cheesborough, *Medical Laboratory Manual for Tropical Countries*, Volume 1. (Volume II in preparation). 1981. Obtainable from M. Cheesborough, 14 Bevills Close, Doddington, Cambs., UK, PE15 0TT.

C.R. Swift, *Mental Health*, African Medical and Research Foundation (AMREF), Nairobi, Kenya, Rural Health Series 6, 1977. A manual for medical assistants and other rural health workers.

D. Werner, *The Village Health Worker, "Lackey" or "Liberator"*, Hesperian Foundation, 1977. 30p.

D.J. Halestrop, *Simple Dental Care for Rural Hospitals*, 40p. Basic knowledge for a medical worker caring for dental conditions. Also available in French. 1975. Medical Missionary Ass., London.

G.J. Ebrahim, *Manuals* designed for use in small hospitals and health centres, Macmillan, London, 1978.

Breast Feeding, the Biological Option, £1.00 *Child Care in the Tropics*, £1.20 *Care of the Newborn in Developing Countries*, £1.40 *Practical Mother and Child Health in Developing Countries*, £1.30 *A Handbook of Tropical Paediatrics*, £1.10

Journals

Contact, issued by Christian Medical Commission, World Council of Churches, 150 route de Ferney, CH-1211 Geneva 20, Switzerland.

Salubritas, quarterly in English, French and Spanish. Published by American Public Health Association, International Health Programme, 1015 Eighteenth Street N.W., Washington D.C. 20036, U.S.A.

Tropical Doctor, a Journal of Modern Medical Practice. Published quarterly by the Royal Society of Medicine. Available from Academic Press Inc. Ltd., 24-28 Oval Road, London NW1 7DX. Subscription £6.00 (UK) £8.80 (overseas).

WHO Chronicle, published by WHO, provides a monthly record of the principal health activities undertaken in various countries with WHO assistance. Subscription Sw.fr.45. Available from HMSO, P.O. Box 569, London SE1 9NH.

World Health, issued monthly by WHO. Available from HMSO World Health: *Traditional Health*, November 1977 World Health: *Primary Health Care*, May 1978

Assignment Children, UNICEF. A journal concerned with children, women and youth in development.

Training of Auxiliaries

AMREF Child Health: a Manual for Medical Assistants and Other Rural Health Workers, African Medical and Research Foundation, Rural Health Series 1, 1975. (P.O. Box 30125, Nairobi, Kenya).

K. Elliott, *The Training of Auxiliaries in Health*, Intermediate Technology Publications Ltd., London, 1975. 75p. A bibliography of material and resources for training auxiliaries.

Dr. J. Everett, *Obstetric Emergencies: A Manual for Rural Health Workers*, African Medical and Research Foundation, Rural Health Series 4. (Address as above).

E. Gally, *Para la Educadora del Hogar*, Editorial Pax-Mexico, 1977. (Rep. Argentina 9, Mexico 1, DF, Mexico)

N. Scotney, *Health Education: A manual for Medical Assistants and other Rural Health Workers*, African Medical and Research Foundation. Rural Health Series 3, 1976. (Address as above).

WHO *Reference Material for Health Auxiliaries and their Teachers*, WHO Offset Publications No. 28, 1976, Geneva. A valuable bibliography

7

of 400 instructional materials on health and nutrition. Annotations are in English and French.

WHO *Training and Utilisation of Auxiliary Personnel for Rural Health Teams in Developing Countries*. WHO Technical Report Series 633, 1979. SW.fr.5.

Community Health Workers Manual, E. Wood, AMREF Rural Health series; Nairobi, 1982. Address as above.

The Primary Health Worker, WHO, Geneva (1977).

Traditional Birth Attendants, WHO Geneva (1975).

The Training of Traditional Birth Attendants, Maureen Williams, CIIR, London.

③ Nutrition and Diarrhoea

Nutrition

M. King, D. Morley and L. Burgess, *Nutrition for Developing Countries*, OUP, 1972, £3.00. Also available in Spanish as *Alimentacion su Ensenanza a Nivel Familiar*, Editorial pax-Mexico 1978 (Rep. Argentina 9, Mexico 1 DF, Mexico).

D. Morley and M. Woodland, *See How They Grow*, MacMillan Tropical Community Health Manuals, London, 1979.

M. Cameron and Y. Hofvander, *Manual on Feeding Infants and Young Children*, (third edition), Oxford University Press, Oxford, 1983. £2.95.

J.E. Brown and R.C. Brown, *Finding the Causes of Child Malnutrition*, Task Force on World Hunger, Atlanta, USA, 1979. A community handbook for developing countries.

A Growth Chart for International Use in Maternal and Child Health Care, WHO Geneva, 1978.

Manual of Nutrition (Reference Book 342), Ministry of Agriculture, Fisheries and Food, HMSO, London (seventh impression 1982).

Energy and Protein Requirements, WHO Technical Report Series No. 522. Geneva 1973.

Measuring Changes in Nutritional Status, WHO Geneva, 1983.

International Code of Marketing of Breast Milk Substitutes, WHO, Geneva, 1981. *Guidelines for Training Community Health Workers in Nutrition*, WHO offset No. 59, 1981. WHO Geneva.

C. de Ville de Goyet, J. Seaman and U. Geijer, *The Management of Nutritional Emergencies in Large Populations*, WHO, 1978. £4.80.

E. Helsing with F. Savage King, *Breast Feeding in Practice*, Oxford University Press, Oxford, 1982. A manual for Health Workers.

J. Koppert, *Nutrition Rehabilitation*, Tri-Med, Blackwells, Oxford, 1977. £1.00.

FAO/UNICEF/WHO, *Methodology of Nutritional Surveillance*. Report of a joint FAO/UNICEF/WHO Expert Committee. WHO Technical Report Series 593. 1976.

FAO Food and Nutrition Paper No. 23, FAO 1982.

Field Guides for the Use of Dried Skim Milk in Nutrition Programmes in Africa, (also available for Asian countries). The Netherlands Nutrition Foundation, The Hague, Holland, 1983. Contains useful nutrition information and recipes when use of dried skim milk is unavoidable.

B.S. Platt, *Tables of Representative Values of Foods Commonly Used in*

Tropical Countries, Medical Research Council, Special Report Series No. 302. Available from HMSO, PO Box 276, London SW8 5DT, (1062 — ninth impression 1980).

Oxfam's Practical Guide to Selective Feeding Programmes, Oxford, 1984.

Diarrhoea

The Management of Diarrhoea and Use of Oral Rehydration Therapy, WHO/UNICEF, Geneva, 1983.

Diarrhoea Dialogue (Journal). Available through AHRTAG, London.

4 Immunisation

J.S. Lloyd, *Improving the Cold Chain for Vaccines* in WHO Chronicle, January 1977.

J. Smith, *Immunise and Protect your Child* in World Health, WHO, Feb/March 1977.

J. Eliford, *How to look after a Refrigerator*, AHRTAG 1980.

A. Battersby, *How to look after a Health Centre Store*, AHRTAG 1983.

U.S. Department of Health and Human Services, *Immunisation — Survey of Recent Research*, C.D.C., London, April 1983.

W.H.O., *Training for Mid-level Managers*, Expanded Programme on Immunisation — series of books from EPI Training Course 1980.

Product Information Sheets, *The Cold Chain*, EPI/WHO/UNICEF, Geneva.

5 Communicable Diseases

General

A.S. Benenson, *Control of Communicable Diseases in Man*. American Public Health Association, Washington, D.C. 13th edition. 1981.

Tuberculosis

K. Toman, *Tuberculosis: Case-Finding and Chemotherapy, Questions and Answers*, WHO, Geneva, 1979, £11.00.

F. Ridehalgh, *Memorandum on Tuberculosis in Developing Countries*, OXFAM 1971 BULLETIN of the International Union Against Tuberculosis (tuberculosis/ respiratory disease/community). Issued monthly from IUAT, 3 rue Georges-Ville, 75116 Paris, France.

Leprosy

WHO *Expert Committee on Leprosy*. Technical Report Series 607, 1977.

WHO *Chemotherapy of Leprosy for control programmes*, 1982.

S.G. Browne, *The Diagnosis and Management of Early Leprosy*. The Leprosy Mission, 50 Portland Place, London W1N 3DG, UK, 1976. Free

Insensitive Feet. Leprosy Mission — as above. Management of foot problems in leprosy. Free.

S.G. Browne, *Drug Resistance in Leprosy*. Free from Leprosy Mission — as above.

Drugs to Combat Dapsone Resistance, Lepra, Fairfax House, Causton Road, Colchester CO1 1P, UK, 1977. Report on meeting held in August 1977.

F.M. Noussitou, *Leprosy in Children* WHO, 1976.

Leprosy "Mini Pack", Health Unit, OXFAM, Oxford.

7

Polio

R.L. Huckstep, *Poliomyelitis*, Churchill Livingstone, 1975. £3.00. Management of severe deformities by surgery and appliances.

Eye Diseases

G.B. Bisley, *The Handbook of Ophthalmology for Developing Countries* O.U.P. 1973. £1.25. Good forward outlook. Includes chapters on how to set up a clinic and on trachoma. Strongly recommended.

P.D. Trevor-Roper, *Lecture Notes on Ophthalmology*, Blackwells Scientific Publications, 5th edition 1974, paperback £2.80.

H.B. Chawla, *Simple Eye Diagnosis*, Churchill Livingstone, 2nd edition 1975, paperback approx. £1.50. Available from Longmans, Harlow, Essex.

J.E.K. Galbraith, *Basic Eye Surgery: A Manual for Surgeons in Developing Countries*, Churchill Livingstone, 1979, paperback approx £5.00. As above.

6 Family Health

Maternal and Child Health/Birth Control

Boston Women's Health Cooperative book, *Our Bodies, Ourselves*, Penguin, Harmondsworth, Middlesex, 1979.

K.L. Michaelson, (ed.), *And The Poor Get Children*, Monthly Preview Press, New York and London, 1981.

A. Aarons, H. Hawes & J. Gayton, *Child to Child*, Macmillan, London, 1979.

Child to Child Activity Sheets, Institute of Health, 30 Guilford Street, London. (Organised by Prof. David Morley and Duncan Guthner.)

The Refugee Child and the Child to Child Programme, above address and Oxfam, Oxford.

D. Maine, *Family Planning — Its impact on the health of women and children*, Columbia University, New York, 1981.

R.L. Kleinman, *Family Planning Handbook for Midwives and Nurses*, IPPF, London, 1977, £1.75.

R.L. Kleinman, *Family Planning Handbook for Doctors*, IPPF (at 18-20 Lower Regent Street, London, SW1Y 4PW, UK), 1977, £3.75.

WHO, *Traditional Birth Attendants*, WHO Offset Publication No. 44, 1979. A field guide to their training, evaluation, and involvement with health services.

WHO, *Education and Training for Family Planning in Health Services*, WHO Technical Report Series No. 508, 1972. Sw.fr.5.

F.P. Mosken and M.L. Williams, *Universal Childbirth Picture Book*, (WIN) 187 Grant Street, Lexington, MA 02173, USA, 1981.

Snowden, William and Hawkins, *The I.U.D.*, Croom Helm, London, £7. A practical guide.

Hawkins and Elder, *Human Fertility Control*, Butterworth, London, £21.

See also Resources for Primary Health Care.

Periodicals

Populi, journal of the United Nations Fund for Population Activities, $2.50 per issue. Information and Public Affairs Division, UN Fund for Population Activities, 485 Lexington Avenue, New York, NY 10017, USA.

UNFPA Newsletter, monthly bulletin of UNFPA activities. Available free, also in French, Spanish or Arabic. Information and Public Affairs Division, UN Fund for Population Activities, 485 Lexington Avenue, New York, NY 10017, USA.

International Family Planning Perspective and Digest, quarterly, published by the Alan Guttmacher Institute, 515 Madison Avenue, New York, NY 10022, USA.

Female Circumcision

A. el Dareer, *Woman, why do you weep?*, Circumcision and its consequences, Zed Press, 1983. Available from J.M. Dent, Letchworth, Herts.

Raquja Haj Dualeh Abdalia, *Sisters in Affliction*, Circumcision and infibulation of women in Africa, Zed Press, 1983. As above.

S. McClean, *Female Circumcision, Excision and Infibulation*, Minority Rights Group, London, Report No. 47. The facts and proposals for change.

Traditional Practices Affecting the Health of Women and Children. WHO/EMO Technical Publication No. 2 Vol 2. Female circumcision, childhood, marriage and traditional taboos etc. — background paper,

F. Hosken, *The Hosken Report — Genital and Sexual Mutiliation of Females*, Women's International Networks (WIN), 187 Grant Street, Lexington, Mass 02173, USA.

Further information and reports:

Forward, Foundation for Women's Health, Research and Development Coordinator, Africa Centre, 38 King Street, London WC2E 8JT.

7 Secondary Health Care

ECHO (previously the Joint Mission Hospital Equipment Board)

ECHO is a non-profit-making organisation which sells and despatches surgical instruments and drugs, dressings etc. at well below current market prices. Initially supplying mission hospitals, ECHO now receives orders from many governments, and is growing rapidly. One new venture is ready-prepared packs for health centres and village clinics. The address of ECHO is: 4 West Street, Ewell, Surrey KT17 1UL, UK.

Working drawings for use by those interested in making their own equipment have been published by the Intermediate Technology Development Group, 9 King Street, London WC2.

Dr D.A. Anderson, *A Model Health Centre*. A Report of the Working Party Appointed in 1972 by the Medical Committee of the Conference of Missionary Societies in Great Britain and Ireland. 1975. £3.00. Available from TALC, Institute of Child Health, 30 Guilford Street, London WC1N 1EH.

Accommodation Standards for Medical Buildings. Volume 1: Dispensaries, Clinics and Health Centres. Volume 2: Hospitals. Medical Architecture Research Unit, Department of Environmental Design, Polytechnic of North London, 1977.

M.P. Mein, *Design for Medical Buildings*. Housing Research and Development Unit, University of Nairobi with the African Medical and Research Foundation, Nairobi, Kenya, 1975.

World Hospitals. Journal of the International Hospital Federation, London. Special Issue on '*Planning and Building Health Care Facilities under Conditions of Limited Resources*'. Volume XI Edition Nos. 2 and 3. Spring/Summer 1975. £4.00.

7

The Role of Hospitals in Primary Health Care. Report of a conference sponsored by the Aga Khan Foundation and WHO, Karachi, Nov. 1981.

The Role of Non- Governmental Organisations in Formulating Strategies for Health for All by the Year 2000. A position paper elaborated by NGO group on PHC., Geneva 1981.

R. England et al, *How to Make Basic Hospital Equipment*, Intermediate Technology Publications Ltd., London, 1979.

Financing of Health Services. WHO Technical Report Series No. 625. WHO, Geneva, 1978.

8 Mental Health

C.R. Swift, *Mental Health*, a manual for medical assistants and other rural health workers, African Medical and Research Foundation, Nairobi.

9 Environmental Public Health

Water

S. Cairncross and R. Feacham, *Small Water Supplies*, Ross Bulletin No. 10, London, 1978.

J. Howard, *Safe Drinking Water*, Oxfam, Oxford, 1979.

A. Pacey, *Hand Pump Maintenance in the context of Community Wells*, I.T. Publications and Oxfam 1981.

J.C. Van Dijk and J.H.C.M. Oomen, *Slow Sand Filtration for Community Water Supply in Developing countries*, IRC, 1982.

E.G. Wagner and J.N. Lanoix, *Water Supply for Rural Areas and Small Communities*, WHO monograph No. 42, 1959.

S.B. Watt and W.E. Wood, *Hand Dug Wells and their Construction*, Intermediate Technology, London, 1979.

Sanitation

R. Feacham and S. Cairncross, *Small Excreta Disposal Systems*, Ross Institute Bulletin No. 8, London, 1978. Excellent guide to all aspects of sanitation.

J. Howard, B. Lloyd, D. Webber, *Oxfam's Sanitation Unit*, Oxfam, Oxford, 1975.

J. M. Kalbermatten et al, A Planners' Guide, Volume 2 of *Appropriate Technology for Water Supply and Sanitation*, World Bank, Washington D.C., 1980.

D. Mara, *Sanitation Alternatives for Low Income Communities*, World Bank, 1982.

D. Mara, *Sewage Treatment in Hot Climates*, J. Wiley and Sons, Chichester, U.K., 1978.

A. Pacey, *Rural Sanitation: Planning and Appraisal*, I.T. Publications and Oxfam, Oxford, 1980.

A. Pacey, *Sanitation in Developing Countries*, Oxfam, Oxford and J. Wiley, 1978.

E.G. Wagner and J.N. Lanoix, *Excreta Disposal for Rural Areas and Small Communities*, WHO monograph, 1958.

U. Winblad and W. Kilame, *Sanitation Without Water*, SIDA, S-105 25 Stockholm, Sweden, Attention Health Division, 1980.

Organisations

The following organisations have extensive lists of publications available:

AMREF, P.O. Box 30125, Nairobi, Kenya.

Hesperian Foundation, Project Piaxtla, Box 1692, Palo Alto, California 94302, U.S.A.

WHO, Distribution and Sales Service, 1211 Geneva 2, Switzerland.

TALC (Teaching Aids at Low Cost), Institute of Child Health, 30, Guildford St., London WC1N 1EH.

7

PART 8 DISASTER GUIDELINES

Mike Goldwater

8

Section 1   Disasters Outline

I   Introduction

The poorest members of a community tend to suffer most from the effects of a disaster, whether natural or unnatural. The poor are already living on the margin of subsistence and lack the reserves to cope with disruption. The suffering caused by disasters is rooted in endemic poverty, and it is the task of development programmes to assist in reducing the level of this poverty and vulnerability to disasters. Disaster response should be, where possible, placed in the context of development and not merely immediate relief (see PART ONE, Section 2). Furthermore, any disaster response must take into account first and foremost the needs and feelings of the victims of disasters, not the logistic requirements of relief agencies and governments. The victims of a disaster should not be viewed just as the passive recipients of relief aid: they are part of a resilient social structure which has suffered a temporary disruption. Relief and reconstruction programmes should take into account the potential strengths and support which the local social structure can offer and the personal potential of the victims themselves. Relief efforts should seek to assist the return to normality as soon as possible, preferably through having a long-term perspective of the development of the physical and human potential of an area and its communities.

To be efficient and effective, the response to a disaster, whether natural or unnatural, must be both timely and appropriate. During the course of Oxfam's wide experience in setting up and running relief programmes, certain basic principles of organisation and strategy have emerged. The purpose of this section is to give an outline of these principles, and to sketch the scope and nature of the kind of aid most likely to be useful in a disaster situation.

The physical phenomena do not constitute a disaster in themselves: disaster is the disruption of people's lives. Disasters can be divided into four basic categories:

- [ ] natural, sudden impact disasters, caused by cyclones, floods, earthquakes;
- [ ] natural, slow onset disasters, caused by drought or famine;
- [ ] unnatural, sudden impact disasters, such as war;
- [ ] unnatural, slow onset or creeping disasters, such as may be caused by chronic political problems developing over a period of time (e.g. build-up of refugees, etc.).

It is a sad geographical reality that some areas are prone to natural hazards. India and the islands of the Caribbean Basin, for example, are prone to cyclones. There is nothing that can be done to prevent them. Their effects, however, can be reduced, often radically, by careful preparedness and mitigation measures.

A sound knowledge of government relief plans in the event of a disaster and the establishment of a good working relationship with the relevant department should be a priority for any field staff working in a hazardous area. Just as important is the need to lobby for the effective application of existing plans.

Preparation also involves making sure whether/where relief supplies, such as medicines, temporary shelter, etc., are available, and having a thorough knowledge of current Oxfam disaster procedures (see below). Other agencies' plans for relief work should also be noted.

8

A disaster can take many forms, but the result is a large number of people whose daily life is severely disrupted and who may be in need of assistance. Disaster response can almost always be divided into three stages:
- [ ] search and rescue;
- [ ] relief;
- [ ] rehabilitation and reconstruction.

1 Search and Rescue

This is almost invariably undertaken by the local population on an *ad hoc* basis. As a rough guide, the appropriateness and usefulness of aid of whatever kind at this particular stage diminishes in direct proportion to the distance between the relief supplier and those affected. That is, only those services with highly specific local knowledge can expect to be of any great help. It is very rare that expatriate agencies can be of use at this stage. There are some notable exceptions — the Swiss dogs trained to locate human beings trapped under buildings, for example.

Field staff on the spot can help simply by making money available for immediate purchases. It is one of the great strengths of NGOs such as Oxfam that they can use small amounts of money quickly and effectively in disasters.

This initial phase lasts between several hours and a number of days, depending on the type and extent of the particular event. After an earthquake, for example, this period is relatively clear-cut. The equivalent stage in a slow onset disaster is that period during which people can still look after themselves and have no need of external assistance. During the initial stage, as the magnitude of the disaster's effects becomes known, constant monitoring is necessary.

2 Relief

The second stage involves the provision of subsistence items such as food, water, shelter, clothing, health care, etc. Which of these is most needed in a particular disaster varies according to the circumstances. An earthquake may leave crops relatively undamaged, but cause great damage to houses. On the other hand, when typhoon Nancy hit Vietnam in 1983, most buildings of importance (e.g. hospitals) were somehow or other repaired, whereas the greatest concern was for the 96,000 tonnes of food which had been lost.

Some voluntary agencies tend to assume that, in a disaster situation, people automatically become demoralised and apathetic. Little mention is made in the media of the fact that in most disasters 95% of the aid required is raised locally. Field staff should support and encourage people's desire and ability to help themselves. In all relief programmes the ultimate objective should be to restore normality as soon as possible, and in the process to ensure that the effects of future disasters are diminished or cushioned as much as possible.

It is impossible to predict precisely which materials will be needed, but subsistence items frequently in demand after a disaster include cooking pots, plates, cups, water carriers, soap, blankets and basic shelter material.

It is in this phase that most of the mistakes occur. For example, in the Guatemalan earthquake of 1976, the following criticisms were made:
- [ ] too much aid was given away;
- [ ] too many of the houses constructed were of a temporary type;
- [ ] some organisations used too many foreign volunteers;
- [ ] too much was done too quickly without proper consultation, so that victims became mere spectators of the work carried out rather than participants.

It is therefore important to assess the extent of locally unmet needs and also:

- [ ] the extent to which the disaster affects the poorer people of the area;
- [ ] the availability of other relief assistance to the affected population;
- [ ] the proximity of an Oxfam field office and Oxfam experience in relief and development in the area in which the disaster has occurred;
- [ ] the effective ability of field staff to visit and work in the affected country, including the acceptability to the Government of Oxfam and its methods;
- [ ] the availability of acceptable channels of relief aid in the affected country.

Only when all available information has been collected can a rational and logical plan be worked out. With the possible exception of small emergency grants which must necessarily be made rapidly, all other relief aid must be carefully considered. A straightforward and methodical approach will guarantee an effective and appropriate application of resources. Speed is of course a priority, but field staff should try to remain detached from the chaos which a disaster brings. Once an ill-planned chain of events has been set in motion, it is extremely easy for the overall aim of the operation to be forgotten or ignored.

It must be remembered that the NGO is usually the junior partner in a post-disaster relief programme, the major burden of expenditure and overall planning being assumed by the large international federations and agencies. Agencies such as the United Nations High Commission for Refugees (UNHCR), the UN Development Programme (UNDP) and the World Food Programme — a department of FAO — have the resources to supply relief materials on a large scale.

It is important for the voluntary agencies to coordinate their operations closely with the UN agencies so that they can avoid duplication and take up the smaller programmes which are outside the UN's mandate or which lack the official Government approval needed for all UN programmes. The considerable flexibility of the NGO enables it to avoid some of the delays usual in the government organisations because of their more cumbersome administrative processes. (More details on the different roles of the the various agencies are given below.) Any disaster relief programme needs financing, and for voluntary agencies like Oxfam, this is dependent upon donor response in the home country. The Disasters Emergency Committee (the British Red Cross Society, the Catholic Fund for Overseas Development, Christian Aid, the Save the Children Fund and Oxfam) arranges for a united appeal (by far the most effective form of publicity) made in the name of the committee on behalf of the member agencies.

Field staff can play a vital role in giving assistance to journalists and television crews in the disaster zone itself. A good press can help fund-raising enormously, and it is therefore important for relief teams to keep reporters informed. Most NGOs have established a good relationship with the media, and this has frequently resulted in sympathetic coverage and considerable help in the field from journalists with information not readily accessible to relief workers. Relief workers must, however, recognise the discrepancy that may exist between the media presentation of a disaster, focusing on the chaos of the situation in the interest of 'newsworthiness', and the reality. Furthermore, the media tend to ignore the efforts of local people on their own behalf.

3 **Rehabilitation and Reconstruction**

The dividing line between this final phase and the relief phase is necessarily imprecise. It is always important to integrate any rehabilitation work with general Oxfam policy on priority groups,

8

methods of project implementation, etc. (See PARTS ONE, TWO and THREE.)

At this stage training in mitigating techniques in areas such as house-building and agriculture should be considered. For example, after the Yemen earthquake of 1982 (YEM 49 B3), Oxfam co-funded a project for educating the local people in new and safer building techniques such as the use of ringbeams and L-shaped steel bars.

It is important that rehabilitation does not consist of handouts, and that self-sufficiency is sustainable in the long term. It is vital to encourage the recovery of local economies as soon as possible. For example, after the tribal clashes in Assam in 1983 (ASM 16 N3), many families were left without looms, one of the major sources of income in the area. New, locally-constructed looms were provided, thus helping to create conditions in which life could begin to get back to normal.

III   Agencies involved in disaster relief

Voluntary agency contributions total a small fraction of the resources available from governments through the UN system; and though voluntary agencies may have a speed and flexibility that enables them to work faster, they cannot supply staple food, shelter materials, clothing or funds on the scale that the UN agencies can. The UN Disaster Relief Office (UNDRO) is not an operational agency in relief work. UNDRO offers technical advice and coordinates information and appeals. It is based in Geneva and works through the local office of the Resident Representative of the UN Development Programme (UNDP). UNDRO channels funds through appropriate government, other UN agencies or occasionally through voluntary agencies (VOLAGS).

1 Government

National governments will to varying degrees bear the brunt of immediate disaster response. Some will work very much on an *ad hoc* basis with the armed forces and police in the forefront of Search and Rescue operations, with the civilian ministries following up with relief and rehabilitation. Other governments now have centralised civilian organisations involved in preparedness, mitigation and disaster response. The Civil Defence will attempt to mobilise and coordinate resources at the time of the emergency: it may also seek to coordinate the work of voluntary agencies. It is important for voluntary agency staff to be in contact with the local government, to avoid duplication of effort, concentration of aid in some districts to the detriment of others, and inefficient use of scarce resources. Evidently some governments will be more efficient and organised than others in their disaster response, and the voluntary agency will have to take this into account.

Many government disaster programmes are unfortunately limited to immediate Search and Rescue and relief operations, and after a few months the programme will finish — just when many unmet rehabilitation needs are becoming evident.

2 The Red Cross

The Red Cross group makes up one sector of the voluntary agencies involved in disaster relief work.

a: The International Committee of the Red Cross (ICRC or CICR) is a specialist agency based in Geneva and undertakes, as a neutral institution, tasks incumbent under the Geneva Conventions, especially related to war and conflicts: the treatment of prisoners of war; protection and assistance to civil and military victims; tracing and reuniting divided families; and the provision of medical assistance to victims of conflicts.

b: The League of Red Cross Societies, also based in Geneva, is the coordinating and operational arm of all national Red Cross and Red Crescent Societies. It is the organisation through which all national Red Cross Societies can channel aid and assistance to the national Red Cross Society of the affected country.

c: The national Red Cross and Red Crescent Societies, each of which has independent status, participate in varying degrees through the League of Red Cross Societies.

Members of each of these Red Cross organisations may be involved in a disaster relief operation. Field staff not familiar with the different sections of the Red Cross group should take care not to confuse them.

3 **Voluntary Agencies**

International voluntary agencies experienced in disaster relief operations fall broadly into two categories — the 'federations' and the 'independent agencies'.

The federations are the central coordinating bodies of nationally based organisations. They include:

a: The World Council of Churches (WCC), Commission on Inter-Church Aid Refugee and World Service (CICARWS), based in Geneva, which represents the Protestant national councils of churches, and usually channels aid through the national council of churches of the affected country. The World Council of Churches includes in its membership *Lutheran World Federation (LWF)*, *Department of World Service (DWS) Church World Service (CWS)*, the *Division of Overseas Ministries of the American Council of Churches*, and *Christian Aid*. Each of these organisations also runs and finances programmes independent of the World Council of Churches.

b: Caritas Internationalis, based in Rome, is the main agency of the Catholic Church in relief work, with counterpart societies often called 'Caritas' in every country where there is a substantial Catholic population: e.g. Deutsche Caritas, Caritas Belgica. In France it is *Secours Catholique*; in Britain the *Catholic Fund for Overseas Development* (CAFOD); in America *Catholic Relief Services* (CRS). CRS operates a large food aid programme using American Government food stuffs.

c: The Licross-Volags Steering Committee for Disasters is made up of the heads of the League of Red Cross Societies, World Council of Churches (CICARWS), Lutheran World Federation, Caritas Internationalis, Catholic Relief Service and Oxfam. The committee meets periodically to coordinate the disaster relief activities of the member agencies. Similar local coordinating committees have been established in various parts of the Third World.

There are a large number of *independent agencies*, including Oxfam, from many countries in Europe, America and Australia, some religiously based, some medical, some technical. The major ones are listed in the address section of this Handbook.

4 **UN Agencies**

Most Third World countries have a United Nations Development Programme (UNDP) office, whose Resident Representative is the senior UN official in the country. The representative has the responsibility for the coordination of all the UN agencies' relief operations at the time of a disaster. The different UN agencies have their specialist fields of work, and experience has shown that UNICEF has the most flexible mandate, enabling it to take on programmes that benefit the population generally; it is not restricted to direct assistance to children. The Food and Agricultural Organisation (FAO) is concerned with the evaluation of the effect of the disaster on agriculture and fisheries and with offering assistance in agricultural reconstruction. The World Food Programme (WFP), a department of FAO, evaluates the food needs and can supply staple food on a large scale. The World Health Organisation takes action

8

to combat epidemics or in eradication programmes, but more usually acts as consultant in health to other UN agencies. The Pan-American Health Organisation (PAHO) operates in the Americas and has an important role in health, nutrition and sanitation matters. The UN High Commission for Refugees (UNHCR) is primarily concerned with the protection of refugees and their material welfare, but is sometimes also involved in work with displaced people within a country.

The UNHCR Emergencies Unit has produced a *Handbook for Emergencies* that summarises UNHCR policies and procedures for field staff. Oxfam field staff facing a refugee emergency, or likely to be cooperating with UNHCR in a refugee relief operation, should use this handbook for reference in dealing with UNHCR and with refugees.

The handbook is divided into two parts. The first, 'Organisation of Responses' (by HCR to refugee emergencies), details the legal definition, protection responsibilities and internal procedure of UNHCR. The second part, 'Sectors of Emergency Assistance', gives policy guidelines on site planning, health, water, supplies, etc.

IV   Disaster preparedness

1 Preparation for Disasters

A sound knowledge of locally available resources, and government and UN disaster plans, is an essential prerequisite to good disaster preparedness. The logistics of preparedness are, however, only secondary and supplementary to an understanding of customs, political structure, economy and history of the society in the hazard-prone area. Disaster response is likely to be more effective where a good knowledge of local affairs can guide the planning and implementation of relief and rehabilitation work. Though it is not always possible, it is desirable that relief programmes are undertaken only by agencies with previous experience in a given area, whose staff understand local conditions.

Preparation for disasters should include access to the information below.

a: The history of past disasters, vulnerability to natural phenomena such as cyclones, earthquakes, climatic change, normal rainfall or regular periodic changes (e.g. incidence of drought and flood), and agricultural patterns (crop, harvest times, etc.).

b: Disaster preparedness by government, local authorities, UN agencies and NGOs. Besides plans for logistics and information flow after a disaster, local authorities should have information on crucial services such as water supplies, food availability, electricity safety circuits, etc.

c: Prediction. Are there early warning or monitoring systems to assist prediction of either long-term problems such as food shortages, or short-term, such as cyclones? In some areas it is advisable to set up programmes monitoring drought, crop failures, prices in local markets in order to predict likely food shortages.

d: Sources of locally obtainable supplies and expertise. If local medical and technical personnel can be recruited quickly after a disaster, they are probably more efficient and cost effective than foreigners, whether staff or volunteers. Where appropriate, some training of local staff might consolidate disaster preparedness. Air freighting supplies is both expensive and potentially disruptive of local markets. Wherever possible, local suppliers and local products should be used.

e: The procedures of all organisations likely to participate in a disaster programme should be ascertained. What are the important addresses, phone and telex numbers? Who is in charge in different organisations? What steps should be followed? Does the organisation have a procedural handbook?

2 Mitigation

The field of disaster mitigation is still developing, and so far it has been easier to recommend mitigation measures in some circumstances than in others. Most mitigation work has concentrated on the high-impact natural disasters which result in physical damage. A great deal of work has been carried out on risk-reducing measures for incorporation into the design and construction of buildings (see Section 4). Examples of measures to reduce the effects of natural disasters on agriculture are also noted below (see Section 2). It is, however, more difficult to plan for the mitigation of the effect of unnatural disasters.

A survey should be made of each area/country to establish the problems which may confront the population and ways of mitigating them. The cost of a well-founded mitigation programme is likely to be far less than that of the effects of an uncontrolled disaster.

The most effective form of disaster mitigation is development. Reducing the vulnerability and poverty of individuals and communities makes them more likely to be able to survive and confront a disaster.

8

Section 2   Disaster Response: Agriculture and Food Supplies

I   Introduction

Disaster relief must be appropriate to the immediate needs of the victims, to their cultural and social norms and to the long-term development of their communities and economy. Often people will be able to articulate their needs in very concrete terms. An attempt must be made to ascertain what *the victims* are saying that they want and need, as opposed to merely listening to what *aid and government* officials are saying that they need. Be wary of artificially compiled shopping lists of demands which bear little relation to the needs and requests of the victims themselves. Some governments and agencies will automatically produce lists of requirements, such as tents and drugs, which have little to do with the actual unmet needs.

Most programmes face a number of obstacles and constraints, ranging from political pressure to simple logistic problems (e.g. shortage of transport); but notwithstanding these, there is no excuse for the expenditure of funds, personnel, time and use of support systems on an inappropriate relief programme. Examples of ill-conceived programmes are unfortunately numerous. At best these responses may be ineffective; at worst they may obstruct other, more appropriate programmes and even cause more harm than good.

II   Agriculture

Farmers, especially the poorer ones in the Third World, are often, through population pressures and social inequalities, found cultivating land in disaster-prone places. The use of flood plains, steep hillsides and regions of marginal and erratic rainfall is common. Hence it is not surprising that poor farmers and landless agricultural labourers usually make up the majority of the victims of natural disasters. The poor are the least capable of responding to a disaster, and find themselves, in the absence of savings or alternative sources of income, further entrenched in debt and dependence.

This sub-section discusses how certain agricultural practices can be used in the case of three types of natural event — drought, flooding and high speed wind — in order to mitigate disaster, or to rehabilitate farm land after a disaster.

1 Drought

a: Preparation and mitigation: nothing can really be done to prevent drought, but the following techniques may help to overcome its impact:
- [ ] irrigation;
- [ ] removal of weeds which successfully compete with the crops for moisture;
- [ ] deep cultivation, if appropriate equipment is available, to allow the roots to enter the sub-soil where moisture may be stored;
- [ ] crop diversification. Some crops, such as sorghum, millet, maize, etc., are comparatively drought resistant. However, altering traditional cropping patterns which have evolved over a great many years and in accord with specific local geographical conditions, should only be attempted with extreme caution. Rather, local information on the best drought-resistant crops should be used;
- [ ] reforestation;

☐ distribution of seeds of local crops for planting can be very important. It is likely that, after a prolonged drought period, available local supplies will have been eaten;
☐ prediction and monitoring.

b: Rehabilitation: in many drought-prone areas rainfall comes in heavy storms, and often in a very short season. Excessive rainfall will exceed the water intake capacity of the soil, and a large proportion of the incoming rain will run off to lower areas and into seasonal streams. The results are both that water is lost from cultivable areas, and also that the run-off causes soil erosion. Programmes for conservation of soil and water in such situations can be most effective and may consist of:

☐ reducing run-off and keeping water on cultivated land, e.g. by construction of small earth embankments at intervals across fields, levelling of sloping land, and other on-field conservation schemes (see PART SIX, Section 2);
☐ collecting excess run-off water and using it for agricultural or domestic/livestock purposes, e.g. run-off catchment schemes or creation of a reservoir with an earth dam in a valley area.

2 Floods

Most of the destruction brought by floods is in the agricultural sector because much of the world's most productive and fertile lands are near rivers and in low-lying delta areas, which are often also densely populated. Floods are the result of heavy rains which exceed the capacity of local rivers. A river's ability to drain excess water can be reduced by silting as a result of deforestation and damming. Cyclonic disturbances, which are always associated with heavy precipitation, are sometimes associated with sea storms and sea-water flooding of the coastal areas.

a: Prevention and mitigation:
☐ natural obstacles, strong bunds along the river and other shoreline protection against sea-water flooding may be the only way of protecting agricultural land;
☐ the best measure is to attempt to grow crops both before and after the flood season, although this will require irrigation facilities;
☐ a good drainage system, involving land levelling if necessary, will also result in a quicker withdrawal of flood water to allow plant recovery.

b: Rehabilitation: most crops, if not at a vulnerable stage of growth, can withstand several hours of submersion in still or slowly moving water. Therefore land reclamation will involve:
☐ draining of the land as soon as possible;
☐ preventing fungus attack on the wet vegetation, by the use of commercial fungicides;
☐ choosing crops which are more tolerant of water-logged conditions, e.g. paddy rice, sugar cane and several perennial grasses;

Any cultivation should be done with care, since working wet soil may cause it to deteriorate. Seed-bed preparation should be attempted only when the top soil is sufficiently dry. Water leaches soluble nutrients from the soil, so the application of nitrate fertiliser in particular will help growth of plants after flood. One of the effects of flooding is the destruction of the next season's seed crops or of seeds in store. Improved seed storage facilities can easily stop this from occurring.

Flooding by sea water may require special measures. Sea water is highly damaging to most cultivated crops, and the possibility of saving crops is therefore minimal. Methods of land reclamation mainly consist of:
☐ leaching with fresh water (irrigation or ponded water). However, leaching will also remove useful salts such as nitrates which should be replaced by green manures or organic fertilisers;
☐ establishing a crop cover as soon as possible. Otherwise, the top soil

8

will become saline again through evaporation and capillary action. Crops should be salt-tolerant (such as suitable varieties of sunflower, barley, sorghum, millet and a number of grasses).

③ High-Speed Winds

High-speed winds, sustained or in gusts, will lay flat or push over any crop which protrudes more than 10 or 20 cms. Unlike floods and droughts, very little can be done about rehabilitation after a disaster. However, a number of measures can be taken to prevent or mitigate crop damage.

Prevention and mitigation:

☐ windbreaks and shelter belts protect houses, farm buildings and crops. Local tree species known to be able to withstand winds (usually a deep rooted type) can be used;

☐ to protect the trees and orchards themselves from destruction, deep ploughing, which loosens the soil, will assist root penetration and deep anchorage. Many farmers tie trees together with rope so that they support each other;

☐ the trees should not be planted in straight rows which offer an unobstructed channel for wind to pass along;

☐ by choosing growing seasons carefully, it is possible to avoid having crops standing in the fields during the months when cyclones are likely to be more frequent or intense. New short-duration paddy rice varieties, for example, can be grown outside the danger months.

④ Aid to Indigenous Groups

It is a difficult and delicate task to provide relief to tribal or nomadic communities, without interfering with their traditional way of life and patterns of social relationships (see general guidelines in PART TWO, Section 5). In many Third World countries rural communities maintaining a traditional tribal way of life, perhaps as nomadic hunter gatherers or pastoralists, are despised by the more Western-orientated, urban people, and by the sedentary peasant farmers. In some countries there is pressure on nomadic groups to give up their way of life in exchange for settled agriculture. Governments may try to use a drought or similar misfortune as an opportunity to force populations to settle as the price of relief. Field staff should be careful to avoid participation in relief schemes involving settlement under duress.

An early priority for field staff faced with a disaster affecting a tribal community is to gain some understanding of the social, political and economic systems by which they live. Social anthropologists and their writings are likely to be most useful in this, and they are usually keen to share their knowledge for the benefit of the affected people. For field staff, action presupposes a reasonable knowledge of the community's traditional ways of life, including 'coping' mechanisms. Field staff should identify university departments and other sources of such information.

⑤ Aid to Pastoralists

Drought has made clear the crucial role played by cattle as a form of famine insurance. They are the single most valuable item in the economy of many herding groups and are seen by the people as their last defence against starvation. They do not serve this function by being consumed but by being exchanged for grain. During drought neighbouring communities with a surplus of grain are able to exact very high prices for it. It would be of great benefit if the exchange could be regulated so that the quantity of grain obtainable for a given type and size of animal is guaranteed. Perhaps this could be done by the setting-up of government or other grain stores and abattoirs.

This action would increase the amount of food available in times of shortage, and would also exploit rather than replace existing social institutions and cultural values. Cattle are an effective means of famine insurance, not only because of their intrinsic economic value and because they are highly mobile forms of wealth, but because of their overwhelming symbolic significance. A large, indeed potentially huge, number of individuals will normally have an interest in the cattle, arising mainly from the complex redistributions of stock that accompany marriage as bride-wealth payments. Thus the use of animals — vital elements in bride-wealth payments — in exchange gives a special quality to the product of the exchange — in this case, grain — and it too becomes something in which many individuals can stake claims.

Field staff should be careful not to adopt the 'rational' position that the extreme symbolic importance given to cattle is an example of uneconomic behaviour. The attitudes of pastoralists towards their cattle can assist the delivery of famine relief.

A programme tried as a drought relief measure in Botswana in recent years seems to have worked well. Government funds were made available to purchase old and weak cattle at a fixed price. These were used to provide meat to drought-relief beneficiaries, as a supplement to other rations. Cattle were purchased in rotation from villages which were offering cattle for sale. The scheme was well received and worked well. It proved to be a relatively cheap method of providing meat, and also helped to maintain the price of livestock, which was to the advantage of herders who might otherwise have been forced into destitution, as prices tend to fall in a drought. Destocking schemes elsewhere have also helped maintain prices received by pastoralists, reduced the number of animals competing for scarce grazing, and culled out weaker animals from herds leaving fitter animals as the basis for recovery.

III   Food aid for disaster relief

1 Oxfam's Position on Food Aid

Oxfam's policy on food aid in general is the result of extensive first-hand experience and research, presented in the publication *Against the Grain* by Tony Jackson, Oxfam, 1982. This book is indispensable for anybody concerned with food aid, as it illustrates vividly the immense gulf between theory and practice in the field. Its principal message is that food aid has too often, even in the recent past, been an inappropriate form of development assistance, and has often proved counter-productive, because:

☐ beneficiaries have not usually been those most in need;

☐ food aid requires considerable administrative resources for handling, storage, distribution and control, and is very costly;

☐ it requires considerable logistical support and absorbs a disproportionate amount of field workers' time and attention;

☐ it creates a relationship of dependence between donors and recipients, disrupts the normal way of life, and very often undermines the community spirit of local people, while encouraging passivity;

☐ perhaps the most crucial medium- and long-term negative effect of food aid is on local food production, which, in the face of large consignments of free or cheap food, can be destroyed altogether.

The above comments apply to what is called Project Food Aid, which absorbs about 30% of all food aid. The remaining 70% is sold or given to Third World governments. Strictly speaking this is not food aid, but rather a form of government budgetary support, or foreign exchange support, and does not concern us here.

The case for food aid can only be made in exceptional circumstances.

8

2 When is Food Aid Needed

Food aid is necessary when a disaster affects food availability. Drought and flooding may disrupt local and regional food production for a time. When food production falls, prices rise, food becomes less accessible to the poor, and seeds are often eaten instead of being planted. Here food aid both in the short and medium term may be required. Other natural hazards, such as hurricanes or earthquakes, may bring a temporary shortage of local foods if crops are damaged or roads are destroyed. A recovery can usually be made soon afterwards. Thus food aid will be needed, if at all, for a limited period only.

In the case of unnatural disasters, such as wars and civil disturbances, which involve refugees, a large influx of people who are not producing food may seriously stretch the local market. At these times, food will have to be imported by the host country; furthermore, these people may have no income with which to purchase food. In such circumstances, food aid will be required.

3 Suitable Food and Distribution

In emergencies, the most pressing need is to distribute the food speedily and equally. The following points raise the issues of what kind of food, where it should come from and how it should be distributed.

a: Once the vulnerable groups have been identified, their requirements must be determined to ensure that the rations suit individual needs, whether the rations are intended to supplement local supply or to be the only food supply.

b: In the past the wrong kind of food has often been sent in disaster relief programmes and in some cases has turned out to be unsafe for local consumption. One way of ensuring the suitability of food is to purchase it as locally as possible. This will also stimulate local or regional production and help to bring conditions back to normal.

c: When food aid is needed, it is important that it arrives on time. Not only is it wasted when it arrives too late, failing to alleviate hardship, but it may arrive at a time when local supplies have been resumed, thereby depressing prices and local production.

d: Distribution can be done through institutions such as schools or hospitals, or through development programmes already in existence. Feeding centres, however, run the risk of becoming permanent institutions which draw people away from their normal pattern of life. (The following section shows how this can be avoided.)

e: A famine does not always mean that there is no local food available, although certain groups may lack the income to buy food. Income-creation projects such as road building may be a suitable alternative to food distribution, providing families with cash. Buying food locally has many advantages, including suitability, speed, cheapness and minimum disruption to the local economy.

f: The option of selling rather than the complex operation of distributing free food should be considered. A shortage of food does not necessarily mean that people have no money, and food can be sold at subsidised prices if need be. Selling also introduces an increased level of accountability and passes the decision on whether or not to take part in the scheme to the people themselves. Sale of food does not compromise the future relationship between the aid workers and the local people, or set up false expectations for subsequent development projects.

g: The sale approach does not preclude the setting up of medically controlled nutrition rehabilitation units for those people who who are in need of special treatment; these clinics would also act as a useful indicator of the food needs of the local community.

4 When should Relief Food Aid Stop?

It is undesirable as well as unnecessary to prolong the distribution of free food after a disaster. Food aid is likely to work against a more self-reliant form of development.

Donors, however, are often keen to integrate emergency food relief with long-term development work, using free food. In most cases it is distributed by the same groups, using the same channels. It is not therefore easy to determine where 'relief' food ends and where 'developmental' food begins. Also, less honourable motives such as convenient disposal of surpluses and competition with other donor governments or agencies may colour the decisions of donors.

Donors should not use emergencies as a way of beginning long-term food aid projects. Once an emergency is over, food aid should be stopped.

8

Section 3   Disaster Response: Health and Nutrition

I   Sudden impact disasters

1 Assessment

While Oxfam has at times made effective immediate responses to sudden impact disasters such as earthquakes, cyclones and bombing of civilian areas, our major contribution has been, and will continue to be, in relief programmes associated with slow onset disasters, whether due to natural causes, primarily drought, or to chronic harassment of civilian populations. It is, however, essential for field staff to be able to make assessments of health and nutrition needs in both types of disasters, and to determine whether an input from Oxfam, in material aid or an operational role, is appropriate. (See also PART SEVEN, Section 3.)

Effective assessment depends primarily on an understanding of the health and nutrition problems that are likely to arise, as indicated in Table 8-1.

Table 8 – 1   Health and Disasters

	Earthquakes	Typhoon/Cyclone Extensive flooding	Floods	Bombing of civilian areas
Deaths	Many	Few	Few	Variable
Severe injuries, requiring major care	Many	Limited	Few	Many
Communicable Diseases	Increased risk in urban areas. Risk in all areas increases with crowding in reception centres, etc.			
Food Scarcity	Rare	Rare	Common	Rare
Major Population Movements	Rare	Rare	Possible	Possible

The second principal component of assessment is an understanding of the time pattern of health and nutrition problems after a physical impact disaster, illustrated in Table 8-2.

Failure to use this simple objective approach leads to many inappropriate responses, such as surgeons arriving a month after the bombing has stopped, or nutritionists looking for malnutrition in the first few days after a sudden disaster.

In the confusion and urgency that surrounds the assessment visit to a new disaster area, an approach based on Tables 8-1 and 8-2 will assist in the objective assessment of the situation.

A third component of assessment relates to what the funding agency can and cannot do. Other, operational agencies are specialists in the first aid management of casualties. Donor agencies may be able to assist with finance or logistics, but it is unlikely that the supply of equipment or personnel is an appropriate use of donor agency resources. It is highly improbable that any external agency will be able to respond to the immediate short-term needs of those injured by a sudden impact disaster.

Table 8 – 2   Time-scale of disasters

	Impact	Week 1	Week 2	Week 3	Week 4
Casualties requiring rescue & resuscitation	▬▬▬				
Casualties requiring major surgery	▬▬▬▬▬▬▬▬				
Casualties requiring long-term nursing		▬▬▬▬▬▬▬▬▬▬▬▬▬▬			
Casualties requiring physiotherapy				▬▬▬▬▬▬▬▬	
Shelter needs	▬▬▬▬▬▬▬▬▬▬▬▬▬▬▬▬▬▬				
Food scarcity				▬▬▬▬▬▬	
Water/sanitation-related disease	▬▬▬▬▬▬▬▬▬▬▬▬▬▬▬▬				

---

2 **Needs and Responses**

a: Management of immediate casualties. The donor agency can play an important role in supporting local organisations, whether government, national Red Cross Society, or local NGOs, usually with immediate finance to mobilise resources. It is rare that outside personnel or supplies can contribute much in this immediate phase, but because of the high media coverage, major responses are often attempted. An important and controversial area is requests for drugs. In the immediate post-disaster phase, a very limited range of drugs is required; most are available locally.

b: Shelter. Simple shelter systems are discussed elsewhere in this manual (PART FIVE, Section 2 and PART EIGHT, Section 4). Lack of shelter in certain climatic areas may be a major cause of ill-health in the post-impact period, and may be an appropriate concern for the funding agency.

c: Communicable disease control. The principal disease problems will be related to *water and sanitation*, either due to breakdown of existing systems (as in the Lebanon, 1982) or due to crowding in camps (e.g. the Burmese refugees in Bangladesh, 1978). The answer to such problems is to improve water and sanitation.

It is essential to attempt to *prevent* water/sanitation related diseases before they reach epidemic proportions. While the major fears are typhoid and cholera, there are few situations where epidemics of these diseases have in fact occurred. In the rare cases where cholera may reach epidemic proportions (as in the Bengal camps of 1971), there may be great demand for special intravenous fluids. Simple rehydration will in fact suffice in most cases. Vaccination against cholera and typhoid has limited value in potential epidemic situations, but may be of use for new arrivals in camps. It must be emphasised that cholera and typhoid vaccines give only *individual* protection; they do not prevent spread of the disease. Field staff must ensure that assistance is not given to such vaccination programmes until adequate action has been taken to improve water supply and sanitation. (For further information on the

8

control of communicable diseases see PART SEVEN, Section 5 and for information on water and sanitation see PART SEVEN, Section 11 and PART EIGHT, Section 4.)

Widespread damage will lead to major environmental pollution from refuse, rotting foods, and in some cases carcases and dead bodies. It is essential that urgent and authoritative action is taken to prevent disease by disposing of these by burying, burning, or use of lime.

Once victims begin crowding into camps and reception centres, health problems will develop as described in the section below on camps.

d: Nutrition. The effect of *sudden impact disasters* on food availability will depend on the type of disaster and the locality. In a disaster affecting a limited geographical area, which is easily accessible and where crops are not lost, it is likely that food will be rapidly moved to the area. In widespread floods occurring during the crop season, both current food stores and the next year's food supply can be destroyed, so that food aid is required. It is unlikely that *malnutrition* will occur in the immediate post-disaster period, as people take more than a month to become malnourished. Consequently, the nutrition response in sudden impact disasters is likely to involve planning food supplies to maintain nutrition rather than emergency programmes to treat malnutrition, a distinction that is not always made. If home food stocks have been destroyed, there may be an immediate need for the most traditional of 'relief kitchens' — almost certainly best run by local groups and preferably using local food.

II Slow onset disasters

Much of Oxfam's relief work over the last few years has been in the chronic relief problems of drought and of political harassment leading to displacement and exodus of refugees. Such disasters are often regarded as 'sudden impact', because they first become apparent to relief agencies when there is a sudden migration of drought victims to towns or mission stations, or a sudden exodus of refugees across a border.

Relief needs in such situations can most usefully be divided into two phases. The first phase is the period of displacement, before refugees arrive at relief centres. While such a period may last only a few days, it can be a very extended period, during which major health and nutrition problems develop. In some situations, this 'displacement phase' may be continuous, as in Eritrea, Tigre and Kurdistan.

Traditionally, relief work is concentrated in the second phase, when displaced people arrive at relief centres, where all the typical health and nutrition problems of relief camps exist.

1 Health in the Displacement Phase

(In this context, displacement includes both periods of physical movement and periods when people remain in their homes prior to the implementation of relief.)

Because in the displacement phase people are usually in small groups, the health problems due to overcrowding and poor sanitation that occur in camps rarely emerge. Forced movement into different ecological zones can introduce new health problems. Changes in the physical environment, such as moving people into higher/colder areas with poor shelter, or to areas where vectors of diseases, malaria, trypanosomiasis, etc., are prevalent, can increase the incidence of disease dramatically.

Political harassment is a major cause of declining health and breakdown of the most basic health care services. Thus if clinics no longer exist to provide even chloroquine and penicillin, diseases which were previously easily treated can become major causes of ill health and death.

It is possible that the funding agency could make a major impact on health care in this phase by the provision of basic drug supplies for use in the displacement area. There will clearly be security and logistic problems but they may not be insurmountable.

[2] Nutrition in the Displacement Phase

Slow onset disasters, with long displacement phases, can lead to severe malnutrition. Malnutrition does not occur 'overnight'. In purely physical terms, it would take ten days of *no* food input for a child to go from 90% weight/height to 80% weight/height, and a further ten days to 70% weight/height, the cut-off point for severe malnutrition. There are of course few situations where there is a sudden, total loss of food input. The usual pattern is a diminishing food input, with long periods of food input below the average minimum of 1,750 calories (7.3 MJ) per person per day. The pattern of malnutrition in the displacement phase will depend on three factors:

☐ nutritional status at the beginning of displacement;

☐ length of displacement;

☐ and the supply of food during displacement. Food may be carried by displaced people. There is also the possibility of harvesting 'famine' crops such as cassava, or of growing short-term crops.

The relationship between time and nutritional status has been mentioned. Clearly one would not expect significant malnutrition during periods of less than three weeks between the displacing event and arrival at a relief centre. However, with up to several months' displacement, widespread malnutrition can occur.

It is useful to consider an average of 10 kg of grain per person as that which can be carried. With a daily food requirement of 0.5 kg/day (= 1,750 cals) this will provide 20 days' ration in addition to use of body reserves.

[3] Food Supplies During Displacement

In the displacement period, a number of situations could arise which would seriously affect nutrition.

a: It could be, for example, that food supplies are interrupted. This will lead to increasing malnutrition, first in vulnerable groups and then in all groups. It is in this situation that the word 'famine' may apply. There are various possible causes for instance:

☐ *food scarcity* which has escaped the knowledge of government and relief agencies — the 'hidden famine'. This is unlikely if political turmoil is involved, as this will have been monitored by the media. It should *not* occur in an area covered by an Oxfam field office if there is a basic monitoring system in operation;

☐ *political factors*, particularly active fighting, can prevent refugees from leaving areas of scarcity or prevent mobilisation of relief into the area: for example, in an 'enclave' of resistance, perhaps as in Biafra; potentially in Tigre/Eritrea, Kurdistan, and parts of Afghanistan.

b: Alternatively, some local food may be available, for instance where people are able to harvest annual grain crops, or perennial root crops such as cassava, and effectively increase their displacement period.

c: Limited ability to move relief food into the area is another possible problem. In zones of political turmoil — especially where the terrain is difficult -both security and logistic problems reduce the amount of relief food that can be delivered. In areas without security problems, ingenuity may be needed to get relief supplies into less accessible regions.

d: It could be that supply is not a problem, but that large sectors of the population lack purchasing power and are unable either to buy food or to draw it into the affected area. Most food relief programmes are oriented towards displaced people who have left the trouble spots, providing food relief in centres located in the safer zones.

8

Thus, NGOs may be able to help in the displacement period, and should seek ways of preventing the mass malnutrition that may already have occurred by the time people arrive at relief centres.

④ The Relief Phase

This is of course the traditional area of work of most agencies. However, the experience of many NGOs over the last few years has emphasised that the traditional approach is in certain respects inappropriate, and there is a need for more objective analysis of the problems and planning of responses.

Two underlying principles must be appreciated:

☐ most people *arriving* at camps have few health problems. It is usually the environment of the camp that causes a deterioration in health. In-camp health programmes must, therefore, be aimed at *prevention* through provision of water, sanitation and immunisation;

☐ a camp should not *produce* malnutrition. There must be both an adequate general ration in the camp to prevent malnutrition, and initial special programmes for the malnourished as appropriate.

Full details on the assessment of health and nutritional needs in newly arrived refugees or in established camps, and the implementation of basic programmes are given in the booklet *Oxfam's Practical Guide to Refugee Health Care*.

The principal responsibility of field staff on first arrival is to undertake a rapid initial assessment of the situation. The data required, which should form the basis of initial appraisal, are:

☐ number of refugees already in camp;
☐ approximate number of new arrivals;
☐ nutritional status of new arrivals;
☐ availability of general food ration in camp;
☐ availability of supplementary food in camp;
☐ principal disease problems of new arrivals;
☐ disease of epidemic proportions in the camp;
☐ present water and sanitation facilities of camp;
☐ shelter facilities of camp;
☐ local agencies present or able to work in camp;
☐ principal immediate needs based on this assessment;
☐ agency or government department responsible.

⑤ Health Care in the Relief Phase

The health programme must be based on three components:

☐ the principal health and nutritional problems of the refugees as they arrive, which have arisen as a result of the event and displacement period;

☐ the factors in the relief camp that are likely to cause a further deterioration in health;

☐ the local resources that exist to meet these needs.

Usually it is the camp itself rather than the event which led to the disruption which is the cause of poor health in relief camp populations. This situation may be understood when we consider the relatively few health risks in the displacement period, and the many health risks that exist in a relief camp. These latter include:

☐ inadequate rations;
☐ poor shelter;
☐ crowding and high potential for disease transmission;
☐ inadequate and polluted water supply;
☐ poor sanitation and environmental health;

The principal diseases are:

☐ related to water/sanitation, e.g. gastro-enteritis, typhoid, cholera, hepatitis, dysentery;

☐ caused by crowding, e.g. respiratory infections, measles, meningitis, typhus, TB;

☐ others, e.g. malaria.

This has been well illustrated in the case of refugees of Rwandan origin leaving Uganda for Rwanda in October 1982. With a short displacement period, the refugees were in a good state of nutrition and health on arrival at the relief camps. The camps displayed all the above risks — the climate was cold and wet, and there was no adequate shelter. When huts were built they were very crowded, the water supply was a muddy stream that rapidly became polluted, and few latrines were constructed. There was a major deterioration in health in the weeks following entry to the camp. Unfortunately, the response of many agencies to all these preventable problems is to provide medicines and to set up curative clinics. Relief agency personnel arriving in an established relief centre may be overwhelmed by the pressing clinical problems and fail to see the health situation in its historical perspective.

There are now few communities, whether local or refugee, without any trained health personnel to contribute to a relief programme. If from the beginning expatriates act as advisers and catalysts, helping to mobilise local resources and avoiding dependence on themselves for day to day running of the programme, there is a good chance that an effective and locally appropriate programme will evolve. If expatriates begin to run programmes themselves, it will be difficult to hand over to local staff, which may cause resentment in the local non-refugee population.

In a rapidly changing relief environment, with new arrivals and changing conditions in the camp, regular disease surveillance and reporting is essential. Such surveillance must be the basis for action and not merely a bureaucratic exercise.

In a given area, standardisation of health programmes in different camps is essential, particularly if different expatriate teams are working in each camp. The example of co-ordination and standardisation set by the Refugee Health Unit in Somalia is instructive.

6 **Nutrition in the Relief Phase**

Two aspects need to be considered. Firstly, an adequate general ration must be provided to ensure that adequately nourished persons arriving at the camp remain reasonably fed. Secondly, it must be determined whether there is an existing problem of malnutrition. If so, it will need to be dealt with, whether it existed on arrival or is due to an inadequate camp ration.

While it may seem obvious that an adequate general ration should be sustained, there are many situations in which the general ration is not planned according to nutritional needs. Recent data from relief camps in the Ethiopian drought indicate deterioration in nutritional status as a result of an inadequate ration. Table 8-3 below shows the position.

Table 8 – 3   Nutrition in Ethiopian Relief Camps

Date of Survey	% Adequately nourished	% Moderately malnourished	% Severely malnourished
March 1983	77%	20%	3%
May 1983	45%	33%	12%

There were no new arrivals during this time. The basic minimum of 1,750 calories per person is provided by ½ kg of grain per day, or 15 kg of grain per month. The general ration in the Ethiopian camps was 25 kg of grain per family per month. For a family of four this gave a daily food energy availability of 730 calories (3 MJ). It is not surprising that there

8

was such a deterioration in nutritional status. The low ration was not due to lack of relief food, but to a local decision to provide a 100 kg bag for four families. At no time had the ration been considered in terms of nutritional requirements.

From the previous sections on nutrition during the displacement phase it should be possible to make some prediction of the likelihood of malnutrition. A survey is essential to determine the magnitude of the problem.

Inaccurate figures are often produced by poor survey techniques. To compare severity in different situations, or different camps in the same area, it is essential that valid surveys are done, so that the most appropriate response is made, and progress, or deterioration, can be monitored.

It is not necessary here to go into details of relief feeding programmes. The principles given in Table 8-4 are generally followed.

Table 8 – 4   Principles of Relief Feeding Programmes

Nutrition Status	Type of feeding	Calories per day
moderately malnourished	supplementary	400-800
severely malnourished	supplementary	1,500-2,000

There are some areas of current debate in feeding programmes that are useful to discuss.

a: Supplementary feeding: In many relief programmes not based on identified nutritional problems is undertaken automatically. There are many disadvantages to supplementary feeding programmes — they take much time and organisation, they may perpetuate a 'relief mentality' and, if not available to the local population, they may cause resentment. If there is a small number of malnourished children, special feeding programmes may not be appropriate. Conversely, if there are many health and nutrition risks, e.g. crowding, poor sanitation, uncertainty over food supply, it may be more appropriate to provide supplementary feeding for all children, including those not malnourished.

The following procedures are suggested:

☐ more than 20% children moderately malnourished — feed all children;

☐ 10% — 20% children moderately malnourished — feed only malnourished;

☐ less than 10% children moderately malnourished — do not start special feeding programmes; treat the small number through clinics.

b: Therapeutic feeding centres must be considered with caution. They take up an inordinate amount of time, concentrate nutritional efforts (and media coverage) on a small number of children when preventive supplementary feeding for a large number may be more useful, and can create possibly inappropriate priorities for a mother who may have several other children to care for.

It is essential that progress in supplementary feeding programmes is monitored, and included in monthly health and nutrition surveillance. Supplementary feeding is often not sufficient to manage a nutritional problem. The reasons for poor weight gain may be not eating food at relief centre, intercurrent illness, or shortage of food in the family, resulting in the supplementary ration becoming the only ration.

It is essential that the relief programme is not prolonged beyond the time of acute relief needs. There are few situations where an acute relief programme should continue for more than six months after the end of the refugee influx. Where this does occur, a continuing expatriate input usually results in a programme that cannot be sustained without outside support, and creates major divisions between refugees and the non-refugee population.

Regular surveillance of changing health and nutritional status should indicate when the acute stage is over.

7 The Post-Relief Phase

After the acute relief phase, the refugees may have several options. If they are 'drought' rather than 'political' refugees, they may be able to move from relief camps back to their villages to begin farming again.

Political refugees may be resettled in agricultural projects in the host country, and then will need only the normal primary health care services of the country, often improved as a by-product of the refugee influx.

For the least fortunate refugees there may be no possibility of return or resettlement, and they may face a bleak future in permanent relief camps.

Relief camps set up in the acute relief stage are rarely ideal for permanent habitation. Temporary shelter, water and sanitation services become over-used, and the camp rapidly develops into a rural shanty town. As the impact of the acute relief stage fades, international agencies, particularly those involved only in relief work, gradually withdraw support, and what may have been a successful relief programme may slide into a new but chronic disaster. Such a decline is not inevitable, but the potential for deterioration exists in any long-term refugee situation.

The state of the refugee camps in Somalia is a useful example. There was a significant deterioration in nutritional status during 1982, as demonstrated by the following figures:

Survey Date	Adequately nourished	Moderately malnourished	Severely malnourished
April 1982	94%	6%	0
January 1983	80%	16%	4%

In addition to this deterioration in quantitative nutritional status, there has been a major problem with scurvy in the camps in Southern Somalia, and other reported vitamin deficiencies in various camps.

The deterioration in nutritional status in the camp surveyed above was due both to an inadequate general ration and to a recent epidemic of gastro-enteritis. The vitamin deficiencies are certainly due to the very limited *range* of food; in January 1983 the ration consisted only of wheat grain and DSM (dried skim milk).

Gastro-enteritis outbreaks are common in such communities, and the problems of environmental health in the crowded camps make it an increasingly serious factor.

There is a very real risk of a serious deterioration in nutrition and health in the Somali camps and in other refugee camps which were not designed for long term habitation.

There is at present little focus on this stage of relief programmes. If we are to avoid secondary disasters in such cases it is essential that greater attention is given to this chronic post-relief stage.

8

Section 4   Disaster Response: Physical Infrastructure

I   Shelter

Shelter and housing are frequent problems for disaster victims. If a disaster destroys people's homes, they will normally provide themselves with temporary shelter as soon as possible, either using the remains of their previous dwellings and any other materials at hand, or by moving in with friends and relatives. Field staff should consider the immediate local purchase of appropriate housing materials or widely available cheap temporary materials such as plastic sheeting, to assist people to shelter themselves and their families. Usually these temporary shelters are adequate for an interim period. The main functions of emergency shelters are protection from heat, cold and rain, plus storage and protection of property. They also provide a base for future action and an address at which to receive other assistance. The emotional security that shelter gives must not be forgotten, nor the privacy it affords. The location must also be appropriate, near employment or farming land, with access to services and utilities. Above all, construction of temporary shelter must not delay the programme of housing rehabilitation and reconstruction.

1 Temporary Shelter

Temporary shelter may be required in certain circumstances, but it is with permanent housing that people most require help. Agencies that supply temporary houses sometimes months after the disaster, and at unit costs as high as permanent local housing, are doing the people a disservice. The houses they provide are, moreover, usually inappropriate in size, style, materials, heating, value and design. They may serve as a disincentive to people who would otherwise start building their own new permanent dwellings.

a: Tents. Tents are to be found in many disaster situations. They are the main form of emergency shelter stockpiled by donor countries and the governments of vulnerable regions (often via the armed forces). They have the 'advantage' of built-in obsolescence, which reduces the risk of permanence, and are liked by governments because they are a visible sign of the effort to help. However, they are costly, usually too small and impossible to extend, unsuitable for housing animals or salvaged goods, and usually involve moving to a camp site away from the damaged property and the human community to which it relates.

b: Plastic sheeting. Oxfam has accumulated considerable experience in the use of plastic sheeting. It can be used in different forms to provide shelter, store property, make ground sheets, etc. For more details see Oxfam's Technical Guide, *Plastic Sheeting*.

Plastic sheeting can provide temporary shelter for those afraid to return to their undamaged homes during the after-shocks of an earthquake.

2 Programme Recommendations

Peasant farmers, shanty town dwellers and other poor people all over the world traditionally build their own houses, and voluntary agencies should not seek to do it for them. The inhabitants should be as involved as possible in any reconstruction programme. Not only does this save scarce financial resources, but more importantly, it gives people an opportunity to practise new building techniques, which can be taught as a part of the programme. Training programmes should particularly be considered in areas vulnerable to cyclones and earthquakes, where simple improved building methods can save houses from total destruction and people from possible loss of life. Courses for builders, and, within a community self-help programme, courses on the use of ring beams and L-shaped steel bars to make houses more stable, are all recommended as a part of a disaster mitigation programme.

Exceptions to the emphasis on a family's own labour and community self-help in housing might need to be made in special hardship cases: for example, a female household head with a large number of small children lacking the labour required for reconstruction, the old, infirm or disabled. A special fund or sub-project could assist these particularly vulnerable people.

Projects should be planned in such a way that when the agency withdraws, the people will be able to carry on this training both technically and financially. As far as possible, the programme should respect traditional techniques and not conflict with existing methods. Use should be made of low-cost, locally available natural or manufactured materials. This is more likely to ensure that after a disaster, materials not destroyed will be re-used in the reconstruction process.

Projects should always be planned with long-term viability and eventual independence in mind. (See PART FIVE, Section 2.)

II   Water supply

In a disaster, water is a greater necessity than food and, except in the most rigorous climates, it is also a greater necessity than shelter.

Human beings require access to some form of water supply on a more or less continuous basis and will consume water that is available to them whether it is polluted or not.

It is therefore essential to establish or re-establish an adequate supply of the safest possible water for the community affected as quickly as possible. A minimum of 3-5 litres of drinking water per person per day for drinking and cooking must be available; in addition, greater quantities of other, reasonable quality water are required for washing clothes and personal hygiene.

Wherever possible a ground water supply (water from underground, i.e. a 'safe' water supply) should be established. In the great majority of cases a ground water supply will need no treatment and can be supplied directly to the population, assuming every effort is made to prevent pollution at the point of collection. These safe water supplies are established through dug wells, tube wells or springs. (See PART SEVEN, Section 11.)

Surface waters should always be treated as suspect, and if they are obviously being polluted, some treatment for disinfection should be arranged.

1 Disinfection of Water

In normal circumstances, there are many criteria by which the wholesomeness of water is judged, such as organic matter, dissolved salts, hardness, matter in suspension, pH and so on. In a disaster, the only criterion which makes a water supply unacceptable is the degree of pathogenic (i.e. disease-forming) material present and the question of whether it poses a serious risk to health.

In emergencies, where people are crowded together, the risk of water being polluted by pathogenic material and being a hazard to health is much greater than in normal circumstances, and the disinfection of water is desirable and often urgently necessary. There are three ways by which this may be done.

a: Boiling the water vigorously for at least 5 minutes, or for longer at high altitudes where boiling point is lower.

b: Use of chemicals, such as proprietary disinfectants (e.g. Milton, Zonite, Javel Water), commercial bleaches (e.g. Chlorox, Dazzle, Regina), bleaching powder and water-sterilising tablets.

Where chemicals are used for disinfecting water, it is important to use the the correct quantity of the chemical. Detailed instructions for doing this are given in the Oxfam Technical Guide, *Safe Drinking Water*, by Burns and Howard.

c: The use of appropriate sand filters to improve water quality.

2 Vessels for Handling and Storing Water

In emergencies, not only is the provision of water a matter of prime importance, but also there may be a lack of pots and pans for cooking, for boiling water and for family storage of water. Also relevant may be the provision of soap and some form of towelling; for boiling water or cooking, fuel and matches may be needed.

3 Oxfam Water Packs

To enable water supplies to be established as quickly as possible, Oxfam has designed a series of water packs which are immediately available for emergency water situations.

The packs already available are for pumping, storage, distribution and water treatment (using the sand filter principle) and a well-digging and hand augering pack.

Fuller details of these water technologies are to be found in PART SEVEN, Section 11.

III   Sanitation

Most people in the Third World are without adequate sanitation, and many have no access to sanitation at all. Much ill-health is directly caused by this. It is a major and serious failure in relief operations, and in development work in general, that many agencies and field workers are very badly informed about the importance of sanitation, particularly in crowded conditions. The low level of competence among many agencies in this subject is a serious problem and one that must be tackled.

1 Diseases Arising from Bad Sanitation

The chief way in which diseases are spread in areas with poor sanitation is by the pollution of food or drinking water with human excreta. This is due to handling of food, spoons and bowls with unwashed hands, flies settling on food, or to direct contamination of water. The diseases transmitted in this way include cholera, typhoid, the dysenteries, polio, hepatitis, and such parasites as roundworm and whipworm. In addition, hookworm is transmitted by direct contact of bare feet with polluted ground, and schistosomiasis (bilharzia) by direct contact of the skin with polluted water.

A human being with cholera can excrete or vomit up to two thousand infective doses of cholera each hour, and such a person may remain infective at this level for up to thirty hours. The job of a sanitation system is first to contain and then to sterilise this material.

A fertile female roundworm that is well established in the human gut will lay as many as a hundred thousand fertile eggs a day, and these are of course passed out in the human excreta. Between 96 and 98% of the population of Bangladesh are estimated to have roundworms. This type of parasite is also a major health problem in India. A handful of such worms lodged in the human gut have first claim on the food consumed and so contribute to malnutrition. (For further information on communicable diseases see PART SEVEN, Section 5.)

2 Some Sanitation Principles

The top priority is to provide a place for people to excrete, and for the excreta to be contained in a chosen location. The benefits of containment are many, but include the control and treatment of the excreta; stopping people from excreting indiscriminately around the site concerned; and stopping rats, insects, vermin and birds having access to the excreta and carrying it off, thereby acting as vectors or distributors of disease. By containing faecal matter, the risk of it being carried into human living quarters on people's feet is also eliminated.

Human excreta under these controlled conditions will over a period of time lose their disease potential. Under normal conditions the pathogens — the disease-forming bacteria — are destroyed or diminished by other bacteria which are harmless to humans. The aim, therefore, is to contain human excreta in a safe location and provide the best possible conditions for these harmless bacteria to do their work.

The benign bacteria which destroy the disease potential of sewage thrive either where plenty of air is available (aerobic bacteria) or under airless conditions (anaerobic bacteria). Both kinds of bacteria are required for good sewage treatment.

The anaerobic stage of treatment is best achieved by passing the sewage through a closed tank. It is also fairly simple to provide the aerobic stage of treatment by allowing the sewage access to plenty of air. This is often achieved by passing the sewage over some form of filter medium, such as broken stone or gravel, allowing the aerobic bacteria to thrive and play their part in the second stage of sewage treatment. Chemical disinfectant usually upsets this natural treatment of sewage and should not be used.

3 Organisation of Public Latrines

Hygienic sanitation is very much more effective in controlling disease than curative medicine. A latrine can save the need for medical treatment, but it must work well and work well all the time. Below are some recommendations for all public sanitation.

a: If sanitation is installed, it must be used by all the population of the community served.

b: Always provide water for washing.

c: The system must be culturally acceptable. Usually people do accept and welcome better sanitation, but it is important to respect any taboos or traditions. Muslim communities, for example, require particular orientation of the latrine area.

d: The latrine must be pleasant to use, and be maintained and cleaned on a full-time basis. Make it a well-paid job for some intelligent person. All too often this daily maintenance and cleaning job is left to someone who is unmotivated and/or untrained, who manages in a short period to bring the latrine into disrepute by allowing it to become smelly and dirty. If sanitation is a key requisite for a community, then the maintenance and servicing of the latrine is a key job and must be made so.

e: The latrine must be within easy walking distance and should be visible to the population it serves.

f: Separate facilities for each sex should be provided, and the facilities must be safe and usable by children. Children are likely to excrete anywhere; they can be as infectious as anybody else, so it is important that they are catered for.

g: Lighting must be provided. Poor or no lighting means that, at best, the facilities will be unusable at night and, at worst, people will defecate on the floor around the latrine, increasing the health risk.

h: Privacy is important for the user of the latrine. All humans prefer to excrete in private, so provide privacy.

i: The latrine should not be sited on low-lying marsh land that floods at the first rains, nor where it may pollute the local water supply.

8

4 Some Simple Sanitation Techniques

If the aim is to establish a long-term sanitation system, field staff should enquire whether there already exists a local sewerage system that could be used. Is there a local sanitation expert? In emergency conditions, or where some kind of sanitation has to be improvised, the expedients listed below are available.

a: Trenches and bored hole latrines. The placing of excreta straight into the ground by trench latrines or bored holes is stll one of the simplest and most successful systems used. The bored hole is to be preferred, as it has some definite advantages, particularly fly control and being able to cover the hole with a squatting plate.

Bored holes are usually made with a hand-operated auger which will cut a hole 24 cm. in diameter and up to 7 metres deep. Augers are also held by Oxfam stores. Bored holes and trench latrines are obviously not feasible where the ground has hard rock near the surface, nor are they suitable in waterlogged ground, nor where high water-tables are present.

When used by the general public, these types of latrine need plenty of supervision on a daily basis, and there should be plans for making new holes or trenches as existing ones are filled. When bored holes are full to within half a metre of the surface, they should be filled in with soil and the squatting plate transferred to a new hole. After a year, it is safe to excavate the contents to use as manure. Bleaching powders and other disinfectants should not be used as they interfere with the natural process by which faeces are turned into a safe manure.

b: Pit latrines are based on exactly the same principle as bored holes except that they are dug by hand instead of being bored with an auger (see PART SEVEN, Section 11 (4a)).

c: Aqua-privies and septic tanks. (See PART SEVEN, Section 11 4b)

d: Emergency excreta disposal. In extreme conditions, bulldozers have been used to clean areas heavily polluted with excreta, by scraping the ground surface and burying the waste material. Under emergency conditions, the military have sometimes gathered excreta into a central location and incinerated it by using liquid or gas fuels. This may sound extreme, but it can be arranged instantly where there is a major disease threat.

5 The Oxfam Sanitation Unit

In answer to the problems and frustrations of trying to cope with massive refugee problems, Oxfam has produced a sanitation unit, which can be moved into a needy situation quickly and assembled to provide a public latrine for around 1,000 people. At the time of writing, 60 of these units have been used in Bangladesh. A booklet, *The Oxfam Sanitation Unit*, available from the Publications Officer, Oxfam, Oxford, gives full constructional details.

The unit can be used as a self-contained packaged system for sanitation and sewage treatment, or it can be looked on as a kit of parts which can be used in several ways, for several different kinds of sanitation. The principal functions of the equipment are as follows:

- ☐ the unit can be used with one tank only as a short-term measure for sewage containment or storage during an emergency;
- ☐ when used with two tanks and a percolating filter, the unit provides continuous treatment of sewage and discharges a safe effluent — it will then provide a long-term public health facility;
- ☐ the squatting plates can be used separately within existing drainage systems to improve the latrines;
- ☐ the tanks can be connected to existing drains to provide sewage treatment where none was previously provided;
- ☐ the squatting plates can be used with bored hole or trench latrines;
- ☐ the flexible tanks can be used for the storage of water if required.

N.B. It is essential that an adequate water supply is available if units are to operate efficiently.

Section 5   Disaster Procedures

I   Information

In all disasters there is a constant demand for reliable, first-hand information.

In the case of slow-onset disaster there should be time to monitor the deteriorating situation and make programme plans. Regular first-hand accounts from the field form an invaluable picture against which to plan a relief programme.

On first news of a high-impact disaster the field staff should:

☐ confirm this information with whatever reliable source they can: the Government, the press, field workers in the area, etc.;

☐ telex or telephone central office, with the information they have received and plans for visiting the area;

In many disasters the international news media are better informed than the national or local media and may be able to add to the field office's picture of events. Reports should also be obtained from UNDRO, the League of Red Cross Societies, the International Committee of the Red Cross and other relief agencies. The field office should be kept informed of the latest news. It is important, therefore, to establish a telex/telephone/telegram link in the field.

1 First Steps

On reaching the disaster area the priorities of the field staff should be as below.

a: To contact the appropriate Government authorities for their assessment of damage and of priority needs, an outline of their relief programme and suggestions as to where help is required (disaster preparedness work will simplify this task).

b: To contact any other international relief agencies to find out their plans, to share information and to liaise about collection of more first-hand information.

c: To contact any local volunteer agencies and community groups, to get their assessment of the situation, to see what they are doing and to find out their plans and needs.

d: To make an independent assessment of the situation, the priority needs, and possible action to meet these needs. Field staff should try to travel around the area, especially, where appropriate, around the poorer rural areas, to assess the situation there.

e: To disseminate this information as quickly as possible to the appropriate agencies. In the first days of a major emergency *communications from the field of operations should be as direct as possible.* Do not use intermediaries if there is any risk of delay or breakdown in communications.

f: To establish channels for funding. It is important for field staff to establish a means of receiving funds, drawing cash and paying grants. This will usually be through a local bank — ideally a branch of a bank with good international links to facilitate international transfers.

2 Assessment

In assessing the impact of the disaster, field staff should give priority attention to:

☐ food stocks held and food supplies expected, with observations of warehouse stocks, markets, prices and normal household stocks;

☐ nutritional status, especially of the most vulnerable categories: children under 5, pregnant women, hospital patients, etc. (for nutritional assessment see Section 3);

☐ the health status of the population and common ailments. This information should be available from hospitals and clinics;

8

477

□ water supplies: quality and quantity;

□ local availability of immediate relief supplies, shelter, building materials, clothes, cloth, cooking utensils, fuel, transport, etc.

③ Communications

One of the first priorities of field staff arriving at a disaster area is to establish a base and set up reliable communications. The first base is usually a modest hotel with a telex if possible. It is important at the outset to establish:

□ a reliable telex number. A hotel or friendly business will usually allow their telex to be used with payment for outgoing messages;

□ a cable address if no telex is available. Cables are extremely expensive compared with telex. Every effort should be made to get a telex link before resorting to cables. Care should be taken to specify how telex or cable messages should be coded or addressed to ensure their arrival at the correct destination.

If field staff are to be in the area for several weeks, they should also establish: a telephone number (and advise the best hours to call); a postal address and, if this is a PO box number, the exact location of the office/base in case people are to be directed to the office.

In Oxfam's experience of disaster relief, two-way radios have not been extensively used. Normal telex and telephone facilities are usually adequate. The operation of radios is a sensitive political issue, and licences are unlikely to be granted easily. However, some UN agencies, national civil defence organisations and voluntary agencies have established radio networks. Field staff should arrange with these organisations to use their circuits if appropriate.

a: Telex/telephone procedures. All telex or cable messages should be numbered sequentially. This provides a reference for the message and a check that none is missed. A careful *record* should be kept of all outgoing and incoming messages, and a *master file* maintained in sequence.

Where there is a third administrative centre involved, a standard procedure must be established determining whether all correspondence between any two offices is to be copied to the third, which can be very expensive, or whether each office is to inform the third only of those items of general importance sent to the second. Failure to establish such a system can result in confusion and misunderstanding, as number sequences are broken and false assumptions are made about who has what information.

Each paragraph of the telex message should start with the project reference number where relevant. In urgent situations when supplies are ordered by telex, a purchase request reference number should be included in the message (see Supplies Procedures, Sub-section II).

b: Telephone. Telephone conversations often include important decisions. A brief file note of the contents of all telephone calls is very important and should either be added to the telex file or filed separately. A telephone log-book is recommended, to record a brief summary of all calls.

④ Reporting

Reporting is vital to enable the head office to keep up to date with changes in the situation and progress in the relief programme and to assist formulation of policies. Many people — broadcasters, journalists, other agencies and the general public — will request reliable and up-to-date information.

In the early days of a relief operation with the situation changing rapidly, lengthy telex reports are the best and most reliable means of reporting, but much of the background information, personal reflections, observations, opinions and confidential material cannot be sent by telex. Weekly reports by field staff, ranging over all aspects of the operation, are of great value to and greatly appreciated by all members of the support team. The topics covered should include (preferably in a consistent framework):

- ☐ the events of the week, with comments on their significance, developing trends, etc.;
- ☐ the political situation, latest figures, estimates of damage, Government's requests to agencies and policies toward them;
- ☐ other agencies and what they are doing;
- ☐ Oxfam programme and tour observations, ideas for future action. Reports of individual projects should be attached separately;
- ☐ morale and distribution of the team;
- ☐ presence of journalists and others;
- ☐ human interest stories and observations;
- ☐ financial situation of the programme.

Field staff should bear in mind that different people in the support team are looking for different information from these reports; for example, the press need to place human interest stories in the media; others need to know whether relief consignments were correctly labelled and consigned and that appropriate goods were sent, etc.

In any large disaster relief operation, events move very fast both in the field and in the home country (with fund-raising campaigns, debates in Parliament, appeals, media pressure, etc.), while communications between the two may be difficult and time-consuming. Experience has shown that each part of the operation takes on a momentum and direction of its own, with the risk that the perceptions of each by the other rapidly become unreal and outdated. For example, the central office is tempted to follow the direction taken by the media and the fund-raising campaign, while the field staff are reorienting the relief programme in response to changing circumstances that are not known to journalists, who then find themselves projecting an inaccurate picture of the operation.

5 Liaison with Other Agencies

It is important in any relief operation to know which other agencies are working in the field and what they are doing, and for field staff to co-ordinate their efforts with other agencies to avoid duplication and confusion.

All voluntary agencies should refer first to Government authorities and take advice or accept the co-ordinating authority of the Government. In many circumstances the voluntary agencies feel frustrated by the delays, apparent confusion and contradiction in Government departments which may be severely disrupted themselves by the disaster. It is therefore important for voluntary agencies to meet to share information, observations and experience, so that each knows what the others are doing. A regular weekly meeting of voluntary agency staff should be attempted.

In the provision of basic services — food, health care, water supplies, housing — there is scope for very wide differences in the standards offered by each agency. Every effort must be made to standardise programmes, ideally in conformity with guidelines from the Government to prevent the confusion that may arise when one group of people receives more food or more sophisticated medical care than another. Without disciplined co-ordination by the agencies, co-operative effort can become competitive effort.

In large disasters there is often an officially designated and generally recognised *'lead agency'*, usually one of the UN agencies, on whom the main burden of agency co-ordination falls. Field staff should support the lead agency in every possible way.

8

The *Emergency Supply Logistics Handbook* — produced by the LICROSS-VOLAG Steering Committee for Disasters — is a useful general guide to procedures in this field.

Every effort must be made to avoid duplication of logistic services by different organisations. Wherever possible, a single, centrally co-ordinated supplies operation should be established in which all involved agencies can participate.

In many situations a unified logistic operation will not be acceptable to all agencies. Although the UN agencies have well-established procedures for supplies, these are not available to all voluntary agencies working in a relief programme, and in any case many would not agree to joining a single UN-operated logistic operation. Speed and independence of action, with their important implications for public relations, are greatly valued by voluntary agencies. If they cannot be unified, supplies and logistic operations must be carefully co-ordinated by all the agencies involved in the operation.

The first priority is to establish a clear understanding of needs, standardise specifications, and decide which agency is to supply what. This is an early priority for inter-agency co-ordinating meetings.

Logistic arrangements should provide for spare capacity, beyond what will suffice if all goes according to plan. There are risks of delays, breakdowns, bottle-necks, red tape and increasing numbers of people requiring assistance as the operation proceeds.

A major trucking operation should, wherever possible, be undertaken by contractors — or in co-operation with the Government or an agency that is normally operating a transport fleet. Field staff should avoid hiring or buying vehicles, employing drivers and mechanics, procuring fuel, arranging maintenance, etc. When a programme of this nature is undertaken, a member of staff with logistic experience should be seconded with special responsibility for this aspect of the work.

Supplies should be purchased locally if possible. The savings that may be made by purchasing in Europe or elsewhere are usually more than made up by the cost of transport, etc., the hours of work and effort in clearing the supplies, and all the uncertainties and risks of importing goods in an emergency. In order to purchase locally, financial arrangements (local bank accounts, etc.) have to be made very early. Standard specifications for commonly needed relief items have been developed on the basis of previous experience (see Appendices I — III).

1 Supplies Procedures

There is always a serious risk in emergencies that speed and urgency may compromise efficiency. Experience has shown that this is particularly true in requesting and purchasing supplies. In emergencies telex requests are sometimes made with poor technical specifications, unclear consignment details, and with inadequate administrative procedures to enable orders to be handled consistently and comprehensively through all their stages.

A standard procedure should be adopted for the purchase of all supplies — whether for administration or for the project itself.

2 Technical Supplies

Field staff have made some serious and expensive mistakes in ordering technical items in disasters, because inadequate technical information is given on the supplies request form, resulting, at worst, in the wrong items being purchased. *In purchasing technical supplies, serious consideration should be given to the possibility of the team member responsible travelling to the supplier and confirming that the purchase is exactly what is required.* This may cost several hundred pounds, but is well worth it if the purchase is of thousands of pounds worth of equipment.

UNIPAC (the UNICEF Packing and Assembly Centre, Copenhagen) stocks a wide range of materials of all sorts, including a selection of relief supplies in an emergency stockpile. Non-governmental organisations may draw on the supplies and services available from UNIPAC on a reimbursement basis. Prices can be obtained from the current UNIPAC catalogue held in UNICEF offices. Purchasing through the regional UNICEF office may be particularly appropriate if UNICEF is also shipping UNIPAC supplies to the same operation. (See Appendices I — IV.)

3 Transport of Relief Supplies

Relief supplies can be transported by: chartered ship, scheduled sea freight, charter flight, scheduled airline flight, rail, road, or in hired trucks. Procedures for the reception and clearance of relief cargoes vary both by country and by means of transport. The following is a guide from Oxfam experience.

a: Sea freight: procedures either by chartered or scheduled ship are dealt with fully in the Licross-Volag *Emergency Supply Logistics Handbook*. In general the reception of sea freight is handled by a shipping agent on behalf of the sender. When shipments by sea are expected, field staff should establish what shipping agency is handling the cargo on arrival and make contact with it. If no agent has been appointed, the field staff should commission a shipping agent to clear the ship's cargo, and work closely with the agent.

b: Air freight by scheduled airline flights: freight is carried by airlines that have established handling and clearance procedures in the airport of arrival. Field staff expecting items by scheduled air freight should contact the airline office for advice on clearance procedures. The efficiency of different airlines in different countries varies widely and in many disasters the routine airport procedures are completely disrupted. Field staff should find out whether it is advisable or necessary for them to meet flights, in order to ensure clearance and safe delivery of the supplies.

c: Charter flights generally require much more involvement of field staff in handling and clearing the cargo, because a charter airline may have no established arrangements for handling at the airport to which it is chartered, and experience shows that shipping agents in the Third World are less familiar with procedures for clearing air freight. The reception of air charters can be a frantic and difficult business. A plane takes only a few hours to load and unload a small cargo. A ship's captain, or scheduled airline, takes more responsibility for cargo and its unloading than the captain of a chartered aircraft, for whom speedy discharge and prompt departure are usually higher priorities. Unlike a ship's cargo for which a negotiable bill of lading has to be countersigned by the consignee or agent, airway bills on chartered aircraft are simply handed over by the captain at the airport to the airport authorities along with the cargo.

Air freight charter flights are often used to deliver relief supplies to a disaster area. Particularly in circumstances where routine airport clearance procedures have broken down, some method of operation has to be worked out by field staff with the responsible government authorities (government ministries, customs, security, airport authority, etc.) as to how these goods are to be unloaded and cleared. In every operation it will depend on local factors. Some balance has to be struck between the urgency of delivering much-needed supplies, in which some normal procedures and regulations are bypassed or disregarded, and the inflexibility of a bureaucracy which demands that some formalities are respected regardless of agencies' ideas about the priority and urgency of humanitarian relief. If a relief operation is receiving

8

frequent relief flights, the field office should draw up appropriate guidelines on local airport procedure so that any member of the office can meet and clear flights.

Field staff may expect telex information of the flight number, estimated time of arrival (ETA) and cargo manifest, and may be asked to telex back acceptance of the flight and cargo. With a charter flight the ETA may change several times, and the final ETA should be taken more as a guide than as the actual arrival time, as charter flights seldom arrive on time.

Copies of the airway bill, cargo manifest and packing lists should be on the flight in a sealed envelope addressed to the field staff. The official papers, carried by the captain and handed to airport officials for distribution to various officials, are unlikely to reach the field staff. These papers are extremely important, as they are the official record of what is on the flight, and may incorporate last-minute changes from the telexed cargo list of several days before.

In most relief operations it is very important that field staff meet all charter flights, and in some circumstances all scheduled flights carrying cargo or passengers, to solve problems over documentation, unloading, customs clearance, warehousing, consignees, passengers, etc. This will involve long hours of waiting in uncomfortable airports. There is no alternative — so always take work or a book to the airport!

d: Rail and road transport can be used, where appropriate, to carry relief supplies. Procedure in these cases depends on local conditions. In recent experience when trucks (lorries) have been hired to carry relief supplies in circumstances where there is political insecurity and considerable risk of delay through lack of fuel, frontier difficulties etc., an Oxfam field staff member has accompanied the vehicles as a 'trouble shooter'. This is a most important role in ensuring that supplies are delivered. It is both difficult and time-consuming, but should not be underestimated.

e: Containers can also be used when practicable — i.e. when there is sufficient freight, and container handling facilities are available at each stage. As containers make strong, secure stores, consideration should be given by field staff to the purchasing or leasing of containers for this purpose.

4 Warehousing and Stock Control

The urgency of the demands for relief items in the early stages of an operation means that goods are often purchased, received and despatched without the routine administration and controls. Speed is important, but it will not be impaired by a simple and readily understood stock control system, which can be set up from the beginning of an operation.

A simple stock control system should use a *stock register* which shows opening stocks, receipts and despatches. Every item entering or leaving the warehouse should be entered, either by items, or chronologically, or both, as a matter of record. 'Bin cards' should duplicate this information and be kept with the stocks in the warehouse. A 'reserve figure' should be indicated on bin cards, so that items may be reordered in good time.

If goods have to be delivered by transporters or drivers to other locations, a *consignment note* should be issued in triplicate, one copy to be retained at the store, one copy to be retained by consignee and one copy to be retained by the driver signed as a receipt by the consignee.

Reasonable details of the consignment should be entered on the consignment note so that pilfering or exchange of goods can be readily identified by the consignee.

Depending on the local situation, the supply of staff, and the value of the cargo, it may be advisable to have a responsible team member accompanying the truck or convoy to ensure that it reaches its destination, or to take appropriate action if it does not.

5 Local Purchase of Relief Supplies

For a foreign relief agency, which is known to be intending to spend large sums, buying goods locally can be a risk. It requires of the buyer honesty, a knowledge of the local market and its conventions, and of the goods needed. Without an experienced buyer in the relief team the alternatives are for one of the field staff to take on the role of buyer (he or she may be honest, but is unlikely to know the local market), for the field office to employ a local buyer who will charge a fixed percentage of the turnover as a fee, or to ask a major company to lend to the field office an experienced buyer on a temporary basis, perhaps at cost.

Orders placed with suppliers should from the first instance be made on a standard printed order form (relating to a supplies request form) which should give clear and precise details of requirements, including where and when goods are to be delivered, and at whose cost, and specifying details of quality of goods supplied.

A system that helps to avoid malpractice by suppliers should be adopted. If two samples are required from the intending supplier, one can remain in the field office for reference, the other should be sent to the storekeeper. The storekeeper may then check the quality of the items delivered against the sample. If they conform, the storekeeper informs the field office, which makes the payment. With this system it is necessary to make clear to the supplier (as the terms on the sample letter do) that payment will not be made earlier than seven days after delivery of the goods. This allows time for checking the consignment.

III Administration

1 Personnel

The use of foreign voluntary or professional staff in the immediate aftermath of a fast-impact disaster is not advised. They are unlikely to arrive in the critical early days after such a disaster. Local people will be the most appropriate during this phase: they are already on the spot, and they know the language, the communities and the people.

In general, volunteers should be used with caution, as they will often lack the experience necessary to operate successfully in an emergency. Wherever possible, experienced local or foreign staff should be used.

Technical consultants are usually available, and where specialised technical advice is required it should be sought from either local people or the list of technical consultants held by most agencies.

The administrative aspects of employing staff should be settled from the beginning. Staff should know their contractual obligations, remuneration, etc.

In circumstances where a large team is fielded, the team leader must be aware of his or her responsibility to employees for their welfare, health, morale and logistic support. This includes feeding staff and providing shelter and possibly rest and recreation.

2 Finance

It is important to establish financial procedures at the beginning of any disaster programme. In a confused situation it is only too easy to lose financial control. A basic accountancy system should be set up, and perhaps even a full-time accountant employed to control finances.

3 Relief Works
(see Food Aid Section 2.II)

Free food distribution encourages a sense of dependence, is damaging to individual self-respect and dignity and can contribute to a reduction in agricultural production. Selecting beneficiaries can be extremely difficult.

The main objectives of relief works are therefore: to provide relief to

8

those who might otherwise be made destitute; to act as a self-selecting mechanism to channel relief to those in need; to enable people to maintain their resources (i.e. avoid selling land or animals); to improve long-term agricultural self-sufficiency and improve community infrastructure (irrigation, schools, etc.).

All able-bodied and many disabled people can join such schemes if they wish, and remuneration should be such as to enable a process of self-selection to operate (i.e. not so high as to attract people away from existing activities, nor too low to permit subsistence).

It is important that people are aware of the scope of the programme, including its short-term nature, registration requirements, levels of remuneration, type of payment (whether in kind or cash), agency responsible, etc.

Ideally, communities will be involved in the selection of projects, although these will have to be screened according to available resources, such as tools, technical advice, etc. The projects should also bring benefits in the long term to the poorer, not the richer members of the community.

There are many arguments for paying in cash rather than food: it provides workers with a choice as to the disposal of their income; it also allows people to buy local food when available, thus maintaining the prices to producers instead of competing with local production. The use of cash also avoids many of the logistic problems incurred when dealing with food, including administration, transport, etc.

4 Distribution of Relief Supplies in Refugee Camps

The distribution of supplies is extremely problematic. Where refugees are not destitute, some items can be sold at a subsidised rate, and the funds reinvested in community projects. However, safeguards should be introduced to enable the sick, old people, etc. to obtain basic supplies. Where possible, individual family assessment should be made to ensure that those most in need receive the required supplies or food. This can lead to a quite complex administrative system which will be more appropriate to some circumstances than to others.

IV Children in emergencies

1 Introduction

In any disaster, whether natural or unnatural, rapid or slow-onset, special attention should always be given to children. They are likely to be far more vulnerable than adults both emotionally and physically; they are more susceptible to disease and malnutrition and are at risk in emergencies of being abandoned by, or separated from, their families.

Since children draw much of their security from their family and look to their parents for guidance, reassurance and love, it is generally preferable that they share the lot of their family in emergencies. Thus, every effort should be made in disasters to safeguard and strengthen family and community structures. The aim should be to address the problems faced by children within the context of a broad integrated programme assisting the community as a whole, rather than devise a specialised intervention aimed exclusively at children. To this end, most emergency programmes designed to help children should follow the broad guidelines outlined elsewhere in this section.

However, sound procedures must be established for assisting children isolated from their families, orphaned, or evacuated during a disaster, and support should only be given to programmes that follow these procedures.

2 Care of Unaccompanied Children

There are various arrangements that can be made to provide for unaccompanied children in emergencies, and these depend very much on the nature of the disaster and the circumstances of each child. Often care must be provided in some form of communal centre. There are a number of different kinds of camp: including sick bays, centres for indefinite stay (or substitute homes) and transit centres geared to providing short-term care while tracing work is carried out and/or other arrangements made.

Children from cultures where living in compounds or large groups is the norm are perhaps better prepared psychologically for life in a centre than those from nuclear families. To help alleviate the children's distress, they should be given as much individual care and attention as possible by people from their own culture while in the centre, and it is vital to have respect for the child's feelings in relation to his or her age and development. In each shelter or centre it should be possible to ascertain:

- [ ] how many children have families/relatives and whether they can return to their care;
- [ ] whether assistance should be provided to the family or community to enable the child's return;
- [ ] for how long the children will need care and how they are to be cared for;
- [ ] whether there are people able to undertake the investigation of the family situation and provide follow-up visits after the children are reunited with their families;
- [ ] whether the people responsible for the children in care and those selected to facilitate their return to their families are sympathetic to the children, and understand their needs, circumstances and the environment from which they have come;
- [ ] whether sufficient funds are available for the work, and facilities for staffing, housing, feeding, educating, transporting, etc. are adequate.

The main aim should be to keep the period in the centre to a minimum by tracing the families, reuniting the children with their families and helping them to readjust to family life. Children who for one reason or another are unable to return to their families may be placed with suitable foster parents, and only when this is not feasible should an institution be considered. There are various kinds of fostering arrangement, the most common probably being an informal one in which no fees are paid or supervision exercised. With the more formal foster homes, the children are usually placed by the Government or a designated agency. Fees may be paid and the home is normally well vetted beforehand and the placing supervised. Often children who are known to have no family links can be placed with the expressed purpose of converting the arrangement from one of fostering into one of legal adoption (although support should only be given to bona fide in-country adoption schemes — for more details see PART TWO, Section 3). In this instance, the home should be supervised until the adoption procedures have been finalised. 'Service' foster homes exist to provide short-term care for children whose families are temporarily unable to do so themselves. Finally, there may be professional foster homes which look after children with special needs — such as handicapped children — and these may exist as an alternative to institutional care.

All fostering agreements — whether designed for evacuees or orphans — should be properly documented and controlled, or there is a risk that the child's rights may be violated. For example, the agreement should specify that the child has a right to retain his/her family name and his/her nationality. There have been cases throughout the world in which foster parents have given their own name to foster children —

8

indicating that they perceive fostering as an arrangement more akin to adoption. In more serious cases, children are 'adopted' de facto by their foster parents, and their natural parents may be unable to secure their return. The right to retain their true nationality may be lost to some children who are removed from, and not returned to, their country of birth.

Long-term institutional care provided for children orphaned during disasters should follow the principles outlined in PART TWO, Section 3. Information should be collected on the traditional ways of caring for orphans in the society concerned. Care should be taken not to introduce alien standards and methods, and children should not be segregated from the rest of the community, for example by placing them in separate villages. They should live in small groups as close as possible to their home of origin. Provision must be made for funding the home after the end of the emergency period.

3 Identification

It is vital not only to identify every unaccompanied child but also to ensure that careful records are kept on each one and the earlier they are taken, for more accurate recall, the better. If the child is to be moved, their records must always accompany them and efforts be made to ensure the records are maintained.

Individual records can be built up by amassing information as below.

☐ Interviewing the child to establish previous experiences, names of relatives, place of residence, etc. Children over the age of six should normally know not only their own and their parents' full names, but also where they come from. However, in certain circumstances they may have been told by relatives to give false information or be too sick or disturbed to remember.

☐ Interviewing other children or adults who might have lived near the child, travelled with him/her or sought help at the same time.

☐ Interviewing staff of shelter or transit centre who remember the circumstances of the child's admission.

☐ Reporting on the health of the child.

☐ Reporting on the child's appearance (noting special characteristics or marks).

Interviews with children should be carried out in an informal and relaxed atmosphere by people who know the language and dialect spoken by the child and are familiar with the environment and circumstances from which he/she comes. It is therefore preferable that the interviewers should be people from the disaster-affected area.

Children under the age of six may have difficulty remembering their identity or where they come from — especially if they are suffering emotional trauma. In this instance a number of special procedures should be followed.

☐ A sympathetic and skilled person should be chosen to spend time with the child, watching out for any clues that might come to light while he/she is at play or talking with other children. Similarly, much can be learnt by monitoring the child during meals, at bath time or during sleep.

☐ Enquiries should be made from anyone who might have recent knowledge of the child.

☐ Photographs should be taken of the child as soon as possible to assist identification. These can be reprinted and circulated publicly, whilst administrative arrangements are made to receive and follow up enquiries from relatives. Young children change appearance very rapidly and may not be recognised by relatives after a long period of separation.

☐ Tooth formation in young children is a fairly reliable indication of age.

□ Babies, toddlers and pre-school children should always wear indestructable identity tags when not with relatives.

Each child's history/record must be reviewed when planning for his/her future, and the best interests of the child should be taken into account when deciding on the type of provision to be made.

4 Establishing the Status of the Child

The status of each unaccompanied child receiving temporary care or shelter should be established before making long-term provision for them. The exact circumstances of the child must be ascertained, whether:

□ he/she has neither parents nor other relatives (this can seldom be established without interviewing people from the child's community);

□ he/she has no immediate family, but does have more distant relatives;

□ he/she is only partially orphaned — with one parent still living;

□ his/her family is still living, but has been separated from him/her;

□ he/she has been rejected by his/her family, temporarily because of hardship, or for more permanent reasons.

It is important to clarify who has legal responsibility for the child before deciding who should care for him/her in the long-term. Unaccompanied children are best taken care of by relatives or other, unrelated families rather than institutions. Plans for their future must be based on a full examination of their circumstances and the motivations and circumstances of those who intend to care for them.

5 Evacuation of Children

This form of intervention is practised on occasion when children's lives are at risk, the aim being to remove them from danger zones to places of safety where their security can be guaranteed. However, only where the threat to life is very great — such as in areas of warfare — should this procedure be considered, although in some cases sick children may be evacuated to places where they can receive proper treatment. To be successful, this type of programme must be both well organised and centralised. Obviously the principles for identifying, establishing the status of, and caring for evacuees should be the same as those applying to unaccompanied children in emergencies. Selection for evacuation must be based on need, and the procedure only undertaken with full documentation and effective supervision. It is preferable that children be placed within their own country, or in neighbouring countries where the same language is spoken, rather than in countries very distant from their home. It is vital to secure the permission of parents before evacuation, since many would prefer that their children remain with them. Parents should be given as much information as possible about the expected destination of their children; many may resent their children being taken to unknown destinations by strangers. Whenever feasible, arrangements should be made to keep parents in regular contact with their children. Care must be taken to ensure that the children are healthy enough to be evacuated and that the procedure will not do them more harm than good. If possible the adults taking responsibility for the children whilst in exile should be of their own ethnic group, speaking the same dialect and able to keep home memories and customs alive.

In many countries evacuation may be seen as a violation of customary practice and national integrity and it is extremely important that no such intervention be supported without prior investigation of policies and attitudes at both local and national level. The policies of the receiving country should be sympathetic to the children's situation and not seek in any way to exploit them, but to protect their interests in both the short and long term.

8

Section 6   Appendices

I   Relief items for which local suppliers should be identified

1. *Food Supplies* Listed by commodities as appropriate locally: Beans, oil, rice, sugar, wheat, etc.

2. *Subsistence Equipment* Blankets, buckets, clothing, housing and construction materials, insecticides and sprayers, kitchen equipment.

3. *Water Supplies* Bleaching powder, drilling rigs, pipes, pumps, tanks.

4. *Medical Supplies* Blood banks, dressings, drugs, equipment.

5. *Agricultural Supplies* Fertiliser, hand tools, pesticides and sprayers, machinery, veterinary supplies.

6. *Logistics* Transport, fuel, warehousing.

7. *Support Equipment* Generators, office supplies.

II   Oxfam Store

Oxfam has a Disaster Store in Bicester, which holds stocks of:

☐ *Blankets.*

☐ *Plastic sheeting:* for use as waterproof shelter — see *Plastic Sheeting — its use for emergency housing and other purposes*, an Oxfam Technical Guide, by J. Howard and R. Spice. The usual stock is: (8mt x 28mt), 375 microns (1,500 gauge) black; (8mt x 28mt), 250 microns (1,000 gauge) black.

☐ *Feeding kits:* for use in nutritional assessment, supplementary and therapeutic feeding. The original Supplementary Feeding Kit (with equipment for feeding 500 children pack in 2 Triwall boxes) and the Therapeutic Feeding Kit (with equipment for 100 children incorporating a nutritional assessment kit packed in one Triwall box) *are being remodelled as 3 separate kits:*

   — *Nutritional Assessment and Surveillance Kit*

   — *Supplementary Feeding Kit (for 250 children)*

   — *Therapeutic Feeding Kit (for 100 children)*

☐ *Sanitation Units:* for use with high density population to provide 'instant' sanitation — see Oxfam Sanitation Unit brochure.

☐ *Water kits:* distribution, pumping, storage etc.

III   Commonly needed relief items

☐ *Tents.* The British Red Cross Society maintains a stock of family-size ridge tents (floor area 4.25 mt x 2.75 mt, height at walls 1.20 mt, at ridge 2.15 mt).

☐ *Family Cooking Sets.* Available from UNIPAC (see standard specifications appended below).

☐ *Soap.* Available at short notice from normal commercial suppliers.

☐ *Drugs and Medical Equipment.* ECHO (The Joint Mission Hospital Equipment Board) exists to supply *bona fide* charities and missions, etc. with drugs and equipment at low cost. Oxfam purchases through them or through similar organisations in Europe. Standard lists of drugs are listed in PART SEVEN of the Handbook.

| IV | **Standard specifications for certain common relief items** |

*Reprint from UNHCR Handbook for Emergencies; UNIPAC numbers in brackets

These specifications have been developed with UNICEF to assist representatives/field staff in drawing up tender requests where local purchase is possible, and to give a clear indication of what could otherwise be supplied at short notice through funding agency headquarters. The UNIPAC catalogue reference is given in brackets where applicable; the actual source of supply through funding agency headquarters would depend on the circumstances and in particular on regional availability.

1. *Blankets, heavy* (similar E500 35 05)

Woven, 30-40% wool, the rest other fibres (cotton, polyester) blanket with stitched ends, size 150 x 200 cm, weight 1.3 kg., packed in pressed bales of 50 items. Each bale of 50 items would be about 0.35 m3 volume and weight 65-70 kg. Large quantities are generally available.

2. *Blankets, light*

Cotton size 140 x 190 cm, weight approx 850 gm, usually packed in pressed bales of 100 items. Each bale of 100 items would be about 0.4m3 volume and weight 85-90 kg. Fairly large quantities generally available ex-stock in Asian region; more limited availability elsewhere.

3. *Buckets, plastic* (217 00 00)

Bucket/pail 10 litre capacity, polyethylene with plated steel-wire bail handle, conical seamless design, suitable for nesting, reinforced or turned lip. Plastic or galvanized buckets are likely to be available locally and are very useful. Plastic are generally to be preferred.

4. *Family cooking sets*, emergency (203 65 10)

12 items aluminium utensils as follows:

Cooking pot, 6 litre, with bail handle and cover Cooking pot, 4 litre, with bail handle Dinner plate, aluminium (4 each) Plastic mug (4 each) Coffee pot, aluminium, 2 litre.

The set is packed in a cardboard carton 25 x 25 x 20 cm, weight 2 kg. The set does not contain cutlery: five stainless steel soup spoons and one stainless steel cook's knife, blade 15-17 cm, can be supplied separately if not available locally. While the set is quite robust, utensils of a heavier gauge aluminium are normally supplied by UNHCR when some delay can be accepted. The advantages of the emergency set are lower weight, packed volume and price. It is therefore particularly suitable when supply by air is necessary.

8

Also recommended by OXFAM:

5. *Plastic sheeting*

Black seamless polyethylene sheeting, 250 microns (1000 gauge), width 5-8m, supplied double-folded in lengths usually of 100-800m, approx. weight 1 kg/4m2. For multipurpose use: roofing, walls, ground sheets, linings, etc. Widely available.

Section 7   Resources

Bibliography

J. Boyden, I. Davis: *Disaster Mitigation* special edition of Reading Rural Development Bulletin 18, Reading University, October 1984.

F. Cuny, *Disasters and Development*, Oxford University Press, New York, 1983.

I. Davis, *Shelter after Disasters*, Oxford Polytechnic Press, Oxford, 1978.

Disasters, The International Journal of Disaster Studies and Practice, International Disasters Institute, London.

LICROSS-VOLAGS Steering Committee for Disasters, *When Disaster Strikes: Emergency Logistics Handbook*, Geneva, 1982.

R. Norton, *Disasters and Settlements*, Disasters Vol. 4, No. 3, 1980, I.D.I., London.

Oxfam, *Practical Guide to Refugee Health Care*, 1983.

Oxfam, *Plastic Sheeting: its use for emergency housing and shelter*, 1981.

Pan-American Health Organisation, *Emergency Health Management after Natural Disasters*, 1981. (Pub. 407)

A. Sen, *Poverty and Famines*, Oxford University Press, Oxford, 1981.

International Refugee Integration Resource Centre, *Refugee Abstracts*, 13 rue Gautier, CH-1201 Geneva, Switzerland.

S. Simmonds, P. Vaughan, S. William-Gunn (eds), *Refugee Community Health Care*, Oxford University Press, Oxford, 1983.

Swedish Red Cross, *Prevention is better than Cure*, Stockholm 1984.

N. Twose, *Why the Poor Suffer Most: Drought and the Sahel*. Oxfam, Oxford, 1984.

UNDRO, *Shelter after Disaster*, Geneva, 1982.

UNHCR, *Handbook for Emergencies*, Geneva, 1982.

UNHCR, *Handbook for Social Services*, Geneva 1984.

A. Wijkman, Lloyd Timberlake, *Natural Disasters: Acts of God or Acts of Man*, Earthscan, London, 1984.

L. Young, *Mitigating Disaster in Agriculture*, in Reading Rural Development Bulletin 18, Reading University, October 1984.

8

ADDRESS
LIST

Philip Wolmuth

ADDRESS LIST

Section 1   Oxfam Field Offices

I   Africa

Burkina Faso
B.P. 489,
Ouagadougou,
Burkina Faso.

Tel: Ouagadougou 336469
Tx: (978) 5403+
A/B AFRICORD 5403 BF
(978) 5255+
A/B SAED 5255 BF
(978) 1111/2+
A/B BCTR 1111 BF*
*Via Assnt Operator 2004

Kenya
P.O. Box 40680,
Nairobi, Kenya.

Tel: 010 2542 47025
Tx: (987) 22924+
A/B 22924 OXFAM
Emerg.(987) 22274+
A/B 22274 EXPRESS

Mali
Oxfam,
B.P. 209,
Bamako, Mali.

Tel: Bamako 223843
Tx: Cabine Publique
BKO-MALI 992/993/994/995
Via Yves Guernard, Euro
Action ACORD
Tel: Bamako 223843

Mozambique
Oxfam,
Avenida Patrice Lumumba 770,
C.P. 2865,
Maputo, Mozambique.

Tel: Maputo 27518
Tx: 6-479 A/B TVTVL MO

Rwanda
B.P. 1298,
Kigali, Rwanda.

Tel: Kigali 4074
Tx: Via (0026) 31971 for
transmission to Oxfam,
Trakig 10, Rwanda.
A/B HRCLS B

Senegal
B.P. 3476,
Dakar, Senegal.

Tel: 010 221 226894
Tx: (906) 671+ for attn.
Oxfam BP 3476 Dakar
A/B 671 PUBLIC SG

**Southern Africa,
Malawi, Zambia**
P.O. Box 35624,
Central Lusaka,
Zambia.

Tel: Lusaka 216742
Tx: (902) 42670+
A/B EAGLE ZA 42670
Emerg.(902) 44560+
A/B CUZAM ZA 44560

Tanzania
P.O. Box 6141,
Arusha, Tanzania.

Tel: Arusha 3697
Tx: (989) 42126+
A/B 42126 CKTVL TZ

Uganda
P.O. Box 280,
Kampala, Uganda.

Tx: (988) 62054
A/B RAPCON UG

Zaire
B.P. 10362,
Kinshasa 1,
Zaire.

Tel: Kinshasa 22248
Tx: (982) 21674
A/B 21674 PTTWC ZR
(982) 21136
A/B 21136 ALAGEM ZR
(982) 21574+
A/B 21574 SITE ZR

Zimbabwe, Botswana, Angola	P.O. Box 4590, Harare, Zimbabwe.	Tel: 010 2630 792610
		Tx: DATAM 4283 ZW
		A/B 4283 ZW

II   Asia

Bangladesh and Burma	Oxfam, 41 Sir Syed Ahmed Rd, (2nd Floor) Block A, Mohammadpur, Dhaka 7, Bangladesh.	Tel: 010 880-2 315386
		Tx: (780) 642940
		A/B 642940 ADABBJ
		Emerg.(780) 655854
		A/B 655854 BDRC BJ

Indonesia

Jl. Galunggung 16,
Candi Baru, P.O. Box 214,
Semarang, Indonesia.

Tel: 010 6224 312956
Tx: (73) 22195+
    A/B 22195 YPDS IN
Telegrams: Oxfam-Semarang
           (Indonesia)

Kampuchea

Oxfam, Monorom Hotel,
Phnom Penh,
Kampuchea.

Telegrams: ARACE+ address

Tx: via (86) 82255+
    A/B 82255 CICR TH*
    *C/o ICRC Delegation,
    20 Sukhumvit − 5014,
    P.O. Box 11-1492,
    Bangkok 10110. Thailand.

Indian Offices

India Central

Oxfam (India) Trust,
P.O. Box 71,
Nagpur 440 001,
Maharashtra, India.

Tel: 010 91712 33737
Tx: (via Tradewings,
    Nagpur, att. Oxfam)
    (81) 715230+
    A/B 715230 TWNG IN

Telegrams: Oxfam-Nagpur
           (India)

India East and Bhutan

Oxfam (India) Trust,
3 Bright Street,
Calcutta 700019,
India.

Tel: 010 9133 41-2563
Tx: (81) 213245+
    A/B 213245 LWSI IN
    Emerg. (81) 213131+
    A/B 213131 BCCA IN

Telegrams: Oxfamindia
           Calcutta (India)

North India and Nepal

Flat No. 314,
Mansarovar Building,
90 Nehru Place,
New Delhi 110019,
India.

Tel: 010 9111 6418591
Tx: (81) 312434+
    A/B 312434 OXFM IN
    (81) 313454+
    A/B 314621 PARS IN

Also for Bridge Oxfam Trading

Tel: New Delhi 682137
Tx: (81) 314621+
    A/B 314621 PARS IN

Telegrams: Oxfam-New Delhi
           110019 (India)

India South	59 Millers Road,
	Benson Town,
	Bangalore 560046,
	India
		Tel: 01091812 565134
		Tx: (81) 845603+
		    A/B 845603 AVTS IN
		
		Telegrams: Oxfam-Bangalore
		    46 (India)

Gujarat and S. Rajasthan	Oxfam (India) Trust, 2 Mangaldeep Society, Near Parikshit Bridge, Gandhi Ashram P.O., Ahmedabad 380027, Gujarat, India.	Tel: Ahmedabad 405890 Tx: None Telegrams: Oxfam-Ahmedabad (India)

Orissa	Oxfam (India) Trust,
	Plot 55A,
	Kharavela Nagar Unit III,
	P.O. Box 170,
	Bhubaneswar 751001,
	Orissa, India.
		Tel: Bhubaneswar 54890
		Tx: (81) 675278+ via
		    Hotel Prachi – attn.
		    OXFAM
		    A/B 675278 PRCH IN
		Telegrams: Oxfam-Bhubaneswar
		    (India)

III   Middle East

Egypt East, Lebanon and Arab Republic	2 Sherif St., Appt. 92, Cairo, Egypt.	Tel: 010 202 754001 Tx: (91) 93445 A/B BARON UN (91) 21451 A/B COSMO UN
Ethiopia, Djibouti	P.O. Box 2333, Addis Ababa, Ethiopia.	Tel: 010 2511 159156 Tx: (980) 21307+ A/B CRDA ADDIS

Lebanon	c/o WSCF,
	P.O. Box 1375,
	Beirut,
	Lebanon
		Tel: 010 961 341902
		Tx: (494) 20949 A/B USI
		    Attn. Mike Scott
		    c/o Yusef Hajjar.
		    Emerg. (494) 23549
		        A/B JOURAS LE
		    Attn. Maia Tabet

Somalia	P.O. Box 2808, Mogadishu, Somalia.	Tel: 010 2521 22637 Tx: (900) 745+ A/B CROCE SUD
Sudan	P.O. Box 3182, Khartoum, Sudan.	Tel: Khartoum 41972

Yemen	c/o British Embassy,
	P.O. Box 1287,
	Sanaa,
	Yemen Arab Republic.
		Tel: 010 9672 73691
		Tx: (895) 2214+
		    A/B RABA YE
		    (895) 2679+
		    A/B YDTEST YE
		    Via Irena for Liz Gascoigne

IV   Latin America and Caribbean

Brazil	Caixa Postal 1987,
	50000 Recife,
	Pernambuco, Brazil.
		Tel: Recife 221 1227
		    231 5449
		Tx: (British Council)
		    814014 A/B BCOU BR

Caribbean	Apartado Postal 20271, Santo Domingo, Dominican Republic.	Tel: 010 1809 5332514 Tx: ITTP, BOOTH 3640001
Mexico, and El Salvador, Guatemala, Honduras, Nicaragua	Apartado Postal No. 5-452, Col. Cuauhtemoc, 06500 Mexico D.F.	Tel: 010 525 514 6043 or 525 2649 Tx: (22) 1771036 + A/B 1771036 BRITME
Peru and Andean countries	Calle Santa Isabel 180, Miraflores, Lima 18, Peru.	Tel: 010 5114 477588 Tx: (36) 25202 + A/B 25202 PE PB HCSAR

NOTE Other areas, e.g. Jordan and the occupied territories, Pakistan, the Philippines, Sri Lanka, Thailand, and refugee relief in S.E. Asia, are covered from Oxfam House, 274 Banbury Road, Oxford, OX2 7DZ, U.K.

Section 2   Oxfam International

Oxfam America,
115 Broadway,
Boston,
Massachusetts 02116, U.S.A.

Tel: (617) 4821211
Tx: 940288 OXFAM
      BSN USA

Oxfam Belgique,
39 Rue de Conseil,
1050 Brussels, Belgium.

Tx: 63939 OXFAM
      BELGIQUE B

Oxfam Canada,
251 Laurier Avenue West, Room 301,
Ottawa,
Ontario K1P 5J6, Canada.

Tx: 543358 OXFAM
      CANADA CN

Oxfam Quebec,
169 rue St. Paul East,
Montreal 127,
Quebec H2Y 1G8, Canada.

Tx: 5560522 OXFAM QUE
      MTL CN

Community Aid Abroad,
75 Brunswick Street,
Fitzroy,
Victoria 3065, Australia.

Tx: 30333 COMCOM AA

Section 3

Resource Centres
Advice Centres and
Research Organisations

I

Agriculture and extension

Agriculture Extension and Rural   Tel: 0734-85234
Development Centre,
University of Reading,
16 London Road,
Reading RG1 5AQ.

Centre for Overseas Pest Research,   Tel: 01-973 8191
College House,
Wrights Lane,
London W8 5SJ.
Pests in growing crops.

Land Resources Division,   Tel: 01-399 5281
Tolworth Tower,
Surbiton,
Surrey KT6 7DY.
Soil survey, water resources,
agronomy.

National Extension College,   Tel: 0223-316644
18 Brooklands Avenue,
Cambridge.

Overseas Liaison Unit,
N.I.A.E.,
Wrest Park,
Silsoe,
Bedford.
Farm mechanisation, oxen, minimum
tillage.

Tropical Store Products Centre,   Tel: 0753-34626
London Road,
Slough,
Bucks SL3 7HL.
Great expertise on crop storage but
take time to answer queries.

II

Appropriate technology, including some agriculture

Agriculture and Water   Tel: 0515-60428
Development Office,
College of Agricultural Engineering,
Silsoe,
Bedford MK45 4DT.

Brace Research Institute, Tel: 010-514457-2000
Macdonald College,
McGill University,
Ste. Anne de Bellevue 800,
Quebec H0A 1C0, Canada.
Windmills, solar power, water supply,
etc.

Groupe de Recherche sur les Tel: 2603680
Techniques Rurales, (GRET),
34 rue Dumont d'Urville,
75116 Paris,
France.

Intermediate Technology Tel: 01-836-9434
Department, (ITDG),
9 King Street,
London WC2E 8HW.

Returned Volunteer Action, (RVA), Tel: 01-278-0804
Technical Consultancy Service,
1 Amwell Street,
London EC1.

T.O.O.L.,
Mauritskade 61a,
Amsterdam,
Netherlands.

Volunteers for International
Technical Assistance, (VITA),
1815 North Lynn Street,
Suite 200,
Arlington,
Virginia 22209-2079,
U.S.A.

III Cooperatives

International Cooperative Alliance, Tel: 01-499-5991
(ICA),
Upper Grosvenor Street,
London W1X 9PA.

Plunkett Foundation for Tel: 0865 53961
Cooperative Studies,
31 St. Giles,
Oxford OX1 3LF.

IV Management

British Executive Service Overseas Tel: 01-839123
(BESO),
116 Pall Mall,
London SW1.
Specialises in providing advisers,
consultants, or temporary managers for
overseas work; costs are low, but not free.

City and Guilds of London Institute,   Tel: 01-580-3050
76 Portland Place,
London W1N 4AA.

Department of Education in   Tel: 01-636-1500
Tropical Areas,
Institute of Education,
20 Bedford Way,
London WC1.

Section 4   Development Agencies

I   British Development Agencies

1 Government and official

Overseas Development Administration,
Foreign and Commonwealth Office,
Eland House,
Stag Place,
London SW1.

Tel: 01-213-3000
Tx: 263907
   MINISTRANT G

The British Council,
10 Spring Gardens,
London SW1A 2BN.

Tel: 01-930-8466

2 Voluntary Development Agencies

British Red Cross Society,
9 Grosvenor Crescent,
London SW1X 7EJ.

Tel: 01-235-5454
Tx: 918657 BRCS G

British Leprosy Relief Association, (LEPRA)
Fairfax House,
Causton Road,
Colchester, Essex.

Tel: 0206-562286

Catholic Fund for Overseas Development, (CAFOD)
2 Garden Close,
Stockwell Road,
London SW9 9TY.

Tel: 01-735-9041

Christian Aid,
P.O. Box 1,
240 Ferndale Road,
Brixton,
London SW9 8BH.

Tel: 01-733-5500
Tx: 916504 CHRAID G

Commonwealth Society for the Deaf,
(Royal National Institute for the Deaf)
105 Gower Street,
London WC1.

Tel: 01-387-8033

Disasters Emergency Committee (DEC),
c/o 9 Grosvenor Crescent,
London SW1X 7EJ.

Tel: 01-235-5454

Royal Commonwealth Society for the Blind,
Commonwealth House,
Heath Road,
Haywards Heath, Sussex.

Tel: 0444-412424

Save the Children Fund,
Mary Datchelor House,
Camberwell Grove,
London SE5.

Tel: 01-703-5400

War on Want,
467 Caledonian Road,
London N7 9BE.

Tel: 01-609-0211

3 British Volunteer Programme and agencies sending volunteers	**British Volunteer Programme (BVP),** 22 Coleman Fields, London N1.	Tel: 01-226-6616
	Catholic Institute for International Relations, (CIIR) 22 Coleman Fields, London N1.	Tel: 01-354-0883
	Friends' Service Council, Friends House, Euston Road, London NW1 2BJ.	Tel: 01-387-3601
	U.N.A. International Service Department, 2 Whitehall Court, London SW1A 2EC.	Tel: 01-930-0679 or 01-930-2931
	International Voluntary Service (IVS) Ceresole House, 53 Regent Road, Leicester.	Tel: 0533-541862
	Voluntary Service Overseas (VSO), 9 Belgrave Square, London SW1X 8PW.	Tel: 01-235-5191
4 Missionary Societies	**Baptist Missionary Society** 93-97 Gloucester Place, London W1H 4AA.	Tel: 01-935-1482
	Church Missionary Society 157 Waterloo Road, London SE1 8UU.	Tel: 01-928-8681
	Church of Scotland, 121 George Street, Edinburgh EH2 4YN.	Tel: 031-225-5722
	Methodist Missionary Society 25 Marylebone Road, London NW1 5JR.	Tel: 01-935-2541
	Salvation Army International Headquarters 101 Queen Victoria Street, P.O. Box 249, London EC4P 4EP.	Tel: 01-236-7020 01-236-5222
	United Society for the Propagation of the Gospel, 15 Tufton Street, London SW1P 3QQ.	Tel: 01-222-4222

II   European Agencies

1 European	**Euro-Action ACORD,** Prins Hendrikkade 48, Amsterdam 1001, Netherlands.	Tx: 23997 IVEF CH

② **Belgium**	**AGENOR,**
	13 rue Hobbema,
	1040 Brussels,
	Belgium.

Association Internationale de Developpement Rural (AIDR),
20 rue de Commerce,
1040 Brussels,
Belgium.

Centre National de Cooperation au Developpement (CNCD),
Rue de Laeken 76,
1000 Brussels,
Belgium.

Tel: 02 218 31 67

Caritas Catholique de Belgique,
(Secours International)
Rue de Commerce 72,
1040 Brussels,
Belgium.

Tel: 02 511 42 55

Freres des Hommes,
Place de Londres 6,
1050 Brussels,
Belgium.

Terre des Hommes,
42-44 Boulevard Brand   Whitlock,
1150 Brussels,
Belgium.

③ **Denmark**   **Danchurchaid,**   Tel: 01 15 28 00
Sankt Peders Straede 3,
1453 Copenhagen K,
Denmark.

④ **France**   **ASCOFAM,**   Tel: 258 68 06
127 Rue Marcadet,
75018 Paris,
France.

Comite Catholique contre la Faim et pour le Developpement,   Tel: 325 31 02
47 Quai des Grands Augustins,
75262 Paris,
France.

Compagnie Internationale de Developpement Rural (CIDR),
B.P.1, Autreches,
60350 Cuise La Motte,
France.

Freres des Hommes,
20 rue de Refuge,
78000 Versailles,
France.

Medecins sans Frontieres (MSF),
19 rue Daviel,
75013 Paris,
France.

Secours Catholique (CARITAS), Tel: 320 14 14
106 rue du Bac,
75341 Paris,
France.

5 Western Germany **Brot fur die Welt,** Tel: 07112 05 11
Stafflenbergstrasse 76, Tx: 723557 DDWS D
7 Stuttgart 1,
W. Germany.

Deutscher Caritasverband Tel: 0761 20 01
(Caritas Germany) Tx: 772475 DEV D
Karlstrasse 40,
P.O. Box 420,
D.7800 Freiburg 1BR,
W. Germany.

Deutsche Welthungerhilfe, Tel: 02221 21 70 18
Adenhaueralle 134, Tx: 8869697 DWHH D
D.300 Bonn,
W. Germany.

**Evangelische Zentralstelle
fur Entwicklungshilfe (EZE),**
Mittlestrasse 37,
5300 Bonne 2 — Bad Godesberg,
W. Germany.

Zentralstelle fur Entwicklungshilfe Tx: 832370 MISA D
(MISEREOR),
Mozartstrasse 11,
5100 Aachen,
W. Germany.

6 Republic of Ireland **Concern,** Tel: 681237
1 Upper Camden Street, Tx: 5596 CERN EI
Dublin 2,
Eire.

Trocaire, Tel: 885173/4 or 887190
169 Booterstown Avenue,
Blackrock,
Co. Dublin,
Eire.

7 Italy **Manitese,**
via Cavenaghi 4,
20149 Milan,
Italy.

8 Netherlands **CEBEMO,** Tel: 070 24 45 94
van Alkemadelaan 1,
The Hague 2077,
Netherlands.

Interchurch Coordination Committee Tel: 030 517704
for Development Projects (ICCO),
Stadhouderslaan 43,
Utrecht,
Netherlands.

| | **NOVIB - Netherlands Organisation for International Development Cooperation,** Amaliastraat 7, 2515JC The Hague, Netherlands. | Tel: 070 624081 |
| | **Stichting HIVOS,** Beeklaan 387, 2652 The Hague, Netherlands. | |

9 **Norway** **Norwegian Refugee Council,** Prof. Dahls Gatel, Oslo 3, Norway. Tel: (472) 60 39 42

Redd Barna *(Norwegian Save the Children)* Jernbanetorget, Oslo 1, Norway. Tel: 02 414635

NORAD, Victoria Terrace 7, Oslo 1, Norway. Tel: 314055

10 **Sweden** **Radda Barnen,** *(Swedish Save the Children)* Box 27320, 5-10254 Stockholm, Sweden.

11 **Switzerland** **Helvetas** *(Swiss Association for Technical Assistance),* St. Moritzstrasse 15, Postfach 8042, Zurich, Switzerland. Tel: 01 6050 60

III American Development Agencies

American Council of Voluntary Agencies for Foreign Service Inc., 200 Park Avenue South, 11th Floor, New York, NY 10003, U.S.A.

American Friends Service Committee, 1501 Cherry Street, Philadelphia, Pennsylvania 19102, U.S.A.

**Cooperative for American Relief
Everywhere (CARE),**
660 First Avenue,
New York,
NY 10016, U.S.A.

Catholic Relief Services (CRS),   Tel: 212 838 4700
1011 First Avenue,
New York,
NY 10022, U.S.A.

(There is also a Geneva office — see under International Agencies)

Church World Service (CWS),   Tel: 212 870 2175
475 Riverside Drive,
Room 620,
New York,
NY 10027, U.S.A.

Inter-American Foundation,
1515 Wilson Boulevard,
Rosslyn,
Virginia 22209, U.S.A.

Lutheran World Relief Inc.,   Tel: 212 532 6350
360 Park Avenue South,
New York,
NY 10010, U.S.A.

**General Conference
Mennonite Church,**
Commission on Overseas Mission,
P. Box 347,
722 Main Street,
Newton,
Kansas 67114, U.S.A.

(This body also has a Canadian affiliate; see also the other
denomination of Mennonites below.)

Mennonite Central Committee,
21 South 12th Street,
Akron,
Pennsylvania 17501,
U.S.A.

(This body also has a Canadian affiliate.)

The Pathfinder Fund,
(a family planning agency)
1330 Boylston Street,
Chestnut Hill,
Massachusetts 02167,
U.S.A.

World Neighbors Inc.,
5116 North Portland Avenue,
Oklahoma City,
Oklahoma 73112, U.S.A.

U.S. Government:

Agency for International Development (USAID),
(including Action/Peace Corps)
Office of Public Affairs,
Department of State,
Washington, D.C. 20523,
U.S.A.

IV   Canadian Agencies

Canadian International Development Agency (CIDA)   Tel: 613 997 5456
Place du Centre,
200 Promenade du Portage,
Hull,
Quebec K1A 0G4,
Canada.

Canadian University Service Overseas (CUSO),
151 Slater Street,
Ottawa,
Ontario K1P 5H5

Development and Peace,   Tel: 9325136
(Developpement et Paix)
2111 Center Avenue,
Montreal,
Canada.

International Development Research Centre (IDRC),
P.O. Box 8500,
Ottawa K1G 3H9,
Canada.

V   International Agencies

Caritas Internationalis,
(Conference of Catholic Charities)
Piazza san Calisto 16,
00153 Rome, Italy.

Catholic Relief Services   Tel: 022 31 46 54
Rue de Cornavin 11,   Tx: 23866 CRS CH
1201 Geneva,
Switzerland.

(see also under American agencies)

Commission of European Communities (CEC/EEC),   Tx: 21877 COMEU B
Directorate General for Development
DG VIII,
rue de la Loi 200,
B-1049 Brussels,
Belgium.

International Committee of the Red Cross (ICRC),
17 avenue de la Paix,
1211 Geneva,
Switzerland.

Tx: 22264 CICR CH

International Council of Voluntary Agencies (ICVA),
13 rue Gautier,
1201 Geneva,
Switzerland.

Tel: 31 66 02
Tx: 22891

International Planned Parenthood Federation (IPPF),
18/20 Lower Regent Street,
London SW1Y 4PW.

Tel: 01 839 2911

International Union of Child Welfare (IUCW),
rue de Varembe,
1211 Geneva 20,
Switzerland.

League of Red Cross Societies (LRCS),
17 Chemin des Crets,
Petit-Saconnex,
Geneva, Switzerland.

Tx: 22555A LRCS CH

LIPCROSS/VOLAGS,
c/o P.O. Box 276,
1211 Geneva 19,
Switzerland.

Lutheran World Federation,
(Department of World Service)
150 route de Ferney,
1211 Geneva 20,
Switzerland.

OECD Development Centre,
2 rue Andre Pascal,
75775 Paris,
Cedex 16, France.

Service Civil International,
(International Voluntary Service)
International Secretariat,
35 Avenue Gaston-Diderich,
Luxembourg-Ville,
Grand Duchy of Luxembourg.

World Council of YMCAs,
37 quai Wilson,
1201 Geneva,
Switzerland.

World Council of Churches,
150 route de Ferney,
P.O. Box 66,
1211 Geneva 20, Switzerland.

Tx: 23423A OIK CH